Lecture Notes in Physics

The Lecture Notes in Physics

The series Lecture Notes in Physics (LNP), founded in 1969, reports new developments in physics research and teaching – quickly and informally, but with a high quality and the explicit aim to summarize and communicate current knowledge in an accessible way. Books published in this series are conceived as bridging material between advanced graduate textbooks and the forefront of research to serve the following purposes:

• to be a compact and modern up-to-date source of reference on a well-defined topic;

• to serve as an accessible introduction to the field to postgraduate students and nonspecialist researchers from related areas;

• to be a source of advanced teaching material for specialized seminars, courses and schools.

Both monographs and multi-author volumes will be considered for publication. Edited volumes should, however, consist of a very limited number of contributions only. Proceedings will not be considered for LNP.

Volumes published in LNP are disseminated both in print and in electronic formats, the electronic archive is available at springerlink.com. The series content is indexed, abstracted and referenced by many abstracting and information services, bibliographic networks, subscription agencies, library networks, and consortia.

Proposals should be sent to a member of the Editorial Board, or directly to the managing editor at Springer:

Dr. Christian Caron
Springer Heidelberg
Physics Editorial Department I
Tiergartenstrasse 17
69121 Heidelberg/Germany
christian.caron@springer.com

Heiner Linke Alf Månsson (Eds.)

Controlled
Nanoscale Motion

Nobel Symposium 131

Springer

Editors

Heiner Linke
Physics Department
University of Oregon
Eugene, OR 97403-1274
USA
E-mail: linke@uoregon.edu

Alf Månsson
Department of Chemistry
and Biomedical Sciences
University of Kalmar
SE-391 82 Kalmar
Sweden
E-mail: alf.mansson@hik.se

H. Linke, A. Månsson (Eds.), *Controlled Nanoscale Motion*, Lect. Notes Phys. 711
(Springer, Berlin Heidelberg 2007), DOI 10.1007/b11823292

Library of Congress Control Number: 2006938044

ISSN 0075-8450
ISBN-10 3-540-49521-5 Springer Berlin Heidelberg New York
ISBN-13 978-3-540-49521-5 Springer Berlin Heidelberg New York

Springer is a part of Springer Science+Business Media
springer.com
© Springer-Verlag Berlin Heidelberg 2007

Typesetting: by the authors and techbooks using a Springer LaTeX macro package
Cover design: WMXDesign GmbH, Heidelberg

Printed on acid-free paper   SPIN: 11823292   54/techbooks   5 4 3 2 1 0

Preface

When the size of a machine approaches the nanometer scale, thermal fluctuations become large compared to the energies that drive the motor. The mechanism and control system for directed nanoscale motion must allow for, or even make use of, this stochastic environment. Controlled motion at the nanoscale therefore requires theoretical descriptions and engineering approaches that are fundamentally different from those that were developed for man-made, macroscopic motors and machines.

Over the past decade, a need to understand and to control directed motion at the nanoscale has arisen in several areas of biology, physics and chemistry. Most notably, the advent of single-molecule techniques in biophysics has given access to detailed information about the performance of molecular motors in biological cells. Combined with a variety of techniques from molecular biology, this information allows conclusions about the physics of biological machines. Even more recently, a variety of approaches including nanofabrication and synthetic chemistry have been used to create artificial nanoscale motors or to control the motion of individual molecules, for example using nanofluidic systems. Many of these approaches were triggered by novel theoretical methods designed to understand how the interplay of stochastic thermal motion and non-equilibrium phenomena can be harnessed to generate an output of useful work.

The present volume is based on selected contributions to the Nobel Symposium 131 on *Controlled Nanoscale Motion in Biological and Artificial Systems*, held on June 13–17, 2005 at Bäckaskog Slott in Sweden. The peer-reviewed chapters in this book are designed to be tutorial and self-contained and provide insight into the state of the art in the following three areas:

Biophysics of molecular motors and single molecules. Molecular motors are proteins or protein complexes that transduce chemical free energy into work through processes generally believed to involve substantial changes in protein structure. This section describes the physical and biochemical principles of molecular motor function together with an account of some important exper-

imental techniques for their study. The section begins with an overview of the
regulation and function of a complex bacterial flagellar motor (Chap. 1). The
focus is then shifted towards molecular motors in eukaryotes and the biophysical principles by which they produce force and linear transport. Chapters
2, 3 and 5 consider the mechanisms of operation of members of the myosin
motor family, which interact with the actin cytoskeleton, and of kinesins and
dyneins, which interact with microtubules. The multitude of biological roles
of motors in living cells include tasks of biomedical relevance, such as axonal transport and embryonal development. Chapters 4 and 5 exemplify these
functions together with accounts of how such diverse tasks can be achieved
by a limited set of motors and cytoskeletal filaments. Chapter 7 describes
the role of molecular motors in nanotube dynamics in living cells, including
a theoretical treatment of the physics of membrane nanotubes. Chapters 6
and 8, finally, consider nanoscale motion in macromolecules not traditionally
counted as molecular motors, including nucleic acid and nucleic acid-binding
proteins (Chap. 6) and polysaccharide modifying enzymes (Chap. 8).

Theory of controlled nanoscale motion. Nanoscale motors and machines typically operate far from thermal equilibrium in an environment characterized by
substantial thermal motion. In addition, thermal fluctuations of the protein
conformational state around a free energy minimum can contribute to the stochastic nature of experimental data. The theory of Brownian motion in and
out of thermal equilibrium therefore plays an important guiding role in the
design of artificial motors and in the analysis of single-molecule experiments.
Chapter 9 describes improved mathematical models of Brownian motion and
their use to calibrate optical tweezers. Chapter 10 represents a tutorial introduction to the Jarzynski equation that allows extraction of information about
equilibrium processes from data taken under non-equilibrium conditions. Finally, Chap. 11 describes theoretical approaches and methods for the accurate
determination of diffusion constants from noisy data.

Controlled motion in nanotechnology. The ability to fabricate and manipulate
nanoscale structures offers an impressive array of methods for the control of
the motion of nanoscale objects, giving access to a new realm of experimental physics. Chapters 12 and 13 provide tutorial introductions to the physics
of nanomechanical and nanofluidic devices for detection and study of single
biomolecules. The subsequent three chapters describe two representative approaches to the construction of artificial molecular motors using self-assembly
techniques, as well as a synthetic nanopore system that allows control of ion
flow similar to a biological ion channel. The final two Chapters (17 and 18)
tie together nanotechnology and biological motors by discussing the physics
and methods of controlling biological motors using nanofabricated structures.

Nobel Symposium 131, on which this volume is based, was sponsored by the Nobel Foundation through its Nobel Symposium Fund. We thank all speakers and participants for their contributions and the Nobel Foundation for generous financial support.

Kalmar and Eugene   *Alf Månsson*
January 2007   *Heiner Linke*

Contents

List of Contributors

J.-F. Allemand
Laboratoire de Physique Statistique
and Dept. Biologie
Ecole Normale Superieure
France
allemand@lps.ens.fr

John Allingham
University of Wisconsin
Department of Biochemistry
USA
jallingham@biochem.wisc.edu

J.L. Arlett
Division of Physics
Mathematics and Astronomy
arlett@cco.caltech.edu

Robert H. Austin
Department of Physics
Princeton University
USA
austin@princeton.edu

D. Bensimon
Laboratoire de Physique Statistique
and Dept. Biologie
Ecole Normale Superieure
France
david.bensimon@lps.ens.fr

Howard C. Berg
Departments of Molecular and
Cellular Biology and of Physics
Harvard University
USA
hberg@mcb.harvard.edu

K. Berg-Sørensen
Department of Physics
Technical University of Denmark
Denmark

G. Charvin
Laboratoire de Physique Statistique
and Dept. Biologie
Ecole Normale Superieure
France
gilles.charvin@lps.ens.fr

Edward C. Cox
Department of Molecular Biology
Princeton University
USA
ecox@princeton.edu

H.G. Craighead
Applied and Engineering Physics
Cornell University
USA
hgc1@cornell.edu

V. Croquette
Laboratoire de Physique Statistique
and Dept. Biologie
Ecole Normale Superieure
France
croquette@lps.ens.fr

M.C. Cross
Division of Physics
Mathematics and Astronomy
California Institute of Technology
USA

András Czövek
Department of Biological Physics
Eötvös University
Hungary
czigor@angel.elte.hu

Gaudenz Danuser
Departments of Cell Biology
The Scripps Research Institute
USA

Imre Derényi
Department of Biological Physics
Eötvös University
Hungary
derenyi@angel.elte.hu

Marileen Dogterom
FOM Institute for Atomic and
Molecular Physics (AMOLF)
The Netherlands
dogterom@amolf.nl

Martijn M. van Duijn
Department of Bioengineering
University of California Berkeley
USA
vanduijn@berkeley.edu

Henrik Flyvbjerg
Biosystems Department
and Danish Polymer Centre
Risø National Laboratory
Denmark
henrik.flyvbjerg@risoe.dk

S.E. Fraser
Kavli Nanoscience Institute
and Division of Biology and
Division of Engineering and
Applied Science
California Institute of Technology
USA

Lawrence S.B. Goldstein
Departments of Cellular and
Molecular Medicine
University of California of San Diego
USA
lgoldstein@ucsd.edu

P.H. Hagedorn
Biosystems Department
Risø National Laboratory
Denmark
peter.hagedorn@risoe.dk

Henry Hess
Department of Materials Science
and Engineering
University of Florida
USA
hhess@mse.ufl.edu

Nobutaka Hirokawa
Department of Cell Biology
and Anatomy
Graduate School of Medicine
University of Tokyo
Japan
hirokawa@m.u-tokyo.ac.jp

Robert D. Horansky
Department of Physics
University of Colorado
USA
Robert.Horansky@Colorado.edu

Christopher Jarzynski
Theoretical Division
Los Alamos National Laboratory
and
Department of Chemistry
and Biochecmistry
Institute for Physical
Science and Technology
University of Maryland
USA
cjarzyns@umd.edu

Gerbrand Koster
Institut Curie
France
and
FOM Institute for Atomic and
Molecular Physics (AMOLF)
The Netherlands
gerbrand.koster@curie.fr

N.B. Larsen
Danish Polymer Centre
Risø National Laboratory
Denmark
and
Biosystems Department
Risø National Laboratory
Denmark
niels.b.larsen@risoe.dk

G. Lia
Harvard University
Chemistry and Chemical Biology
USA
lia@fas.harvard.edu

T. Lionnet
Laboratoire de Physique Statistique
and Dept. Biologie
Ecole Normale Superieure
France

Thomas F. Magnera
Department of Chemistry
and Biochemistry
University of Colorado
USA
magnera@eefus.colorado.edu

Alf Månsson
School of Pure and Applied
Natural Sciences
University of Kalmar
Sweden
alf.mansson@hik.se

Dietmar J. Manstein
Institute for Biophysical Chemistry
Germany
manstein@bpc.mh-hannover.de

Charles R. Martin
Department of Chemistry
University of Florida
USA
crmartin@chem.ufl.edu

Josef Michl
Department of Chemistry
and Biochemistry
University of Colorado
USA
michl@eefus.colorado.edu

Lars Montelius
Division of Solid State Physics and
The Nanometer Consortium
University of Lund
Sweden
lars.montelius@ftf.lth.se

S. Mosler
Danish Polymer Centre
Risø National Laboratory
Denmark

K.C. Neuman
Laboratoire de Physique Statistique
and Dept. Biologie
Ecole Normale Superieure
France

Ian A. Nicholls
School of Pure and
Applied Natural Sciences
University of Kalmar
Sweden
ian.nicholls@hik.se

Kazuhiro Oiwa
Kobe Advanced CT Center
(KARC)
National Institute of Information
and Communications Technology
(NICT)
Japan
oiwa@nict.go.jp

Pär Omling
Division of Solid State
Physics and The Nanometer
Consortium
University of Lund
Sweden
par.omling@vr.se

M.R. Paul
Department of Mechanical
Engineering
Virginia Polytechnic Institute
and State University
USA
mrp@vt.edu

John C. Price
Department of Physics
University of Colorado
USA
john.price@colorado.edu

Jacques Prost
Institut Curie
France
and
ESPCI
France
jacques.prost@curie.fr

Ivan Rayment
Department of Biochemistry
University of Wisconsin
USA
Ivan_Rayment@biochem.wisc.edu

M.L. Roukes
Kavli Nanoscience Institute
and Division of Physics
Mathematics and Astronomy and
Division of Engineering and
Applied Science
California Institute of Technology
USA
roukes@caltech.edu

O.A. Saleh
Materials Department and
Biomolecular Science and
Engineering Program
University of California
USA
saleh@engineering.ucsb.edu

K.T. Samiee
Applied and Engineering Physics
Cornell University
USA
kts3@cornell.edu

E. Schäffer
Center of Biotechnology
Technical University
Germany

D. Selmeczi
Danish Polymer Centre
Risø National Laboratory
Denmark
and
Department of Biological Physics
Eötvös University
Hungary
david.selmeczi@risoe.dk

Sameer B. Shah
Department of Bioengineering
University of Maryland
USA
sameer@umd.edu

Zuzanna S. Siwy
Department of Physics and
Astronomy
University of California
USA
and
Department of Chemistry
Silesian University of Technology
Poland
zsiwy@uci.edu

Gudmund Skjåk-Bræk
Department of Biotechnology
The Norwegian University of Science
and Technology
Norway
gudmund.skjaax-braek
@biotechntnu.no

Marit Sletmoen
Biophysics and Medical Technology
Department of Physics
The Norwegian University of Science
and Technology
Norway
marit.sletmoen@phys.ntnu.no

J.E. Solomon
Division of Physics
Mathematics and Astronomy
California Institute of Technology
USA

S.M. Stavis
Applied and Engineering Physics
Cornell University
USA
sstavis@gmail.com

Bjørn Torger Stokke
Biophysics and Medical Technology
Department of Physics
The Norwegian University of Science
and Technology
Norway
bjorn.stokke@phys.ntnu.no

Sven Tågerud
School of Pure and
Applied Natural Sciences
University of Kalmar
Sweden
sven.tagerud@hik.se

Reiko Takemura
Okinaka Memorial Institute for
Medical Research
Japan

S. Tolić-Nørrelykke
Max Planck Institute for the Physics
of Complex Systems
Germany
tolic@nbi.dk

Viola Vogel
Department of Materials
Swiss Federal Institute of Technology
(ETH)
Switzerland
viola.vogel@mat.ethz.ch

Y.M. Wang
Department of Physics
Princeton University
USA
ymwang@wuphys.wustl.edu

Ge Yang
Departments of Cell Biology
The Scripps Research Institute
USA

H. Yokota
Department of Molecular Physiology
The Tokyo Metropolitan Institute
of Medical Science
Japan
hiroaki_yokota@rinshoken.or.jp

Bernard Yurke
Bell Laboratories
USA
yurke@lucent.com

1

Navigation on a Micron Scale

H.C. Berg

Departments of Molecular and Cellular Biology and of Physics,
Harvard University, Cambridge, Massachusetts 02138, USA

Abstract. *E. coli* is a bacterium 1 μm in diameter. It swims in a nutrient medium, counting molecules of interest as it goes along. On the basis of these counts, it accumulates in regions that it deems more favorable. How does nature design, construct, and operate such a nanomachine?

Modern work on the motile behavior of bacteria began in 1965, when Julius Adler published a symposium paper "Chemotaxis in *Escherichia coli*" [1]. An electron micrograph of a negatively-stained cell that appeared there is shown in Fig. 1.1. By chemotaxis, Adler meant the ability of cells to move toward a source of chemicals that they deemed favorable (called attractants) or to move away from the source of chemicals that they deemed unfavorable (called repellents). *E. coli* was the organism of choice for studies of the molecular biology of behavior, because more was known about *E. coli* than any other free-living thing. Earlier work in the field of bacterial motility goes all the way back to van Leeuwenhoek, who described swimming bacteria in 1676. More than 200 years passed before Theodor Engelmann (in Utrecht) and Wilhelm Pfeffer (in Tübingen) began systematic studies of responses of bacteria (different species isolated from the wild) to oxygen and a variety of other chemicals. Pfeffer used the word "chemotaxis," because he thought that bacteria could steer; the mechanism turned out to be different, but the name stuck. For reviews that include this history, see [2,3].

A schematic drawing of *E. coli* is shown in Fig. 1.2. The cell has a 3-layered wall. The outermost layer is penetrated by proteins, called porins, that allow the passage of water-soluble molecules of low molecular weight (the size of sucrose or less). The peptidoglycan layer is a quasi-rigid network that gives the cell its non-spherical shape. The inner membrane is a lipid bilayer, like the plasma membrane of a human cell. There are two kinds of external organelles, pili, involved in adhesion, and flagella, that enable cells to swim. Several flagella (about 4 to 6, on average) arise at random points on the sides of the cell. Each is driven at its base by a reversible rotary motor, embedded in the cell wall [4].

H.C. Berg: *Navigation on a Micron Scale*, Lect. Notes Phys. **711**, 1–13 (2007)
DOI 10.1007/3-540-49522-3_1 © Springer-Verlag Berlin Heidelberg 2007

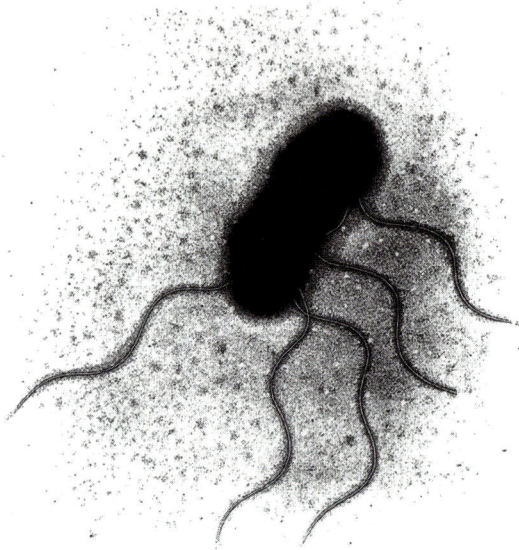

Fig. 1.1. An electron micrograph of *E. coli* published by Julius Adler [1]. This cell is negatively stained with the salt of an element of high atomic number, which has spread only a few micrometers out from the cell body. As a result, the flagellar filaments appear truncated. The diameter of the cell body is about 1 μm, and the wavelength of the filaments is about 2.3 μm

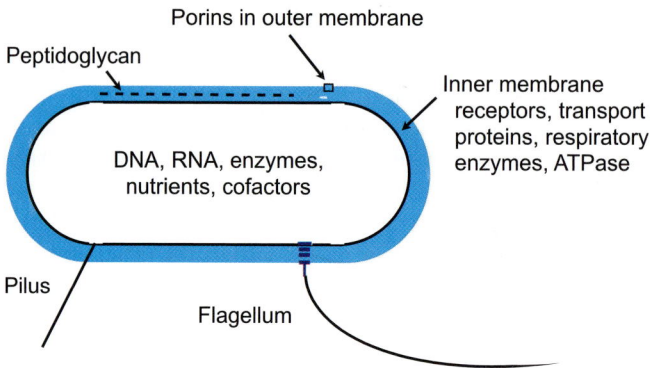

E. coli K12:   4,286 genes   ≈ 1 μm = 10^{-4} cm in diameter

H. sapiens: ≈ 25,000 genes

Fig. 1.2. *E. coli* is a Gram-negative bacterium with a 3-layered cell wall, comprising an outer membrane (with porins), a peptidoglycan layer, and an inner membrane. There are cytoskeletal elements attached to the inner membrane (not shown) but none within the cytoplasm itself, which is devoid of membranes and other structural elements present in eukaryotic cells. The external organelles shown include a pilus and a flagellum. The pilus is straight, the flagellum helical

Adler improved upon an assay developed by Pfeffer, in which cells in a dilute suspension swim up spatial gradients of a chemical diffusing from the tip of a capillary tube and then swim inside. After about an hour, the tube is withdrawn and a fraction of its contents spread on a nutrient agar plate, which on the following day, yields one colony per bacterium plated. By varying the initial concentrations of chemicals in the tube and counting colonies, Adler produced dose-response curves [5]. Remarkably, he found that cells respond strongly to certain chemicals that they can neither transport across the inner cell membrane nor metabolize [6]. So chemotaxis is a matter of taste rather than material gain: cells have specific chemoreceptors [6].

When this behavior was studied with a tracking microscope [7], it was found that a cell picks a direction at random, swims in a nearly straight line, and then abruptly changes its mind, as shown in Fig. 1.3. Intervals during which the cell swims steadily are called runs, while those during which it changes its mind are called tumbles. Runs are relatively long, about 1 sec on average, and tumbles are relatively short, about 0.1 sec, on average; both run and tumble intervals are distributed exponentially. Some of the dots in the track (Fig. 1.3) are close together, but these represent runs in a direction in or out of the plane of the figure. If one places one of Adler's capillary tubes containing an attractant at the right edge of the figure, runs to the right get longer; however, runs to the left stay about as long as they are in the absence of a stimulus. That is, the random walk becomes biased, and the bias is positive. Put colloquially, *E. coli* is an optimist: if life gets better, it enjoys it more; if life gets worse, it doesn't worry about it!

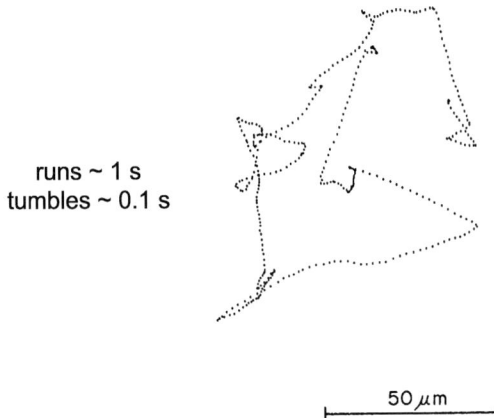

runs ~ 1 s
tumbles ~ 0.1 s

50 μm

Fig. 1.3. Thirty seconds in the life of one wildtype *E. coli* cell, tracked in a homogeneous isotropic medium, in the absence of any chemical attractants [7]. This is a 2-dimensional projection of a 3-dimensional track. The interval between dots is ~0.08 sec. The cell swam about 20 diameters per second and executed 26 runs and tumbles

It wasn't long before we learned that bacteria swim by rotating their flagellar filaments [4]. If the direction of rotation is counter-clockwise (CCW) as viewed from behind the cell, the several filaments form a bundle that pushes the cell forward [8]. If one or more filaments turn clockwise (CW), the bundle fragments, and the cell moves erratically with little net displacement [9]. So bacterial behavior depends upon the direction of flagellar rotation [10]. When runs get longer, as during a chemotactic response, cells spend more time rotating their flagella CCW.

If one looks carefully at the runs in Fig. 1.3, it is evident that cells cannot swim in a straight line. They are subject to rotational Brownian motion. This problem was discussed by Einstein in 1905, who found that the root-mean-square angular displacement of a small particle (here, the cell) increases as the square-root of the time. If one puts in numbers, one finds that an *E. coli* cell will wander by a root-mean-square angle of 90° in about 10 sec. Therefore, after 10 sec, it has forgotten where it has been. So measurements of concentrations made more than 10 sec ago are no longer relevant. This sets an absolute limit on the time that cells have to decide whether life is getting better or worse.

We found that *E. coli* understands this problem by studying the chemotactic response of *E. coli* at the level of a single flagellar motor, as shown in Fig. 1.4 [11, 12]. We used the tethered-cell technique developed by Silverman and Simon [13], as shown in the inset. Break most of the flagella off of cells by viscous shear and attach the remaining flagellar stubs (ideally, no more than one per cell) to a glass slide using an anti-filament antibody. Now, the flagellar motor, rather than driving the filament, turns the cell body. If a micropipette containing a negatively-charged attractant (e.g., the amino acid aspartate) is brought to within a few micrometers of such a cell, the attractant can be

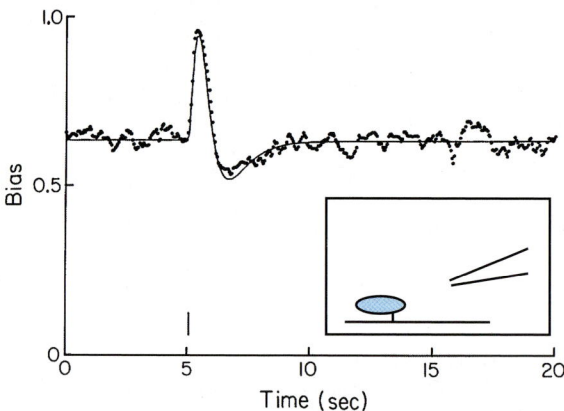

Fig. 1.4. The response of *E. coli* to a short pulse of attractant, delivered at 5.06 seconds. Tethered cells were stimulated by iontophoretic pipettes, as shown in the inset

ejected in a controlled manner by passage of an electrical current. The plot shows the result obtained when tethered cells were exposed to a short pulse applied at 5.06 seconds. The ordinate is the bias of the cell, i.e., the fraction of time that it spins CCW. The effect of the short pulse is a biphasic response lasting about 4 sec, called an impulse response. The bias approaches 1 within about 0.5 sec, returns to the baseline within about 1 sec, and remains below the baseline for the following 3 sec. The areas of the two lobes of the response are nearly the same. The argument is rather elaborate, but what this experiment shows is that cells compare counts of attractant molecules made over the past second with counts made over the previous 3 seconds and respond to the difference. So, cells make temporal comparisons over a time span substantially less than 10 sec: they have a short-term memory spanning 4 sec [11, 12].

One of *E. coli*'s flagellar motors is shown in Fig. 1.5. A motor is made of about 20 different kinds of parts. There is a ring in the cytoplasm called the

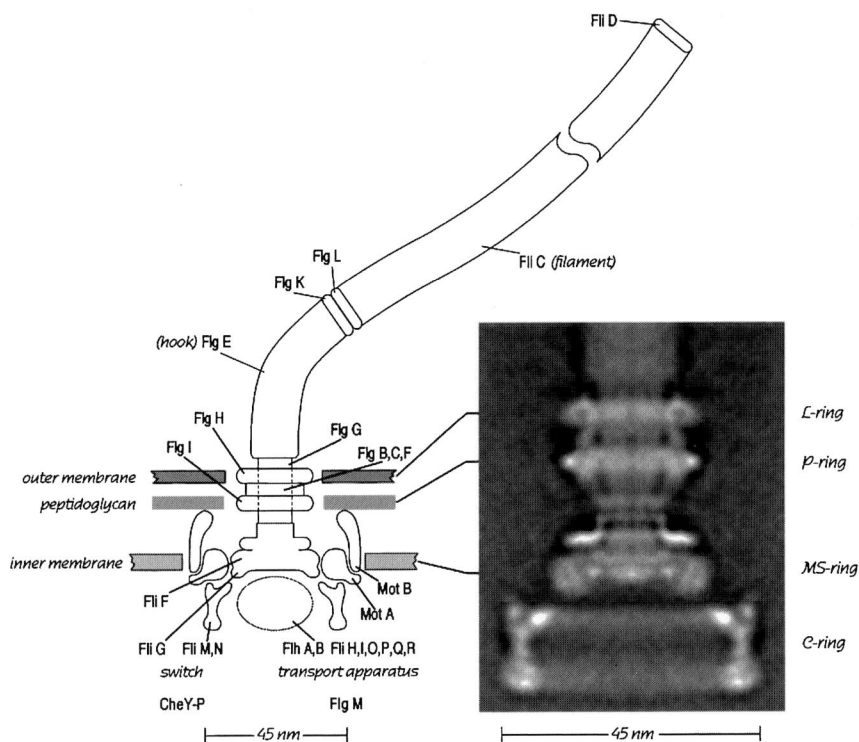

Fig. 1.5. The flagellar rotary motor. The inset is a reconstruction of cryo-electron micrographs assembled by David DeRosier. It is what one would see looking through a spinning rotor. Most proteins are labeled Flg, Flh, or Fli, depending upon where their genes are located on the *E. coli* chromosome. Null mutants of any of these genes result in cells without flagellar filaments. Mutations in *mot* genes result in paralyzed flagella

C-ring, a set of rings in the inner membrane called the MS-ring, a drive shaft that extends from the MS-ring to a flexible coupling (or universal joint) called the hook, a bushing that gets the drive-shaft through the outer layers of the cell wall, called P and L rings, two adapter proteins, and finally a long helical filament terminated by a cap. The motor is driven by sets of proteins that span the inner membrane but are bolted down to the peptidoglycan layer, called Mot proteins. MotA (4 copies) and MotB (2 copies) comprise force-generating units (at least 8) powered by a transmembrane proton flux. This is an ion-driven machine, not an ATP-driven machine. The motor is assembled from the inside out in a highly regulated manner. Surprisingly, the filament grows at its distal end rather than at the proximal end.

A great deal is known about the assembly and function of the flagellar motor, but not much about its structure at the atomic level. As a result, we do not really understand how it works, i.e., how proton flow makes it turn, or what the events are that cause it to switch (change its direction of rotation). However, as judged by its torque-speed relationship, shown here in Fig. 1.6, the motor utilizes a power-stroke mechanism; it does not behave as a thermal ratchet. Were it a thermal ratchet, there would be a barrier to backwards rotation, and the torque-speed curve would be concave downwards rather than convex [14]. For recent reviews, see [15–17].

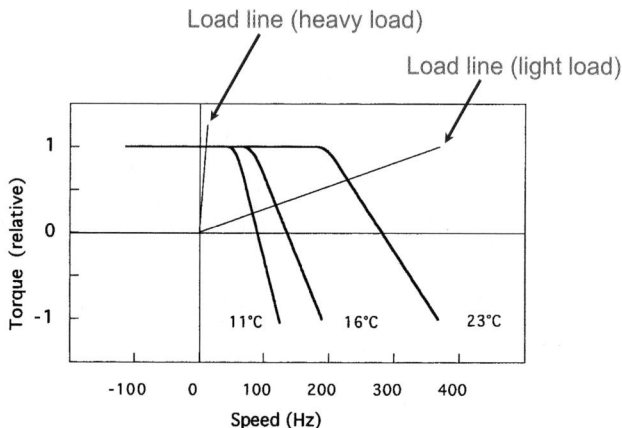

Fig. 1.6. The torque-speed curve for the flagellar motor of *E. coli* shown at three temperatures (*thick lines*), together with two load lines (*thin lines*), one for a heavy load (*left*) and the other for a light load (*right*). The slopes of the load lines are the rotational viscous drag coefficients for the objects being spun, e.g., the body of a tethered cell (a heavy load) or a 0.3 μm-dia latex bead (a light load). The plateau of the torque-speed curve is not quite as flat as shown; at 23°C, the torque drops about 10% between stall (0 Hz) and the knee (∼200 Hz). The torque-speed curves were obtained by electrorotation – a complicated story; see [18] – or by increasing the slopes of load lines by addition of a viscous agent [19]

The main quest of most of the people studying bacterial chemotaxis is to understand the signal transduction pathway. What is the machinery that enables cells to swim in a purposeful manner, i.e., what controls the direction of flagellar rotation? Signal processing involves interactions between different kinds of proteins, beginning with receptors that bind specific chemical attractants, shown in Fig. 1.7. The receptors are long α-helical molecules sensitive

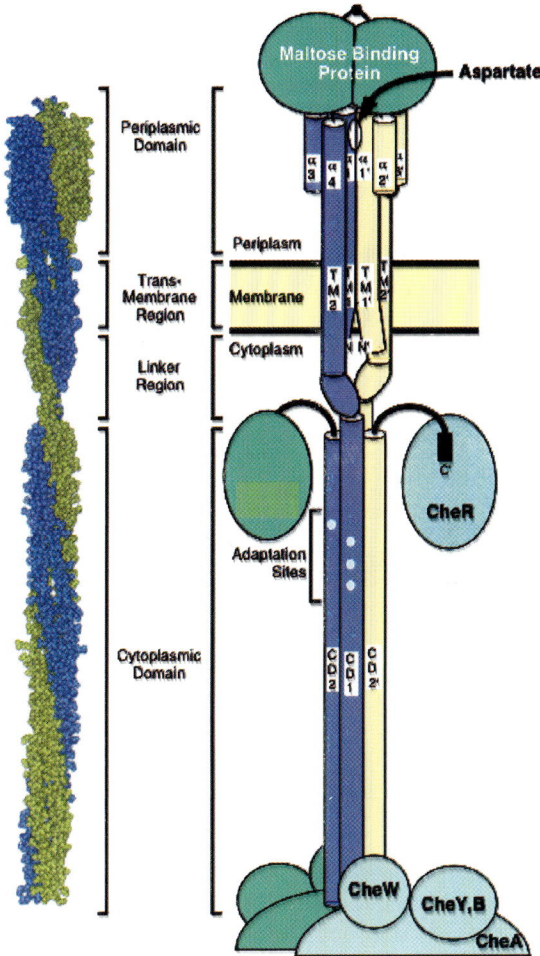

Fig. 1.7. A space-filling model of the aspartate receptor dimer (*left*) and a cartoon, drawn on the same scale, of its complex with other chemotaxis proteins (*right*). The other proteins are described in Fig. 1.8. The dimer is 38 nm long. Aspartate binds directly to a cleft in the periplasmic domain. Maltose binds indirectly via a binding protein that binds to the tip of the periplasmic domain. The labels α (for α-helical), TM (for transmembrane), and CD (for cytoplasmic domain) refer to different regions of the protein. Figure courtesy of Joseph Falke, who used the space-filling model of [20]

to serine (Tsr), to aspartate or maltose (Tar), to ribose or galactose (Trg), or to certain dipeptides (Tap). These ligands bind to a domain in the periplasm, outside the cell's inner membrane, or they bind to a small binding protein that, in turn, binds to this receptor domain.

The rest of the signal-transduction machinery is shown schematically in Fig. 1.8. The receptors control the activity of a kinase, CheA, a molecule that phosphorylates (adds phosphate to) a small signaling molecule called CheY. CheY-P diffuses to the flagellar motors, where it binds and increases the probability of CW rotation. If one adds an attractant, say aspartate, the kinase activity goes down, less CheY-P is made, less binds to the flagellar motors, and runs are extended. However, there is a problem, because the lifetime of CheY-P is too long. So the cell has another molecule, a phosphatase called CheZ, that removes the phosphate. Thus, when the cell is suddenly exposed to aspartate, its motors respond within a few 10ths of a second.

The rapid change in the kinase activity following a large step-addition of an attractant is shown in the diagram at the right (Fig. 1.8). After a few minutes, the cells adapt, i.e., the kinase activity returns to its initial value, even though the attractant is still there. This process involves the methylation of the cytoplasmic domain of the receptor, catalyzed by a methyltransferase, CheR. The methylation sites (adaptation sites) are shown by the 4 grey dots

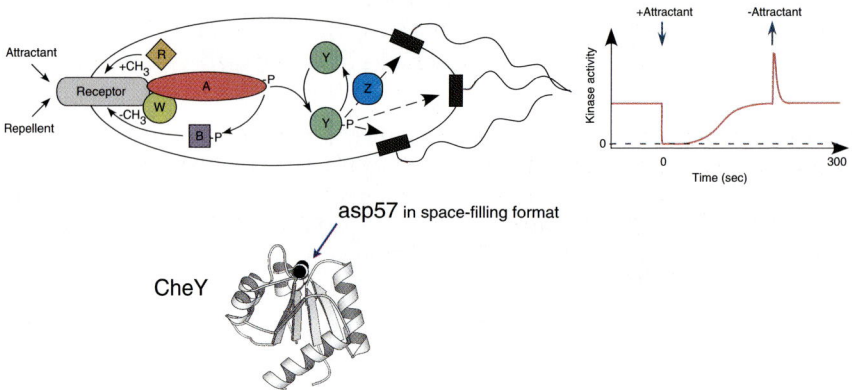

Fig. 1.8. *E. coli's* signal transduction pathway. Receptors in clusters activate a kinase, CheA. When activated, the kinase phosphorylates a small signaling protein called CheY, shown at the bottom in a ribbon diagram. A phosphate group is transferred from his48 of CheA (not shown) to asp57 of CheY (*highlighted*). Copies of CheY-P diffuse to the base of each flagellar motor where they bind, increasing the likelihood of CW rotation. The phosphate is removed by a phosphatase, CheZ. The kinase is linked to the receptor via a coupling factor, CheW. The cytoplasmic domain of the receptor is methylated by a methyltransferase, CheR, and demethylated by the methylesterase CheB, which is also activated by the kinase. The diagram at the right shows changes in kinase activity generated by the sudden addition and eventual removal of a large amount of attractant. Figure courtesy of Victor Sourjik [26]

in Fig. 1.7. When the attractant is removed, these methyl groups are clipped off by a methylesterase, CheB, which is activated by the kinase. Adaptation allows cells to sense changes in their environment without worrying about ambient concentrations. This increases the range of concentrations over which they can respond.

We have been studying interactions of these proteins by fluorescence resonance energy transfer (FRET), as illustrated in Fig. 1.9. An interesting result is shown in Fig. 1.10, in which the change in kinase activity is plotted as a function of the amount of attractant added. On the left are shown a set of dose-response curves obtained by addition of the non-metabolizable aspartate analog, α-methylaspartate. The four sets of curves were obtained at different ambient concentrations of α-methylaspartate, as noted in the figure legend. All of these kinds of data can be condensed into one figure, as shown on the right, if the fractional change in kinase activity is plotted as a function of the fractional change in receptor occupancy. The receptors show prodigious gain: when their occupancy changes by 1%, the kinase activity changes by 35%. Evidently, this amplification occurs via the interactions of adjacent receptors, which are arranged in tight clusters, as shown in Fig. 1.11.

In more recent work, we have studied changes in sensitivity generated by changes in receptor composition [21], imaged the CheZ/CheY-P interaction in different parts of the cell [22], and studied the binding of CheY-P to FliM, its target at the base of the flagellar motor [23]. It turns out that most of the chemotaxis proteins enumerated in Fig. 1.8 are concentrated in the receptor clusters. So much of the current work on bacterial chemotaxis focuses on

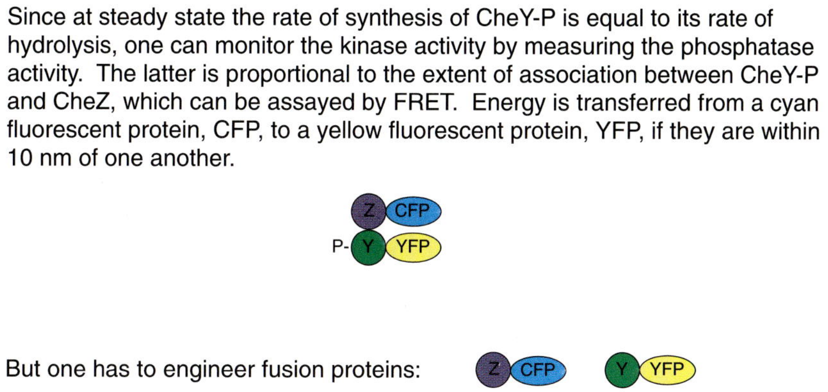

Since at steady state the rate of synthesis of CheY-P is equal to its rate of hydrolysis, one can monitor the kinase activity by measuring the phosphatase activity. The latter is proportional to the extent of association between CheY-P and CheZ, which can be assayed by FRET. Energy is transferred from a cyan fluorescent protein, CFP, to a yellow fluorescent protein, YFP, if they are within 10 nm of one another.

But one has to engineer fusion proteins:

Fig. 1.9. An illustration of how the kinase activity can be monitored by fluorescence resonance energy transfer (FRET). A fusion of one protein is constructed with a cyan fluorescent protein, and a fusion of a second protein is constructed with a yellow fluorescent protein. When the two proteins interact (bind to one another), fluorescence energy can transfer from the cyan to the yellow fluorescent protein. So the cyan fluorescence goes down and the yellow fluorescence goes up. These changes are measured

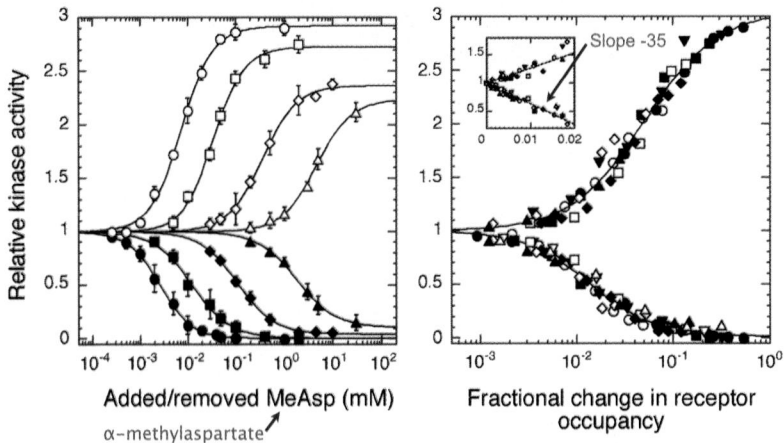

Fig. 1.10. Dose-response curves obtained by the FRET technique. The curves on the left were obtained with cells at different ambient concentrations of the non-metabolizable attractant α-methylaspartate (0, 0.1, 0.5, and 5 mM; circles, squares, diamonds, and triangles, respectively). At a given ambient concentration, more α-methylaspartate was added, as indicated on the abscissa, and the kinase activity was measured (*closed symbol*). After this addition, the cells were allowed to adapt, the concentration of α-methylaspartate was returned to the ambient level, and the kinase activity was measured again (*open symbol*). The cells were allowed to adapt yet again, and the process was repeated with a different increment of α-methylaspartate, until the full dose-response curve was obtained (6 to 10 measurements following the addition and removal of attractant). Finally, a different ambient concentration was chosen, and the entire process was repeated. When the fractional change in kinase activity is plotted as a function of the fractional change in receptor occupancy, all of such data fall on two curves, as shown on the right [24]. Upon addition of attractant, the fractional change in kinase activity is 35 times larger than the fractional change in receptor occupancy. Upon removal of attractant, a similar enhancement is observed (slope not indicated). This amplification occurs via receptor-receptor interactions

attempts to understand how these clusters are assembled and how they function; see the reviews by [25, 26].

Allosteric models for motor swiching, e.g. [27], predict that substantially more CheY-P should bind to FliM when the motor spins CW than when it spins CCW. This is something that we can test by the FRET technique, by looking at motors driving tethered cells. Figure 1.12 shows a cell exposed to a saturating dose of the amino acid serine, that drastically lowers the CheY-P concentration. Under the conditions of this experiment (very little CheZ), more CheY binds to the cluster than does CheY-P, so the cluster brightens. Since the affinity of the motors for CheY is much smaller than that for CheY-P, the motors darken. This experiment shows that one can monitor the binding of CheY-P to single motors.

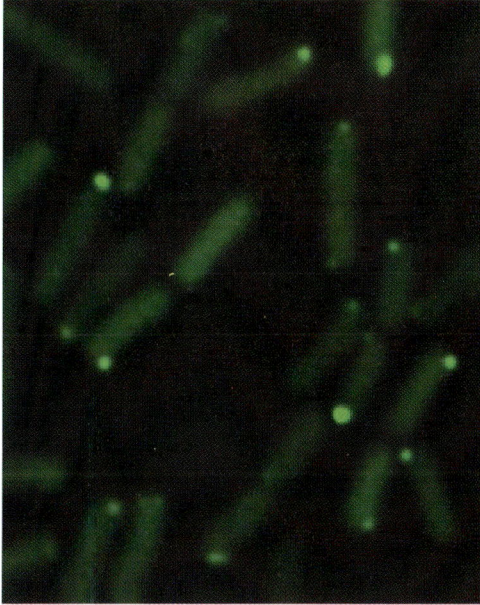

Fig. 1.11. Receptor clusters tagged by YFP-CheR and visualized by fluorescence. The clusters appear as diffraction-limited spots about 0.2 μm in diameter near the cell poles. Photograph courtesy of Victor Sourjik

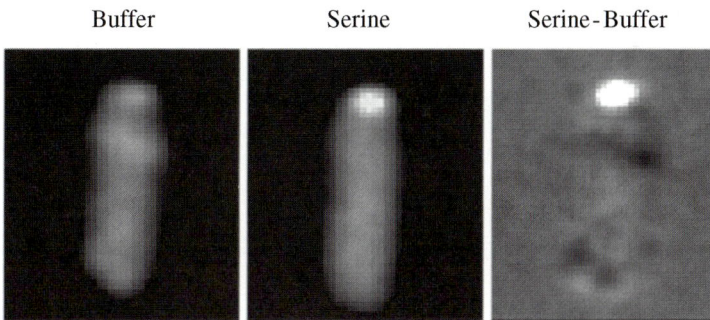

Buffer   Serine   Serine-Buffer

Fig. 1.12. A cell expressing CheY-YFP imaged in a phosphate buffer, before (*left*) and after (*middle*) addition of a saturating dose of the amino acid serine (an attractant). The image on the right is the difference of these two images. The only CheZ expressed by this cell is a variant that does not bind to the cluster; therefore, CheY binds to the cluster (to the kinase CheA) but CheY-P does not. The fluorescence of the fusion protein is the same whether it is phosphorylated or not. The serine response lowers the kinase activity, so CheY-P is converted to CheY. Thus, the receptor cluster brightens as more CheY binds to CheA, and several motors darken, as CheY-P dissociates from FliM. Photograph courtesy of Ady Vaknin

To summarize: *E. coli* is a remarkable nanomachine. It is a self-replicating autonomous robot, about 1 μm in diameter, that lives in an aqueous medium and swims many diameters per second by rotating long helical filaments, each driven at its base by a reversible rotary motor, only 45 nm in diameter. The direction of rotation of these motors, and hence the behavior of the cell, is controlled by receptors in clusters, usually at a cell pole, that count molecules of interest in the surrounding medium and activate a small signaling protein that diffuses through the cytoplasm and binds to the flagellar motors, changing the likelihood that they spin clockwise or counterclockwise. All of this behavioral machinery is built from proteins, objects of exquisite specificity, interacting in different ways, enabling the cell to navigate in a microscopic world.

References

1. J. Adler (1965). Chemotaxis in *Escherichia coli. Cold Spring Harbor Symp. Quant. Biol.*, **30**, pp. 289–292.
2. H.C. Berg (1975). Chemotaxis in bacteria. *Ann. Rev. Biophys. Bioeng.*, **4**, pp. 119–136.
3. H.C. Berg (2004). *E. coli in Motion.* New York: Springer-Verlag.
4. H.C. Berg and R.A. Anderson (1973). Bacteria swim by rotating their flagellar filaments. *Nature*, **245**, pp. 380–382.
5. J. Adler (1973). A method for measuring chemotaxis and use of the method to determine optimum conditions for chemotaxis by *Escherichia coli. J. Gen. Microbiol.*, **74**, pp. 77–91.
6. J. Adler (1969). Chemoreceptors in bacteria. *Science*, **166**, pp. 1588–1597.
7. H.C. Berg and D.A. Brown (1972). Chemotaxis in *Escherichia coli* analysed by three-dimensional tracking. *Nature*, **239**, pp. 500–504.
8. R.M. Macnab, and M.K. Ornston (1977). Normal-to-curly flagellar transitions and their role in bacterial tumbling: Stabilization of an alternative quaternary structure by mechanical force. *J. Mol. Biol.*, **112**, pp. 1–30.
9. L. Turner, W.S. Ryu, and H.C. Berg (2000). Real-time imaging of fluorescent flagellar filaments. *J. Bacteriol.*, **182**, pp. 2793–2801.
10. S.H. Larsen, R.W. Reader, E.N. Kort, W. Tso, and J. Adler (1974). Change in direction of flagellar rotation is the basis of the chemotactic response in *Escherichia coli. Nature*, **249**, pp. 74–77.
11. S.M. Block, J.E. Segall, and H.C. Berg (1982). Impulse responses in bacterial chemotaxis. *Cell*, **31**, pp. 215–226.
12. J.E. Segall, S.M. Block, and H.C. Berg (1986). Temporal comparisons in bacterial chemotaxis. *Proc. Natl. Acad. Sci. USA*, **83**, pp. 8987–8991.
13. M. Silverman and M. Simon (1974). Flagellar rotation and the mechanism of bacterial motility. *Nature*, **249**, pp. 73–74.
14. R.M. Berry and H.C. Berg (1999). Torque generated by the flagellar motor of *Escherichia coli* while driven backward. *Biophys. J.*, **76**, pp. 580–587.
15. H.C. Berg (2003). The rotary motor of bacterial flagella. *Annu. Rev. Biochem.*, **72**, pp. 19–54.

16. D.F. Blair (2003). Flagellar movement driven by proton translocation. *FEBS Lett.*, **545**, pp. 86–95.
17. R.M. Macnab (2004). Type III flagellar protein export and flagellar assembly. *Biochim. Biophys. Acta*, **1694**, pp. 207–217.
18. H.C. Berg and L. Turner (1993). Torque generated by the flagellar motor of *Escherichia coli. Biophys. J.*, **65**, pp. 2201–2216.
19. X. Chen and H.C. Berg (2000). Torque-speed relationship of the flagellar rotary motor of *Escherichia coli. Biophys. J.*, **78**, pp. 1036–1041.
20. K.K. Kim, H. Yokota, and S.-H. Kim (1999). Four helical bundle structure of the cytoplasmic doman of a serine chemotaxis receptor. *Nature*, **400**, pp. 787–792.
21. V. Sourjik and H.C. Berg (2004). Functional interactions between receptors in bacterial chemotaxis. *Nature*, **428**, pp. 437–441.
22. A. Vaknin and H.C. Berg (2004). Single-cell FRET imaging of phosphatase activity in the *Escherichia coli* chemotaxis system. *Proc. Natl. Acad. Sci. USA*, **101**, pp. 17072–17077.
23. V. Sourjik and H.C. Berg (2002a). Binding of the *Escherichia coli* response regulator CheY to its target measured *in vivo* by fluorescence resonance energy transfer. *Proc. Natl. Acad. Sci. USA*, **99**, pp. 12669–12674.
24. V. Sourjik and H.C. Berg (2002b). Receptor sensitivity in bacterial chemotaxis. *Proc. Natl. Acad. Sci. USA*, **99**, pp. 123–127.
25. J.S. Parkinson, P. Ames, and C.A. Studdert (2005). Collaborative signaling by bacterial chemoreceptors. *Curr. Opin. Microbiol.*, **8**, pp. 1–6.
26. V. Sourjik (2004). Receptor clustering and signal processing in *E. coli* chemotaxis. *Trends Microbiol.*, **12**, pp. 569–576.
27. T.A.J. Duke, N. Le Novère, and D. Bray (2001). Conformational spread in a ring of proteins: a stochastic approach to allostery. *J. Mol. Biol.*, **308**, pp. 541–553.

2

Myosin Motors: The Chemical Restraints Imposed by ATP

I. Rayment[‡] and J. Allingham

University of Wisconsin, Department of Biochemistry, Madison, Wisconsin, USA
jallingham@biochem.wisc.edu

Most molecular motors use ATP. This is not surprising since this is the universal currency of energy for most of life's processes. The question to be addressed in this chapter is how does the fundamental chemistry of ATP hydrolysis influence the observed organization of linear molecular motors seen today with a focus on myosin. This chapter is written to ask in simple terms what can be gained by reconsidering the chemistry of ATP hydrolysis.

2.1 Chemistry and Thermodynamics of ATP Hydrolysis

Adenosine triphosphate is a simple molecule whose hydrolysis to either ADP and inorganic phosphate or AMP and pyrophosphate is used to drive many otherwise thermodynamically unfavorable reactions. Its value arises from the large exergonic free energy associated with these hydrolysis reactions ($\Delta G^{\circ\prime} = -32.2$ and -30.5 kJ $\cdot$ mol^{-1} respectively [1]). Compounds such as ATP are often referred to as "high energy" intermediates, which is a qualitatively convenient way of accounting for its role in biological processes. However, such terminology is inadequate for explaining the molecular basis for its exergonic properties, or more importantly, accounting for the chemical and atomic details that are seen in the enzymes that use this substrate.

From the chemical perspective the phosphoester bond joining the β and γ phosphoryl groups is no different from any other single phosphorous oxygen linkage, so that the high energy is not associated with the linkage itself. Indeed, ATP is a surprisingly stable molecule in aqueous solution, which contributes to its value as a participant in biological transformations. The source of the large free energy of hydrolysis arises from differences in electronic and electrostatic properties of the reactants and products. One source of the free

---

[‡] Address correspondence to: Ivan Rayment, Department of Biochemistry, University of Wisconsin, 433 Babcock Drive, Madison, WI 53706, Ivan_Rayment@biochem.wisc.edu

I. Rayment and J. Allingham: *Myosin Motors: The Chemical Restraints Imposed by ATP*, Lect. Notes Phys. **711**, 15–40 (2007)
DOI 10.1007/3-540-49522-3_2   © Springer-Verlag Berlin Heidelberg 2007

energy change is the reduction in the electrostatic repulsion between the negatively charged groups of the phosphoanhydride on hydrolysis. Likewise there is greater resonance stabilization for the products. Another source is increased solvation of the products relative to the reactants. These details influence how ATP is accommodated by all enzymes that utilize ATP but are particularly important where the energy of hydrolysis is utilized to generate work, because they dictate when and how the free energy is released. It is important to realize that the free energy of hydrolysis cannot be directly transferred from the bond cleavage event, since in solution this energy would be lost immediately to the solvent. Instead, the conversion of the free energy of hydrolysis into work must be coupled to differences in binding energy for the reactants and products by the motors during their motile cycles.

The actual free energy released upon ATP hydrolysis depends on the concentration of the reactants and products as well as the pH, divalent metal ion concentration, and ionic strength. These variables are difficult to measure accurately and vary depending on time and location within an organism, but under normal physiological conditions ΔG is estimated to be ≈ 50 kJ $\cdot$ mol^{-1}. Importantly, divalent cations such as Mg^{2+} have a high affinity for ATP and are always involved in phosphoryl transfer reactions, so that in all instances the true substrate is M^{2+} ATP rather than the nucleotide alone. These variables have a profound influence on myosin's hydrolytic and motor function where parameters such as ADP and phosphate concentration are well known to influence and in many cases regulate the activity of myosin [2–4].

2.2 Hydrolysis of MgATP

The hydrolysis of ATP is fundamentally a phosphoryl transfer reaction which has been studied extensively both in biological and chemical systems. From the latter it is widely accepted that the transfer of the phosphoryl moiety from ATP to water is predominantly a dissociative process that proceeds through a metaphosphate-like transition state with little bond order to the attacking nucleophile (water) or leaving ADP (Fig. 2.1) [5]. There has been extensive discussion over the nature of the transition state for phosphoryl transfer within an enzyme active site. Most of the questions focus on whether the observed rate enhancement is the result of transition state stabilization of a fundamentally metaphosphate-like transition state or whether it is the result of a shift to a more associative mechanism where there is expected to be substantial bond formation between the attacking water molecule and the phosphorus atom. This is a very difficult question to address experimentally for myosin due to the complexity of its kinetic cycle. Techniques such as kinetic isotope effects, which allow one to probe the nature of the transition state, are difficult to apply because the bond cleavage event is not the rate-limiting step.

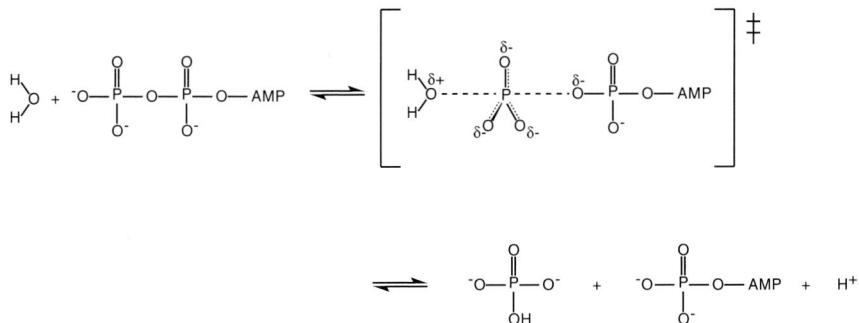

Fig. 2.1. Hydrolysis of ATP. The exact charges and resonance structure of the meta-phosphate like transition state are not defined since this depends on the extent of bond cleavage and formation

Regardless of the ambiguity surrounding the nature of the transition state for enzymatically catalyzed systems, it is clear that these catalysts dramatically increase the rate of hydrolysis relative to that which happens in solution. The first order uncatalyzed rate in water at pH 7 and 25 °C is $\sim 5 \times 10^{-9} \cdot \sec^{-1}$ (Calculated from [5] and [6]), whereas the bond-splitting rate of ATP hydrolysis by myosin at 25°C is $> 100 \cdot \sec^{-1}$ which represents a rate enhancement of $\sim 2 \times 10^{10}$. Indeed, based on the Eyring equation it can be estimated that a rate enhancement of 2×10^{10} demands a reduction in free energy of the transition state for hydrolysis of $\sim 57 \text{ kJ} \cdot \text{mol}^{-1}$. Although this rate enhancement sounds impressive, it is fairly modest compared to many other enzymatically catalyzed reactions [7]. Given the intrinsic stability of ATP in solution, the evolution of an effective way of reducing the free energy of the transition state for the hydrolytic step is central to the use of phosphoanhydrides as an effective energy source. In the absence of a dramatic rate enhancement these compounds would be kinetically useless as energy sources. This rate enhancement is a fundamental property of the active site of myosin that must be explained in molecular terms before a full understanding of its chemical mechanism is achieved.

2.3 Kinetic Cycle for Myosin

The fundamental kinetic cycle of myosin is well established and although there are important differences between myosin isoforms they all appear to follow the general cycle shown in Fig. 2.2 (For excellent reviews of the kinetics of myosin and its isoforms see [8–12]).

The cycle is often viewed as starting at a point where myosin is bound to actin in the absence of ATP (the rigor state), though this is not the most populated state for most myosins. The first step in the cycle is the binding of MgATP to myosin which induces a series of conformational changes

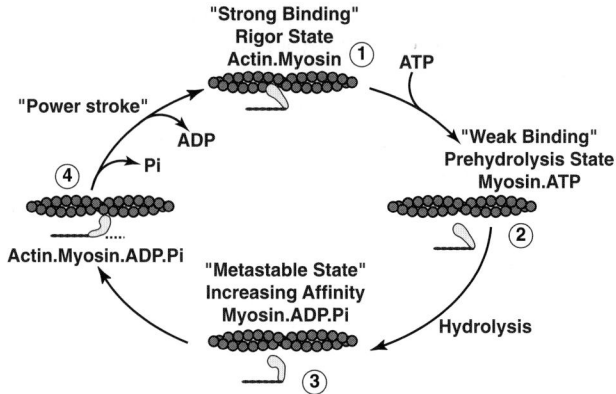

Fig. 2.2. Traditional kinetic cycle for myosin. A schematic of the alternation between the strongly and weakly bound states depending on the nucleotide bound to myosin

in myosin that lower the binding affinity of myosin for actin from $\sim 2 \times 10^7$ M^{-1} to $\sim 1.5 \times 10^2$ for skeletal muscle myosin [13, 14]. In the second stage, also accompanied by a conformational change in the myosin head, ATP is hydrolyzed but not released from the active site. This is often described as the metastable state because bond cleavage has occurred, yet the hydrolysis products have not been released. The conformational change at this stage may be viewed as "priming" the molecule for the powerstroke. Thereafter myosin acquires a renewed affinity for actin and during rebinding to actin, phosphate is released and the powerstroke initiated. The release of phosphate is seen as the key step in the transition from the weakly bound non-stereospecific states to the strongly bound stereospecific states. The final kinetic step is the release of MgADP. The scheme shown here implies at least four distinct conformations for myosin around the pathway.

At first sight, the hydrolytic cycle utilized by myosin appears counterintuitive since the hydrolysis event is uncoupled from the force generating step, but on more careful examination, it is a scheme in which the fundamental chemistry associated with the hydrolysis of ATP can be accommodated efficiently within a biological machine. The issue here is one of rate constants and efficiency.

Even though it is comfortable to feel that biological motors are fast, in molecular terms they are quite slow, both in the speed with which they move and their kinetic turnover number. Typically the fastest myosin motor, such as the unconventional myosin from characean alga [15], moves at ~ 50 $\mu m \cdot sec^{-1}$ whereas the typical turnover number of actin-activated skeletal myosin is $\sim 10 \cdot sec^{-1}$ though many are much slower. In contrast, the intrinsic bond cleavage rate for ATP hydrolysis by myosin is $>100 \cdot sec^{-1}$, where the lower value for the overall turnover number reflects that the slow step is product release [16,17]. This fundamental paradox is true for all molecular motors. Thus, in all motors there must be some form of modulation of the rate constants for binding and release of product in order to accommodate the fundamental differences in rate between the chemical and biological processes. Remarkably, these adaptations allow biological motors to operate with outstanding thermodynamic efficiency (upwards of 50% for some myosin-based motors).

The scheme that has evolved for myosin reflects that for most isoforms the motor spends only a limited time associated with actin. That is, myosin has a low duty ratio, which is defined as the fraction of time spent on actin [18]. In those cases where myosin spends a considerable time bound to actin (as with myosin V for example), modulation of the release rate of ADP changes the motor from a low to high ratio motor [19, 20].

2.4 Structures of Myosin

At this time there is a wealth of structural information available for myosin from a wide variety of classes of myosin. These form the foundation of a sound model for the contractile cycle, which though not complete, provides insight into the utilization of ATP as an energy source. A summary of the high-resolution structures that have been deposited in the Protein Data Bank is given in Table 2.1. The current structures can be divided into roughly three classes: the prehydrolysis state, metastable state, and almost rigor. At present major structural questions that need to be addressed are the structure of the actomyosin complex both in the presence and absence of ADP.

The first molecular model for myosin was derived for chicken skeletal myosin subfragment-1 which revealed the overall architecture of the myosin head [21], much of which is shared by all myosin isoforms. On the basis of this structure a model for the contractile cycle was proposed [22], much of which has been shown to be correct. As seen in Fig. 2.3a, the myosin head is characterized by a motor domain attached to an extended α-helical segment that serves as a binding site for two light chains (in skeletal muscle myosin). The globular motor domain is split by two clefts. One of these constitutes the nucleotide-binding pocket, whereas the other divides the distal part of the molecule into an upper and lower domain (often referred to as the 50 kD domains reflecting earlier proteolytic studies on myosin.) The second cleft is now understood to mediate the conformational changes in the nucleotide-binding site with the actin binding interface. It is now well established that the motor domain alone is sufficient to generate force through its hydrolysis of ATP and interaction with actin, where the extended helical segment that binds the light chains serves to amplify the conformational changes and increase the size of the power stroke.

Initially it was unknown where the structure of chicken skeletal myosin subfragment-1 fit into the contractile cycle for myosin since the structure was solved in the absence of nucleotide. The approximate fit of the tail and motor domain into the image reconstruction suggested that the overall arrangement

Table 2.1. Conformations of Current High-Resolution Myosin X-ray Crystal Structures

Myosin	PDB code	Nucleotide	50 kDa cleft	Conformation	Reference
Gg SkII	2MYS	MgSO$_4$	open	Pre-hydrolysis	[21]
Dd II	1FMW	MgATP	open	Pre-hydrolysis	[26]
Dd II	1FMV	none	open	Pre-hydrolysis	[26]
Dd II	1MMD	MgADP.Bef$_x$	open	Pre-hydrolysis	[23]
Sc II	1B7T	MgADP	open	Pre-hydrolysis	[27]
Gg V	1W7J	MgADP.Bef$_x$	open	Pre-hydrolysis	[30]
Dd II	1MND	MgADP.Alf$_4$	partially closed	Metastable	[23]
Dd II	1VOM	MgADP.VO$_4$	partially closed	Metastable	[24]
Gg SmII	1BR1, 1BR2	MgADP.Alf$_4$	partially closed	Metastable	[28]
Sc II	1QVI	MgADP.VO$_4$	partially closed	Metastable	[65]
Dd I (MyoE)	1LKX	MgADP.VO$_4$	partially closed	Metastable	[66]
Dd II	1Q5G	none	partially closed	Near-rigor	[67]
Sc II	1S5G	MgADP	open??	Near-rigor	[68]
Sc II	1SR6	none	open??	Near-rigor	[68]
Gg V	1OE9, 1W8J	none	closed	Rigor-like	[29, 30]
Pig VI	2BKH,	none	closed	Rigor-like	[34]

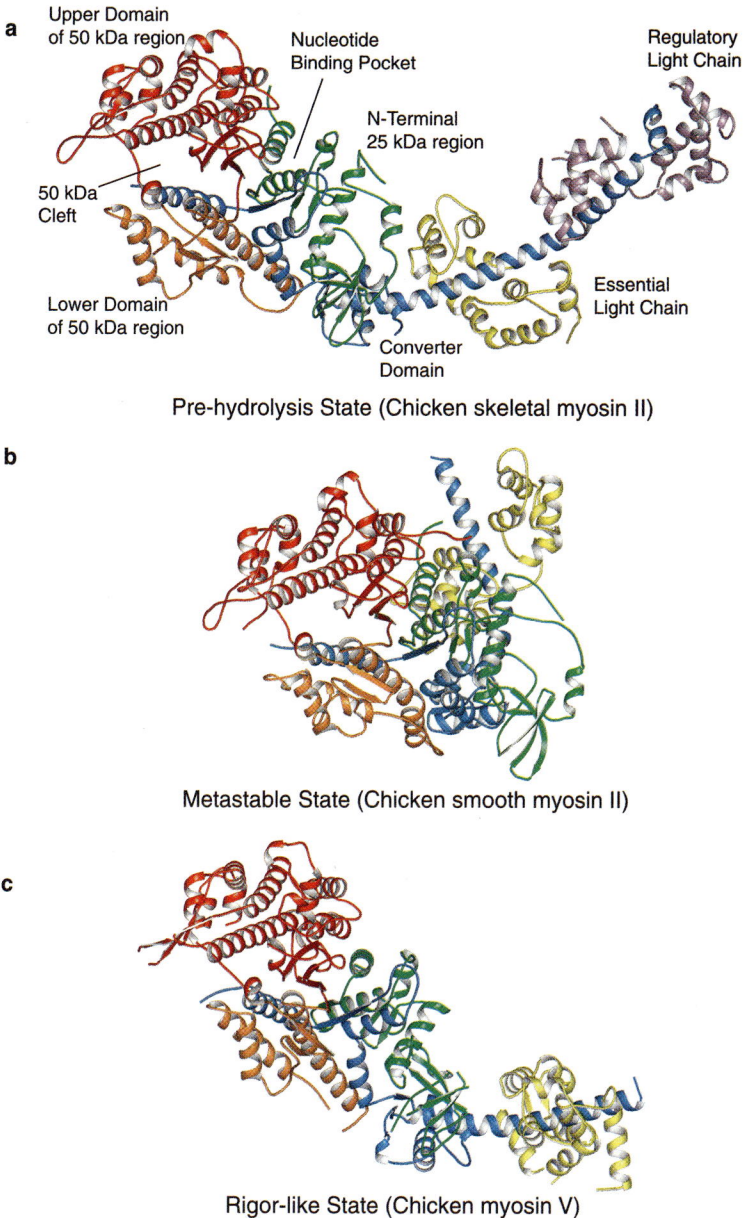

a

Upper Domain
of 50 kDa region

Nucleotide
Binding Pocket

N-Terminal
25 kDa region

Regulatory
Light Chain

50 kDa
Cleft

Lower Domain
of 50 kDa region

Essential
Light Chain

Converter
Domain

Pre-hydrolysis State (Chicken skeletal myosin II)

b

Metastable State (Chicken smooth myosin II)

c

Rigor-like State (Chicken myosin V)

Fig. 2.3. Major conformational states of myosin. The pre-hydrolysis state (*top*) is represented by the structure of the S1 portion of chicken skeletal myosin II (PDB code 2MYS) [21]. The major subdomains are labeled and have been colored similarly for each myosin isoform shown. Light chains have been included in their respective conformations bound to the lever arm of each model. The metastable state (*middle*) is represented by the MgADP.Alf$_4$ form of chicken smooth muscle myosin II (PDB code 1BR1) [28]. The rigor-like state (*bottom*) is represented by the apo form of chicken myosin V (PDB code 1OE9) [29]

of domains might be similar to the rigor state, however an unfavorable interaction between the lower 50 kD domain of myosin and actin implied that the cleft should close when myosin binds to actin [22]. To examine this question, the structure of the motor domain of myosin II from *Dictyostelium discoideum* (S1dC) was determined in the presence of nucleotide analogs that mimicked the prehydrolysis and transition state for hydrolysis [23–26]. This revealed that the structure observed in chicken skeletal myosin subfragment-1 represents a prehydrolysis state (post-rigor) where the cleft that splits the 50 kD region is open and is configured in such a way that it is unable to hydrolyze ATP. In particular, the hydrogen bonding network at the apex of the nucleotide-binding pocket is unable to support the in-line nucleophilic attack of a water molecule on the γ-phosphate.

The open nature of the 50 kD cleft appears to be the defining feature of the prehydrolysis state, rather than the relative orientation of the light chain binding domain since structural studies of scallop myosin reveal that this section of the molecule can adopt a wide range of conformations [27]. This is consistent with the energetics of the contractile cycle. The key feature of the prehydrolysis state is that it should bind ATP tightly and cause a reduction in the affinity of myosin for actin. The determinant for this property resides in the motor domain, whereas the position of the tail is unimportant immediately after release from actin. Indeed, it could be argued that a single well defined position for the tail in the prehydrolysis state would be energetically unfavorable since it would reduce the off-rate of myosin from actin and would slow the conformational change to the metastable state.

In the contractile cycle, understanding the peculiarity of the metastable state is central to establishing the molecular connection between ATP hydrolysis and energy transduction. Structural studies with transition state analogs, first with *Dictyostelium* S1dC [23, 24] and then with chicken smooth muscle myosin motor domain with its essential light chain [28], revealed how the active site is configured for ATP hydrolysis and provided insight into the molecular basis for retention of hydrolysis products. Importantly, the structural studies of smooth muscle myosin motor domain with its essential light chain show a major change in orientation of the light chain-binding region relative to the motor domain, which is communicated from the nucleotide-binding site via the converter domain (Fig. 2.3b). It is widely accepted that this conformation is representative of the start of the powerstroke (pre-powerstroke conformation) and is the best model for the metastable state.

The final stage in the contractile cycle involves the rebinding of myosin to actin and release of first phosphate and then MgADP. Phosphate release is known to be the committing step in initiating the powerstroke. This is clearly a multi-step process and at this time there are still questions about the final structure of acto-myosin. Fortunately, the recent structures of myosin V have provided insight into the conformational changes that are associated with this process [29, 30]. The original studies on chicken skeletal myosin subfragment-1 suggested that the large cleft in the 50 kD region of the motor domain might close when myosin binds to actin. Indirect evidence for this has been provided through the use of spectroscopic probes [31] and electron microscopy [32, 33], however the most definitive evidence has come from the study of myosin V [29] and more recently from the structures of myosin VI [34].

Structural studies of myosin V show that for this myosin isoform the large cleft is closed when the protein is in the apo form and the lever arm is in a position characteristic of that in the actomyosin rigor complex (Fig. 2.3c) (PDB accession numbers 1OE9, 1W8J) [29]. Furthermore, the nucleotide-binding elements (P-loop, Switch I and II) within the nucleotide-binding pocket, (including the β-sheet), have adopted previously unseen conformations that interfere with ATP binding, demonstrating the structural communication between myosin's actin-binding surface and nucleotide-binding site. This communication is transmitted through the coordinated movement of subdomains that are interconnected at conserved points within the motor domain of all myosins. The β-sheet of the motor domain is an essential component of this communication network.

In the apo structure of myosin V, closure of the cleft involves repositioning the upper 50-kDa subdomain relative to the N-terminal subdomain, producing a much closer alignment of the upper and lower 50-kDa subdomains. This results in exclusion of a substantial amount of water relative to the solvent content observed in open cleft myosin structures. In addition, the conformation of the nucleotide-binding pocket is altered relative to the weak actin-binding pre-hydrolysis state such that the distance between the P-loop and switch I has increased sufficiently to disable their ability to coordinate Mg^{2+} and for switch I to interact with the nucleotide. Also, the position of the P-loop observed within the active site appears to obstruct nucleotide entry. Switch II has adopted a new conformation as well. This element is one of the interconnection points described above that is intimately involved in positioning the subdomains that are involved in cleft closure and movement of the lever arm.

The structure of myosin V rigor-like crystals soaked with MgADP showed a slight movement of the P-loop to accommodate the nucleotide, which was held weakly (PDB accession numbers 1W7I [30]). However, no additional rearrangements of the switch I element were observed that would enable coordination of Mg^{2+} or the γ-phosphate. This provides strong evidence for the order of products release as $Pi > Mg^{2+} > ADP$. Based on the conformational changes spanning the pre-hydrolysis state for myosin II to those observed in the rigor-like state of myosin V, it is predicted that the sequence of events following the initial binding of the phosphate chain of ATP to the P-loop would involve an inward movement of switch I, through flexion of the β-sheet, in order to stabilize the γ and β-phosphates and Mg^{2+} [30]. As part of the communication network of the motor domain, repositioning of switch II during these events would then stimulate opening of the cleft and progression to the low actin-affinity pre-hydrolysis state [35].

While these structures emphasize the connection between the actin-binding interface and nucleotide-binding site during the contractile cycle, they also reveal the gap in our knowledge of the structural states that embody a tight actin-binding state with tightly bound MgADP, as well as a state showing the release of phosphate.

2.5 Active Site of Myosin

The structural studies of the metastable state, as mimicked by the structures with transition state analogs, provide insight into how the chemical properties of ATP are accommodated by this family of molecular motors. As noted earlier, myosin provides dramatic rate enhancement for the hydrolysis of ATP compared to the uncatalyzed rate in water. Indeed, the bond splitting rate achieved is far higher than the stepping rate of this family of molecular motors. It appears that this is achieved first and foremost by providing a framework that positions the hydrolytic water in the correct location for in-line acceptance of the metaphosphate-like species present at the transition state. Partial closure of the 50 kDa cleft and the associated rearrangement of the hydrogen-bonding network around the γ-phosphoryl moiety is a central component of this framework. This conformational change also serves to isolate the active site from bulk solvent, which aids in retention of the hydrolysis product (Fig. 2.4). Of course the movements of the loops associated with the active site are also coupled to large conformational changes that serve to prime the molecule for the powerstroke.

In earlier years it was suggested that myosin and most other phosphoryl transferases would require or utilize a base to catalyze the removal of the proton from the "attacking" nucleophile. This suggestion has profoundly influenced progress in the literature and discussion of ATP and GTP dependent hydrolytic processes. Indeed, the reaction scheme shown in Fig. 2.1 is highly unfavorable unless the proton on the attacking water molecule is removed. The question is when is the proton removed and where does it go? Is it abstracted by a base (side chain in the active site) or does it depart later in the reaction pathway? Under normal circumstances the pKa of a proton on water or an alcohol is $\sim$14. Furthermore, in efficient acid/base catalysis the pKa's of the catalytic groups generally match the pKa of the proton that is removed.

The structures of the MgADP $\cdot$ VO$_4$ complexes indicate that no catalytic base is present in the active site of myosin, which suggests that the water molecule is not deprotonated prior to its attack on the γ-phosphorous atom. Overall this arrangement is consistent with the solution studies of phosphoryl transfer, which indicate that the transition state should be metaphosphate-like and would predict that there is little bond formation to the nucleophile. If this is true, then it implies that formation of the bond to the attacking water occurs after the transition state is reached and that bond formation will lead to a decrease in the pKa of the attached protons, which will facilitate proton transfer. Thus the question becomes where does the proton on the

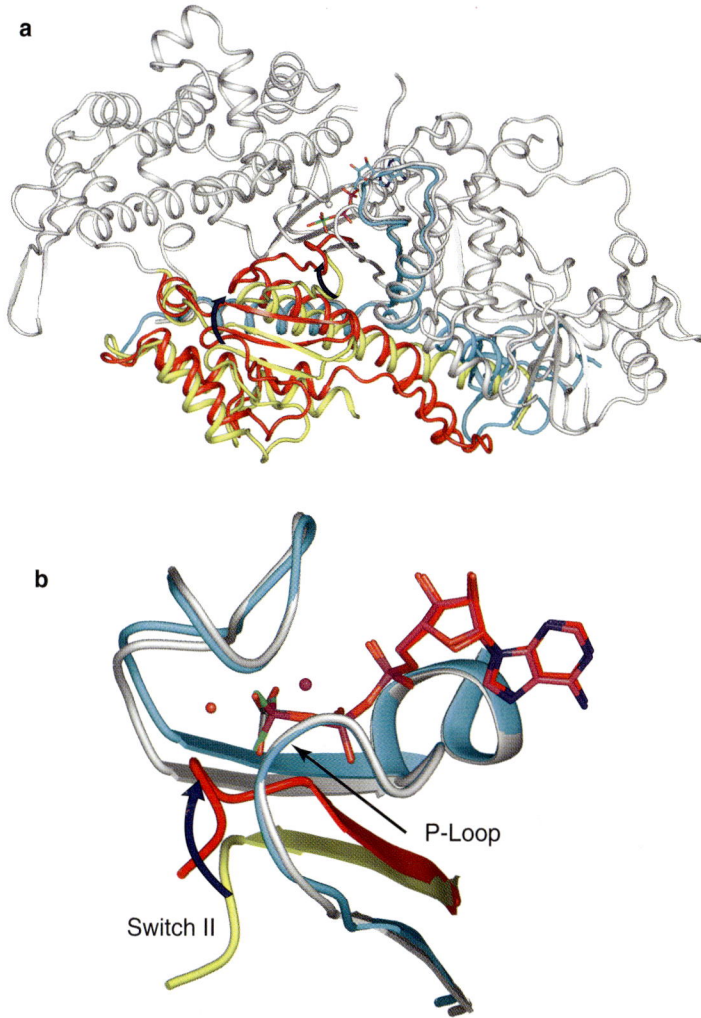

Fig. 2.4. Superposition of *Dictyostelium* S1dC in the prehydrolysis and metastable state as mimicked by the complexes with $MgADP \cdot BeF_x$ and $MgADP \cdot VO_4$. (**a**) shows an overview of the molecules to reveal the long range effects of binding a transition state analog in the active site. (**b**) shows the local changes associated with the γ-phosphate binding pocket. In particular the altered conformation of Switch II serves to close-off the active site from bulk solvent and position the water for nucleophilic attack on the γ-phosphoryl moiety. In this figure the entire polypeptide chain for the $MgADP \cdot BeF_x$ complex is included where the lower domain of its 50 kD region is depicted in yellow. For clarity only lower domain of the 50 kD region and C-terminal segment are shown for the $MgADP \cdot VO_4$ complex. These are depicted in red and cyan respectively. The figure was prepared with the program Pymol from coordinates with the accession numbers 1MMD and 1VOM respectively [23, 24]

water molecule (now seen as being attacked by metaphosphate) go? This is an important issue, since early kinetic studies show that the proton produced in the hydrolysis of ATP is released at the same time as the phosphate [36], which implies that the proton from the nucleophilic water remains in the active site in the metastable state.

Computational and experimental studies on small G-proteins, which utilize a similar nucleotide binding motif, suggest that the ultimate base in their hydrolytic reactions is the γ-phosphate [37, 38] which raised the possibility that these enzymes and myosin proceed via a mechanism that incorporates substrate-assisted general base catalysis. At one extreme this would entail transfer of the proton from the attacking water molecule to the γ-phosphate followed by nucleophilic attack by a hydroxyl anion where this would imply an associative mechanism. Solution kinetic measurements do not support this mechanism [39], but together the results suggest that the proton from the hydrolytic water is eventually transferred to the γ-phosphate. How this is achieved is subject to discussion. Direct transfer of the proton from the water molecule to a phosphoryl oxygen would require a four-center transfer which is stereochemically unfavorable. In myosin, a six center transfer has been proposed in which the proton is shuttled by hydrogen exchange from the hydrolytic water to the γ-phosphate via Ser236 [24].

The structures of the transition state complexes provide a satisfactory explanation for how myosin catalyzes the hydrolysis of ATP, but there remains the question of what prevents the release of inorganic phosphate once the hydrolytic event has occurred. This is important since the stability of this state is a central feature of the motile cycle of myosin.

Insight into the stability of the metastable state can be gained by simply considering the chemistry of ATP hydrolysis. Based on the earlier description of the source of the free energy, the metastable state must accomplish the following. It must minimize the loss of free energy due to reduction in the electrostatic repulsion between the negatively charged groups of the phosphoanhydride upon hydrolysis. In addition, it must prevent the resonance stabilization and solvation of the products relative to the reactants. Although no structure is available for the true metastable state, calculations which place the phosphate into the active site provide insight into how these molecular restraints are accommodated by the active site (Fig. 2.5) [40].

The issue of electrostatic repulsion appears to have been solved by the polar/ionic environment that neutralizes the charge on the γ- and β-phosphates. Every lone pair on the phosphoryl oxygens participates fully in hydrogen bonding or ionic interactions, the latter of which include contributions from the magnesium ion and Lys185. This same set of interactions prevents the phosphate ion (in the model) from being solvated. The few water molecules that are observed in the $ADP \cdot VO_4$ complex participate in a well-ordered hydrogen bonding network. Finally the ionic environment serves to localize the protons on the phosphate and eliminates resonance stabilization and prevents their loss to solvent.

a

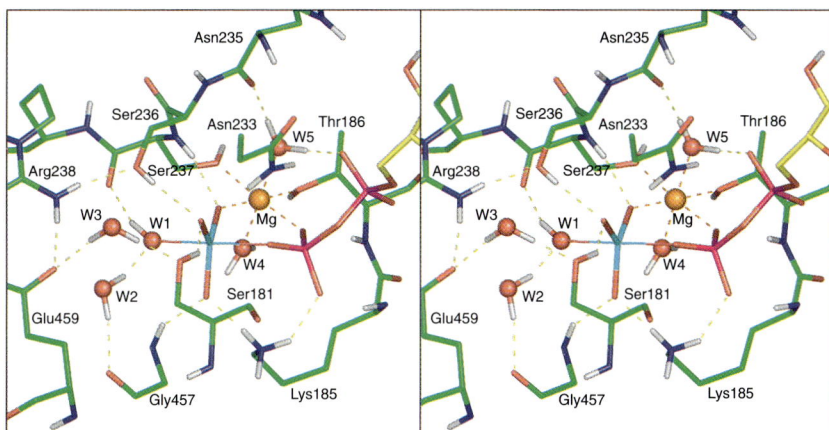

b

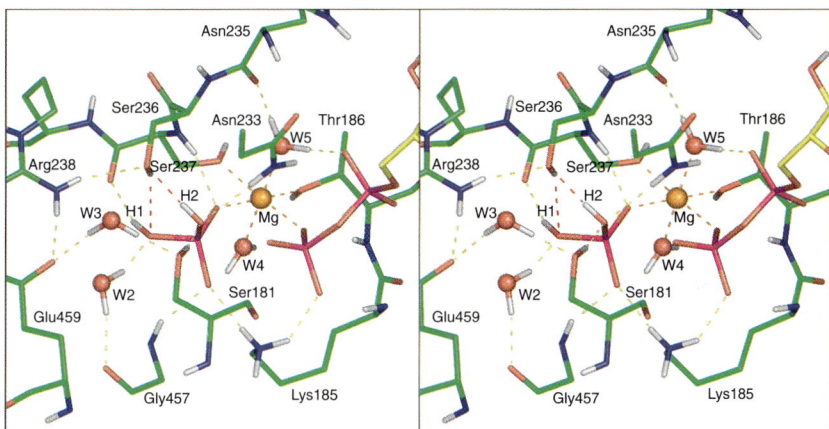

Fig. 2.5. Model for the interaction of phosphate with the metastable state. (**a**) shows a stereo view of the γ-phosphate pocket in the MgADP $\cdot$ VO$_4$ form of S1Dc myosin II. Hydrogen bonds are shown as dashed yellow lines. Mg^{2+} and its coordinating bonds are shown in orange. Waters involved in the coordination of the γ-phosphate and Mg^{2+} are labeled W1 to W5. W1 is depicted as the attacking nucleophile partially bound to the vanadate moiety. (**b**) shows a stereo view of the MgADP.VO$_4$ form of S1Dc with phosphate modeled in place of vanadate. Modeling of phosphate into the active site was done by superimposing the coordinates for the phosphoanhydride chain from the molecular dynamics and combined QM/MM reaction path calculations described in reference [40] onto those of the MgADP.VO$_4$ form of S1Dc (PDB accession number 1VOM, [24]). The putative positions of the protons acquired by phosphate during hydrolysis are shown as H1 and H2. The path of a portion of the proton shuttling route is shown as dashed red lines. The phosphate atoms are shown in magenta, oxygen in red, vanadate in cyan, nitrogen in blue, hydrogen in white, carbon atoms of myosin in green, and the nucleotide in yellow. The figures were prepared with the program Pymol

From the perspective of this model, release of phosphate, which occurs at the start of the power stroke, might easily be triggered by a small change in the ionic interactions within the γ-phosphate binding pocket. Although the exact nature of this signal is unknown, release of the hydrolysis product in an ordered manner is essential for establishing a conformational cycle in which energy can be converted into work. It is also clear from the structural studies to date that phosphate must depart via a different route from which it entered [41].

2.6 Comparison with G-proteins: Molecular Switches

The active site for myosin shares considerable structural similarities with that of the G-proteins so that developments in this field have profoundly influenced our understanding of myosin [42]. The widely used "switch I" and "switch II" nomenclature to describe conformational switches associated with the magnesium and γ-phosphate binding sites, respectively, arose from studies of the G-proteins [43]. From a functional view-point, both of these groups of proteins utilize a hydrolytic event to change the binding affinity of the protein for its ligands (both proteins and nucleotides). G-proteins can be viewed as molecular switches that alternate between GTP and GDP forms, which differ in their biological function. In general terms the GDP form is inactive in signaling whereas the GTP form is active. In the absence of other factors, the rate of hydrolysis of GTP by G-proteins is exceedingly slow (for in-depth reviews see [44] and [45]). The close similarity between the active sites of the G-proteins and myosin suggests that there should be strong mechanistic similarities (Fig. 2.6).

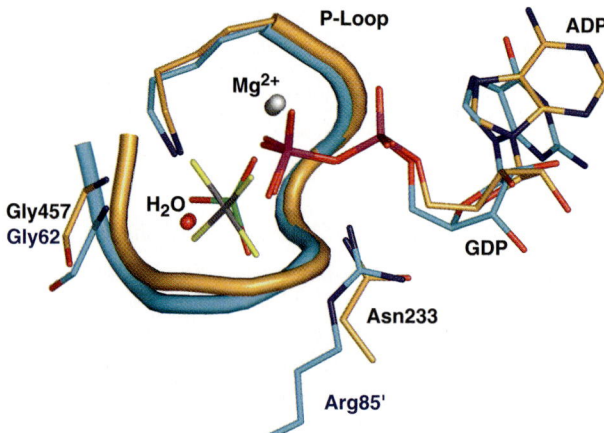

Fig. 2.6. Comparison of the ATP and GTP γ-phosphate binding pocket in myosin and the G-proteins. Myosin and RhoA are depicted in yellow and cyan respectively. The figure for the myosin S1dC·MgADP·VO$_4$ and RhoA·GDP·AlF$_4^-$ complexes was prepared with the program Pymol from coordinates with the accession numbers 1VOM and 1TX4 respectively [24, 48]

As noted above, G-proteins by themselves are exceedingly poor GTPases, whereas when bound to G-protein activating proteins (GAPs) they hydrolyze GTP at ~15 s^{-1}, which is ~10^5 faster than the uncatalyzed rate [46]. Structural studies of Ras and Rho, two small G-proteins, bound to their respective GAPs in the presence of MgGDP · AlF$_x$ which mimics the transition state for hydrolysis, show how catalysis is achieved [47, 48]. In the activated state, most but not all, GAPs insert an arginine residue into the active site [49]. In those proteins that utilize an arginine, this is an essential part of the catalytic machinery [50]. It has been proposed that, the arginine stabilizes the negative charge on the transition state in an associative mechanism, however its role in a more dissociative mechanism is less clear.

A major functional difference between the G-proteins and myosin is the maintenance of the metastable state in myosin, which allows for a three state system. In the G-proteins, which alternate between two conformations, there is no requirement for controlling the release of the free energy of hydrolysis once the bond-cleavage event has occurred. Careful examination of the structure of RhoA and its GTPase-activating protein complexed with MgGDP · AlF$_4$ together with a model of how phosphate might bind indicates that in this active site the phosphate released from the hydrolysis of GTP would not exist as H$_2$PO$_4^-$ (Fig. 2.7). Rather, the ionic environment surrounding the terminal AlF$_4^-$ or phosphate (as modeled) demands that at least one proton must be lost after the initial bond cleavage and transfer of the proton from the attacking water molecule to the γ-phosphate must occur. This is due to the presence of the highly conserved glutamine and arginine residue, which when phosphate is modeled in the active site are expected to form hydrogen bonding and ionic interactions with the same phosphoryl oxygen. This is in contrast to myosin where the same phosphoryl oxygen serves as a hydrogen bond donor to Ser236 and acceptor to Ser181 (Fig. 2.8). Release of a proton will result in a decrease in free energy due to resonance stabilization. This will also encourage release of the phosphate from the active site due to charge repulsion between the terminal phosphoryl group of GDP and inorganic phosphate. Kinetic studies reveal that phosphate release is fast and not the rate limiting step in the GAP-activated hydrolytic process [46].

Comparison of the active site of G-proteins with that of myosin suggests that the catalytic arginine is not strictly required for hydrolysis, since myosin does not have such a residue and it is not observed in all activated complexes [49]. In myosin, the side chain of Asn233 takes the place of the arginine guanidinium group. The role of the arginine might be to stabilize the product and *prevent* phosphate from being retained in the active site. Certainly the presence of an additional positive charge in the active site close to the γ-phosphoryl moiety will lower the pKa of the phosphate and encourage the release of protons.

It is of interest that no structures of GDP · VO$_4$ bound to a G-protein have been reported [51]. Rather, complexes have only been seen with AlF$_4^-$, which clearly cannot be protonated. The absence of vanadate complexes is

a

b

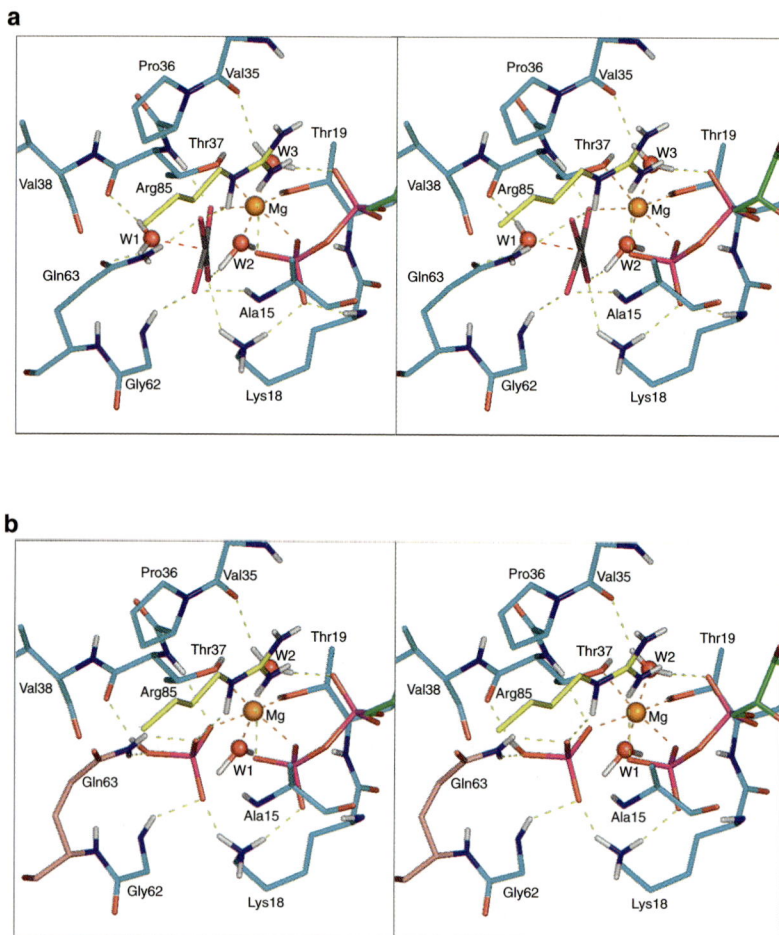

Fig. 2.7. Model for the interaction of phosphate with the metastable state of the Rho/GAP GTPase. (**a**) Shows a stereo view of the γ-phosphate pocket in the MgGDP. AlF_4^- form of RhoA and its GTPase-activating protein. Hydrogen bonds are shown as dashed yellow lines. Mg^{2+} and its coordinating bonds are shown in orange. Waters involved in the coordination of the γ-phosphate and Mg^{2+} are labeled W1 to W3. W1 is depicted as the attacking nucleophile bound (*red dashed line*) to the AlF_4^- moiety. (**b**) Shows a stereo view of the MgGDP. AlF_4^- form with phosphate modeled in place of AlF_4^-. Modeling of phosphate was done by superimposing the coordinates for the phosphoanhydride chain from the molecular dynamics and combined QM/MM reaction path calculations described in reference [40] onto those of the MgGDP. AlF_4^- form of the Rho/GAP GTPase (PDB accession number 1TX4). The orientation of Gln63 has been modified to illustrate a potential structural transition and hydrogen bond formation upon GTP hydrolysis. The phosphate atoms are shown in magenta, oxygen in red, aluminum in grey, fluorine in pink, nitrogen in blue, hydrogen in white, carbon atoms of Rho/GAP GTPase in cyan and yellow, and the nucleotide in green. The figures were prepared with the program Pymol

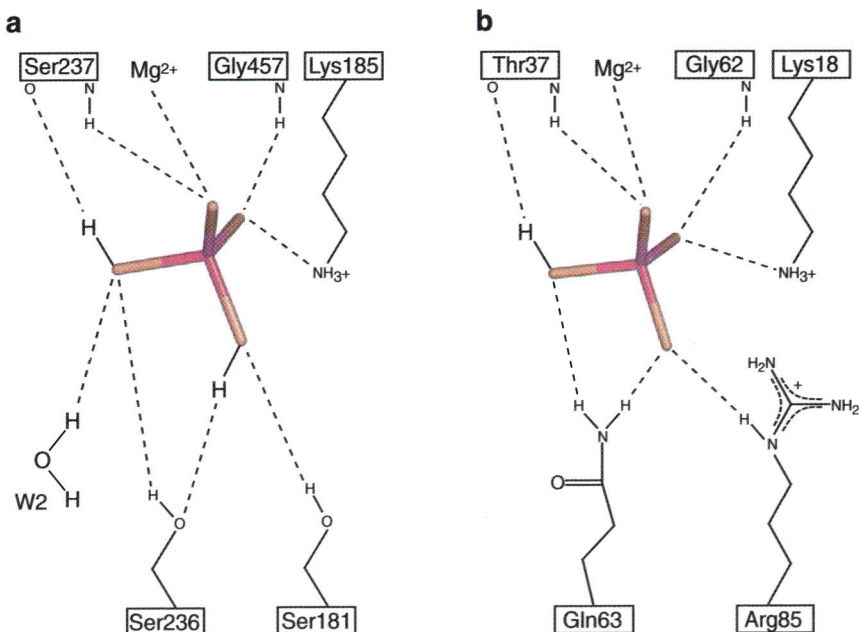

Fig. 2.8. Schematic representations of the coordination of the γ-phosphate modeled into myosin S1dC and the Rho/GAP GTPase. (**a**) Shows the predicted proton positions and the hydrogen bonds and Mg^{2+} interaction for the γ-phosphate in S1Dc myosin. (**b**) Shows the interactions in the γ-phosphate pocket of Rho/GAP GTPase. In this figure phosphate atoms are shown in magenta and oxygen atoms (including W2 from Fig. 2.7) are in red

consistent with the hypothesis that G-proteins have an active site that is complementary to a deprotonated phosphate, which is expected to shift the equilibrium towards hydrolysis and release of phosphate.

Comparison of the active sites of myosin and the G-proteins provides an explanation for why myosin is able to maintain the metastable state and control the release of free energy from the hydrolysis of ATP. It is of interest then, whether the same principles are maintained in other molecular motors such the kinesin superfamily which move along microtubules.

2.7 Kinesin Based Motors

The kinesin family of molecular motors has more members than the myosin superfamily and in many ways is intellectually more challenging because many kinesins are intrinsically processive motors. The study of kinesins has a much shorter history, but fortunately, developments over the last several years have established a sound kinetic model for its motile cycle [52]. A brief summary

of the kinetic model for conventional kinesins is presented in order to allow
for a discussion of the nucleotide-binding site in the molecular motor.

This group of proteins is characterized by their ability to travel long distances along microtubules, which serve as their tracks, without dissociating.
These motors are defined as having a very high duty ratio [18]. As with many
members of the myosin superfamily, conventional kinesins are dimeric molecules where normal movement depends on interactions between the two motor
domains.

Kinesins come in many shapes and sizes, but are all characterized by having
a segment of their sequence that can be identified as a motor domain. This
motor domain contains a core of typically ~330 amino acids which is much
smaller than the motor domain of myosin (~760 amino acids). Also, in contrast
to the myosin superfamily which predominately places its motor domain at
the N-terminus, the motor domain in the kinesin superfamily may be located
at either end or in the middle of the polypeptide. In keeping with their major
function as machines for moving organelles, all kinesins have sections that have
been implicated in cargo-binding, where in many cases this includes additional
proteins (light chains). Also, in most cases the polypeptide chain that connects
the motor domain to the rest of the molecule includes an α-helical segment
that allows the motor domains to homodimerize.

As noted, kinesins move their cargoes along microtubules where most kinesins appear to move in a processive hand-over-hand manner along the filament (Fig. 2.9). Each step corresponds to a longitudinal translation of ~80 Å
which is equivalent to the distance between equivalent sites on adjacent tubulin heterodimers. Thus, the fundamental question is how is the hydrolysis of
ATP coupled to a unitary step along the filament and how is futile loss of
hydrolysis products prevented.

The kinetic cycle for conventional kinesin is complex. A recent model for
the stepping activity of kinesin is shown in Fig. 2.9 [53, 54]. A central component of this model is that the two motor domains are enzymatically out of
phase, which ensures that only one head is firmly attached to the microtubule
at a time. This scheme starts with head 1 tightly bound with the other in a
dissociated state coordinated to ADP. Binding of ATP to head 1 prompts a
conformational change that causes the head 2 to swing forward to the next microtubule binding site where head 2 enters a weakly bound state which rapidly
releases ADP. Hydrolysis of ATP in head 1 is followed by tight binding of head
2 to the leading site. Loss of phosphate from head 1 allows detachment of the
now trailing head. A key component of this model is the strain dependent
communication between the two heads which provides directionality to the
cycle as well as providing a resolution to the earlier confusion associated with
the solution studies [54]. It also places the hydrolysis and phosphate release at
a specific location in the cycle, which is thermodynamically essential for an
energy transducing system. Significantly, both for myosin and kinesin phosphate release is coupled to a rate-limiting step in the hydrolytic cycle. Therefore, as with myosin, the structural states of kinesin that lead up to and

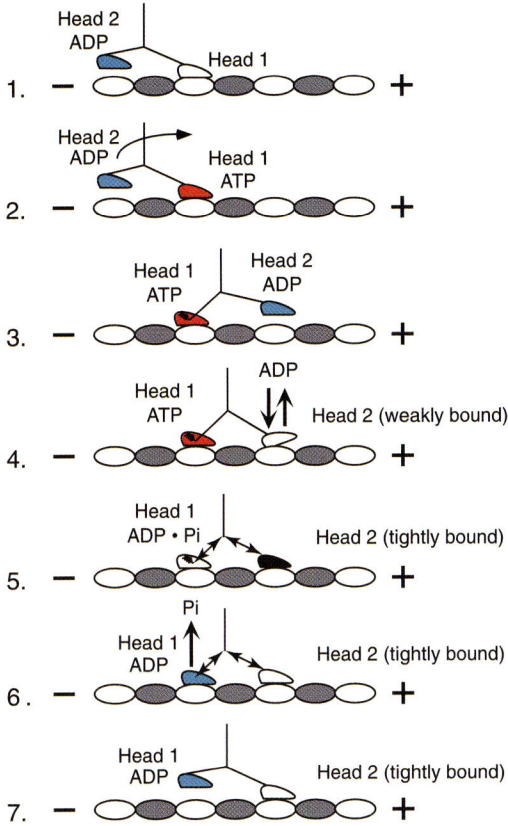

Fig. 2.9. Kinetic cycle for kinesin-based motors. Adapted from Fig. 2.1 in reference [53] and [54]

follow the hydrolysis of ATP are central to understanding the molecular basis of energy transduction.

A large number of high-resolution structures are available for motor domains of members of the kinesin superfamily together with a few structures of dimeric species. These include several structures of kinesin motor domains bound to non-hydrolysable and transition-state analogs. The implications of these structures in developing a model for the stepping cycle have been recently reviewed [55].

As with myosin, there is no high resolution structure for kinesin bound to microtubules, however an image reconstruction of the motor domain of KIF1A in its ADP and ADPPCP state bound to microtubules [56] has provided insight into the overall arrangement of the active site of kinesin relative to the α/β subunits of the tubulin monomers. As shown in Fig. 2.10, the nucleotide-binding site lies in close proximity to an α/β tubulin dimer. This reconstruction in conjunction with the more recent structures of KIF1A

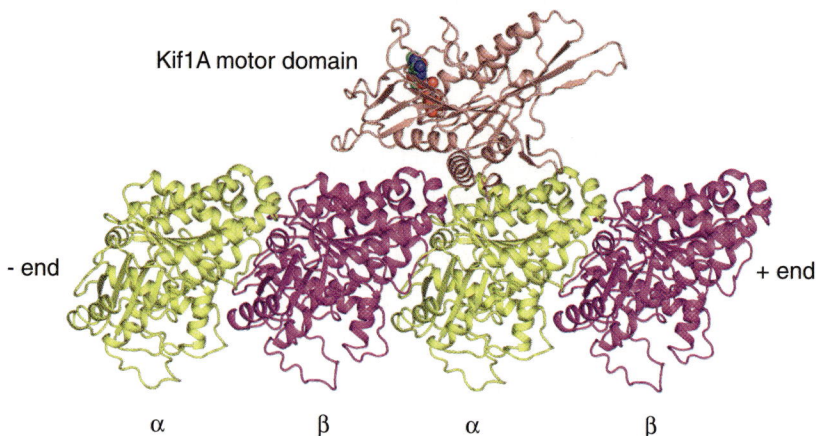

Fig. 2.10. Model for the interaction of one motor domain of kinesin with two α/β heterodimers of tubulin based on an image reconstruction of the motor domain from KIF1A bound to microtubules [56]. The orientation was chosen to show the close relationship between the nucleotide binding site and the microtubule-binding interface. The figure was prepared with the program Pymol from coordinates with the accession number 1IA0 [56]

bound to transition state analogs forms the basis of the current model for the stepping cycle of the kinesin superfamily of molecular motors [57].

The structure of the kinesin motor domain bears many topological similarities to myosin, even though it is considerably smaller (Fig. 2.11) [58–60]. It utilizes a P-loop motif to coordinate the nucleotide and contains the same switch I and II elements common to the G-proteins [61]. As such the requirements for hydrolyzing ATP and retaining the hydrolysis products seen in myosin should be observed in these structures. The initial structures of members of the kinesin superfamily all contained ADP reflecting the high affinity of kinesin for the hydrolysis product. The second phase of structures incorporated non-hydrolyzable analogs and complexes of vanadate and metalofluoride with MgADP all bound to the motor domain of KIF1A [56,57]. It should be noted that KIF1A is a monomeric kinesin and has been associated with a diffusive Brownian mechanism [62,63], however it has also been demonstrated that this motor can dimerize and under these circumstances functions as a processive motor [64]. Given the sequence and structural similarity between KIF1A and conventional kinesins it is likely that this motor shares the same molecular mechanism for energy transduction described above.

Examination of the MgADP · AlF$_x$ complex with KIF1A motor domain suggests, based on the discussion developed earlier for myosin and the G-proteins, that the observed complex does not represent either the phosphate pre-release structure or a model for the transition state for hydrolysis. As can be seen, the active site is quite open and there are limited contacts

Chicken Skeletal Myosin S1

Human Kinesin Motor Domain (1BG2)

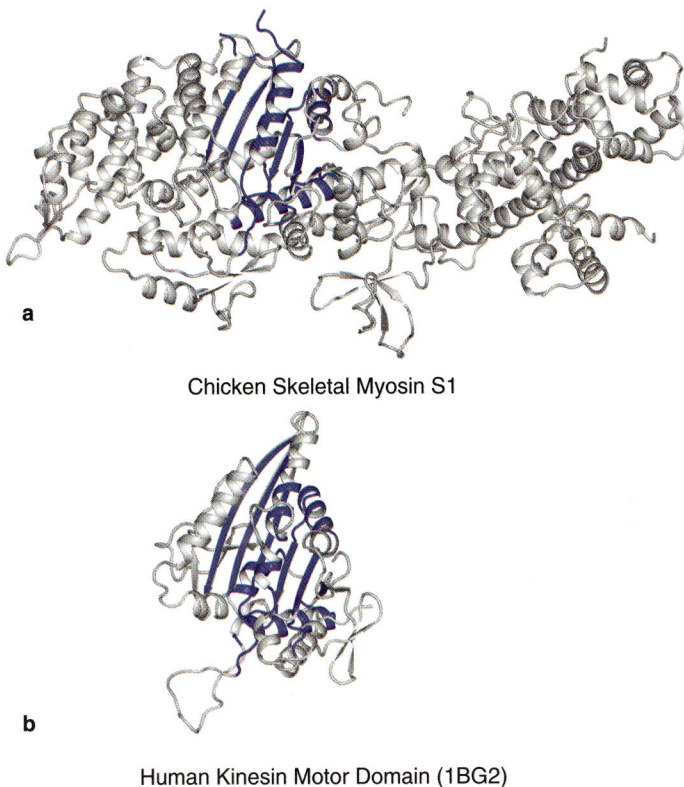

Fig. 2.11. Comparison of myosin subfragment 1 and a kinesin motor domain. This shows the topological similarities surrounding the nucleotide-binding site these two classes of motor domain. The secondary structural elements that are shared by both motor domains are depicted in blue. The figure was prepared with the program Pymol from coordinates with the accession numbers 2MYS and 1BG2 [21, 58]

between the protein and the terminal AlF_x moiety (Fig. 2.12). In addition the absolutely conserved glycine residue in Switch II is not coordinated directly to the AlF_x moiety as observed in myosin or the G-proteins, which suggests that the transition state analog has not bound in the location expected for the γ-phosphoryl group. Likewise, the components of Switch I are not coordinated to the magnesium ion as found in all other P-loop containing proteins. It can be readily envisaged that in its current form, this conformation could not retain phosphate since there is nothing that will prevent the loss of protons or provide charge neutralization necessary to retain the product in the active site. Simple chemical consideration of the requirements for hydrolyzing ATP and retention of the hydrolysis product, in conjunction with the kinetic model, suggests that there is much more to be learned about the active site of the kinesin superfamily. It can be stated with certainty that the none of

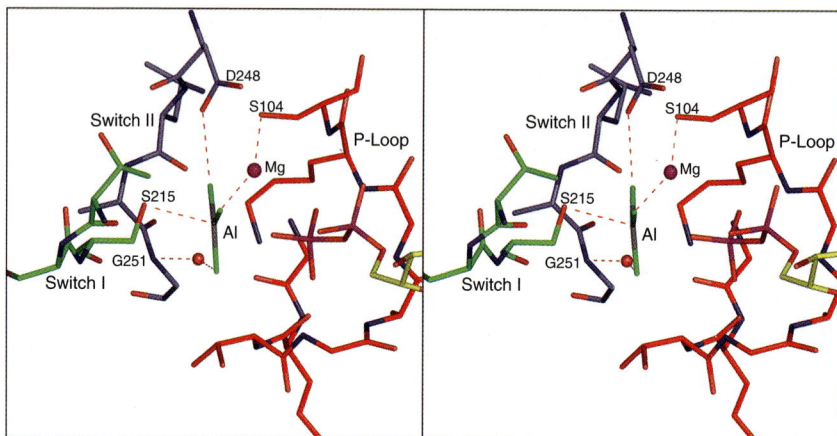

Fig. 2.12. Stereo view of the coordination of AlF$_3$ to the kinesin motor domain of KIF1A. The P-Loop, Switch I, and Switch II are depicted with red, green, and cyan carbon atoms, respectively. The hydrogen bonding and ionic interactions of the AlF$_3$ moiety with the contents of the active site are shown in *dashed red lines*. As can be seen, the coordination sphere of this phosphate analog is far from complete. The figure was prepared with Pymol with coordinates with the accession number 1VFX [57]

the current high-resolution models reflect the true state for the active site of kinesin when it is bound to microtubules prior to or after the committing chemical step.

2.8 Conclusions

The study of molecular motors draws on disciplines that cover all fields of scientific endeavor extending from cell biology, through chemistry and biochemistry, into physics and mathematics. Progress is made when the understanding gained from one area is applied across interdisciplinary boundaries. Ironically the fundamental constraints observed and accepted within one discipline are often unknown to others. The purpose of this review has been to reconsider the fundamental chemical features of nucleotide hydrolysis within the context of recent developments in our structural and kinetic knowledge of the motile cycles of myosin and kinesin.

Simple consideration of the source of free energy from the hydrolysis of a phosphodiester bond provides immediate insight into the observed organization of the kinetic cycles of linear motors. Because of the great differences in kinetic rate of bond hydrolysis and cycle time of the motors, both myosin and kinesin have evolved mechanisms for coupling the release of product with a fundamental step in the motor cycle. In turn, this demands that the motor proteins themselves must have a structural framework or defined conformation that serves to retain the hydrolysis products. While it is clear how this is achieved for myosin, it is less clear that this state has been defined for the kinesin superfamily. For both systems, questions of communication between interacting components still remain. In the case of myosin it is unknown how the interactions between actin and myosin serve to trigger the release of phosphate, whereas for the kinesin superfamily, the exact nature of the signal from one head to the other which triggers the release of phosphate remains to be discovered.

Acknowledgements

We thank Dr. Qiang Cui (Department of Chemistry, University of Wisconsin) for stimulating discussions. This work was supported by a research grant to IR from the National Institutes of Health (AR35186) and by a fellowship to JSA from the Canadian Institutes of Health Research (64606).

References

1. W. P. Jenks (1976). Handbook of biochemistry and molecular biology: proteins. G. D. Fasman. Cleveland, Ohio, CRC Press. **1**, pp. 296–304.
2. M. Nyitrai and M. A. Geeves (2004). Adenosine diphosphate and strain sensitivity in myosin motors. *Philos Trans R Soc Lond B Biol Sci* **359**, pp. 1867–77.
3. Y. Takagi, H. Shuman, et al. (2004). Coupling between phosphate release and force generation in muscle actomyosin. *Philos Trans R Soc Lond B Biol Sci* **359**, pp. 1913–20.
4. A. V. Somlyo, A. S. Khromov, et al. (2004). Smooth muscle myosin: regulation and properties. *Philos Trans R Soc Lond B Biol Sci* **359**, pp. 1921–30.
5. S. J. Admiraal and D. Herschlag (1995). Mapping the Transition State for ATP Hydrolysis: Implications for Enzymatic Catalysis. *Chemistry and Biology* **2**, pp. 729–739.
6. S. J. Admiraal and D. Herschlag (1999). Catalysis of phosphoryl transfer from ATP by amine nucleophiles. *Journal of the American Chemical Society* **121**, pp. 5837–45.
7. R. Wolfenden (2003). Thermodynamic and extrathermodynamic requirements of enzyme catalysis. *Biophys Chem* **105**, pp. 559–72.
8. M. A. Geeves and K. C. Holmes (1999). Structural mechanism of muscle contraction. *Annu Rev Biochem* **68**, pp. 687–728.
9. J. Howard (2001). Mechanics of motor proteins and the cytoskeleton. Sunderland, Mass., Sinauer Associates, Publishers.
10. S. Wakelin, P. B. Conibear, et al. (2002). Engineering Dictyostelium discoideum myosin II for the introduction of site-specific fluorescence probes. *J Muscle Res Cell Motil* **23**, pp. 673–83.
11. E. M. De La Cruz and E. M. Ostap (2004). Relating biochemistry and function in the myosin superfamily. *Curr Opin Cell Biol* **16**, pp. 61–7.

12. W. Zeng, P. B. Conibear, et al. (2004). Dynamics of actomyosin interactions in relation to the cross-bridge cycle. *Philos Trans R Soc Lond B Biol Sci* **359**, pp. 1843–55.

13. L. E. Greene and E. Eisenberg (1978). Formation of a ternary complex: actin, 5′-adenylyl imidodiphosphate, and the subfragments of myosin. *Proc Natl Acad Sci USA* **75**, pp. 54–8.

14. J. A. Sleep and R. L. Hutton (1978). Actin mediated release of ATP from a myosin-ATP complex. *Biochemistry* **17**, pp. 5423–30.

15. T. Kashiyama, N. Kimura, et al. (2000). Cloning and characterization of a myosin from characean alga, the fastest motor protein in the world. *J Biochem (Tokyo)* **127** pp. 1065–70.

16. C. R. Bagshaw and D. R. Trentham (1973). Reversibility of Adenosine-Triphosphate Cleavage by Myosin. *Biochemical Journal* **133**, pp. 323–328.

17. H. D. White, B. Belknap, et al. (1997). Kinetics of nucleoside triphosphate cleavage and phosphate release steps by associated rabbit skeletal actomyosin, measured using a novel fluorescent probe for phosphate. *Biochemistry* **36** pp. 11828–36.

18. J. Howard (1997). Molecular motors: structural adaptations to cellular functions. *Nature* **389**, pp. 561–7.

19. M. Nyitrai, G. Hild, et al. (2000). Flexibility of myosin-subfragment-1 in its complex with actin as revealed by fluorescence resonance energy transfer. *Eur J Biochem* **267**, pp. 4334–8.

20. S. S. Rosenfeld, A. Houdusse, et al. (2005). Magnesium regulates ADP dissociation from myosin V. *J Biol Chem* **280**, pp. 6072–9.

21. I. Rayment, W. R. Rypniewski, et al. (1993). Three-Dimensional Structure of Myosin Subfragment-1: A Molecular Motor. *Science* **261**, pp. 50–58.

22. I. Rayment, H. M. Holden, et al. (1993). Structure of the Actin-Myosin Complex and its Implications for Muscle Contraction. *Science* **261**, pp. 58–65.

23. A. J. Fisher, C. A. Smith, et al. (1995). X-ray Structures of the Myosin Motor Domain of *Dictyostellium discoideum* Complexed with MgADP · BeF$_x$ and MgADP · AlF$_4^-$. *Biochemistry* **34**, pp. 8960–72.

24. C. A. Smith and I. Rayment (1996). X-ray structure of the Magnesium(II).ADP.Vanadate Complex of the *Dictyostelium discoideum* Myosin Motor Domain to 1.9 Å Resolution. *Biochemistry* **35**, pp. 5404–17.

25. A. M. Gulick, C. B. Bauer, et al. (1997). X-ray structures of the MgADP, MgATPgammaS, and MgAMPPNP complexes of the Dictyostelium discoideum myosin motor domain. *Biochemistry* **36**, pp. 11619–28.

26. C. B. Bauer, H. M. Holden, et al. (2000). X-ray structures of the apo and MgATP-bound states of Dictyostelium discoideum myosin motor domain. *J Biol Chem* **275** pp. 38494–9.

27. A. Houdusse, V. N. Kalabokis, et al. (1999). Atomic structure of scallop myosin subfragment S1 complexed with MgADP: a novel conformation of the myosin head. *Cell* **97**, pp. 459–70.

28. R. Dominguez, Y. Freyzon, et al. (1998). Crystal structure of a vertebrate smooth muscle myosin motor domain and its complex with the essential light chain: visualization of the pre-power stroke state. *Cell* **94**, pp. 559–71.

29. P. D. Coureux, A. L. Wells, et al. (2003). A structural state of the myosin V motor without bound nucleotide. *Nature* **425**, pp. 419–423

30. P. D. Coureux, H. L. Sweeney, et al. (2004). Three myosin V structures delineate essential features of chemo-mechanical transduction. *Embo J* **23**, pp. 4527–37.

31. C. M. Yengo, E. M. De La Cruz, et al. (2002). Actin-induced closure of the actin-binding cleft of smooth muscle myosin. *J Biol Chem* **16**, p. 16.
32. N. Volkmann, D. Hanein, et al. (2000). Evidence for cleft closure in actomyosin upon ADP release. *Nat Struct Biol* **7**, pp. 1147–55.
33. N. Volkmann, G. Ouyang, et al. (2003). Myosin isoforms show unique conformations in the actin-bound state. *Proc Natl Acad Sci USA* **100**, pp. 3227–32.
34. J. Menetrey, A. Bahloul, et al. (2005). The structure of the myosin VI motor reveals the mechanism of directionality reversal. *Nature* **435**, pp. 779–85.
35. K. C. Holmes, R. R. Schroder, et al. (2004). The structure of the rigor complex and its implications for the power stroke. *Philos Trans R Soc Lond B Biol Sci* **359**, pp. 1819–28.
36. C. R. Bagshaw and D. R. Trentham (1974). Characterization of Myosin-Product Complexes and of Product-Release Steps During Magnesium Ion-Dependent Adenosine-Triphosphatase Reaction. *Biochemical Journal* **141**, pp. 331–49.
37. T. Schweins, M. Geyer, et al. (1995). Substrate-assisted catalysis as a mechanism for GTP hydrolysis of p21ras and other GTP-binding proteins. *Nat Struct Biol* **2**, pp. 36–44.
38. T. Schweins and A. Warshel (1996). Mechanistic analysis of the observed linear free energy relationships in p21ras and related systems. *Biochemistry*, **35** pp. 14232–43.
39. S. J. Admiraal and D. Herschlag (2000). The substrate-assisted general base catalysis model for phosphate monoester hydrolysis: Evaluation using reactivity comparisons. *Journal of the American Chemical Society* **122**, pp. 2145–48.
40. G. H. Li and Q. Cui (2004). Mechanochemical coupling in myosin: A theoretical analysis with molecular dynamics and combined QM/MM reaction path calculations. *Journal of Physical Chemistry B* **108**, pp. 3342–57.
41. R. G. Yount, D. Lawson, et al. (1995). Is Myosin a "Back Door" Enzyme? *Biophysical Journal* **68**, pp. 44s–49s.
42. C. A. Smith and I. Rayment (1996). Active Site Comparisons Highlight Structural Similarities Between Myosin and Other P-Loop Proteins. *Biophysical Journal* **70**, pp. 1590–602.
43. M. V. Milburn, L. Tong, et al. (1990). Molecular Switch for Signal Transduction: Structural Differences Between Active and Inactive Forms of Protooncogenic *ras* Proteins. *Science* **247**, pp. 939–45.
44. S. R. Sprang (1997). G protein mechanisms: insights from structural analysis. *Annu Rev Biochem* **66**, pp. 639–78.
45. I. R. Vetter and A. Wittinghofer (2001). The guanine nucleotide-binding switch in three dimensions. *Science* **294**, pp. 1299–304.
46. A. E. Nixon, M. Brune, et al. (1995). Kinetics of inorganic phosphate release during the interaction of p21ras with the GTPase-activating proteins, p120-GAP and neurofibromin. *Biochemistry* **34**, pp. 15592–8.
47. K. Scheffzek, M. R. Ahmadian, et al. (1997). The Ras-RasGAP complex: structural basis for GTPase activation and its loss in oncogenic Ras mutants. *Science* **277**, pp. 333–8.
48. K. Rittinger, P. A. Walker, et al. (1997). Structure at 1.65 A of RhoA and its GTPase-activating protein in complex with a transition-state analogue. *Nature* **389**, pp. 758–62.
49. M. J. Seewald, C. Korner, et al. (2002). RanGAP mediates GTP hydrolysis without an arginine finger. *Nature* **415**, pp. 662–6.

50. M. R. Ahmadian, P. Stege, et al. (1997). Confirmation of the arginine-finger hypothesis for the GAP-stimulated GTP-hydrolysis reaction of Ras. *Nat Struct Biol* **4**, pp. 686–9.
51. D. R. Davies and W. G. Hol (2004). The power of vanadate in crystallographic investigations of phosphoryl transfer enzymes. *FEBS Lett* **577**, pp. 315–21.
52. R. A. Cross (2004). The kinetic mechanism of kinesin. *Trends Biochem Sci* **29**, pp. 301–9.
53. L. M. Klumpp, A. Hoenger, et al. (2004). Kinesin's second step. *Proc Natl Acad Sci USA* **101**, pp. 3444–9.
54. S. S. Rosenfeld, P. M. Fordyce, et al. (2003). Stepping and stretching. How kinesin uses internal strain to walk processively. *J Biol Chem* **278**, pp. 18550–6.
55. E. P. Sablin and R. J. Fletterick (2004). Coordination between motor domains in processive kinesins. *J Biol Chem* **279**, pp. 15707–10.
56. M. Kikkawa, E. P. Sablin, et al. (2001). Switch-based mechanism of kinesin motors. *Nature* **411**, pp. 439–45.
57. R. Nitta, M. Kikkawa, et al. (2004). KIF1A alternately uses two loops to bind microtubules. *Science* **305**, pp. 678–83.
58. F. J. Kull, E. P. Sablin, et al. (1996). Crystal Structure of the Kinesin Motor Domain Reveals a Structural Similarity to Myosin. *Nature* **380**, pp. 550–5.
59. E. P. Sablin, J. F. Kull, et al. (1996). Three-Dimensional Structure of the Motor Domain of NCD, a Kinesin-Related Motor with Reversed Polarity of Movement. *Nature* **380**, pp. 555–9.
60. F. J. Kull, R. D. Vale, et al. (1998). The case for a common ancestor: kinesin and myosin motor proteins and G proteins. *J Muscle Res Cell Motil* **19**, pp. 877–86.
61. E. P. Sablin and R. J. Fletterick (2001). Nucleotide switches in molecular motors: structural analysis of kinesins and myosins. *Curr Opin Struct Biol* **11**, pp. 716–24.
62. Y. Okada and N. Hirokawa (2000). Mechanism of the single-headed processivity: diffusional anchoring between the K-loop of kinesin and the C terminus of tubulin. *Proc Natl Acad Sci USA* **97**, pp. 640–5.
63. Y. Okada and N. Hirokawa (1999). A processive single-headed motor: kinesin superfamily protein KIF1A. *Science* **283**, pp. 1152–7.
64. M. Tomishige, D. R. Klopfenstein, et al. (2002). Conversion of Unc104/KIF1A kinesin into a processive motor after dimerization. *Science* **297**, pp. 2263–7.
65. S. Gourinath, D. M. Himmel, et al. (2003). Crystal structure of scallop Myosin s1 in the pre-power stroke state to 2.6 a resolution: flexibility and function in the head. *Structure (Camb)* **11**, pp. 1621–7.
66. M. Kollmar, U. Durrwang, et al. (2002). Crystal structure of the motor domain of a class-I myosin. *Embo J* **21**, pp. 2517–25.
67. T. F. Reubold, S. Eschenburg, et al. (2003). A structural model for actin-induced nucleotide release in myosin. *Nat Struct Biol* **10**, pp. 826–30.
68. D. Risal, S. Gourinath, et al. (2004). Myosin subfragment 1 structures reveal a partially bound nucleotide and a complex salt bridge that helps couple nucleotide and actin binding. *Proc Natl Acad Sci USA* **101**, pp. 8930–5.

3

How Linear Motor Proteins Work

K. Oiwa[1] and D.J. Manstein[2]

[1] Kobe Advanced ICT Research Center (KARC), National Institute of Information and Communications Technology, 588-2 Iwaoka, Nishi-ku, Kobe 651-2492, Japan
[2] Institutes for Biophysical Chemistry and Structure Analysis, Hannover Medical School, OE4350, Carl-Neuberg-Straße 1, D-30623 Hannover, Germany

3.1 Introduction

Most animals perform sophisticated forms of movement such as walking, running, flying and swimming using their skeletal muscles. Although directed movement is not generally associated with plants, cytoplasmic streaming in plant cells can reach velocities greater than $50\,\mu m/s$ and thus constitutes one of the fastest forms of directed movement. Unicellular eukaryotic organisms and prokaryotes display diverse mechanisms by which they are able to actively move towards a food source, light or other sensory stimuli. On the cellular level active transport of vesicles and organelles is required, since the cytoplasm resembles a gel with a mesh size of approximately 50 nm, which makes the passive transport of organelle-sized particles impossible. For elongated cells such as neurons, even proteins and small metabolites have to be actively transported.

Linear motor proteins, moving on cytoskeletal filaments such as actin filaments and microtubules, are predominantly responsible for the motile activity in eukaryotic cells. They are chemo-mechanical enzymes that use the chemical energy from adenosinetriphosphate (ATP) hydrolysis to generate force and to move cargoes along their filament tracks. Under physiological conditions, the energy input per molecule of ATP corresponds to the chemical free energy liberated by its hydrolysis to adenosinediphosphate (ADP) and inorganic phosphate (P_i), ca. 10^{-19} J (100 pNnm). The thermodynamic efficiency of motor proteins varies between 30 and 60%. As machines, motor proteins are unique since they convert chemical energy to mechanical work directly, rather than through an intermediate such as heat or electrical energy.

3.2 Structural Features of Cytoskeletal Motor Proteins

Three families of linear, cytoskeletal motor proteins have been described: kinesin, dynein, and myosin (Fig. 3.1). Kinesin and dynein family members

K. Oiwa and D.J. Manstein: *How Linear Motor Proteins Work*, Lect. Notes Phys. **711**, 41–63 (2007)
DOI 10.1007/3-540-49522-3_3

move along microtubules while the members of the extended myosin superfamily move along actin filaments. The number of molecular engines and motor proteins has greatly increased with the advance of the various genome
projects. We now know more than eighteen myosin subfamilies or classes.
Thirty-nine myosin heavy chain genes have been found in the human genome;
nine of these are expressed in muscle tissues, while the remainders, the so-
called unconventional myosins, are responsible for cell motilities other than
muscle contraction. Fourteen subfamilies are known for kinesins and three for
dyneins.

Comparisons of the available full-length sequences of dynein heavy chains
have shown that dynein is a member of the AAA$^+$ (ATPase Associated
with various cellular Activities) protein family [1]. So far, more than 200
AAA$^+$ family members have been identified that participate in diverse cellular processes. Dyneins are further divided into two groups: cytoplasmic dynein
and axonemal dynein. Cytoplasmic dynein is composed of two identical heavy
chains of about 530 kDa each. Additionally, cytoplasmic dyneins consist of
two 74 kDa intermediate chains, about four 53–59 kDa intermediate chains,
and several light chains. Axonemal dynein shares the same heavy chain structure but its overall structure is far more complicated and will not be discussed
here in detail (reviewed in [2]).

Dynein heavy chains consist of the C-terminal head with two elongated
flexible structures called the stalk (microtubule-binding domain) and the N-
terminal tail (cargo-binding domain). The head and the stalk form a motor domain (Fig. 3.1d). The motor domains of all dyneins are generally well
conserved in sequence (with 40–80% similarity) [3] and are indistinguishable
by electron microscopy at the single-particle level (reviewed in [4]). A single
dynein motor domain has a mass of almost 380 kDa. Most of this mass is contained in seven protein densities that encircle a cavity to produce a ring-like
architecture, named a head ring. At least six of the seven densities probably
correspond to the highly conserved AAA-modules containing specialized motifs for ATP binding (e.g. P-loop motif). The first four of six AAA-modules
are predicted to bind nucleotides [5]. The first P-loop (P1) has the most highly
conserved sequence among dyneins and was previously assigned as the principal site of ATP hydrolysis by vanadate-mediated photocleavage of the dynein
heavy chain. Further strong support for a functional role of P1 has been provided by molecular dissection of cytoplasmic dyneins in which mutation of
this P-loop abolished motor activitiy [4,6]. The microtubule binding domain
named the stalk, is flanked by two relatively long coiled-coil regions and is
structurally conserved among all dyneins. This stalk extends from the head
between AAA-modules #4 and #5. Its tip interacts with microtubules in an
ATP-sensitive manner.

Based on electron microscopic observation of an axonemal dynein, our
group proposed a model in which multiple conformational changes are coordinated in such a way that the changes between the AAA domain #1 and
its neighbors are amplified by the docking of the head ring onto a linker that

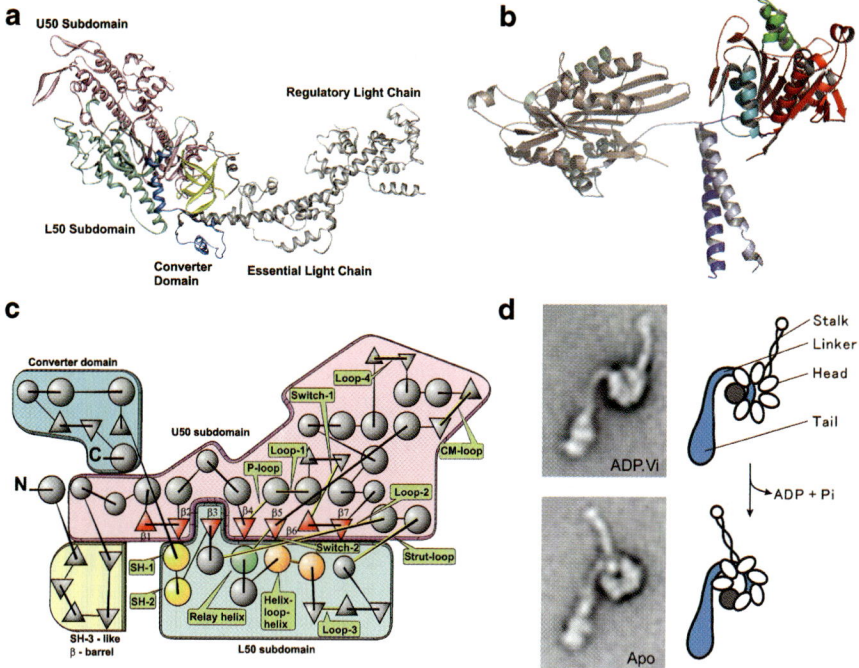

Fig. 3.1. Structure and topology of molecular motors. (**a**) Structure of the myosin motor domain with light chain binding region. N-terminal domain, green; L50 or actin binding domain, cyan; U50, red; C-terminal domain, blue. The essential and regulatory light chains are colored yellow and magenta, respectively. (**b**) Structure of the kinesin 1 motor domain dimer. Switch-1 region, green; switch-2 region, cyan; neck linker and neck helix, blue. (**c**) Topological map of the myosin motor domain. In addition to the domains and subdomains shown, crystallographic results reveal that the segment between $\beta7$ and switch-2 moves as a solid body and can be regarded as an independent subdomain. Helices are shown as *circles* and β-strands as *triangles*. The background colours are: N-terminal SH3-like β-barrel, yellow; U50 subdomain, pink; L50 subdomain, cyan; converter domain, light-blue. The 7-stranded central β-sheet is shown in red ($\beta1$ 116–119; $\beta2$ 122–126; $\beta3$ 649–656; $\beta4$ 173–178; $\beta5$ 448–454; $\beta6$ 240–247; $\beta7$ 253–261) (modified from [7]). (**d**) Isolated molecules of a monomeric flagellar inner arm dynein (subspecies c), imaged in two biochemical states using negative staining electron microscopy. These images show that the ADP.Vi-molecule has the same general form as the apo-molecule, but the latter is more compact. Schematic diagram of the power stroke of axonemal dynein. Rigid coupling between AAA domain #1 and a linker causes a rolling of the head towards the tail which translates the microtubule by 15 nm under zero load conditions. The distance between the tip of the stalk in these two conformations is approximately 15 nm [8]

connects the tail and the head ring [8] (Fig. 3.1d). When the stalk tip binds tightly to a microtubule, this may promote a concerted conformational change in AAA-modules #4 to #1, leading to activation of release of ADP and phosphate from AAA-module #1. Rigid coupling between AAA-module #1 and linker causes a rolling of the head towards the stem. The linker docks on the head ring and the resultant rotation of the head ring swings the stalk. Judging from the sequence and structural similarities between axonemal and cytoplasmic dyneins, it appears reasonable to assume that the two dynein subfamilies adopt the same mechanism for force generation.

Kinesin [9,10] and myosin [11] motor domains share a common core structure even though they move on different filament tracks and the myosin motor domain is twice as large as the ~350 residue kinesin motor domains (Fig. 3.1a, b). The core structure of kinesin and myosin motor domains is distantly related to the GTPase subunit of heterotrimeric G proteins and small G-proteins of the Ras family [12, 13]. The core structure includes three conserved sequence motifs, termed P-loop, switch-1 and switch-2 at the nucleotide-binding pocket, which act as γ-phosphate sensors. Their switching between ATP and ADP states is associated with specific intramolecular movements, analogous to the nucleotide dependent conformational transitions in G-proteins, which are central to the mechanism of kinesin and myosin family motors. To carry out directed movements, motor proteins must be able to associate with and dissociate from their filament tracks. All motor proteins have at least one force-producing motor domain that contains in addition to the active site, which is responsible for ATP binding and hydrolysis, a binding site for the cytoskeletal filament. In myosins and conventional kinesin, a neck domain connects the motor domain to the tail region. The neck region of myosins is formed by one or more so-called IQ motifs serving as binding sites for calmodulin or calmodulin-like light chains (Fig. 3.1a). The resulting complex of extended α-helical heavy chain and tightly bound light chains serves as a lever arm amplifying and redirecting smaller conformational changes within the myosin motor domain that occur during the interaction with nucleotide and filamentous actin (F-actin) [14, 15]. The cargo-binding tail domain shows high diversity both between motor families and within a single motor family. This enables different motors to bind different cargoes and thus to perform a wide variety of cellular functions.

3.3 In Vitro Motility Assays: A Link between Physiology and Biochemistry

A key issue in motor protein research is the mechanism of chemomechanical coupling. For myosin, we would like to understand how a series of chemical events such as ATP binding, hydrolysis of ATP, release of phosphate, and ADP release induce changes in the ATP binding site, and how the changes are coupled to events at the actin binding site and transmitted into large scale structural changes leading to the production of working strokes of 5 to 50 nm.

Detailed structural information is required to unravel the mechanism by which force and movement are produced. In addition, to link the enzyme kinetics of the motor protein ATPase in solution with the mechanics and energetics of motor proteins, it is essential to use experimental systems in which both ATP-turnover and mechanical parameters can be accurately measured. Here, motor protein studies were advanced greatly by the development of in vitro motility assays. In such assays, the motility of a system consisting only of the purified motor protein, F-actin or microtubules, ATP and buffer solution can be measured under well-defined conditions. The in vitro assay systems are useful in filling the gap between the physiology and biochemistry of motor proteins because these systems enable us to directly observe force generation and movement involving only a very small number of protein molecules.

Two typical geometries are used for in vitro motility assays: bead assays and surface assays (Fig. 3.2). In the former, filaments are fixed to a substrate, such as a microscope slide, and motors are attached to small plastic beads, typically 1 μm in diameter, or to the tip of a fine glass needle. The motion of the beads or of the needle along the filaments in the presence of ATP is visualized using a light microscope. Position and movement of the beads or the needle are measured photo-electrically and can be determined with a resolution on the order of nanometers and sub-millisecond time-response. In the surface assays, the motors themselves are fixed to the substrate, and filaments are observed to diffuse down from solution, attach to and glide over the motor-coated surface. Visualization of the filaments is readily accomplished using dark-field or fluorescence microscopy. Measurements performed on large numbers of enzyme molecules frequently hide important details of their mechanism. Similarly, the discrete actions performed by individual motor proteins are buried in the average when the net output from a large number of asynchronous motors is monitored. In recent decades, rapid progress in a number of technologies such as atomic force microscopy, optical trap nanometry and fluorescence microscopy has provided us with tools to follow the dynamics of single-molecules in situ with spatial and temporal resolution extending to the Å and μs ranges. This allows the direct observation of the dynamic properties of molecular motors, which macroscopic ensemble-averaged measurements cannot detect. Two fundamental motor protein parameters, coupling efficiency and step-size, that can only be indirectly inferred from conventional in vitro motility assays are now accessible by single molecule approaches, permitting the direct and simultaneous observation of ATP-turnover and force production.

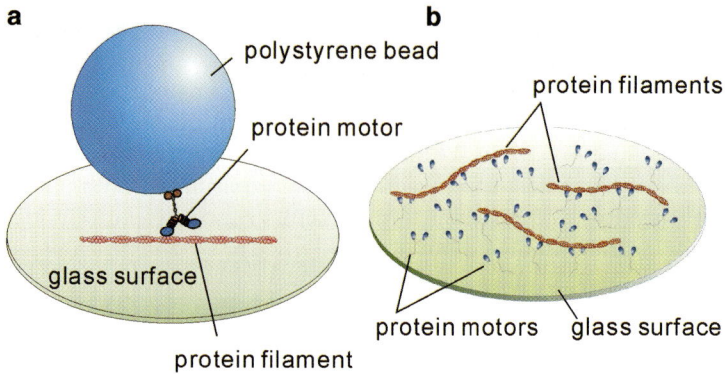

Fig. 3.2. Geometries for quantitative in vitro motility assays. (**a**) Motile activity can be detected and quantified by attaching the motor to a bead and allowing it to interact in an ATP-dependent manner with microtubules or oriented actin filaments on the cover glass surface. (**b**) Alternatively, individual fluorescence labelled and phalloidin-stabilized actin filaments can be observed moving over a lawn of myosins and microtubules can be observed moving over a lawn of dynein or kinesin motors

3.4 Structural Features of the Myosin Motor Domain

The core of the myosin motor domain is formed by a central, 7-stranded β-sheet and is surrounded by α-helices (Fig. 3.1a). The N-terminal 30 residues of the myosin heavy chain extended between the motor domain and the neck region. Unless otherwise stated, sequence numbering refers to the *Dictyostelium* myosin-II heavy chain. Residues 30–80 form a protruding SH3-like β-barrel domain. The function of this small domain is unknown; however, it is absent in class-I myosins and thus appears not to be essential for motor activity. A large structural domain, which accounts for 6 of the 7 strands of the central β-sheet, is formed by residues 81–454 and 594–629 (Fig. 3.1c). This domain is usually referred to as the upper 50K domain (U50). A large cleft divides the U50 from the lower 50K domain (L50), a well defined structural domain formed by residues 465–590. The actin binding region and nucleotide binding site of myosin are on opposite sides of the seven-stranded β-sheet and separated by 40–50 Å. The P-loop and the switch motifs are located in the U50 domain and form part of the ATP binding site. Switch-1 and switch-2 contact the nucleotide at the rear of the nucleotide binding pocket. The switch motifs move towards each other when ATP is bound and move away from each other in its absence. Conformational changes during the transition between different nucleotide states do mostly correspond to rigid-body rotations of secondary and tertiary structure elements. The motor domain can thus be regarded as consisting of communicating functional units with substantial movement occurring in only a few residues.

Residues 630–670 form a long helix that runs from the actin binding region at the tip of the large cleft to the 5th strand of the central β-sheet. A turn and a broken helix are formed by residues 671–699. The segments of the broken helix are frequently referred to as SH1 and SH2-helices. The converter domain formed by residues 700–765 functions as a socket for the C-terminal light chain binding domain and plays a key role in the communication between the active site and neck region.

3.5 Amplification of the Working Stroke by a Lever Arm Mechanism

According to the lever arm model, the step size of myosins is predicted to be proportional to the length of the neck region. An important feature of the lever arm model is that it has the neck region rigidly attached to the converter domain. During the power stroke, converter domain and lever arm swing together in a rigid body motion (Fig. 3.3a). The axis of rotation lies close to and is oriented almost parallel to a long α-helix formed by residues 466–496, which is usually referred to as the relay helix (Fig. 3.1c). The swinging movement, from an initial *UP* position to a *DOWN* position at the end of the working-stroke, is accompanied by the release of the hydrolysis products inorganic phosphate and ADP. The actomyosin ATPase cycle (Fig. 3.4) can be described by a number of intermediate states, which have ATP, its hydrolysis products or no nucleotide bound at the active site and display high or low affinity for F-actin.

The ATP-bound and nucleotide-free states are structurally and biochemically well-defined. They define extreme positions in regard to nucleotide and actin affinity as well as lever arm position. In the ATP-bound state, the affinity of myosin for F-actin is 10,000-fold reduced compared to the nucleotide-free protein. Conversely, strong binding to F-actin reduces the affinity for nucleotide 10,000-fold. The following sequence of events explains the reciprocal relationship between actin and nucleotide affinity. The switch-1 loop preceding β2, the switch-2 loop following β3, and the P-loop following β4 undergo large conformational changes upon ATP-binding. These movements of the active site loops stabilize γ-phosphate binding and the coordination of the Mg^{2+} ion. Additionally, the three edge β-strands: β5, β6, and β7 change their orientation, which leads to large changes in the relative position of the U50 and L50 subdomains and opening of the large cleft between them. Changes of nucleotide binding loop structures are thereby coupled to large movements of the actin binding region. The central role played by the interaction between γ-phosphate and nucleotide-binding loops explains why ATP but not ADP can induce the changes leading to low actin affinity. The establishment of a tight network of hydrogen bonds in the final ATP-myosin complex is concomitant with the formation of a proper active site. This allows the ATPase cycle to advance to the myosin-ADP-P_i intermediate. However, ATP hydrolysis is not the key step that drives the cycle forward. In fact the equilibrium constant for the hydrolysis step is close to unity. What makes the cycle unidirectional

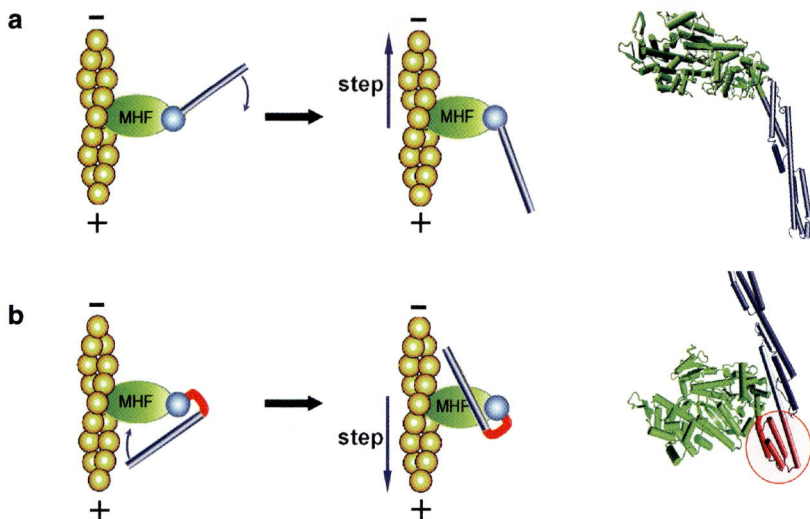

Fig. 3.3. Mechanical models for myosin-based forward and backward movement. (**a**) Power-stroke of a conventional barbed plus-end directed myosin. (**b**) The insertion of a domain (*red*) between the converter region (*blue*) and the lever arm (*orange*) reverses the direction of the power-stroke and produces a minus-end directed myosin. The lever arm moves tangential to the long axis of the actin filament. The right hand panels show the myosin motor domain in the post power-stroke state (*green*) with an artificial lever arm consisting of two α-actinin repeats (*blue*) or two α-actinin repeats and a directional inverted that is derived from the hGBP 4-helix bundle (*red*)

is the irreversibility of ATP binding. Binding of the myosin motor domain to actin induces a reversal of the sequence of conformational changes that are induced by ATP binding. Strong actin binding induces closing of the large cleft between U50 and L50 [16]; this leads to a distortion of the central β-sheet, the outward movement of the nucleotide binding loops disrupts the coordination of the Mg^{2+} ion and thereby ADP binding [17,18]. The loss of the Mg^{2+}-ion coordination induced by actin-binding is similar to the effect of GTPase exchange factors on the release of GDP by small G-proteins. Therefore, actin can be viewed as an ADP-exchange factor for myosin [12,18]. Concomitant with the transition from the ATP-bound state to the rigor-like state, the lever arm swings from its *UP* position to the *DOWN* position [19].

The various genome projects have led to the identification of a large number of new myosins and myosin classes. However, their characterisation is greatly impeded by the fact that the sequencing projects cannot identify the light chains that are associated with the individual myosin heavy chains. In the case of class-I myosins, analysis of the motor activity of the native protein is further hampered by the presence of an ATP-insensitive actin-binding site in the tail region [20]. Here, the direct fusion of an artificial lever arm to

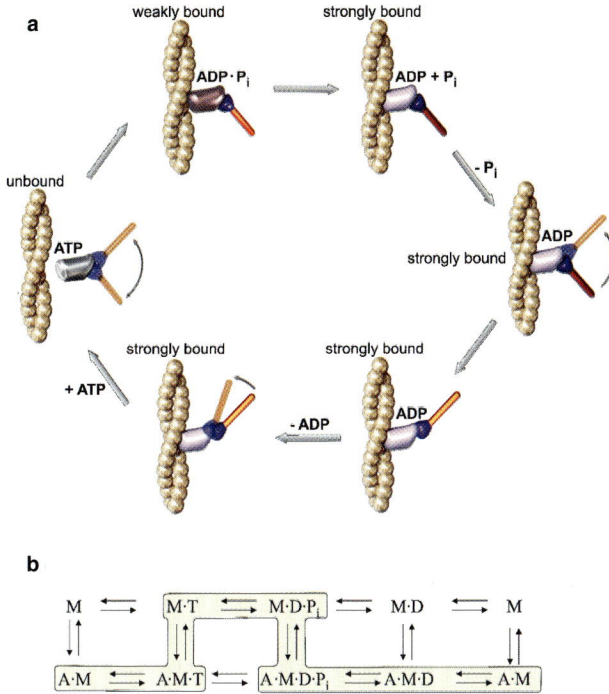

a

weakly bound   strongly bound

ADP·P$_i$ ADP + P$_i$

-P$_i$

unbound

ATP   ADP

strongly bound

+ ATP   strongly bound   strongly bound

- ADP   ADP

b

$$M \rightleftarrows M{\cdot}T \rightleftarrows M{\cdot}D{\cdot}P_i \rightleftarrows M{\cdot}D \rightleftarrows M$$
$$A{\cdot}M \rightleftarrows A{\cdot}M{\cdot}T \rightleftarrows A{\cdot}M{\cdot}D{\cdot}P_i \rightleftarrows A{\cdot}M{\cdot}D \rightleftarrows A{\cdot}M$$

Fig. 3.4. Actomyosin ATPase cycle. (*Upper panel*) Mechanochemical scheme of the actomyosin ATPase cycle. Actin monomers are shown as golden spheres. The motor domain is colored metallic grey for the free form, purple for the weakly-bound form, and violet for the strongly-bound form. The converter is shown in blue and the lever arm in orange. Starting from the *top-right* the following sequence of events is shown: ATP binding induces dissociation of the actomyosin complex. The lever arm returns in the pre-powerstroke position and ATP hydrolysis occurs. Actin-binding can be described in terms of a three-state docking model. The initial formation of a weakly bound collision complex between the myosin head and F-actin is governed by long range ionic interactions and is strongly dependent on ionic strength but independent of temperature. Strong binding is initiated by the following isomerisation to the A-state, leading to the formation of stereospecific hydrophobic interactions. The A-state is affected by organic solvent and temperature. Ionic strength has a comparatively small effect on the formation of this state. The following stepwise transition upon P$_i$ release and then ADP release to a strongly bound state (R-state) involves major structural rearrangements with formation of additional A-M contacts. The conformational changes involve both hydrophobic and ionic interactions. In many myosins both P$_i$ and ADP release result in movement of the lever arm and contribute to the working stroke. The A-to-R transition is dependent on the effective concentration of F-actin and hydrolysis products, as the reaction sequence is readily reversible [21]. (*Lower panel*) Minimal description of the myosin and actomyosin ATPase as defined in solution. *Vertical arrows* indicate the actin association and dissociation from each myosin complex. In every case the events shown can be broken down into a series of sub steps involving one or more identifiable protein conformational changes. The states with a *yellow* background represent the predominant pathway for the actomyosin ATPase (modified from [7])

the motor domain greatly facilitates the production and characterization of recombinant myosin motors displaying full motile activity. It could be demonstrated that the kinetic properties of myosin motor domains from various classes are not affected by the fusion with an artificial lever arm [22, 23]. The replacement of the native neck region with artificial lever arms is facilitated by the limited number of contacts between the motor domain and the neck region. The main technical problems that need to be overcome are the creation of a stiff junction and tight control over the direction in which the lever projects away from the motor domain. Spectrin-like repeats have been used to produce constructs with artificial lever arms of different length for a variety of unconventional myosins.

We applied this approach to characterize the motor properties of a myosin XI, which drives cytoplasmic streaming [22]. Cytoplasmic streaming, as found in plant cells and algae, belongs with velocities of up to 100 μm s^{-1} to the fastest forms of actin-based motility. Electron microscopy of rotary shadowed native plant myosin XI show two head domains with elongated neck regions that are much larger than those observed with myosin II. The neck regions join in a thin stalk of 25 nm length, which has two smaller globular domains attached to its distal end. This structural organization is similar to that of myosin V, whose mechanical properties have been extensively studied [24]. A *Chara corallina* myosin XI motor domain fused to two α-actinin repeats, corresponding to an artificial lever arm of approximately 12 nm length, moves actin filaments with a mean velocity of 16.2 μm/s in the in vitro motility assay [22, 25].

3.6 Backwards Directed Movement

Actin filaments are formed from G-actin monomers and microtubules from α/β-tubulin dimers. Due to the head-to-tail arrangement of these constituent building blocks, the resulting filaments are polar structures. The inherent polarity of the filaments and the stereospecificity of their interactions with motor proteins form the basis for directional movement in biological systems. Each individual type of motor protein moves towards either the plus or the minus end of its respective filament. The difference in the polymerization rates distinguish the fast-growing plus-ends from the slower-growing minus-ends. The directionality of a motor protein can be readily determined by in vitro motility assays. The minus-end of an actin filament or a microtubule can be fluorescently labeled to distinguish it from the plus-end. Using this technique, it was discovered that, different from conventional kinesin, the kinesin-related protein motors ncd [26] and Kar3 [27] are minus-end-directed microtubule-based motors and, different from conventional myosin, myosin VI [28, 29] and myosin IX [30] are minus-end directed actin-based motors.

Structural studies have shown that the myosin power-stroke occurs tangentially to a circle that is defined by the circumference of the actin filament

(reviewed in [31]). As described above, the origin of the translational movement of the lever arm is the rotation of the converter domain that results from
the conformational changes associated with ATP-binding, ATP-hydrolysis,
release of products and the interactions with F-actin. A lever arm mechanism
implies that a reversal of the direction of movement can be achieved simply by
rotating the attachment point of the lever arm through 180°. If the lever arm
points in the opposite direction, the same rotation of the converter region
that produces plus-end directed movement in the native myosin will lead to
the opposite translational movement of the tip of the lever arm and therefore
minus-end directed movement in the mutant constructs (Fig. 3.3) [32].

The following points need to be considered in engineering a construct with
a lever arm rotated by 180°. First, a suitable protein or protein domain needs
to be identified whose insertion leads to a near 180° rotation of the lever
arm. Secondly, steric clashes between the lever arm and the motor domain
must be avoided. Finally, rigid junctions have to be created and the individual building blocks joined in the proper orientation. The following molecular
building blocks were used for the generation of an artificial minus-end directed
myosin: a directional inverter formed by a 4-helix bundle motif derived from
human guanylate binding protein-1 (hGBP), an artificial lever arm formed by
Dictyostelium α-actinin repeats 1 and 2, and a plus-end directed class I myosin
motor domain derived from *Dictyostelium* MyoE [32]. The design of a functional construct was considerably facilitated by the fact that the structures
of all three building blocks used for the generation of the backwards-moving
motor were known [33–35].

3.7 Surface-Alignment of Motor Proteins and their Tracks

The controlled attachment of motor proteins and their protein tracks to well-
defined surface areas offers a potential route to the production of functional
nanomachines. To this end, effective and non-destructive methods have been
investigated for immobilizing these proteins on surfaces and for steering the
resulting output in the form of force and movement in defined directions.
Additionally, the motion of the protein filaments or beads needs to be under
tight directional control and not random as in the standard in vitro assay
system [36].

Microtubules have a low affinity for clean glass, but modifications of the
glass surface can provide selective attachment of microtubules. Immobilizing
microtubules selectively on lithographically patterned silane surfaces was first
reported by [70]. They found that microtubules bound strongly to amine-
terminal silanes while retaining the ability to act as active rails for kinesin
motility. By exposing the silane surface to light from a deep UV laser, they
produced aminosilane patterns lithographically with line widths varying from
1 to 50 µm, and used these patterns for selective adhesion of microtubules.

Using microtubules oriented by buffer flow and immobilized with aminosilane, [37] demonstrated that kinesin-coated silicon microchips can move across the microtubule surface. In these experiments microtubules were aligned along the patterns but not with equal polarity.

Alignment of protein tracks, microtubules or actin filaments with the same polarity is fundamental to applications of motor proteins as elementary force generators in nanotechnology. When driven by very small number of kinesin molecules, microtubules aligned under continuous buffer flow with the same polarity during active sliding [38]. However, the unipolar alignment of microtubules is quickly abolished by thermal agitation after cessation of the flow. Improving this technique, Böhm and coworkers (2001) immobilized microtubules aligned in buffer flow on kinesin-coated surface with gentle treatment with 0.1% glutardialdehyde. Even with this glutardialdehyde treatment, microtubules retained their activities as substrate for kinesin motility. Kinesin-coated beads (glass, gold and polystyrene) with 1–10 μm diameters moved unidirectional on the microtubules with average velocity of 0.3–1.0 μm s^{-1} over a distance up to 2.2 mm [39].

A simple and versatile method used in our laboratory for arranging microtubules on a glass surface in a defined array with uniform polarity uses positively charged nanometer-scale polyacrylamide beads and directional buffer flow [11]. Polarity-marked microtubules attached via their seed-end to the bead-surface even at high ionic strength. The seed-ends, which are located at the minus-end of the microtubules, contain ethylene glycol bis [succinimidyl-succinate] (EGS) cross-links and tetramethylrhodamine at a much higher concentration than the rest of the microtubule. Without EGS-crosslinking, the seed part does not bind to surfaces at high ionic strength. Low ionic strength solution is then introduced in the flow-cell and microtubules are directionally aligned in the direction of buffer flow (Fig. 3.5). Microtubules bind to beads tightly at low ionic strength. Generally, surface binding of microtubules is ionic strength-dependent and fully reversible. Surface-bound microtubules support normal movement of kinesin-coated beads in one direction, indicating that microtubules remain intact even after binding to the surface. The convenience of this procedure for orienting microtubules with uniform polarity makes these surfaces useful not only for powering nanometer-scale devices but also for measuring spectroscopic properties of microtubule motors, such as kinesin and dynein.

3.8 Controlling the Direction of Protein Filament Movement Using MEMS Techniques

The gliding movement of microtubules or actin filaments is now spatially controllable. A number of methods for the control of filament movement in defined directions have been developed. In these methods, chemical modifications and micro-fabrications of the surface have been used. In order to control the track

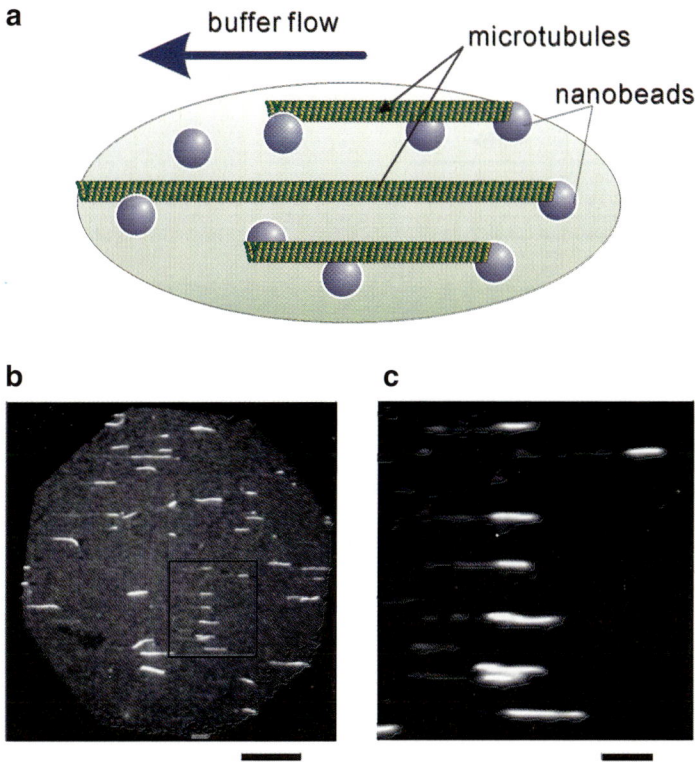

Fig. 3.5. Polarity-marked microtubules attached on positively-charged glass surface. (**a**) Illustration of experimental arrangement of nanobeads and seeds of microtubules. The glass surface was first coated with positively-charged polyacrylamide nanobeads. Since the EGS-cross linked seeds of microtubules (*bright fluorescence*) had negative charges, microtubules selectively attached on the surface at their minus ends. Buffer flow oriented these microtubules on the surface. (**b, c**) Fluorescence microscopic observation of aligned microtubules. The polarity of microtubules can be easily identified from the position of the brightly labeled seeds. The images demonstrate how microtubules can be efficiently aligned in a unipolar fashion by buffer flow. Scale bar, 15 μm in b, 4 μm in c

along which filaments glide, it is necessary to restrict the location of active motors to specific regions of the surface. While the detailed interactions of motor proteins with surfaces are not well understood, it has been observed that myosin motility is primarily restricted to the more hydrophobic resist surfaces [40, 41]. Thus, myosin and proteolytic fragments of myosin can be readily aligned on polytetrafluoroethylene (PTFE)-deposited surfaces, resulting in the movement of actin filaments being restricted to the well-defined fabricated tracks [42]. PTFE thin films are readily generated by rubbing a heated glass-coverslip surface with a PTFE rod while applying a defined

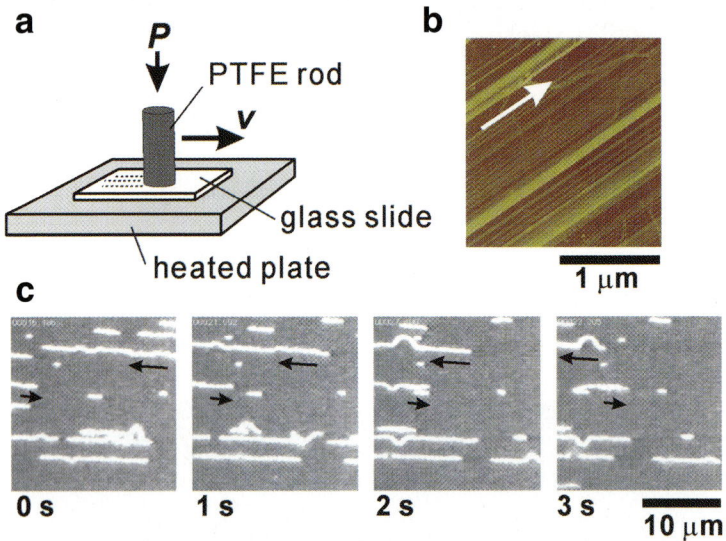

Fig. 3.6. Alignment of myosin or proteolytic fragments of myosin on poly-tetrafluoroethylene (PTFE)-deposited surfaces. (**a**) Mechanical deposition of PTFE thin film on glass surface. A rod of solid PTFE is moved against a coverslip on a hot plate at a constant rate of 50 mm/s and constant pressure (7×10^4 Pa). (**b**) Atomic force microscopic observation of the PTFE thin film deposited on the glass surface. (**c**) A sequence of fluorescence microscopy images showing the movement of fluorescent actin filaments on a myosin S1-coated PTFE-thin film. Actin filaments are moving on the ridges of the PTFE deposit in a bi-directional fashion. Reproduced with permission from [43]

pressure [44]. The PTFE thin films were used for alignment or graphoepitaxial crystal growth of a variety of substances [44, 45]. The resulting films on the coverslip surface consist of many linear and parallel PTFE ridges of 10–100 nm width (Fig. 3.6). Myosin or its proteolytic fragments, subfragment 1 (S1) or heavy meromyosin (HMM) bind to the ridges without losing activity and actin filaments move on these ridges with the speed that is typical for in vitro motility of myosin or myosin-fragment. This PTFE technique was also used for the kinesin-microtubule system [46].

However, the widths, heights and shapes of the PTFE ridges are difficult to control. To overcome this difficulty, various polymers were examined for use as surface coating, and several effective polymers that maintain the activity of motor proteins have been reported. Thus methods have been developed in which a glass or silicone surface is coated with resist polymers such as polymethylmethacrylate (PMMA) or SAL601. UV radiation, electron beams or soft lithography are then used to remove resist from defined regions and to draw specific patterns on the substrate. With careful selection of the buffer solutions [e.g. pH, ionic strength, concentration of motors, choice of blocking substances such as casein and bovine serum albumin (BSA) and/or detergents], motility can be restricted either to the unexposed, resist polymer surface or to the exposed underlying substrate.

PMMA was the first resist polymer found to be useful for immobilizing myosin molecules while retaining their abilities to support the movement of actin [42]. Patterned surfaces were prepared by photolithography with PMMA-coated glass coverslip. Various patterns of tracks of PMMA were fabricated on coverslips, and HMM was introduced and immobilized on the patterns. Fluorescent actin filaments were then added in the presence of ATP. Their movements were confined to the PMMA tracks (Fig. 3.7). Through the study of the behavior of actin filaments moving in PMMA tracks, we found that the probability of an actin filament making a U-turn is low within a track of a few micrometers width. In addition, actin filaments often moved along the edges of the tracks when they approached the edge at low angles instead of escaping from the tracks. Thus, simple patterns can effectively bias the movement of actin filaments, confining them to unidirectional movement (Fig. 3.7c) [47].

In the experiments described above, the PMMA tracks are raised above the surrounding glass surface. This leads to actin filaments running off the tracks and their number gradually decreasing over time. Given the potential applications of this system, it is thus necessary to develop a way to restrict the movement of filaments to one dimension along linear tracks for extended periods of time. Restricting kinesin-driven movement of microtubules along linear tracks was achieved by using micrometer-scaled grooves lithographically fabricated on glass surfaces [48–50]. In the presence of detergent, kinesin selectively adsorbed onto a glass surface from which the photoresist polymer has been removed, not on the photoresist polymer itself [50]. The tracks thus have a reversed geometry as compared with those used previously i.e. they were channels bordered by walls of the resist material. Microtubules rarely climbed up the walls and moved away from the track. Therefore this method allowed us to limit the kinesin-driven movement of microtubules effectively to one dimension along a linear track. The sidewall collisions described above and the subsequent guidance of microtubules along the sidewall were well characterized by [49]. Similar nano-structured surfaces were also used for the actomyosin system [40] although actin filaments often climbed up the wall and escaped from tracks owing to their lower flexural rigidity compared with microtubules. This limitation has been overcome by shaping the surface morphology with nanometer precision, which forces the filaments to move exclusively on the tracks [51].

While it is possible to use chemical and topographical patterning to guide protein filaments and restrict their movement to particular tracks, it is more difficult to control the direction of movement along the track. The difficulty arises because the orientation in which motors bind to a uniform surface is not well controlled. The conversion of bidirectional movement into unidirectional movement along the linear tracks was finally accomplished by simply

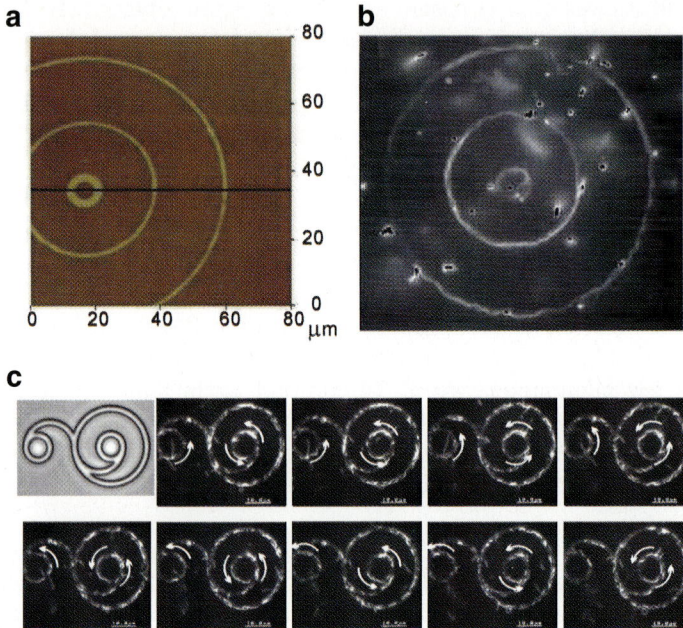

Fig. 3.7. (**a**) Atomic force micrograph of triple concentric circular PMMA tracks. (**b**) Fluorescence microscope images of actin filaments moving on PMMA tracks coated with myosin HMM molecules. The superpositioning of image at different time points shows that the movement of actin filaments is restricted to the tracks. (**c**) Extraction of unidirectional movement of actin filaments on HMM-coated PMMA tracks. The *first panel* shows a bright-field image of circular PMMA tracks. The fluorescence microscope images show actin filaments moving counter-clockwise along the tracks and being directed towards the smaller circular tracks

adding arrowhead patterns on the tracks [50]. Most microtubules entering the arrowheads against the direction in which the arrowheads points are induced to make an 180° turn (Fig. 3.8). As a result, unidirectional movement is generated by the rectifying action of the arrow-headed pattern. Precise analyses and design of these rectifiers has been carried out by [71]. Arrowhead rectifiers have enabled us to construct microminiaturized circulators, in which populations of microtubules rotate in one direction and transport materials on the micrometer scale in a predefined fashion.

In addition to the spatial control of the movement of protein filaments, the temporal control of motor protein activity has been investigated. Rapidly chasing the buffer solution with a new one is the simplest way to control the activity of motor proteins. Flushing out ATP induces rapid cessation of protein filament movement. To control the concentration of ATP, photo-activatable ATP is an alternative method. Kaplan and coworkers showed that light-induced activation of caged complexes can control the activity of

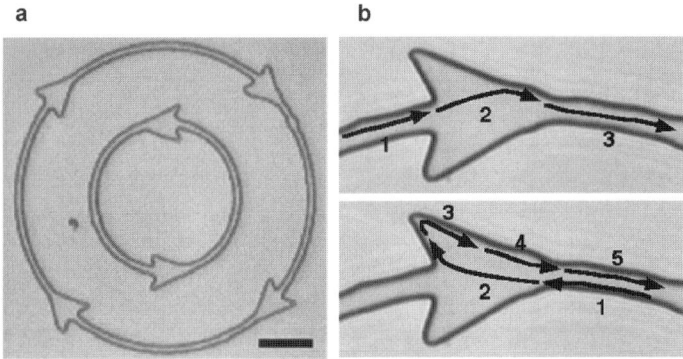

Fig. 3.8. Micro-fabricated circular grooves with arrowhead patterns to extract unidirectional movement from bi-directional, rotational movement of microtubules. (**a**) An optical microscopic image of the grooves. With this pattern, microtubules in the *outer circle* are moving clockwise, while those in the *inner circle* are moving counter-clockwise. (**b**) Schematic diagram of the arrowhead functioning as a rectifier of microtubule movement

proteins [52]. Caged nucleotides have been commonly used not only for the study of motor protein function [53] but also for controlling motor proteins in nanomachine-development [54]. One promising approach to controlling the activity of motor proteins is the development of caged proteins pioneered by G. Marriott [55, 56]. Caged heavy meromyosin was prepared by conjugating the thiol reactive reagent 1-(bromomethyl)-2-nitro-4, 5-dimethoxybenzene with the critical thiol group in the so-called SH1-helix. This treatment renders the molecule inactive. It can be reactivated by a pulse of near-ultraviolet light. Following irradiation with UV light, actin filaments on HMM-coated surface concomitantly start to move with velocities comparable to those of unmodified HMM [55].

On the other hand, the fast and reversible on- and off-switching of the motile activity of motor proteins needs to be investigated for the application of motor proteins to nanomachines. Rapid perturbation such as a temperature jump can be used to control movement of protein filaments. Kato et al., developed a temperature-pulse microscope in which an IR laser beam locally illuminated an aggregate of metal particles bound on a surface [57]. Using this system, the temperature of a microscopic region of ca. 10 μm in diameters was raised reversibly in a square-wave fashion with rise and fall times of several ms with a temperature gradient up to 2 degrees C/micrometer. Using an in vitro motility assay, they showed that the motor functions can be thermally and reversibly activated even at temperatures that are high enough to normally damage the proteins. By combining directional control of movement of protein filaments with this temperature jump method or application of light, external electric and magnetic fields, it should be possible to control cargo loading and unloading as well as the motor protein activities.

Controlling the position and orientation of motor proteins with sub-nanometer precision constitutes another key technology: motor proteins need to be placed with nanometer accuracy on a surface and their orientation controlled within a few degrees. Spudich and coworkers demonstrated that HMM molecules, sterospecifically bound to a single actin filament in rigor, could be transferred to nitrocellulose-coated surface by addition of ATP and that transferred HMM supported motility of actin filaments [58]. Combining this technique to filament-alignment techniques may provide a surface on which motor proteins are aligned with high spatial precisions and orientation.

Several methods for coupling motor proteins to the surface have been reported. Fusion motor proteins with bacterial biotin-binding proteins can bind specifically to streptavidin-coated cargoes or surfaces. Many peptide tags fused to expressed proteins have been commonly used to make the purifications easy. Some of these tags were used to couple the proteins to surfaces coated with the complementary ligand or antibodies [59–61].

3.9 Conclusions and Perspectives

Here, we have described the basic properties of motor proteins and how molecular biological techniques can be used to generate recombinant motors with well defined properties. The alignment of motor proteins and cytoskeletal filaments while maintaining their functions has been achieved by the use of nano- and micro-fabrication techniques. The methods described here are useful for establishing micrometer- or nanometer-scale arrays of motor proteins and filaments, and straightforward in their application. The use of motor proteins in nanometric actuators is moving a step closer towards realization. The generation of backwards- and forwards-moving motors that display increased thermal stability, optimized kinetic properties, and tight regulation by external signals will play an important part in the integration of biomolecules into nanotechnological devices.

The mechanochemical coupling in myosin, as described here, is a paradigm for linear motor proteins in general and suggests that the activity of these nanomachines can be mediated or regulated by divers mechanisms. The occurrence of myosins with lever arms of different length constitutes a simple means of increasing the size of the working stroke and thereby the velocity [23]. Similarly, the angle through which the lever arm swings affects the size of the working stroke and velocity. It has been shown that the lever arm of class I myosins swings though a ~30° larger angle than in conventional myosin [34]. Fine-tuning of the rate of ATP turnover provides another way to modulate the velocity of motor proteins. This can be done by changing the rate of the ATP hydrolysis step or by modification of the rate of product release [62]. As the release of the hydrolysis products is greatly accelerated by actin-binding, modulation of the interaction with actin provides one means to affect motor function. In the case of some unconventional myosins, phosphorylation of a so-called "TEDS-residue" is required for normal coupling between the actin and nucleotide bindings sites [63–65]. The negative charge introduced by the phosphate group stabilizes the rigor complex by reducing the dissociation rate constant more than 30-fold. Product inhibition by ADP provides another means to reduce the velocity of the motor protein. Mg^{2+}-ions, which act more like catalysts during the ATPase cycle, can affect the rate of ADP-release. For class-I and class-V myosins, it has been shown that changes in the concentration of free Mg^{2+}-ions within the physiological range affect velocity [63,66]. Motor activity can also be modulated by changes in the stiffness of the lever arm. The Ca^{2+}-ion dependent binding of light chains can induce such changes. Direct mechanical coupling between the heads of kinesin or myosin heavy chain dimers provides a further way to modulate motor activity and, with appropriate tuning of the hydrolysis and product release steps, can play a key role in the generation of processive motors and the directional movement of motor proteins in the absence of a stiff lever arm [67–69].

Acknowledgements

We thank E. Mandelkow and A. Marx for providing Fig. 3.1b. The work was supported by Special Coordination Funds for Promoting Science and Technology, the Ministry of Education, Culture, Sports, Science and Technology (K.O.); Fond der Chemischen Industrie and DFG grants MA1081/5-3 and MA1081/6-1 (D.J.M.).

References

1. R. D. Vale (2000). AAA proteins. Lords of the ring. *J Cell Biol*, **150**, pp. 13–20.
2. L. M. DiBella and S. M. King (2001). Dynein motors of the Chlamydomonas flagellum. *Int Rev Cytol*, **210**, pp. 227–268.
3. I. Milisav (1998). Dynein and dynein-related genes. *Cell Motil Cytoskeleton*, **39**, pp. 261–272.
4. M. P. Koonce and M. Samso (2004). Of rings and levers: the dynein motor comes of age. *Trends Cell Biol*, **14**, pp. 612–619.
5. D. J. Asai and M. P. Koonce (2001). The dynein heavy chain: structure, mechanics and evolution. *Trends Cell Biol*, **11**, pp. 196–202.
6. K. Oiwa and H. Sakakibara (2005). Recent progress in dynein structure and mechanism. *Curr Opin Cell Biol*, **17**, pp. 98–103.
7. M. A. Geeves, R. Fedorov, and D. J. Manstein (2005). Molecular mechanism of actomyosin-based motility. *Cell Mol Life Sci*, **62**, pp. 1462–1477.
8. S. A. Burgess, M. L. Walker, H. Sakakibara, P. J. Knight, and K. Oiwa (2003). Dynein structure and power stroke. *Nature*, **421**, pp. 715–718.
9. F. Kozielski, S. Sack, A. Marx, M. Thormahlen, E. Schonbrunn, V. Biou, A. Thompson, E. M. Mandelkow, and E. Mandelkow (1997). The crystal structure of dimeric kinesin and implications for microtubule-dependent motility. *Cell*, **91**, pp. 985–994.

10. F. J. Kull, E. P. Sablin, R. Lau, R. J. Fletterick, and R. D. Vale (1996). Crystal structure of the kinesin motor domain reveals a structural similarity to myosin. *Nature*, **380**, pp. 550–555.

11. I. Rayment, W. R. Rypniewski, K. Schmidt-Base, R. Smith, D. R. Tomchick, M. M. Benning, D. A. Winkelmann, G. Wesenberg, and H. M. Holden (1993). Three-dimensional structure of myosin subfragment-1: a molecular motor. *Science*, **261**, pp. 50–58.

12. R. S. Goody and W. Hofmann-Goody (2002). Exchange factors, effectors, GAPs and motor proteins: common thermodynamic and kinetic principles for different functions. *Eur Biophys J*, **31**, pp. 268–274.

13. R. D. Vale (1996). Switches, latches, and amplifiers: common themes of G proteins and molecular motors. *J Cell Biol*, **135**, pp. 291–302.

14. M. Anson, M. A. Geeves, S. E. Kurzawa, and D. J. Manstein (1996). Myosin motors with artificial lever arms. *EMBO J*, **15**, pp. 6069–6074.

15. T. Q. Uyeda, P. D. Abramson, and J. A. Spudich (1996). The neck region of the myosin motor domain acts as a lever arm to generate movement. *Proc Natl Acad Sci USA*, **93**, pp. 4459–4464.

16. P. B. Conibear, C. R. Bagshaw, P. G. Fajer, M. Kovacs, and A. Malnasi-Csizmadia (2003). Myosin cleft movement and its coupling to actomyosin dissociation. *Nat Struct Biol*, **10**, pp. 831–835.

17. P. D. Coureux, A. L. Wells, J. Menetrey, C. M. Yengo, C. A. Morris, H. L. Sweeney, and A. Houdusse (2003). A structural state of the myosin V motor without bound nucleotide. *Nature*, **425**, pp. 419–423.

18. T. F. Reubold, S. Eschenburg, A. Becker, F. J. Kull, and D. J. Manstein (2003). A structural model for actin-induced nucleotide release in myosin. *Nat Struct Biol*, **10**, pp. 826–830.

19. K. C. Holmes, I. Angert, F. J. Kull, W. Jahn, and R. R. Schroder (2003). Electron cryo-microscopy shows how strong binding of myosin to actin releases nucleotide. *Nature*, **425**, pp. 423–427.

20. H. Brzeska, T. J. Lynch, and E. D. Korn (1988). Localization of the actin-binding sites of Acanthamoeba myosin IB and effect of limited proteolysis on its actin-activated Mg2+-ATPase activity. *J Biol Chem*, **263**, pp. 427–435.

21. M. A. Geeves and P. B. Conibear (1995). The role of three-state docking of myosin S1 with actin in force generation. *Biophys J*, **68**, pp. 194S–199S; discussion 199S–201S.

22. K. Ito, T. Kashiyama, K. Shimada, A. Yamaguchi, J. Awata, Y. Hachikubo, D. J. Manstein, and K. Yamamoto (2003). Recombinant motor domain constructs of Chara corallina myosin display fast motility and high ATPase activity. *Biochem Biophys Res Commun*, **312**, pp. 958–964.

23. C. Ruff, M. Furch, B. Brenner, D. J. Manstein, and E. Meyhofer (2001). Single-molecule tracking of myosins with genetically engineered amplifier domains. *Nat Struct Biol*, **8**, pp. 226–229.

24. M. Tominaga, H. Kojima, E. Yokota, H. Orii, R. Nakamori, E. Katayama, M. Anson, T. Shimmen, and K. Oiwa (2003). Higher plant myosin XI moves processively on actin with 35 nm steps at high velocity. *EMBO J*, **22**, pp. 1263–1272.

25. S. Higashi-Fujime, R. Ishikawa, H. Iwasawa, O. Kagami, E. Kurimoto, K. Kohama, and T. Hozumi (1995). The fastest actin-based motor protein from the green algae, Chara, and its distinct mode of interaction with actin. *FEBS Lett*, **375**, pp. 151–154.

26. R. A. Walker (1995). ncd and kinesin motor domains interact with both alpha- and beta-tubulin. *Proc Natl Acad Sci USA*, **92**, pp. 5960–5964.

27. S. A. Endow, S. J. Kang, L. L. Satterwhite, M. D. Rose, V. P. Skeen, and E. D. Salmon (1994). Yeast Kar3 is a minus-end microtubule motor protein that destabilizes microtubules preferentially at the minus ends. *EMBO J*, **13**, pp. 2708–2713.

28. J. Menetrey, A. Bahloul, A. L. Wells, C. M. Yengo, C. A. Morris, H. L. Sweeney, and A. Houdusse (2005). The structure of the myosin VI motor reveals the mechanism of directionality reversal. *Nature*, **435**, pp. 779–785.

29. A. L. Wells, A. W. Lin, L. Q. Chen, D. Safer, S. M. Cain, T. Hasson, B. O. Carragher, R. A. Milligan, and H. L. Sweeney (1999). Myosin VI is an actin-based motor that moves backwards. *Nature*, **401**, pp. 505–508.

30. A. Inoue, J. Saito, R. Ikebe, and M. Ikebe (2002). Myosin IXb is a single-headed minus-end-directed processive motor. *Nat Cell Biol*, **4**, pp. 302–306.

31. M. A. Geeves, and K. C. Holmes (1999). Structural mechanism of muscle contraction. *Ann Rev Biochem*, **68**, pp. 687–728.

32. G. Tsiavaliaris, S. Fujita-Becker, and D. J. Manstein (2004). Molecular engineering of a backwards-moving myosin motor. *Nature*, **427**, pp. 558–561.

33. W. Kliche, S. Fujita-Becker, M. Kollmar, D. J. Manstein, and F. J. Kull (2001). Structure of a genetically engineered molecular motor. *EMBO J*, **20**, pp. 40–46.

34. M. Kollmar, U. Dürrwang, W. Kliche, D. J. Manstein, and F. J. Kull (2002). Crystal structure of the motor domain of a class-I myosin. *EMBO J*, **21**, pp. 2517–2525.

35. B. Prakash, L. Renault, G. J. Praefcke, C. Herrmann, and A. Wittinghofer (2000). Triphosphate structure of guanylate-binding protein 1 and implications for nucleotide binding and GTPase mechanism. *EMBO J*, **19**, pp. 4555–4564.

36. S. J. Kron and J. A. Spudich (1986). Fluorescent actin filaments move on myosin fixed to a glass surface. *Proc Natl Acad Sci USA*, **83**, pp. 6272–6276.

37. L. Limberis and R. J. Stewart (2000). Toward kinesin-powered microdevices. *Nanotechnol*, **11**, pp. 47–51.

38. R. Stracke, K. J. Bohm, J. Burgold, H. J. Schacht, and E. Unger (2000). Physical and technical parameters determining the functioning of a kinesin-based cell-free motor system. *Nanotechnol*, **11**, pp. 52–56.

39. K. J. Böhm, R. Stracke, P. Muhlig, and E. Unger (2001). Motor protein-driven unidirectional transport of micrometer-sized cargoes across isopolar microtubule arrays. *Nanotechnol*, **12**, pp. 238–244.

40. R. Bunk, J. Klinth, L. Montelius, I. A. Nicholls, P. Omling, S. Tagerud, and A. Mansson (2003). Actomyosin motility on nanostructured surfaces. *Biochem Biophys Res Commun*, **301**, pp. 783–788.

41. D. V. Nicolau, H. Suzuki, S. Mashiko, T. Taguchi, and S. Yoshikawa (1999). Actin motion on microlithographically functionalized myosin surfaces and tracks. *Biophys J*, **77**, pp. 1126–1134.

42. H. Suzuki, A. Yamada, K. Oiwa, H. Nakayama, and S. Mashiko (1997). Control of actin moving trajectory by patterned poly(methylmethacrylate) tracks. *Biophys J*, **72**, pp. 1997–2001.

43. H. Suzuki, K. Oiwa, A. Yamada, H. Sakakibara, H. Nakayama, and S. Mashiko (1995). Linear Arrangement of Motor Protein on a Mechanically Deposited Fluoropolymer Thin Film. *Jpn J Appl Phys*, **34**, pp. 3937–3941.

44. J. C. Wittmann and P. Smith (1991). Highly oriented thin-films of poly(tetrafluoroethylene) as a substrate for oriented growth of materials. *Nature*, **352**, pp. 414–417.
45. D. Fenwick, P. Smith, and J. C. Wittmann (1996). Epitaxial and graphoepitaxial growth of materials on highly orientated PTFE substrates. *J Mat Science*, **31**, pp. 128–131.
46. J. R. Dennis, J. Howard, and V. Vogel (1999). Molecular shuttles: directed motion of microtubules slang nanoscale kinesin tracks. *Nanotechnol*, **10**, pp. 232–236.
47. K. Oiwa, D. M. Jameson, J. C. Croney, C. T. Davis, J. F. Eccleston, and M. Anson (2003). The 2'-O- and 3'-O-Cy3-EDA-ATP(ADP) complexes with myosin subfragment-1 are spectroscopically distinct. *Biophys J*, **84**, pp. 634–642.
48. J. Clemmens, H. Hess, J. Howard, and V. Vogel (2003a). Analysis of microtubule guidance in open microfabricated channels coated with the motor protein kinesin. *Langmuir*, **19**, pp. 1738–1744.
49. J. Clemmens, H. Hess, R. Lipscomb, Y. Hanein, K. F. Bohringer, C. M. Matzke, G. D. Bachand, B. C. Bunker, and V. Vogel (2003b). Mechanisms of microtubule guiding on microfabricated kinesin-coated surfaces: Chemical and topographic surface patterns. *Langmuir*, **19**, pp. 10967–10974.
50. Y. Hiratsuka, T. Tada, K. Oiwa, T. Kanayama, and T. Q. Uyeda (2001). Controlling the direction of kinesin-driven microtubule movements along microlithographic tracks. *Biophys J*, **81**, pp. 1555–1561.
51. R. Bunk, M. Sundberg, A. Mansson, I. A. Nicholls, P. Omling, S. Tagerud, and L. Montelius (2005). Guiding motor-propelled molecules with nanoscale precision through silanized bi-channel structures. *Nanotechnol*, **16**, pp. 710–717.
52. J. H. Kaplan, B. Forbush, 3rd, and J. F. Hoffman (1978). Rapid photolytic release of adenosine 5'-triphosphate from a protected analogue: utilization by the Na:K pump of human red blood cell ghosts. *Biochemistry*, **17**, pp. 1929–1935.
53. J. A. McCray, L. Herbette, T. Kihara, and D. R. Trentham (1980). A new approach to time-resolved studies of ATP-requiring biological systems; laser flash photolysis of caged ATP. *Proc Natl Acad Sci USA*, **77**, pp. 7237–7241.
54. H. Hess, J. Clemmens, D. Qin, J. Howard, and V. Vogel (2001). Light-controlled molecular shuttles made from motor proteins carrying cargo on engineered surfaces. *Nano Letters*, **1**, pp. 235–239.
55. G. Marriott and M. Heidecker (1996). Light-directed generation of the actin-activated ATPase activity of caged heavy meromyosin. *Biochemistry*, **35**, pp. 3170–3174.
56. G. Marriott, P. Roy, and K. Jacobson (2003). Preparation and light-directed activation of caged proteins. *Methods Enzymol*, **360**, pp. 274–288.
57. H. Kato, T. Nishizaka, T. Iga, K. Kinosita, Jr., and S. Ishiwata (1999). Imaging of thermal activation of actomyosin motors. *Proc Natl Acad Sci USA*, **96**, pp. 9602–9606.
58. Y. Y. Toyoshima, C. Toyoshima, and J. A. Spudich (1989). Bidirectional movement of actin filaments along tracks of myosin heads. *Nature*, **341**, pp. 154–156.
59. Y. Inoue, Y. Y. Toyoshima, A. H. Iwane, S. Morimoto, H. Higuchi, and T. Yanagida (1997). Movements of truncated kinesin fragments with a short or an artificial flexible neck. *Proc Natl Acad Sci USA*, **94**, pp. 7275–7280.

60. S. Itakura, H. Yamakawa, Y. Y. Toyoshima, A. Ishijima, T. Kojima, Y. Harada, T. Yanagida, T. Wakabayashi, and K. Sutoh (1993). Force-generating domain of myosin motor. *Biochem Biophys Res Commun*, **196**, pp. 1504–1510.

61. A. H. Iwane, K. Kitamura, M. Tokunaga, and T. Yanagida (1997). Myosin subfragment-1 is fully equipped with factors essential for motor function. *Biochem Biophys Res Commun*, **230**, pp. 76–80.

62. R. W. Lymn and E. W. Taylor (1970). Transient state phosphate production in the hydrolysis of nucleoside triphosphates by myosin. *Biochemistry*, **9**, pp. 2975–2983.

63. S. Fujita-Becker, U. Dürrwang, M. Erent, R. J. Clark, M. A. Geeves, and D. J. Manstein (2005). Changes in Mg^{2+} ion concentration and heavy chain phosphorylation regulate the motor activity of a class I myosin. *J Biol Chem*, **280**, pp. 6064–6071.

64. E. M. Ostap, T. Lin, S. S. Rosenfeld, and N. Tang (2002). Mechanism of regulation of Acanthamoeba myosin-IC by heavy-chain phosphorylation. *Biochemistry*, **41**, pp. 12450–12456.

65. Z. Y. Wang, F. Wang, J. R. Sellers, E. D. Korn, and J. A. Hammer, 3rd (1998). Analysis of the regulatory phosphorylation site in Acanthamoeba myosin IC by using site-directed mutagenesis. *Proc Natl Acad Sci USA*, **95**, pp. 15200–15205.

66. S. S. Rosenfeld, A. Houdusse, and H. L. Sweeney (2005). Magnesium regulates ADP dissociation from myosin V. *J Biol Chem*, **280**, pp. 6072–6079.

67. N. J. Carter and R. A. Cross (2005). Mechanics of the kinesin step. *Nature*, **435**, pp. 308–12.

68. P. B. Conibear and M. A. Geeves (1998). Cooperativity between the two heads of rabbit skeletal muscle heavy meromyosin in binding to actin. *Biophys J*, **75**, pp. 926–937.

69. M. Nyitrai and M. A. Geeves (2004). Adenosine diphosphate and strain sensitivity in myosin motors. *Philos Trans R Soc Lond B Biol Sci*, **359**, pp. 1867–1877.

70. D. C. Turner, C. Chang, K. Fang, S. L. Brandow, and D. B. Murphy (1995). Selective adhesion of functional microtubules to patterned silane surfaces. *Biophys. J*, **69**, pp. 2782–2789.

71. M. G. L. van den Heuvel, C. T. Butcher, R. M. M. Smeets, S. Diez, and C. Dekker (2005). High rectifying efficiencies of microtubule motility on kinesin-coated gold nanostructures. *Nano Letters*, **5**, pp. 1117–1122.

4

Axonal Transport: Imaging and Modeling of a Neuronal Process

S.B. Shah[1], G. Yang[3], G. Danuser[3], and L.S.B. Goldstein[2]

[1] Dept. of Bioengineering, University of Maryland, College Park, MD
`sameer@umd.edu`
[2] Dept. of Cellular and Molecular Medicine, University of California of San Diego, La Jolla, CA
`lgoldstein@ucsd.edu`
[3] Dept. of Cell Biology, The Scripps Research Institute, La Jolla, CA

Abstract. Owing to their unusual geometry and polarity, neurons face a tremendous transport challenge. In particular, the bi-directional movement of many cargoes between cell body and synapse that takes place within extremely long, narrow axons requires motor-driven active transport along polarized microtubules. We summarize some imaging and theoretical modeling strategies recently developed to better understand axonal transport and neuronal function. Our approaches are motivated by three questions: (1) Can we predict the response of a complex trafficking system to perturbations of various components, either alone, or in combination? (2) What is the relationship between in vitro measurements of single motor properties and the movement of motor-cargo complexes in vivo? (3) What key principles govern the operation of the neuronal transport system? We discuss the imaging of vesicular transport in *Drosophila melanogaster* larval axons, and the development of quantitative schemes to define transport function via the extraction of kinematic parameters from these images. The application of these schemes to images from wild-type larvae and larvae expressing mutations in specific transport proteins allows rigorous quantification of transport kinematics in functional and dysfunctional neurons. Finally, we present some strategies and results for the theoretical modeling of axonal transport, and discuss the integration of these results with experimental data.

4.1 Neuronal Function: A Tremendous Transport Challenge

Owing to their polarity and unusual geometry, neurons face a complex transport challenge. Many proteins and organelles are synthesized and assembled in neuronal cell bodies, which are typically no more than 10–30 µm in diameter. These cargoes are then transported by molecular motors along extremely

S.B. Shah et al.: *Axonal Transport: Imaging and Modeling of a Neuronal Process*, Lect. Notes Phys. **711**, 65–84 (2007)
DOI 10.1007/3-540-49522-3_4

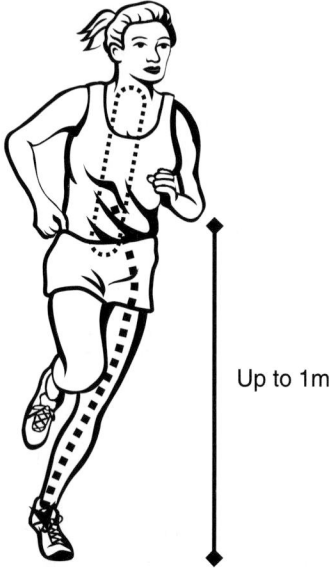

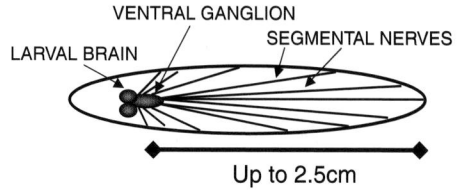

Fig. 4.1. (**a**) In a human, single axons (*heavy dotted line*) can span a distance of up to 1 meter, from cell bodies in the spinal cord (*dotted oval*) to neuromuscular junctions in the leg musculature. (**b**) In a fruit fly larva, the nervous system also faces a complex transport challenge. Axons within segmental nerves originate in cell bodies located in the ventral ganglia and terminate at neuromuscular junctions

long, narrow axons towards the presynaptic nerve termini. In addition, motor-cargo complexes are transported in the opposite direction, sending packets of information or recycled materials from synapses towards the cell body. Axons may be up to one meter long in humans, with volumes up to, and in excess of, 1000 times that of the supporting cell body (Fig. 4.1). The sheer volume and diversity of cargoes being transported simultaneously within an axon are staggering; motors must move a host of cargoes varying in size and function, ranging from vesicles to mitochondria to signaling complexes to entire segments of the cytoskeleton. To add to the challenge, axonal calibers may be as narrow as 200 nm – barely wider than the dimensions of the transported cargoes themselves!

4.2 Meeting the Challenge: Key Players in the Neuronal Transport System

Much of the long-distance transport within an axon takes place along polarized microtubule tracks, which are oriented with their plus (polymerizing) end pointing away from the cell body. On these microtubules, ATP-dependent motor proteins shuttle their cargoes bi-directionally. Kinesin motors are primar-

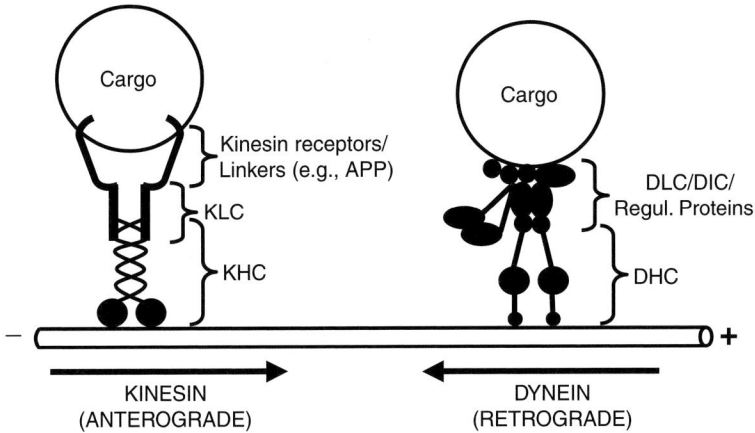

Fig. 4.2. Kinesin (*left*) consists of dimeric heavy chains (KHC) and light chains (KLC), and binds cargo via receptors, or linker proteins, such as Amyloid Precursor Protein (APP). Dynein (*right*) also consists of dimeric heavy and light chains (DHC, DLC) as well as several regulatory subunits including intermediate chains (DIC) and the dynactin complex. Most kinesin isoforms move towards the plus end of a microtubule and dynein moves towards the minus end

ily responsible for moving their cargo in the anterograde direction (away from the cell body), while dynein motors, and a few minus-end oriented kinesins, are responsible for retrograde transport (towards the cell body; Fig. 4.2). The biophysical properties of these motors and also the mechanisms of force generation by which they move along microtubules have been studied extensively, using a variety of elegant in vitro biophysical and biochemical assays (e.g., Vale et al. 1992 [1–7]).

Transport is generally categorized as fast or slow based on the average velocities of a protein or cargo of interest. Fast cargoes move an average of 20–400 mm/day (1–10 μm/sec), while slow cargoes move an average of 0.2–4 mm/day (0.01–0.1 μm/sec). Imaging of selected cargoes in vivo has shed some light on differences between these different classes of cargo Brady et al., 1982 [8–10]. Fast and slow average velocities might result not from constant cargo motion at fast and slow speeds, but rather, via bursts of high velocities punctuated by pauses of varying duration. These discrepancies between average cargo velocity and single-motor or instantaneous velocity suggest that the motion of motor-cargo complexes requires regulation of the motors and their associated proteins and cargoes.

Proposed interactions between an assortment of motors and a plethora of associated cargoes and regulatory proteins, have been reviewed extensively [11–13]. Several lines of evidence suggest that the tail region of kinesin, containing various combinations of kinesin heavy chains (KHC) and kinesin light chains (KLC), plays an important role in binding specific cargoes [14, 15].

In addition, cargo attachment and detachment may regulate kinesin switching from activated and inactivated states respectively [16, 17]. In the case of dynein, the dynactin complex, in addition to several dynein light and intermediate chains (DLC, DIC), are crucial for the regulation of motor activity [18,19]. Although we have only covered some basic features of axonal transport biology in this chapter, further details on neuronal transport, transport proteins, and their relevance to neurodegeneration may be found in several recent reviews Gunawardena et al., 2004 [20–22].

4.3 Unraveling Mechanism: Using Imaging and Modeling

Despite considerable information about the individual components of the neuronal transport system, there are several gaps in our understanding. First, what is the relationship between the biophysical properties of individual motors determined in vitro and the movement of motor-cargo complexes in vivo? Second, how are interactions between specific proteins integrated to generate a complex trafficking system and to respond to perturbations? Third and finally, what key principles allow the neuron to take advantage of trafficking to support its functions?

In this chapter, we summarize some new strategies to better understand axonal transport. We discuss the imaging of vesicular transport in *Drosophila melanogaster* larval axons, and the development of quantitative metrics of transport function via the extraction of kinematic parameters from images. The application of these schemes to images from wild-type larvae provides a rigorous quantification of normal transport kinematic properties, and allows insights into properties of dysfunctional neurons in larvae expressing mutations altering specific transport proteins. Finally, we present some approaches to theoretical modeling of axonal transport and discuss the potential for such modeling not only to describe transport, but to generate new hypotheses about mechanisms of vesicle transport.

4.4 In Vivo Traffic Cameras: Imaging of Vesicles in Larval Segmental Nerves

Fluorescent proteins with a range of spectral properties enable the tagging and imaging of specific proteins in vivo [23, 24]. For illustrative purposes within this chapter, we focus on the analysis of vesicles containing Amyloid Precursor Protein (APP), a kinesin-associated cargo protein, fused to YFP Kamal et al. 2000 [25, 26]. Expression of APP-YFP in a select few segmental neurons, by using the SG26.1/62B driver [27], allows the imaging of one distinct axon over a very dark background. Introduction of mutations affecting transport

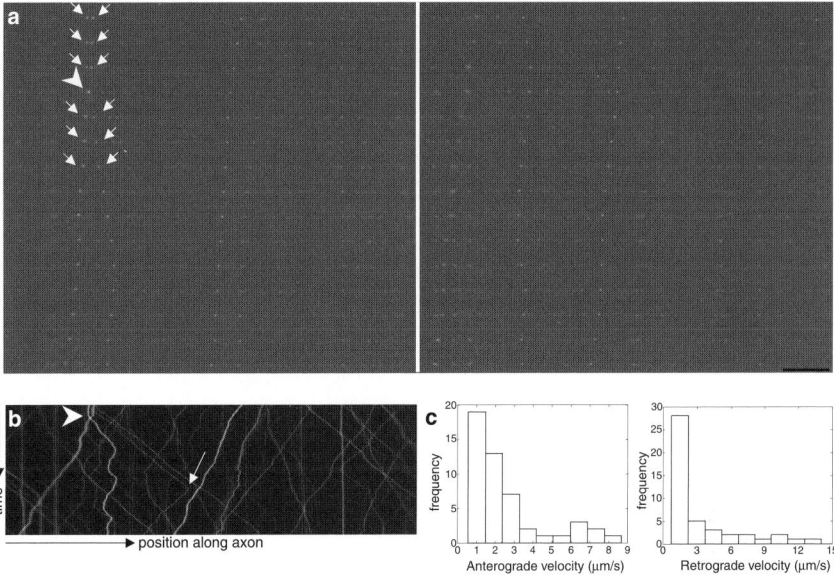

Fig. 4.3. (a) Sample panel of images from a time-lapse movie of wild-type fruit fly expressing APP-YFP. Axons were imaged using a 100x/NA 1.4 oil-immersion objective, with a capture rate of 10 Hz and 2×2 binning. This resulted in a pixel size of approximately $0.13\,\mu$m. Every fifth frame is shown, for a total of thirty frames. Frames corresponding to $t = 0\,\sec - 7.5\,\sec$ are shown in the first column, and $t = 7.5\,\sec - 15\,\sec$ in the second column. Two vesicles (*arrows*) are shown moving in opposite directions, and crossing (*arrowhead*). Bar $= 10\,\mu$m. (b) Kymograph corresponding to images in 3a. Position along the length of the imaged region of the axon is plotted on the x-axis. Elapsed time increases downward along the y-axis. Note the range of intensities, and the frequent superposition of tracks. Crossing vesicles from a are again indicated by an *arrowhead*. Reversing vesicle is shown with an *asterisk*. (c) Distributions of anterograde and retrograde velocities for single runs (terminated by a pause or a reversal) for vesicles shown in the sample movie. Note the considerable heterogeneity of velocities in both directions

proteins (e.g., KLC, DLC, members of the dynactin complex, etc.) then allows imaging and analysis of transport in the context of defects in specific transport components.

To image nerves and axons routinely, segmental nerves of third-instar larvae of a desired genotype are exposed by fine dissection. Animals are then pinned down, and inverted onto a coverslip for imaging. Cell bodies expressing APP-YFP are conspicuous in the ventral ganglia, and fluorescing axons emerging from these cell bodies are readily located. Axons are imaged using an inverted wide-field fluorescence microscope (Nikon TE2000-U), with a capture rate of 10 Hz, suitable for detecting vesicles being transported at greater than 5 microns/second. Specific sample imaging parameters are shown in the legend for Fig. 4.3, showing a montage of APP-YFP transport in wild-type larval axons. These parameters were used to image a mutant that has

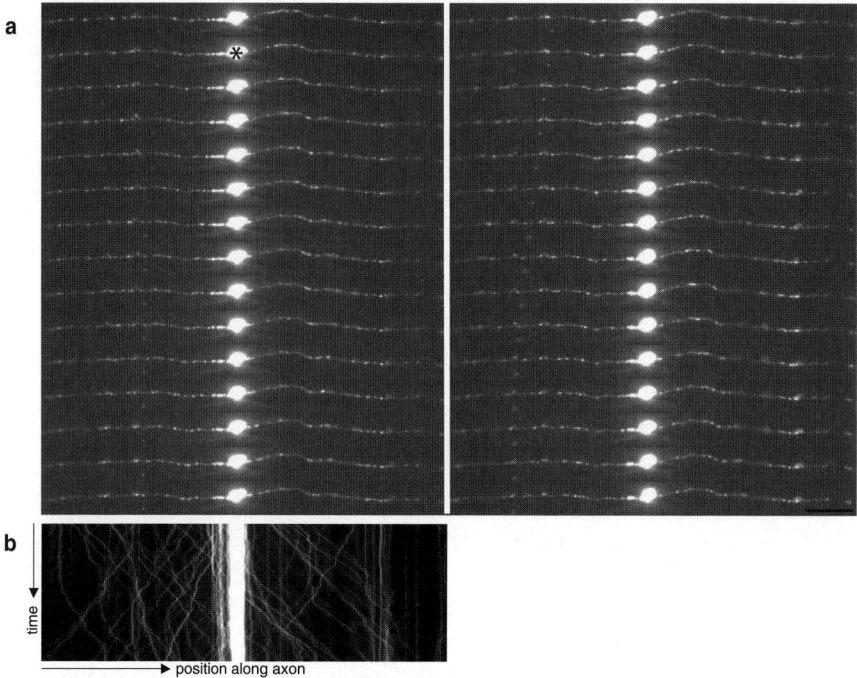

Fig. 4.4. (a) Sample panel of images from a time-lapse movie of a mutant fruit fly under-expressing DLC and expressing APP-YFP. Axons were imaged using the same configuration as in Fig. 4.3. Every fifth frame is shown, for a total of thirty frames. Frames corresponding to $t = 0\,\mathrm{sec} - 7.5\,\mathrm{sec}$ are shown in the first column, and $t = 7.5\,\mathrm{sec} - 15\,\mathrm{sec}$ in the second column. A large accumulation of vesicles is indicated by an *asterisk*. Bar $= 10\ \mu$m. (**b**) Kymograph corresponding to images in 4a

50% of the normal amount of a dynein light chain. Massive accumulations of fluorescently labeled vesicles, or clogs, were observed in some mutant axons, suggesting a transport defect (Fig. 4.4).

Though we have discussed the imaging of proteins *in vivo*, specific components of the transport machinery may also be imaged in a cell culture system. This can be done by transfecting primary neurons with constructs expressing fluorescently tagged proteins in combination with constructs causing either increased or reduced production of specific proteins [25]. This approach has the advantage of flexibility with respect to the targeted protein, and also offers the advantages of imaging vesicle motion on a flat surface. On the other hand, the loss of a true physiological environment and variable protein expression levels affect the interpretation of results.

4.5 Breaking Down the Film: Vesicle Tracking and Parameter Extraction

Several features are apparent when viewing the movies. First, due to the geometry of the axon, bi-directional movement is essentially constrained to one dimension. On the other hand, instances of vesicle superposition (e.g., merging, splitting, and crossing, as shown in Fig. 4.3, arrowhead) cause significant complexity. In addition, vesicles are observed with a range of intensities. While a general sense of transport may be obtained by simply viewing the movies (e.g., bi-directional movement; high vesicle density within the imaging field, etc.), rigorous analysis of transport kinematics requires the tracking of each individual vesicle over the course of the movie. This tracking has traditionally been performed by kymograph analysis (e.g., [28,29]). By drawing a line along the length of the imaged axon and plotting the intensity profile along that line at each time point, one can visualize the tracks of individual vesicles spatially and temporally (Fig. 4.3b; 4.4b). By tracing the path of each vesicle, one can then extract velocities, pause frequencies, reversal frequencies, and directions of vesicle motion.

While the kymograph offers the ability to quantify some aspects of vesicle transport, it has several drawbacks. Rigorous quantification of kinematic parameters is rather tedious and inefficient, due to the necessity for manually tracing each individual line on a given kymograph. In fact, a quick calculation shows that for a simple statistical comparison between two groups of movies, assuming 30–50 vesicles per movie and 10–20 movies per group, 600–2000 lines must be traced! Attempts to reduce the effort by measuring only a subset of lines tend to bias the data set, and can mask important differences in kinematic properties. In addition, vesicles of weaker fluorescent intensity are not easily resolved, and the properties of vesicles varying in size and intensity are not easily separated. To resolve some of these issues, we have applied methods in particle tracking to automate the extraction of kinematic parameters for individual vesicles from these movies.

The basic approach to tracking imaged vesicles involves four steps (details are published elsewhere; [30]). The first step uses well-characterized techniques in fluorescence speckle microscopy to detect vesicles as points [31]. The identity of these points is based on the detection of local maxima above the local background, which is defined by Delauny triangulation of local minima in the fluorescent signal. Sometimes, multiple local maxima are found for larger vesicles, resulting in some errors in vesicle localization. These vesicles can be defined as two-dimensional blobs using multi-resolution wavelet analysis and local thresholding techniques [32], and are localized on the basis of their area and shape. Tracks for each vesicle are then generated by utilizing a global optimization scheme based on bipartite correspondence between points in consecutive frames. This correspondence is evaluated both by position (as done in most existing nearest-neighbor particle tracking schemes), and by vesicle fluorescence intensity and velocity.

This tracking scheme accurately joins vesicles from frame to frame, however, tracks are often broken at sites of vesicle merging, splitting, and crossing. The many modes of track breakage pose a significant challenge, and confound traditional particle-tracking software, which are usually based on nearest-neighbor linkage of particles from frame to frame. The solution to this problem required the development of gap-closing algorithms, which piece together track fragments and fill in gaps between broken tracks; for a given track, all candidate tracks within a search radius are screened and joined to ensure continuity in direction, velocity, and intensity values. Tracks terminating at another track with a spike in fluorescence intensity and no obvious reinitiation of movement are considered merges. Splits are detected by running the same algorithm in the backward direction. Tracks with interruptions due to their merging with other tracks are then completed by duplicating the overlapping portion. Finally, incomplete tracks (that disappear completely during the course of imaging) are not considered. The generation of a complete track for each vesicle allows definition of a complete set of transport metrics and a thorough description of a vesicle's kinematic properties. Parameters defined include instantaneous and average vesicle velocities in each direction, vesicle run lengths, pause and reversal frequencies, flux, and vesicle spacing and distribution. Interestingly, heterogeneity, potentially masked with a less thorough analysis, is evident even within a single movie, as seen in the bimodal velocity distribution for a sample wild-type movie (Fig. 4.3c).

4.6 Understanding the Data: Theoretical Modeling of Axonal Transport

Given our ability to observe and quantify the movement of individual vesicles, it would initially appear that a stochastic, single-particle, microscopic model would ideally simulate the data. A microscopic model would also allow the simulation of events such as vesicle collision, merging, and splitting that we observe in vivo. Indeed we are currently working on the development of a microscopic model to understand cooperation and competition between different motors and physical interactions between cargoes.

Alternatively, a macroscopic model would be useful to evaluate transport on a more global scale. Such a model would provide specific parameters, corresponding to specific physical principles (e.g., diffusion vs. directional motor-driven transport), that govern the organization and spatial distribution of cargoes. Our vesicle tracking algorithms would then allow us to verify such a model by quantifying vesicle organization, vesicle flux, and spatial profiles of vesicle density and type. A macroscopic model would also define the sensitivity of the transport system to local perturbations. An example of this type of analysis is to predict the effect of vesicle accumulations (e.g., the clog in Fig. 4.4) on the distribution of cargoes at varying distances from the clog.

Finally, assessing the response of a transport system at varying scales is critical to linking microscopic and macroscopic models, so that we can determine
how specific parameters governing bulk transport are related to parameters
governing the behavior of individual cargoes.

We will devote the remainder of this chapter to the macroscopic modeling of axonal transport. We will discuss the simulation of axonal transport
using basic principles of convection, diffusion, and reaction. In addition, we
also propose an approach to model vesicle accumulations (axonal clogs) and
resultant changes in vesicle distributions. Finally, we present preliminary efforts to predict the response of experimental systems with our model.

4.6.1 Numerical Solution
of Convection-Diffusion-Reaction Equations

Reference [37] recently developed an elegant theoretical treatment of intracellular microtubule-based transport of organelles, based on fundamental
convection-diffusion-reaction equations. In this model, each motor-cargo complex, or vesicle, was modeled as a "particle" that existed in a microtubule-
bound or unbound state. Particle density at each state was determined based
on binding rate constants, which allowed transitions between unbound and
bound states (Fig. 4.5, solid lines). Bound particles moved either anterogradely
or retrogradely in one dimension (along an axon) at a steady velocity, while
unbound particles diffused freely. Details of their analytical solutions are found
in their original paper, but we will highlight some relevant observations and
limitations of their analysis.

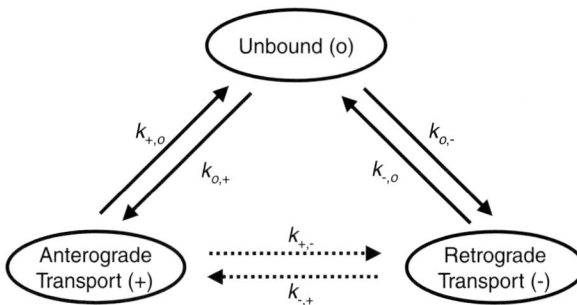

Fig. 4.5. Particles (*vesicles*) exist in one of three states. Unbound particles (o)
move via diffusion and bound particles move anterogradely (+) or retrogradely (−).
The original Smith and Simmons model did not allow for direct transitions between
bound anterograde and bound retrograde states (*dotted line*)

To enable a straight-forward analytic solution, the authors made several
assumptions. First, only a symmetric case was considered, with equal bind-
ing/unbinding rates and equal velocities in each direction. This provided an

adequate model for a single type of motor traveling on microtubules with equally mixed polarity, but is not ideal for the more biologically relevant case with two distinct classes of motors moving across microtubules of uniform polarity (cf. Fig. 4.3c). Initial particle distributions were also limited to step rises in particle concentration at the cell body or dispersion from a point distribution of particles. In addition, boundary conditions required constant reservoirs of particles at the "cell body" and at the "nerve terminal." Consequently, the temporal response of more complicated spatial profiles, such as those found experimentally, could not be solved. Nevertheless, solutions constrained by initial values and boundary conditions predicted the basic kinetics of bulk particle dispersion. On the other hand, due to the severe limitations on initial and boundary conditions, transient theoretical responses could not be readily determined, and furthermore, could not be linked to experimental data. We used a 1D macroscopic reaction-diffusion-transport model based on that proposed by Smith and Simmons, to describe bulk bidirectional vesicle transport:

$$\frac{\partial d_0}{\partial t} - D\frac{\partial^2 d_0}{\partial x^2} = -\left(k_{0,+} + k_{0,-}\right)d_0 + k_{+,0}d_+ + k_{-,0}d_- \qquad (4.1)$$

$$\frac{\partial d_+}{\partial t} + v_+\frac{\partial d_+}{\partial x} = k_{0,+}d_0 + k_{-,+}d_- - \left(k_{+,0} + k_{+,-}\right)d_+ \qquad (4.2)$$

$$\frac{\partial d_-}{\partial t} + v_-\frac{\partial d_-}{\partial x} = k_{0,-}d_0 + k_{+,-}d_+ - \left(k_{-,0} + k_{-,+}\right)d_- \qquad (4.3)$$

Table 4.1 provides details on the mathematical notation. This model differs from the original by allowing for a direct state transition between anterograde and retrograde states ($k_{+,-}$ and $k_{-,+}$; Fig. 4.5, dotted lines). From a biological perspective, these coefficients could describe a direct interaction between dynein-driven and kinesin-driven vesicle transport, such as that proposed by [33]. Reference [34] developed a similar model recently, which may be studied for further theoretical details. This model allows for an additional motor to bind a given particle, potentially affecting microtubule binding kinetics. However, because we are unable to directly count the number of bound motors experimentally, we use Eqs. (4.1)–(4.3).

To address some of the limitations resulting from the requirement for an analytical solution to the system of partial differential equations, we elected to solve the set of reaction-diffusion-transport equations numerically. Numerical finite-difference modeling allowed us to consider a range of boundary conditions – both constant and dynamic – as well as the more complex initial particle distributions taken directly from our experimental observations. In addition, solving these equations numerically eased the requirement for constant and/or symmetric velocities and reaction parameters.

We solved Eqs. (4.1)–(4.3) using the Crank-Nicholson scheme [35]. Some instability has been documented for numerical solutions to convection-diffusion -reaction equations [36], so we tested several sets of initial density profiles and

Table 4.1. Mathematical Notation. Note that vesicles are treated as particles

t	Time coordinate
x	Space coordinate along 1D axonal microtubule
d_o	Distribution density of particles that are detached from axonal microtubules (in number of particles per unit volume)
d_+	Distribution density of bound anterograde transport particles
d_-	Distribution density of bound retrograde transport particles
D	Diffusion rate coefficient of detached particles
$k_{o,+}, k_{+,o}$	Transition rate coefficients of detached particles to and from anterograde transport particles, respectively
$k_{o,-}, k_{-,o}$	Transition rate coefficients of detached particles to and from retrograde transport particles, respectively
$k_{+,-}, k_{-,+}$	Transition rate coefficients of anterograde transport particles to and from retrograde transport particles, respectively
v_+	Anterograde velocity
v_-	Retrograde velocity

boundary conditions. To display salient features of the system response to changes in parameters, we show a profile that simulates the dispersion of particles from a central peak (Fig. 4.6a). Responses to other initial profiles may be found elsewhere ([34, 37]). Setting constant values at the spatial boundaries (Dirichlet boundary conditions) resulted in instability at the edges of the spatial profile (data not shown). This was expected given the dynamic nature of the density spatial profiles, particularly for extreme parameter values. To remedy this situation, reflective boundary conditions were imposed at each step, such that the edge density values for each particle distribution ($d_{x=0}$ and $d_{x=M}$) at timestep t_n were used as boundary values $d_{x=0}$ and $d_{x=M}$ for the next timestep t_{n+1}. Experimentally, these boundary conditions would correspond to a no-flux condition, or an absence of particles entering or leaving the imaging field. Solutions were obtained using customized implementation of the Crank-Nicholson scheme in MATLAB 7.0.4.

4.6.2 System Response to Perturbations of Modeling Parameters

Representative temporal and spatial system responses for each particle state are shown in Fig. 4.6 for the parameters listed in the figure legend. To focus on the most dynamic region of each profile, a region of 20 μm is shown for each response. The unbound diffusing particles initially disperse rapidly, as seen by flattening and widening of the initial central peak during the first half-second (Fig. 4.6c). This is followed by a more gradual dispersion. The bound particle peak densities (Figs. 4.6d–e) initially increase, in synch with the rapid reduction in unbound particle peak density, and then gradually decrease. As expected, bound anterograde particles migrate anterogradely (to the right) and bound retrograde particles migrate retrogradely (to the left).

The dispersion of bound particles is coupled to the dispersion of unbound particles, as seen by the spreading out of all three particle profiles over time (Figs. 4.6b-e).

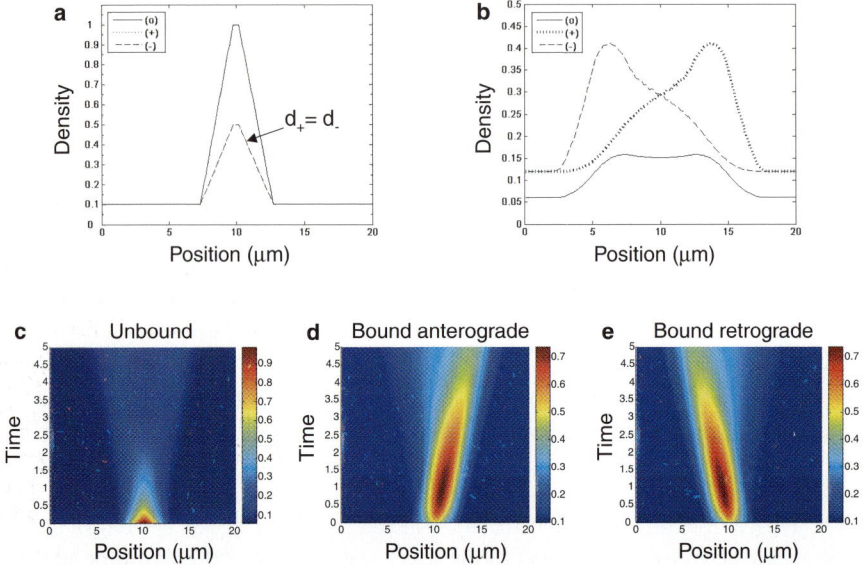

Fig. 4.6. Theoretical initial profiles for unbound particles (d_o, *solid line*), bound anterograde particles (d_+, *dotted line*) and bound retrograde particles (d_-, *dashed line*). (**a**) Particles disperse from a central peak; profiles for d_+ and d_- overlap. (***b-e***) Sample response to initial conditions shown in a. (**b**) Spatial profile of all three particle states at $t = 5$ sec. (**c**) Sample response of unbound particle spatial profile over position and time for $t = 5$ sec. (**d-e**) Sample response of bound anterograde particle distribution and bound retrograde particle distribution over position and time for $t = 5$ sec. The following parameters, which could be calibrated within the model, were used: $D = 0.1\,\mu\text{m}^2/\text{sec}$; $v_+ = v_- = 1\,\mu\text{m/sec}$; $k_{o,+} = k_{o,-} = 1\,\text{sec}^{-1}$; $k_{+,o} = k_{-,o} = 0.5\,\text{sec}^{-1}$; $k_{+,-} = k_{-,+} = 0\,\text{sec}^{-1}$. Densities of each particle state are given by the colormap

To determine the critical parameters leading to this type of system response, individual parameters are varied one at a time (Fig. 4.7). Increasing diffusivity five-fold, to the diffusivity of a 1 µm sphere diffusing in water, has very little effect on the system response, though a slight reduction in bound particle peaks was observed (Fig. 4.7a). However, changes in velocities and reaction coefficients have dramatic effects on spatial distributions. Reducing retrograde velocity (Fig. 4.7b) results in asymmetry of both bound and unbound particle populations, with unbound particle dispersion biased towards the side of the faster anterograde velocity. Reducing binding rates symmetrically (reduction in $k_{o,-}$ and $k_{o,+}$, Fig. 4.7c) results in a reduction of the bound

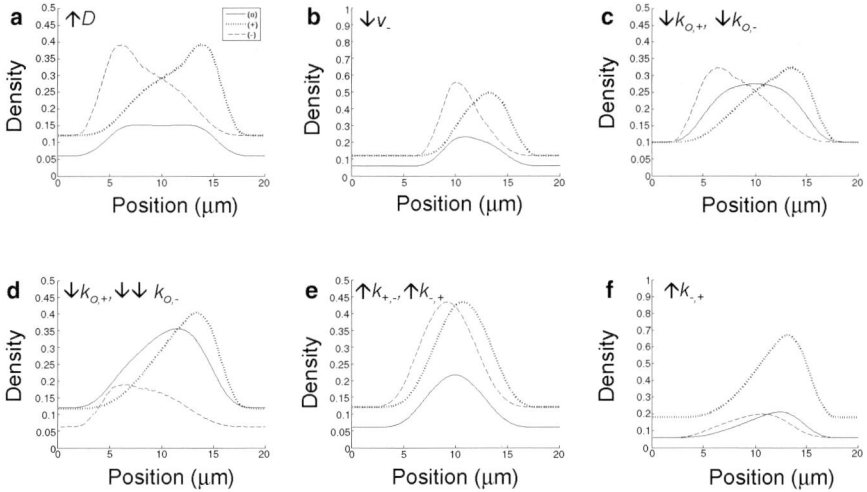

Fig. 4.7. Variations in individual parameters: Displayed are the final spatial profiles of all three particle states at $t = 5$ sec for the same parameters as shown in Fig. 4.6, with a change in (**a**) an increase in diffusivity ($D = 0.5$ mm^2/sec); (**b**) a decrease in retrograde velocity ($v_- = 0.1$ mm/sec); (**c**) a decrease in binding reaction constants ($k_{o,+} = 0.5$ sec^{-1}; $k_{o,-} = 0.5$ sec^{-1}); (**d**) an unequal decrease in binding reaction constants ($k_{o,+} = 0.5$ sec^{-1}; $k_{o,-} = 0.2$ sec^{-1}) (**e**) an increase in direction change reaction constants ($k_{+,-} = k_{-,+} = 0.5$ sec^{-1}); (**f**) an increase in retrograde to anterograde reaction constant ($k_{+,-} = 0$ sec^{-1}; $k_{-,+} = 0.5$ sec^{-1})

particle peaks compared to the unbound particle peaks, though the net range over which bound particles dispersed does not change. When binding rates are reduced asymmetrically (lower $k_{o,-}$ compared to $k_{o,+}$, Fig. 4.7d), both bound anterograde and unbound particle density peaks increase, at the expense of retrograde particles. Again, the range of dispersion does not change. When symmetric transitions from bound anterograde to bound retrograde states are allowed (increases in $k_{+,-}$ and $k_{-,+}$, Fig. 4.7e), bound particle densities do not disperse as rapidly. However, if only a one-sided transition from bound to unbound state is allowed (lower $k_{+,-}$ compared to $k_{-,+}$, Fig. 4.7f), a dramatic increase in the anterograde bound peak density is observed, as well as a skewing and slight increase in the unbound peak density. These results collectively reveal a strong relationship between velocity and range of active dispersion of bound particles. Additionally, the coupling between the distributions of bound and unbound particles highlights the significant influence of the reaction coefficients on system response. Our results are in agreement with the numerical system responses of [34], which were obtained with an unknown numerical solution scheme.

4.6.3 Simulation of an Axonal Traffic Jam

To simulate defects in axonal transport, such as those observed in the dynein light chain mutant (Fig. 4.4), we modified the spatial profile of critical parameters to simulate an axonal blockage. There may be multiple mechanisms leading to these axonal traffic jams, however, most blockages display disorganization of the cytoskeleton and/or accumulation of organelles. Consequently, we speculate that near the blockage location, moving particles would either become detached, or perhaps decrease in velocity due to physical blockage. Therefore, we modified the binding and unbinding rates as well as velocities in the vicinity of a "blockage." For example, a particle moving anterogradely is more likely to detach in the vicinity of a blockage (Fig. 4.8a, dotted line), while

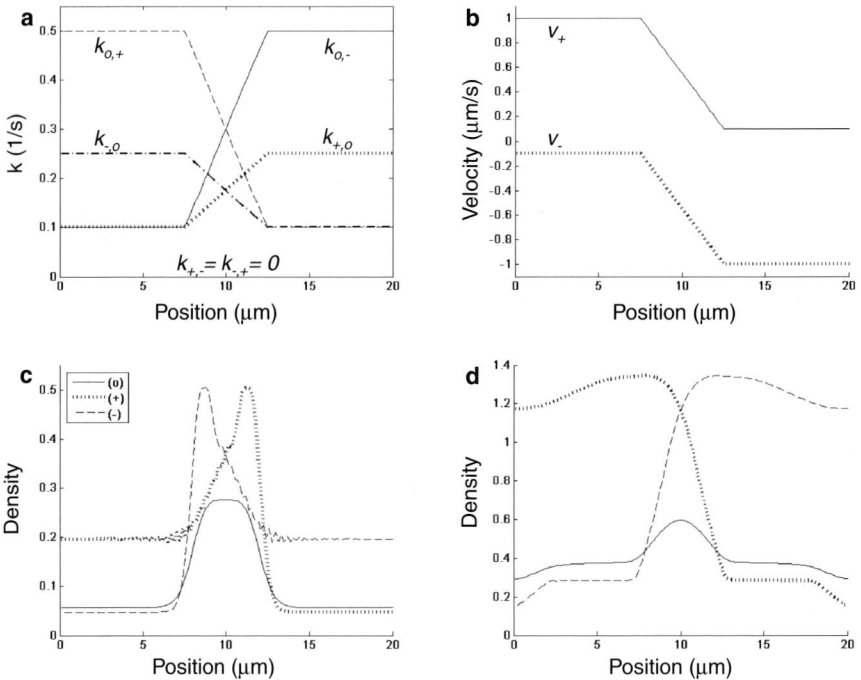

Fig. 4.8. Congestion, or blockage, was generated by changes in model parameters. As bound particles approached the "blockage" in the middle of the profile, they were more likely to slow down and more likely to detach. Unbound particles near the clog were less likely to attach. (**a**) Reaction constant profiles ($k_{o,+}$: dashed; $k_{-,o}$: *dashed-dotted*; $k_{+,o}$: *dotted*; $k_{o,-}$: *solid lines*) and (**b**). Velocity as a function of position (v_+: *solid*; v_-: *dotted lines*). Note that retrograde velocity is plotted as a negative value. (**c**) Response after $t = 5$ sec of initial central peak profile (Fig. 4.6a) to this perturbation. (**d**) Response after $t = 5$ sec of constant initial density profiles (initial bound anterograde and retrograde particle densities $= 0.5$, unbound $= 1.0$) to the same perturbation

a particle that is detached in the vicinity of a blockage is less likely to attach and move anterogradely into the blockage (Fig. 4.8a, dashed line). Similarly, an anterogradely moving particle approaching a blockage from its proximal (left) side would slow down (Fig. 4.8b, solid line), as would a retrogradely moving particle approaching a blockage from its distal (right) side (Fig. 4.8b, dotted line).

Imposition of these spatially changing parameters has dramatic effects on the particle density profiles. For initial distributions of particles in a center peak (i.e., within the blockage, cf. Fig. 4.6a), particles are confined to the vicinity of the blockage (Fig. 4.8c). On the other hand, for initial distributions of moving particles that are constant through the blockage region, particles accumulate at the edges of the blockage. Bound anterograde particles accumulate at the proximal edge of the clog, while bound retrograde particles accumulate at the distal edge of the clog (Fig. 4.8d).

4.6.4 Linking Theory to Experiment

An important question is whether macroscopic modeling is appropriate for understanding the distributions of small groups of individual particles, such as those imaged in the *Drosophila* larvae. To do this, we determined spatial and temporal density profiles for stationary particles (modeled by particles in an unattached state), particles moving anterogradely (modeled by particles in bound, anterograde state) and particles moving retrogradely (modeled by particles in bound, retrograde state) by calculating position and direction of individual particles over periods of 1.5 seconds (15 frames, or ten time points per 150 frame movie). Spatially, the 696 pixel-wide images were divided into 24 bins, and Gaussian smoothing ($\sigma = 2$) was used to make spatial profiles continuous.

Model parameters were optimized to match the experimental profile at specific time points. Assuming time-invariant and space-invariant diffusion rate, transition rates, and transport velocities, the model has a total of 9 parameters. However, by making some assumptions based on our experimental data, we simplified the optimization. First, for the time scales over which our movies were generated, vesicles were either observed to move at a rate consistent with active transport, or remain stationary. Consequently, the diffusivity was set to a negligible value ($D = 0.01\mu m^2/sec$). In addition, through dimensional analysis, we were able to further reduce the number of independent parameters to 6. It should be noted that while the values for several reaction parameters will likely be negligible when fitting model predictions to distributions from wild-type movies, we predict an increase in these parameter values when fitting data from movies of motor mutants.

Figure 4.9 shows a sample initial spatial profile of densities in each of the three particle states (Fig. 4.9a) as well as the profiles of each state after 1.5 seconds, for experimental data (lines) and model data (symbols) (Fig. 4.9b). For the parameter set selected, the model fits experimental data surprisingly well. The selected anterograde velocity ($v_+ = 1.6\,\mu\text{m/s}$) and retrograde velocity ($v_- = -1.1\,\mu\text{m/s}$) accurately predicted the appropriate translation of the peak densities of bound anterograde and retrograde densities (arrowheads, Fig. 4.9b). These velocities were comparable, but not identical, to the average of individual particle velocities measured with our tracking program ($v_+ = 2.1\,\mu\text{m/sec}$ and $v_- = -1.8\,\mu\text{m/sec}$). By selecting appropriate binding and unbinding constants ($k_{o,+} = 0.13\,\text{sec}^{-1}, k_{o,-} = 0.12\,\text{sec}^{-1}, k_{+,o} = k_{+,-} = k_{-,+} = k_{-,o} = 0\,\text{sec}^{-1}$), increases in anterograde and retrograde density peaks were also accurately predicted (arrowhead and asterisk, Fig. 4.9), as was the trend of fewer stationary particles in the proximal half of the profile (arrow, Fig. 4.9b).

On the other hand, a few features were not accurately predicted. The reduction and dispersion of diffusing particles was grossly overestimated, despite negligible diffusivity. This error was likely a result of the imposition of reflective boundary conditions, when in reality, there is a measurable influx and efflux of particles of various states over time. Another inaccuracy was the failure to simulate the emergence of new retrograde clusters of particles (double arrows, Fig. 4.9b). This was primarily a result of our assuming a spatially constant rate of attachment and detachment. By increasing (in a piece-wise fashion) $k_{o,-}$ to $0.20\,\text{sec}^{-1}$ and $k_{+,-}$ to $0.12\,\text{sec}^{-1}$ for the distal (right) one-fifth of the spatial field, and $k_{o,-} = 0.25\,\text{sec}^{-1}$ for the proximal (left) one-third of the spatial field, the model more accurately predicted experimental results (Fig. 4.9c).

4.7 Conclusions and Future Directions

Our in vivo imaging system allows the measurement of kinematic parameters of individual vesicles during axonal transport. Our newly developed algorithms to measure cargo distributions from these images provide input parameters and experimental verification of a macroscopic model of the axonal transport system. Our model accurately predicts several features of the spatial and temporal response of vesicle density distributions to initial distributions measured from experiment. We expect greater accuracy once we impose more realistic boundary conditions. By introducing random fluctuations in model parameters at certain time points, it may be possible to simulate the "noise" inherent to biological systems within the framework of our macroscopic model. However, given the requirement for spatial variability in macroscopic model parameters (Fig. 4.9c), it may be more realistic to describe the stochastic dynamic behavior of individual vesicles via microscopic vesicle transport models. Microscopic models would also allow the simulation of observed events such as vesicle collision, merging, and splitting. Currently, we are developing microscopic vesicle transport models to understand the cooperation and competition between different motors (cf. [38]), and the influence of the shape of vesicles and organelle accumulations on vesicle transport. The linking of

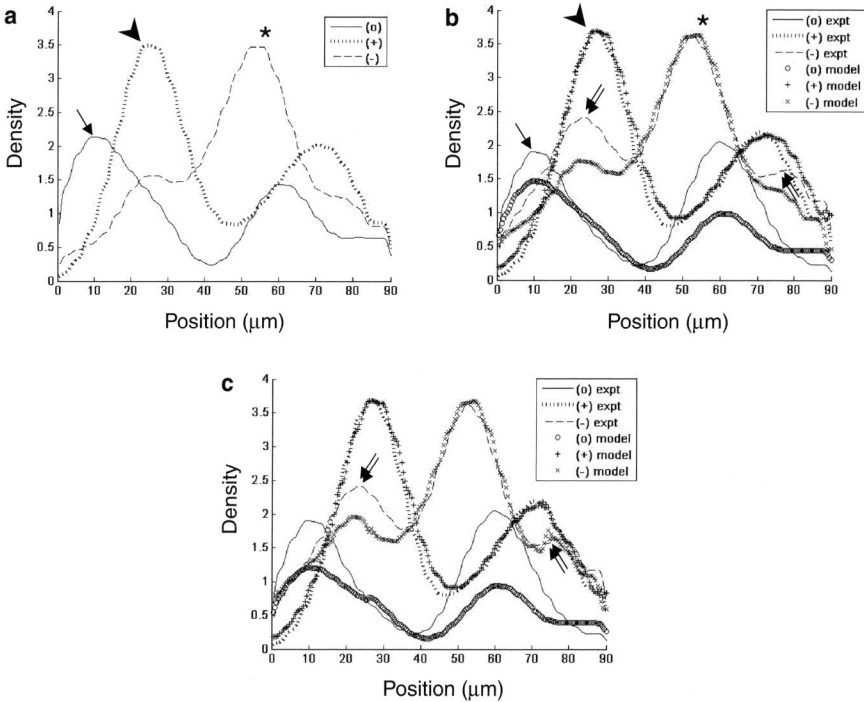

Fig. 4.9. (a) Sample initial experimental spatial profiles for unbound particles (d_o, *solid line*), bound anterograde particle (d_+, *dotted line*) and bound retrograde particles (d_-, *dashed line*) from movie shown in Fig. 4.3. (b) Spatial profile after 1.5 seconds from experiment (*lines*) and from model stationary (d_o, "o"), bound anterograde particle (d_+, "+") and bound retrograde particles (d_-, "x"). Parameters were spatially constant, and are given in the text. (c) Comparison between experimental spatial profiles (*lines*) and model profiles (*symbols*). Several parameters, given in the text, were varied in step-wise manner, and resulted in a better fit. *Arrowhead* indicates an anterograde peak, *asterisk* indicates a retrograde peak, *single arrow* indicates a stationary peak, and *double arrows* indicate two newly developing retrograde peaks in the experimental profiles

such stochastic microscopic model predictions to experimental data will not be trivial. However, we will benefit greatly from our ability to directly measure transport parameters in vivo, allowing us further insight into motor cooperativity and cargo interactions.

It is important to ultimately bridge the gap between between macroscopic and microscopic models; for example, state transition probabilities of multiple individual motors bound to a cargo could be correlated directly to the transition rate coefficients in the macroscopic vesicle transport model. Recently, there have been many significant advances in high-resolution imaging and also quantification of critical features of dynamic biological processes, including our own axonal imaging assays and tracking software. Given these advances, the time is right for integrating increasingly accurate experimental readouts into more meaningful theoretical models.

Acknowledgements

We acknowledge Shermali Gunawardena for the schematic drawing of *Drosophila* anatomy, Richard Brusch for technical assistance with imaging, and the UCSD Department of Bioengineering for MATLAB software. This work is supported in part by NIH Ruth-Kirschstein Fellowship GM071145 (to S.B.S.), NIH grant GM35252 (to L.S.B.G.), and NIH grant R01 GM60678 (to G.D.). L.S.B.G. is an investigator of the Howard Hughes Medical Institute.

References

1. C.M. Coppin, J.T. Finer, J.A. Spudich, R.D. Vale (1996). Detection of sub-8-nm movements of kinesin by high-resolution optical-trap microscopy. *Proc Natl Acad Sci USA*, **93**(5), pp. 1913–7.
2. J. Howard, A.J. Hudspeth, R.D. Vale (1989). Movement of microtubules by single kinesin molecules. *Nature*, **342**(6246), pp. 154–8.
3. K. Kawaguchi, S. Ishiwata (2000). Temperature dependence of force, velocity, and processivity of single kinesin molecules. *Biochem Biophys Res Commun*, **272**(4.3), pp. 895–9.
4. S.C. Kuo, M.P. Sheetz (1993). Force of single kinesin molecules measured with optical tweezers. *Science*, **260**(5105), pp. 232–4.
5. R. Mallik, B.C. Carter, S.A. Lex, S.J. King, S.P. Gross (2004). Cytoplasmic dynein functions as a gear in response to load. *Nature*, **427**(6975), pp. 649–52.
6. R.D. Vale (1993). Measuring single protein motors at work. *Science*, **260**(5105), pp. 169–70.
7. R.D. Vale, F. Malik, D. Brown (1992). Directional instability of microtubule transport in the presence of kinesin and dynein, two opposite polarity motor proteins. *J Cell Biol*, **119**(6), pp. 1589–96.
8. R.D. Allen, J. Metuzals, I. Tasaki, S.T. Brady, S.P. Gilbert (1982). Fast axonal transport in squid giant axon. *Science*, **218**(4577), pp. 1127–9.
9. L. Wang, A. Brown (2001). Rapid intermittent movement of axonal neurofilaments observed by fluorescence photobleaching. *Mol Biol Cell*, **12**(10), pp. 3257–67.
10. S.T. Brady, R.J. Lasek, R.D. Allen (1982). Fast axonal transport in extruded axoplasm from squid giant axon. *Science*, **218**(4577), pp. 1129–31.
11. B.W. Guzik, L.S. Goldstein (2004). Microtubule-dependent transport in neurons: steps towards an understanding of regulation, function and dysfunction. *Curr Opin Cell Biol*, **16**(4), pp. 443–50.
12. N. Hirokawa, R. Takemura (2005). Molecular motors and mechanisms of directional transport in neurons. *Nat Rev Neurosci*, **6**(4.3), pp. 201–14.
13. V. Muresan (2000). One axon, many kinesins: What's the logic? *J Neurocytol*, **29**(11–12), pp. 799–818.

14. A.B. Bowman, A. Kamal, B.W. Ritchings, A.V. Philp, M. McGrail, J.G. Gindhart, L.S Goldstein (2000). Kinesin-dependent axonal transport is mediated by the sunday driver (SYD) protein. *Cell,* **103**(4), pp. 583–94.

15. K.J. Verhey, D. Meyer, R. Deehan, J. Blenis, B.J. Schnapp, T.A. Rapoport, B. Margolis (2001). Cargo of kinesin identified as JIP scaffolding proteins and associated signaling molecules. *J Cell Biol,* **152**(5), pp. 959–70.

16. R. Cross, J. Scholey (1999). Kinesin: the tail unfolds. *Nat Cell Biol,* **1**(5), pp. E119–2.

17. D.D. Hackney, J.D. Levitt, J. Suhan (1992). Kinesin undergoes a 9 S to 6 S conformational transition. *J Biol Chem,* **267**(12), pp. 8696–701.

18. S.J. King, T.A. Schroer (2000). Dynactin increases the processivity of the cytoplasmic dynein motor. *Nat Cell Biol,* **2**(4.1), pp. 20–4.

19. A.W. Tai, J.Z. Chuang, C.H. Sung (2001). Cytoplasmic dynein regulation by subunit heterogeneity and its role in apical transport. *J Cell Biol,* **153**(7), pp. 1499–509.

20. N. Hirokawa, R. Takemura (2003). Biochemical and molecular characterization of diseases linked to motor proteins. *Trends Biochem Sci,* **28**(10), pp. 558–65.

21. S. Roy, B. Zhang, V.M. Lee, J.Q. Trojanowski (2005). Axonal transport defects: a common theme in neurodegenerative diseases. *Acta Neuropathol (Berl),* **109**(4.1), pp. 5–13.

22. S. Gunawardena, L.S. Goldstein (2004).Cargo-carrying motor vehicles on the neuronal highway: transport pathways and neurodegenerative disease. *J Neurobiol,* **58**(4.2), pp. 258–71.

23. A.B. Cubitt R. Heim, S.R. Adams A.E. Boyd L.A. Gross R.Y. Tsien (1995). Understanding, improving and using green fluorescent proteins. *Trends Biochem Sci,* **20**(11), pp. 448–55.

24. M.M. Falk, U. Lauf (2001). High resolution, fluorescence deconvolution microscopy and tagging with the autofluorescent tracers CFP, GFP, and YFP to study the structural composition of gap junctions in living cells. *Microsc Res Tech,* **52**(4.3), pp. 251–62.

25. C. Kaether, P. Skehel, C.G. Dotti (2000). Axonal membrane proteins are transported in distinct carriers: a two-color video microscopy study in cultured hippocampal neurons. *Mol Biol Cell,* **11**(4), pp. 1213–24.

26. A. Kamal, G.B. Stokin, Z. Yang, C.H. Xia, L.S. Goldstein (2000). Axonal transport of amyloid precursor protein is mediated by direct binding to the kinesin light chain subunit of kinesin-I. *Neuron,* **28**(4.2), pp. 449–59.

27. G.B. Stokin, C. Lillo, T.L. Falzone, R.G. Brusch, E. Rockenstein, S.L. Mount, R. Raman, P. Davies, E. Masliah, D.S. Williams, L.S. Goldstein (2005). Axonopathy and transport deficits early in the pathogenesis of Alzheimer's disease. *Science,* **307**(5713), pp. 1282–8.

28. S. Gunawardena, L.S. Her, R.G. Brusch, R.A. Laymon, I.R. Niesman, B. Gordesky-Gold, L. Sintasath, N.M. Bonini, L.S. Goldstein (2003). Disruption of axonal transport by loss of huntingtin or expression of pathogenic polyQ proteins in Drosophila. *Neuron,* **40**(4.1), pp. 25–40.

29. H.M. Zhou, I. Brust-Mascher J.M. Scholey (2001). Direct visualization of the movement of the monomeric axonal transport motor UNC-104 along neuronal processes in living Caenorhabditis elegans. *J Neurosci,* **21**(11), pp. 3749–55.

30. G. Yang, A. Matov, G. Danuser (2005). Reliable Tracking of Large Scale Dense Antiparallel Particle Motion for Fluorescent Live Cell Imaging, in *Proc. IEEE*

Workshop on Computer Vision Methods for Bioinformatics at CVPR., San Diego, CA, USA.

31. A. Ponti, P. Vallotton, W.C. Salmon, C.M. Waterman-Storer, G. Danuser (2003). Computational analysis of F-actin turnover in cortical actin meshworks using fluorescent speckle microscopy. *Biophys J*, **84**(5), pp. 3336–52.

32. J-C. Olivo-Marin (2002). Extraction of spots in biological images using multi-scale products. *Pattern Recognition*, **35**, pp. 1989–1996.

33. S.P. Gross, M.A. Welte, S.M. Block E.F. Wieschaus (2002). Coordination of opposite-polarity microtubule motors. *J Cell Biol*, **156**(4), pp. 715–24.

34. A. Friedman, G. Craciun (2005). A model of intracellular transport of particles in an axon. *J Math Biol*, **51**(4.2), pp. 217–46.

35. J. Hoffman (1992). *Numerical methods for engineers and scientists.*, McGraw-Hill, Inc.

36. W. Hundsdorfer, J.G. Verwer (2003). *Numerical solution of time-dependent advection-diffusion-reaction equations.*, Springer.

37. D.A. Smith and R.M. Simmons (2001). Models of Motor-Assisted Transport of Intracellular Particles. *Biophys J*, **80**(1) pp. 45–68.

38. M. Badoual, F. Jülicher, J. Prost (2002). Bidirectional Cooperative Motion of Molecular Motors. *Proc Natl Acad Sci*, **99**(10), pp. 6696–6701.

5

Intracellular Transport and Kinesin Superfamily Proteins: Structure, Function and Dynamics

N. Hirokawa[1] and R. Takemura[2]

[1] Department of Cell Biology and Anatomy, Graduate School of Medicine,
University of Tokyo, Tokyo, Japan
hirokawa@m.u-tokyo.ac.jp
[2] Okinaka Memorial Institute for Medical Research, Tokyo, Japan

Abstract. Using various molecular cell biological and molecular genetic approaches, we identified kinesin superfamily proteins (KIFs) and characterized their significant functions in intracellular transport, which is fundamental for cellular morphogenesis, functioning, and survival. We showed that KIFs not only transport various membranous organelles, proteins complexes and mRNAs fundamental for cellular functions but also play significant roles in higher brain functions such as memory and learning, determination of important developmental processes such as left-right asymmetry formation and brain wiring. We also elucidated that KIFs recognize and bind to their specific cargoes using scaffolding or adaptor protein complexes. Concerning the mechanism of motility, we discovered the simplest unique monomeric motor KIF1A and determined by molecular biophysics, cryoelectron microscopy and X-ray crystallography that KIF1A can move on a microtubule processively as a monomer by biased Brownian motion and by hydolyzing ATP.

Key words: Kinesin superfamily proteins, KIFs, Intracellular Transport, Microtubule, Biased Brownian Movement

5.1 Introduction

A neuron has a highly polarized structure, and stimuli received at the dendrites are directionally transmitted to the axon through the cell body [1]. Axons are quite often one meter long, but most of the proteins in axons and synaptic terminals where neuron transmits signals to the next neuron or target cell are synthesized in the cell body first and then delivered over a long distance by a process called axonal transport. Thus, axonal transport plays significant roles in neuronal morphogenesis, function, and survival. However, dendritic transport also plays important roles in neurons. Furthermore, intracellular transport plays essential roles in many other types of cell. It is becoming increasingly clear that the basic principles of intracellular transport

N. Hirokawa and R. Takemura: *Intracellular Transport and Kinesin Superfamily Proteins: Structure, Function and Dynamics*, Lect. Notes Phys. **711**, 85–121 (2007)
DOI 10.1007/3-540-49522-3_5   © Springer-Verlag Berlin Heidelberg 2007

are conserved, and a common molecular machinery is used for intracellular transport in neurons and other types of cell. Because of this, axonal transport and dendritic transport serve as a good model system for elucidating one of the fundamental mechanisms of sustaining life in organisms.

Cells transport different cargoes to different destinations at different velocities. This is similar to the transport of merchandises in our society. Fruit, vegetables or grains grown in farms are transported to metropolitan areas using different means of transportation, such as airplanes, trains, or trucks at different speeds. Likewise, computers and electric appliances are assembled in industrial areas and transported by various means to different regions of a country. The survival of a civilized community depends on proper transportation on a daily basis. Cells also have very intricate mechanisms of achieving efficient intracellular transport using many different molecular motors. Maintaining proper intracellular transport is essential for any cellular functions to be achieved properly.

In the axon, proteins synthesized in the cell body are transported either in the form of membrane organelles or protein complexes. When directly observed by high-resolution optical microscopy, many membrane organelles of different shapes are transported at different speeds in axons. In addition, it can be observed that membrane organelles are transported bidirectionally, from the cell body to the periphery and from the periphery to the cell body. Transport in the direction of the cell body to the periphery is called anterograde transport, and transport in the opposite direction is called retrograde transport.

A quick-freeze, deep-etch electron microscopy of axons revealed that membranous organelles are transported along microtubules, which are 25-nm-diameter tubules and are actually a polymer composed of heterodimers of α- and β-tubulins [2, 3]. Short crossbridges are noted between membranous organelles and microtubules and it is very likely that these crossbridges correspond to molecular motors [2–5] (Fig. 5.1). On close examination, one notes that there are different shapes of crossbridges [2,4,5]. Some crossbridges have a stalk and two globular heads (Fig. 5.1a); the stalk is on the side of membranous organelles, and the globular heads are on the side of microtubules. Some crossbridges are simply globular heads and lack stalks (Fig. 5.1b). Crossbridges composed of small globular heads and short stalks are also seen with mitochondria (Fig. 5.1c). These observations suggest that there might be several forms of molecular motors. Many different types of membrane organelle are also transported in axons, such as the precursors of axonal membranes, synaptic vesicles, or presynaptic plasma membranes, and mitochondria. Therefore, we hypothesized that many motors must be required for the highly complex transport process in axons to proceed properly. We searched the molecular motor gene family using the mouse brain cDNA library, and identified the first ten kinesin superfamily proteins "(KIFs)" [6].

KIFs have a motor domain, which has a conserved ATP-binding sequence and a conserved microtubule-binding sequence [1, 3, 43] (Fig. 5.2). Amino

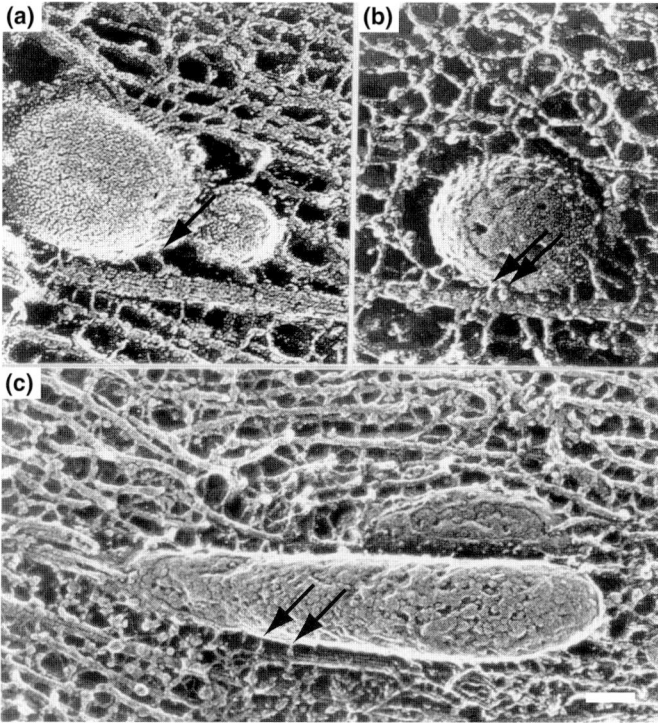

Fig. 5.1. Quick-frozen, deep-etched axons in which various types of crossbridge (*arrows*) are identified between membranous organelles and microtubules. (**a**) A crossbridge (*arrow*) composed of two heads and a long stalk (∼25 nm). (**b**) Only globular heads (*arrows*) are recognizable between a membranous organelle and a microtubule. (**c**) Crossbridges (*arrows*) between mitochondria and microtubule are short and have small heads. Bar, 50 nm. Reproduced with permission from [5]

acid sequence homologies in the motor domains are 30–60%, however other regions are quite variable. KIFs can be grouped into three types depending on the position of the motor domain. N-kinesins have a motor domain in the N-terminal region, M-kinesins in the middle, and C-kinesins in the C-terminal region (Fig. 5.2). KIFs are also of various molecular shapes; some are monomers, and some form heterodimers or homodimers with or without light chains or associated proteins (Fig. 5.3).

Recently, we have identified all KIF genes in the mammalian and human genomes [32]. There are 45 genes in total. Two to three mRNAs are sometimes formed by alternative splicing; therefore, the number of KIF proteins is probably twice that of genes or perhaps even larger. We have studied many of these genes and clarified their structures and functions by molecular cell biology and molecular genetics. We have also clarified the mechanisms by which motor proteins move along microtubules by single-molecule biophysics and cryoelectron microscopy, and X-ray crystallography.

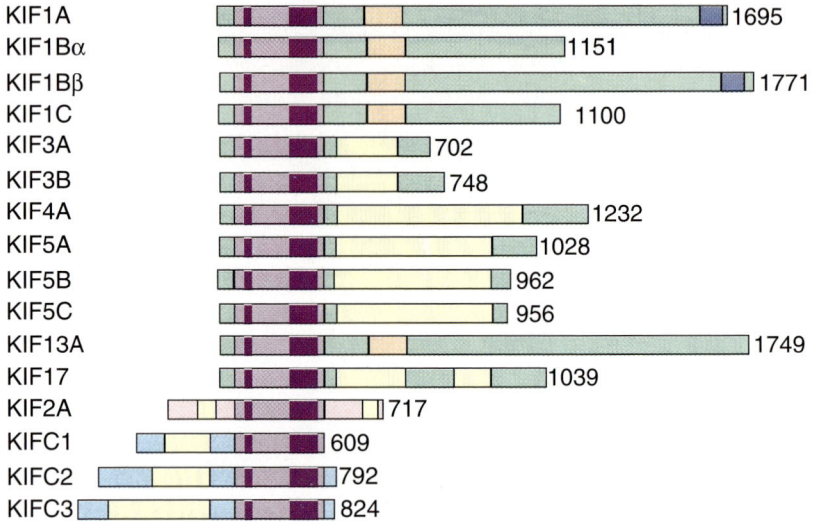

Fig. 5.2. The domain structures of principal KIFs. The motor domains are shown in purple. The ATP-binding consensus sequence is indicated by a *thin red line* and the microtubule-binding consensus sequence by a *thick red line*. The dimerization domains, forkhead-associated domains and pleckstrin homology domains are shown in *yellow, orange,* and *dark blue* respectively. The number of amino acids in each molecule is shown on the *right*. N-kinesins are shown in *green*, M-kinesin in *pink* and C-kinesin in *pale blue*. Relationship of these KIFs and the new standardized kinesin family names [64] is following; KIF5 belongs to Kinesin-1 family, KIF3 and KIF17 to Kinesin-2 family, KIF1 and KIF13 to Kinesin-3 family, KIF4 to Kinesin-4 family, KIF2 to Kinesin-13 family, KIFC1, KIFC2, KIFC3 to Kinesin-14 family

Before starting to discuss about molecular motors, one important point has to be explained, that is, the polarity of microtubules. Microtubules serve as rails, and a microtubule has a plus end and a minus end. The plus end is where the microtubule polymerizes faster; therefore, it is the fast growing end. The minus end is opposite to the plus end. In axons, microtubules are unipolar, so that the plus ends point to the periphery. Because of this, anterograde motors are plus-end-directed and retrograde motors minus-end-directed. In epithelial cells, the minus ends of microtubules are directed towards the apical surface. In fibroblasts, microtubules radiate in various directions from the microtubule-organizing center near the nucleus and their plus ends are directed towards the periphery. Thus, cells have very highly organized microtubule rails.

5.2 Monomeric Motors and Their Functions

First, we would like to discuss the monomeric motors of the KIF1 family, KIF1A, KIF1Bα, and KIF1Bβ. The discovery of monomeric motors is also

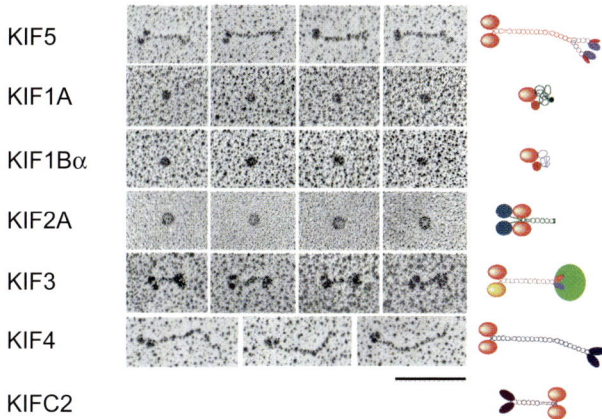

Fig. 5.3. Principal members of KIFs observed by low-angle rotary shadowing. Diagrams, constructed on the basis of electron microscopy or predicted from the analysis of their primary structures, are shown on the *right* (the *large red ovals* in each diagram indicate motor domains). Reproduced with permission from [3]. Scale bar, 100 nm

very important in terms of the mechanism underlying the motility of molecular motors. However, we will discuss this mechanism later, and, in this section, the biology of KIF1 motors is discussed.

5.2.1 KIF1A – Monomeric Motor for Synaptic Vesicle Precursor Transport

KIF1A has its motor domain in the N-terminal region, therefore, a N-kinesin, and has 1695 amino acid residues [6,13] (Fig. 5.2). A unique property of KIF1A is that experimental data strongly suggest that it exists as a monomer [13]. Sucrose velocity gradient, native polyacrylamide gel electrophoresis (PAGE) and differential laser light scattering analyses suggest that the native molecular weight of KIF1A in solution is 200 kD, which closely agrees with the molecular weight of 190 kD estimated from the amino acid sequence. Therefore, findings strongly suggest that KIF1A exists as a monomer. Low-angle rotary-shadowing electron microscopy also revealed that KIF1A is a single globular molecule (Fig. 5.3).

A motility assay using microtubules demonstrated that KIF1A moves towards the plus end of microtubules at a velocity of 1.2 μm/s [13]. Because microtubules in axons are oriented with the plus ends directed towards the periphery, the plus-end direction of KIF1A suggests it to be an anterograde motor. Whether this is the case in vivo can be confirmed by nerve ligation. When peripheral nerve axons are ligated in the middle, anterogradely transported membrane organelles accumulate in the region proximal to the ligation site, whereas retrogradely transported membrane organelles accumulate in the region distal to the ligation site. When a sciatic nerve was ligated and stained with an anti-KIF1A antibody, a strong immunofluorescence intensity was observed in the proximal region, but not in the distal region. This observation strongly suggests that KIF1A is associated with anterogradely transported membrane vesicles.

The next question was what is the cargo of KIF1A? To answer this question, first, the ligated nerve was stained with antibodies specific to various proteins that are transported in axons [13]. It was shown that KIF1A colocalized with synaptophysin, a synaptic vesicle protein, but not with syntaxin 1A, a synaptic plasma membrane protein, suggesting that KIF1A transports membrane vesicles containing synaptic vesicle proteins, but not synaptic plasma membrane proteins (Fig. 5.4a). Results also suggest that synaptic vesicle proteins and synaptic plasma membrane proteins are transported in different types of membrane vesicle. Another approach we undertook was immuno-

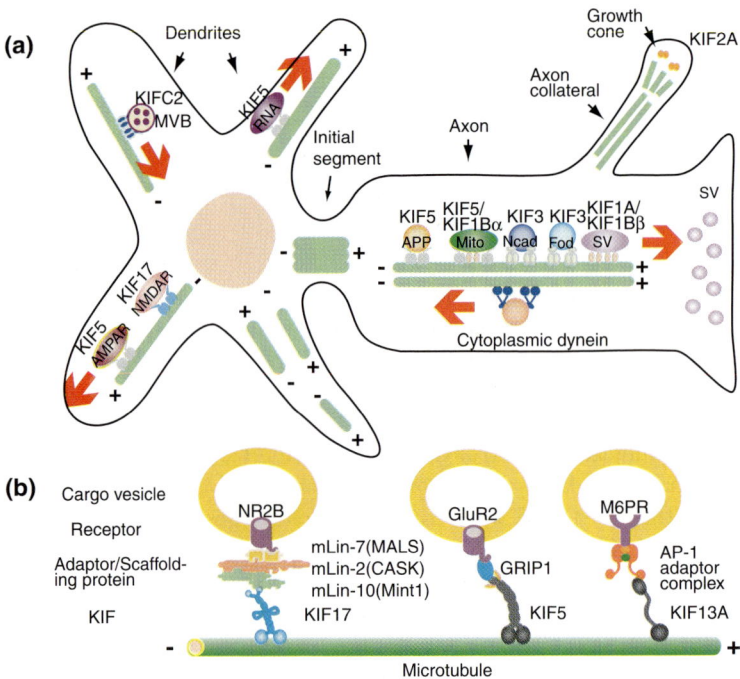

Fig. 5.4. (a) Motors and cargoes in axonal and dendritic transports. In the axon, microtubules (shown in *green*) are unipolar and the plus end (+) points to the synaptic terminal. In dendrites, microtubules are of mixed polarity in the proximal regions, although it becomes unipolar at the distal regions. KIFs transport various cargoes along microtubules in axon and dendrites. See texts for detail. The nucleus is shown in *pink*. (b) Molecular details of binding of cargo vesicles to KIFs. KIFs tend to recognize and bind receptor molecules in cargo vesicles via adaptor/scaffolding proteins. (Modified with permission from [7])

precipitation. The membrane vesicle fraction was first isolated from axons, and then the KIF1A-containing vesicle fraction was immunoisolated using beads conjugated with the anti-KIF1A antibody. The immunoblotting of this KIF1A-containing vesicle fraction revealed that conventional kinesin (KIF5, see Sect. 3.2 for detail) and KIF3 are not present in this vesicle fraction; suggesting that KIF1A transports membrane vesicles different from those transported by conventional kinesin or KIF3. The fraction contained synaptic vesicle proteins such as synatophysin, synaptotagmin, or Rab3A, but, not synaptic plasma membrane proteins such as syntaxin 1A or SNAP25 in agreement with the results of nerve ligation.

The results mentioned up to this point demonstrate that KIF1A is a monomeric, plus-end-directed fast motor and participates in the anterograde axonal transport of synaptic vesicle precursors containing synaptophysin, synaptotagmin, and Rab3A. To further clarify the functional and biological significance of this motor protein in vivo, we next generated a KIF1A knockout mouse [20].

Mice homozygous for the mutant KIF1A gene were born alive, but all died within 24 hr after birth. The mice showed motor and sensory abnormalities. Therefore, their nervous system was studied. First, it was found that the number of synaptic terminals per unit area decreased to 50–60% of that of wild-type mice [20]. Second, the density of synaptic vesicles at synaptic terminals also decreased to 50–60% of that of wild-type mice [20]. However, please note that the density did not decrease to zero. Third, when various regions of the central nervous system were examined, focal neuronal deaths were observed in the cerebrum, rhinencephalon, and amygloid areas. Axonal degeneration was also widespread, but it was suggested to be a change secondary to neuronal death [20]. These observations indicate that the lack of KIF1A results in neuronal death.

To examine the process of neuronal death, we cultured hippocampal neurons [20]. The survival curve showed that the hippocampal neurons of mutant mice started to die after 6 days in culture, and most died after 13 days. In the wild-type neurons, the synthesis of KIF1A started to increase after 6 days.

In summary, KIF1A is an anterograde fast motor that transports synaptic vesicle precursors and is vital in neuronal function and survival. However, because the synaptic vesicle density of the mutant mice did not become zero, it was suggested that there must be another motor that transports synaptic vesicle precursors. This last point will be dealt with later.

5.2.2 KIF1Bα – Monomeric Motor for Mitochondrial Transport

We will now discuss a new motor that we discovered following KIF1A, that is, KIF1B [11]. KIF1B is also a N-terminal motor, and it is now called KIF1Bα [31] (Fig. 5.2). Experimental data support that KIF1Bα also exists as monomer, and by low-angle rotary-shadowing electron microscopy, KIF1Bα is observed as a single-head globular molecule, but sometimes it has a short tail [11] (Fig. 5.3).

To examine what types of cargo are transported by KIF1Bα, we stained cultured Neuro2A cells with an anti-KIF1Bα antibody [11]. The anti-KIF1Bα antibody immunoreacted with large dotlike structures, indicating that KIF1Bα transports large membranous organelles. To characterize these organelles, various markers and the anti-KIF1Bα antibody were double-stained, and it was found that MitoTracker, a marker of mitochondria, colocalized with KIF1Bα. This demonstrated that KIF1Bα transports mitochondria [11] (Fig. 5.4a).

To confirm this, the mitochondrial fraction was isolated from the mouse brain, and whether KIF1Bα was concentrated in the fraction was examined [11]. Cytochrome oxidase, a mitochondrial marker, was highly enriched in the high-centrifugation membrane vesicle (P3) fraction, and KIF1Bα was also abundant in this fraction. The results further support the notion that KIF1Bα transports mitochondria.

We further examined whether KIF1Bα could transport mitochondria along microtubules in vitro [11]. For this, axonemes, which are microtubule bundles, isolated from the *Chlamydomonous* flagella were firmly fixed to a coverglass to prevent gliding, and purified mitochondria and KIF1Bα were added in the presence of ATP. The mitochondria moved along microtubules at a velocity of 0.5 μm/s. Therefore, KIF1Bα is a motor that transports mitochondria. We will not describe in detail, but we generated knockout mice of KIF5B and KIF5C, two closely related conventional kinesins [21, 44]. The analysis of the KIF5B or KIF5C knockout mice revealed that KIF5s are also motors for mitochondrial transport. Thus, mitochondria are transported by two plus-end-directed microtubule-dependent motors, KIF1Bα and KIF5s (Fig. 5.4a).

5.2.3 KIF1Bβ – Second Monomeric Motor for Transport of Synaptic Vesicle Precursors

Next, we will describe aother motor that transports synaptic vesicle precursors [31]. In the analysis of KIF1A knockout mice, we realized that there must be another motor for transport of synaptic vesicles precursors. Because the transport of synaptic vesicle precursors is particularly important for neuronal function, we searched for this motor. We indeed found one, KIF1Bβ [31]. KIF1Bβ is a new isoform formed by alternative splicing from the gene for KIF1Bα, which we have described above. The motor domains of KIF1Bα and KIF1Bβ were identical, but C-terminal tail regions were completely different (Fig. 5.2). Because motors bind to microtubules at motor domains and bind to cargoes at tail domains, having different tails indicates that the motors transport different cargoes.

We then analyzed the KIF1Bβ cargo. Biochemical analyses such as subcellular fractionation or flotation assay, GST-pulldown using the tail region of KIF1Bβ, vesicle immunoprecipitation using an anti-KIF1Bβ antibody, and immunoelectron microscopy using antibodies specific to various axonally transported membrane proteins and KIF1Bβ all demonstrated that KIF1Bβ transports synaptic vesicle precursors containing synaptotagmin, synaptophysin, and SV2 [31] (Fig. 5.4a).

To study the in vivo function of KIF1Bβ, we then generated KIF1Bβ knockout mice [31]. KIF1Bβ knockout mice were born alive, but die within 30 min of birth due to apnea. A histological examination of the lungs showed no visible signs of abnormality, but pulmonary alveoli did not expand properly. We then examined the muscles involved in respiration, but these as well did not reveal any significant defects. Therefore, it was most likely that the apnea was of a neurological origin, and we examined the central nervous system. In the KIF1Bβ knockout mice, the number of neuronal cell bodies was less than 25% of that in the control. Neuronal loss in the knockout mouse respiratory center was severe, and this was likely the cause of neonatal apnea. The density of synaptic vesicles also decreased to 50–60%.

When hippocampal neurons are cultured, mutant cells start to die after 3 days in culture, and a significant proportion of cells die after 1 week in culture [31]. We would like to point out some details here; the KIF1Bβ knockout mice were generated such that the KIF1B motor domain was disrupted; therefore, both KIF1Bα and KIF1Bβ were not expressed in knockout mice. To assess which isoform of KIF1B is responsible for the neuronal loss phenotype, a rescue experiment was performed. When exogenous KIF1Bβ or KIF1Bα was introduced to the mutant culture, only KIF1Bβ rescued the neuronal death. Therefore, KIF1Bβ was mainly responsible for the mutant phenotype. This seemed to be reasonable because no changes in mitochondrial distribution were observed in this mutant mouse neuron, and mitochondria have another motor, KIF5s, for their transport as we discussed above. Therefore, the phenotype appeared to be caused by lack of KIF1Bβ, not that of KIF1Bα.

5.2.4 Charcot-Marie-Tooth Disease Type 2A Caused by Mutation in KIF1Bβ

We made one crucial observation. The mutant mice died shortly after birth, but heterozygotes survived and appeared normal. However, these 50% normal mice showed progressive muscle weakness and motor discoordination after one year of age [31]. We performed a series of behavioral tests; 1-year-old heterozygotes showed a staggering gate and shorter retention times in fixed bar and rotarod tests than wild types. These behavioral tests indicated muscle weakness and motor discoordination, and these phenotypes were similar to those of human neuropathy. Therefore, we first examined whether synaptic vesicle precursors decreases in the heterozygotes. In the peripheral axons of heterozygous mice, the levels of KIF1Bβ and synaptic vesicle proteins, such as synaptotagmin or SV2, were one-half those in wild-type mice. However, similar changes were not observed for nonsynaptic vesicle proteins such as SNAP-25 or syntaxin 1A. These results indicate that in the heterozygotes, the extent of transport of synaptic vesicle precursors specifically decreases to one-half that of the wild type.

We found that murine KIF1Bβ is mapped on chromosome 4E. The Charcot-Marie-Tooth disease type 2A (CMT2A) had been mapped to the overlapping human chromosome region 1p35–36, although the causal gene had not been identified. CMT is the most common inherited neuropathy in humans with a prevalence of 1/2500 and has several subtypes [45]. To examine whether KIF1Bβ mutation is the cause of the disease, we analyzed a pedigree of CMT2A [31]. In all affected family members in this pedigree, but not in unaffected siblings or healthy controls, a heterozygous Q-to-L (glutamine-to-leucine) mutation was found in the consensus ATP-binding site of the KIF1Bβ motor domain. We also found that this mutation caused a significant decrease in microtubule-dependent ATPase activity. In addition, when the wild-type KIF1Bβ and mutant KIF1Bβ were exogenously expressed in fibroblasts, wild-type KIF1Bβ moved to the plus ends of microtubules and accumulated at the cell periphery, whereas mutant KIF1Bβ remained and aggregated in the perinuclear region. In summary, we have found a new motor that transports synaptic vesicle precursors. KIF1Bβ plays essential roles in neuronal function and survival. We also found that a KIF1Bβ gene mutation is a cause of hereditary human neuropathy [31, 45].

5.3 Dendritic Transport and Mechanisms of Cargo Recognition

Here, we will discuss dendritic transport, in terms of motors and cargoes, and the physiological significance of the transport. A series of experiments have also demonstrated a principle of how cargoes are recognized by motors and how their selective transport to dendrites are achieved [1]. Therefore, we will also discuss these points.

5.3.1 Transport of NMDA Receptor by KIF17

We found a new motor, KIF17, a N-terminal motor domain-type motor with 1038 amino acids; it is a plus-end-directed motor with a velocity of 1.2 μm/s [27]. A unique feature of KIF17 is its localization; that is, it mainly localizes in the cell body and dendrites, as was clearly demonstrated in the cerebral cortex by immunofluorescence microscopy. The same localization was confirmed in cultured hippocampal neurons. In these experiments, axons can be distinguished from dendrites by staining with an anti-neurofilament antibody.

The next important question is what is the cargo of KIF17. Immunoisolation using beads revealed that the cargoes are membranous organelles [27]. To analyze the types of membranous organelle, proteins that bind to the C-terminal tail of KIF17 were searched by yeast two-hybrid assay, and mLin-10 (mouse homologue of *Caenorhabditis elegans* LIN-10), also known as Mint1, was identified as the binding partner. mLin-10 has two PDZ domains, and our data indicate that the KIF17 tail domain binds to the first PDZ domain. The binding was also confirmed by immunoprecipitation using an anti-mLin-10 antibody and a BIAcore system. This finding is very exciting because other laboratories have shown that mLin-10 binds to N-methyl-D-asparatate (NMDA)-type glutamate receptors via mLin-2 (CASK) and mLin-7 (MALS/ Velis) [46], suggesting that KIF17 transports NMDA-type glutamate receptors. We therefore examined whether it was indeed the case.

The floating assay showed that the peak fraction of KIF17 coincided with the peak fractions of mLin-10, mLin-2, and mLin-7, and the NR2B subunit of NMDA receptor [27]. The peak fraction was then used in immunoprecipitation as the membrane-solubilized lysate or the vesicle fraction. The results directly demonstrated that KIF17 binds to the NR2B subunit via the mLin-10, mLin-2, and mLin-7 complex. When the membrane fraction associated with KIF17 was isolated and dropped onto *Chlamydomonas* flagellum microtubules with KIF17 and ATP, the cargo vesicles moved to the plus ends [27]. By immunofluorescence microscopy, the cargoes moving along microtubules were found to contain NR2B. These results indicate that KIF17 transports the NMDA receptor subunit NR2B via the interaction with the tripartite scaffolding protein complex containing mLin-10, mLin-2 and mLin-7 (Fig. 5.4b, left). However, up to this point we have presented in vitro data; therefore, we needed to determine whether KIF17 indeed transports NMDA receptors in neurons in vivo.

We first expressed the yellow fluorescent protein (YFP)-KIF17 in cultured hippocampal neurons [38]. YFP-KIF17 distributed mainly in the cell body and dendrites, not in axons. We then examined whether the NR2B subunit colocalized with YFP-KIF17-binding membrane vesicles by immunofluorescent antibody staining. Many of the NR2B staining represented clusters of NMDA receptors that had been already incorporated into postsynaptic sites that were costained by the anti-PSD-95 antibody. However, NR2B staining that colocalized with YFP-KIF17 was not costained by the anti-PSD-95 antibody, suggesting that the NR2B staining that colocalized with YFP-KIF17 were in the process of active transport in dendrites. A statistical analysis showed that 45% of YFP-KIF17 colocalized with NR2B subunits. The results strongly suggest that KIF17 transports NMDA receptors in neurons in vivo. To further verify this, we also performed functional blocking of KIF17 in hippocampal neurons [38]. We used two strategies; one was expressing antisense oligonucleotides and another was expressing dominant-negative mutants that express only the C-terminal tails. Both strategies showed that blocking KIF17 resulted in a significant decrease in the density of NMDA receptor clusters. This result strongly supported the idea that KIF17 transports NMDA receptors (Fig. 5.4a).

When the NMDA receptor is functionally blocked by an AP-V antagonist, the expression of NR2B subunits is upregulated. When cultured hippocampal neurons are treated with AP-V, the expression of KIF17 is also upregulated [38]. Therefore, it is suggested that KIF17 and NR2B are coregulated at the transcriptional level, which is also supported by the analysis of transgenic mice, which will be discussed later.

5.3.2 Transport of AMPA Receptor by KIF5

As shown above, we have demonstrated that KIF17 transports one type of glutamate receptor, the NMDA-type receptors [27]. We have also identified the motor that transports α-amino-3-hydroxy-5-methylisoxazole-4-propionate (AMPA)-type receptors, another type of glutamate receptor [34]. We were searching the cargoes of KIF5, a conventional kinesin. Conventional kinesin is composed of heavy chains, that is, KIF5 (5A, 5B, 5C), and light chains. Light chains bind membrane vesicles containing the amyloid precursor protein (APP) and other molecules, therefore, they are among the cargoes of KIF5 (Fig. 5.4a). However, we were searching cargoes that bind to the tail region of heavy chains using the yeast two-hybrid method. We found glutamate receptor-interacting protein 1 (GRIP1), which binds to the GluR2 subunit of AMPA receptor [34]. Immunoprecipitation showed that GRIP1 indeed binds to both KIF5, the heavy chain of conventional kinesin, and the GluR2 subunit in neurons (Fig. 5.4b, center). In addition, when a KIF5 dominant-negative construct, which contains only the GRIP1-binding site but lacks the motor domain, was expressed in cultured neurons, the density of GluR2 clusters in dendrites significantly decreased, suggesting that GluR2 is normally transported to dendrites by KIF5 via the interaction with GRIP1 [34] (Fig. 5.4a).

Furthermore, when the minimal kinesin-binding domain of GRIP1 is expressed in hippocampal neurons, conventional kinesins are recruited to dendrites [34]. In contrast, when the JNK/stress-activated protein kinase-associated protein 1 (JSAP1), which functions as a binding partner of kinesin via light chains, is expressed, conventional kinesins predominantly localize to axons [34]. These results suggest that binding of cargoes to conventional kinesins either via heavy chains or light chains determines the polarity of transport, that is, the transport of cargoes to axons or dendrites.

In summary, we have shown that the C-terminal tail of KIF5 binds to GRIP1 and transports vesicles containing the GluR2 subunit of AMPA receptor in dendrites. We have also shown that the binding of GRIP1 to heavy chains tends to steer KIF5 to dendrites rather than to axons, although the mechanism is as yet to be elucidated.

5.3.3 Use of Adaptor/Scaffolding Protein Complexes

We will not go into detail, but we have also identified a new motor, KIF13A, and showed that KIF13A transports vesicles containing the mannose-6-phosphate receptor from the Golgi apparatus to the plasma membrane. In this transport, the tail region of KIF13A binds to the AP-1 adaptor protein complex, which then recognizes the mannose-6-phosphate receptor enabling its transport from the Golgi apparatus to the plasma membrane (Fig. 5.4b, right) [30].

This series of experiments has demonstrated that KIF17 transports the NMDA-type glutamate receptors, which is vital for important neuronal functions such as memory and learning, and that KIF5 transports AMPA-type receptors, which are also very important for neuronal functions. In addition, this series of experiments has also established the mechanism by which motor molecules recognize and bind to their cargoes (Fig. 5.4b). In the case of KIF17, it binds to mLin-10, and via the mLin-10, mLin-2, and mLin-7 scaffolding protein complex to the NMDA receptor NR2B subunit (Fig. 5.4b, left). In the case of KIF13A, it binds to β-1 adaptin and via the AP-1 adaptor complex to the mannose-6-phosphate receptor (Fig. 5.4b, right). KIF5 binds to GRIP1 at the heavy chains and via GRIP1 to the GluR2 subunit of the AMPA receptor (Fig. 5.4b, center). Therefore, the tail region of motor molecules recognizes and binds to the adaptor or scaffolding proteins in the transport of cargo membrane organelles containing functional membrane proteins. The use of adaptor and scaffolding proteins appears to be one of the basic mechanisms for the recognition and transport of cargoes [1,7].

5.3.4 Enhanced Spatial and Working Memory in Transgenic Mice Overexpressing KIF17

We next examined the in vivo role of KIF17. We generated a gain-of-function model, that is, a transgenic mouse overexpressing KIF17, using the calmodulin-dependent kinase II (CaMKII) promoter [36]. The mouse is apparently normal, but overexpresses KIF17.

We then performed behavioral tests [36]. In the delayed matching-to-place version of the Morris water maze task, which measures working/episodelike memory, the swimming speeds were not significantly different, but the working memory of transgenic mice overexpressing KIF17 was significantly better than the wild type. In the standard hidden-platform version of the Morris water maze task, the transgenic mice demonstrated a better spatial learning and memory than the wild type. It means that we generated a smart mouse.

We then investigated what was happening in the smart mouse's brain. In the hippocampus and cerebral cortex, the expression level of the KIF17 protein was high [36]. In addition, the mRNA expression levels of KIF17 and the NR2B subunit of NMDA receptors were high. NR2B transcription is regulated by the phosphorylation of cyclic AMP responsive element-binding protein (CREB). Therefore, we compared the expression levels of phosphorylated CREB in the wild-type and transgenic mice. Although the expression levels of CREB were the same, that of phosphorylated CREB was significantly up-regulated in the transgenic mice compared with that in the wild type [36]. The promoter region of NR2B has a CREB consensus sequence, which is regulated by phosphorylated CREB. Interestingly, we found that KIF17 also has a CREB consensus sequence in the promoter region [36].

From these results we propose the following as a potential mechanism by which the overexpression of KIF17 leads to the improvement of learning and memory. First, the overexpression of KIF17 leads to an increase in the dendritic transport of vesicles carrying the NR2B subunit of the NMDA receptor. When the receptor is incorporated in the postsynaptic site, synaptic transmission efficiency increases. As a result, calcium influx at the postsynaptic site increases leading to the activation of signal transduction cascades such as cyclic AMP-dependent protein kinase (protein kinase A, PKA), mitogen-activated protein kinase (MAP kinase), and calmodulin-dependent protein kinase, and culminating in the enhancement of the phosphorylation of CREB. Then, mRNA transcription of KIF17 and NR2B is enhanced, which in turn enhances protein translation. This will lead to a positive feedback of enhanced dendritic transport of the NR2B subunit, and ultimately learning and memory improve, that is, a smart mouse is generated. In summary, the transgenic mouse experiments showed that motor proteins play significant roles in forming a basis for higher-order brain function.

5.3.5 Transport of RNA-Containing Granules by KIF5

Recently, we have also clarified a mechanism by which mRNA is transported by molecular motors [28]. As we discussed previously, KIF5, the conventional kinesin, is composed of heavy chains and light chains. We searched for a cargo that binds via heavy chains, and identified one that binds to the amino acid 865–923 of the tail domain, which we named the minimum binding site (MBS) [28]. The cargo is a large granule containing various RNA-associated proteins such as hnRNP-U, Purα, Purβ, PSF, DDX1, DDX3, SYNCRIP, TLS, and NonO. KIF5 has three closely related genes, *kif5a*, *kif5b*, and *kif5c*, and the granule binds to all the three KIF5s. Of the granular components, we have examined 17 proteins in detail, and immunoprecipitation experiments showed that all of these proteins are consistently contained in the granule. The granule also contains several specific mRNAs such as that for CAMKIIα or activity-regulated cytoskeleton-associated protein (Arc), but not other mRNAs such as that for tubulin. Therefore, there is a specificity in which mRNAs are transported in the granule.

We also generated antibodies to these 17 proteins, and demonstrated that all these proteins and KIF5 colocalize with the Purα-containing granule [28]. We also showed that mRNAs for CaMKII and Arc colocalize with the granule. Furthermore, we visualized the movement of Purα-containing granules in dendrites [28]. The movement velocity was $\sim$0.1 μm/s.

When RNA interference (RNAi) is used to suppress the expression of the component proteins of RNA-containing granules, such as hnRNP-U, Purα, PSF, or staufen, mRNA transport is suppressed [28]. These observations confirm that these proteins play significant roles in mRNA transport. However, the suppression of other proteins such as DDX3 or SYNCRIP does not affect mRNA transport, suggesting that some proteins are not directly involved in transport processes [28].

We have performed proteomic analysis of the transporting granule and identified at least 42 proteins [28]. The proteins include those involved in RNA transport, protein synthesis, RNA helicases, hnRNPs, and other RNA-associated proteins. In this way, the mechanism by which mRNAs are transported is becoming clear.

5.4 KIF3, Left–Right Determination and Development

Here, we will discuss about a heterotrimeric motor, KIF3, which has a unique physiological role not observed in other KIFs. KIF3 forms a heterodimer composed of KIF3A and KIF3B, which are derived from separate genes [15]. The heterodimer associates with kinesin superfamily-associated protein 3 (KAP3) on its C-terminal tail, and therefore, KIF3A, KIF3B, and KAP3 form the KIF3 motor (Fig. 5.3) [17]. KIF3A and KIF3B are abundantly expressed in neurons, but they are ubiquitously expressed in other tissues as well [15,17].

In neurons, the KIF3 motor transports membrane vesicles associating with α-fodrin and these vesicles are important for neurite elongation [26]. The KIF3 motor is bound to α-fodrin via KAP3. To examine role of the KIF3 motor in vivo, we generated KIF3A and KIF3B knockout mice [22,24]. These knockout mice showed very similar phenotypes; therefore, we will mainly discuss the KIF3B knockout mice [22].

5.4.1 Mice Lacking KIF3 Shows Randomization of Left–Right Asymmetry

The lack of KIF3A and KIF3B is embryonically lethal [22, 24]. Therefore, we analyzed the embryos. The embryos showed various phenotypes including an open brain, pericardial swelling and abnormality in turning. Among these various phenotypes, we focused our attention on the randomization of left–right determination [22, 24].

In humans, the Kartagener syndrome has been described, and 50% of patients with this disease have a normal internal organization, that is, the heart on the left, the liver on the right, and the spleen and pancreas on the left. However, the remaining 50% of patients have a mirror image distribution, that is, *situs inversus* viscera. The Kartagener syndrome patients also show male infertility and respiratory failure, but the reason for the randomization of left–right determination was not known. The first visible sign in the left–right determination in embryogenesis is cardiac looping. Initially, the heart is a straight tube, but it will fold to form an atrium and a ventricle. Normally it will form the D-loop, but in *situs inversus*, the L-loop. One hundred percent of wild-type and heterozygous mice showed the D-loop, whereas 50% of KIF3B mutant mice showed the D-loop and the remaining 50% the L-loop.

5.4.2 KIF3 is Essential for Ciliogenesis of the Nodal Cells

The left–right determination is the focus of much research in developmental biology. A triangle ventral dent in an early embryo called the "node" plays an important role. The node regulates the expression of a series of genes that are expressed specifically on the left side, for example, *nodal, lefty1, lefty2, and Pitx2*. However, the upstream events were not known. The whole-mount *in situ* hybridization of wild-type and knockout mice revealed that the expression of the most upstream gene, *lefty2*, is randomized in the knockout mice. The results indicate that events upstream of the expression of these genes are regulated by KIF3.

We then observed the node. Scanning electron microscopy revealed that monocilia are present on nodal cells in the wild-type mice, but are missing in the knockout mice. The wild-type nodal cilia have microtubule doublets in a "9 + 0" arrangement, instead of the usual "9 + 2" arrangement. In the 9 + 0 arrangement, the central pair is missing, and usually this type of cilia are considered to be immotile. In the knockout mice, although the cilia are missing, basal bodies are present. The first obvious question is how the absence of KIF3 causes the absence of cilia. We demonstrated that KIF3 is present in the nodal cilia of wild-type mice by immunofluorescence and immunoelectron microscopies. When this result is combined with data from other laboratories, it indicates that the KIF3 heterodimer transports protein complexes carrying the ciliary components from the base to the tips of cilia along microtubules [22, 24, 47]. In the knockout mice, cilia are not formed because ciliary component proteins are not transported in the cilia because of the absence of the KIF3 motor.

5.4.3 Nodal Cilia Rotate to Produce a Constant Leftward Flow

Why does the left–right determination becomes randomized when nodal cilia are not present? We next decided to approach this question [22,24]. In ordinary cilia or flagella, nine pairs of microtubule doublets composed of the "A" and "B" subunits are circularly arranged and in the center a pair of microtubules is present [48]. In most of these 9 + 2 type of cilia or flagella, beating movements are observed. In contrast, nodal cilia do not have the central pair and were, therefore, thought to be immotile. We first examined whether nodal cilia are indeed immotile by video microscopy using live embryos. Surprisingly, the cilia were rotating clockwise at about 600 cycles/min. However, it remain unclarified how the motility could be involved in the left–right determination. Therefore, we attempted to determine the total effect of individual rotations. We added fluorescent-dye-labeled beads to the extraembryonic fluid in the node. To our surprise, the beads moved from the right to left; therefore, in wild-type mice, there is a constant leftward flow of extraembryonic fluid in the node. We named this flow "nodal flow". In the node of heterozygous mice, in which nodal cilia are present and motile, the beads showed a leftward flow.

However, in the node of knockout mice, in which nodal cilia are missing, the beads only show the Brownian movement, suggesting that the nodal flow is absent.

In summary, the KIF3 motor transports protein complexes in nodal cilia from the base to the tip in the wild-type mice. The nodal cilia rotate in a clockwise manner at about 600 cycles/min, and this rotation generates a flow of extraembryonic fluid in the node, the nodal flow. The leftward nodal flow generates a concentration gradient of a putative morphogen X on the left side of the node, which we will discuss later. The knockout mice do not have cilia due to the absence of KIF3 and thus no ciliary movements. There would then be no nodal flow, and the secreted morphogen X would simply diffuse, switching on left-right-determining gene cascades randomly. Thus, it was shown that the KIF3 motor regulates important events in development.

5.4.4 How is Nodal Flow Generated

We are left with two more questions: how do individual rotation movements generate the net effect of a linear leftward flow, and what is the putative morphogen. Regarding the first question, we first examined whether the same nodal flow that we have demonstrated in mice is observed in other vertebrates such as rabbits, whose embryos are closer to those of humans, or medaka fish, which are similar to zebrafish [49]. We found that cilia are present in a region of the rabbit notochord that corresponds to the murine node and in medaka Kupffer's vesicles, which play an important role in the left–right determination. We also found that most of these cilia have the 9 + 0 arrangement, and also rotate clockwise. In addition, these rotation movements generate the leftward nodal flow. Therefore, we found that the presence of cilia, rotation movements of cilia and nodal flow are mechanisms conserved in vertebrates.

We have also established a video system with a high temporal resolution and observed ciliary movements in detail. The cilia rotate with their axis tilted at an angle of 40 degrees posteriorly [49]. Because of this tilt, when a cilium rotates to the left, that is, movement from the right to the left, it moves in a nearly vertical plane and from the hydrodynamic point of view, the movement effectively generates the nodal flow because of shear resistance. However, in returning, that is, movement from the right to the left, it moves near the surface and, therefore, cannot effectively generate the nodal flow. Thus, because of the tilting of the cilia posteriorly at an angle of 40 degrees, the leftward flow is generated effectively, but not the rightward flow.

In conclusion, three elements, the structural assymmetry within cilia, which creates clockwise movements, the presence of the node on the ventral side of an embryo, and the tilting of the axis of rotation toward the posterior at an angle of 40 degrees, combined together produce the leftward flow generated by rotation movements. These observations also explain the pathogenesis of Kartagener's syndrome. The causal genes for Kartagener's syndrome are the components of the dynein motor, which is responsible for the movements of cilia and flagella; and its mutations mainly occur in light chains and intermediate chains. The *iv* mouse is a model of Kartagener's syndrome [50]. Nodal cilia are present, but immotile due to the defect in dynein motor activity. Immotile cilia do not create nodal flow and cause the same phenotype as that of KIF3 knockout mouse, that is, the randomization of the left–right determination. Dynein motor defects also cause the immotility of sperm flagella, and thus male infertility. The immotility of cilia in respiratory epithelial cells underlies the inability of coughing up sputum and thus chronic respiratory complications. Therefore, why one-half of the patients with Kartagener's syndrome have *situs inversus* and how the *situs inversus* is related to other symptoms are now understood.

5.4.5 Leftward Nodal Flow Transports Nodal Vesicular Parcels Carrying Sonic Hedgehog and Retinoic Acid

We hypothesized that the leftward nodal flow creates a concentration gradient of a putative morphogen X, which is likely a secreted protein. However, what exactly is the signal transduction mechanism of this flow? In search of morphogens and the signal transduction mechanism, we have found a new mode of extracellular transport of morphogens in the node [51].

Because the findings of other laboratories suggest that fibroblast growth factor (FGF) signaling is associated with left determination [52], we examined the distribution of FGF receptors in the node. Immunofluorescent staining showed that FGF receptors are localized on the nodal cilia and perinodal cell surface. Thus, it appears that FGF proteins do not generate a left-right concentration gradient by themselves. We then examined the effect of blocking FGF signaling on the left-determination process. In control embryos, the static elevation of Ca^{2+} level starts from the left margin of the node and laterally propagates towards the left lateral plate mesoderm as shown by the use of calcium-sensitive fluorophores [53]. However, when FGF signaling is blocked by a specific inhibitor of the FGF receptor tyrosine kinase, SU5402, or a dominant-negative recombinant peptide of an extracellular domain of mouse FGF receptor, FGFR-DN, the Ca^{2+} level elevation is significantly suppressed. The suppressed Ca^{2+} signals are rescued by the supplementation of downstream morphogen candidates, namely, Sonic hedgehog (SHH) or retinoic acid (RA), suggesting that FGF, SHH and RA are involved in an event specifying laterality.

We then labeled membrane lipids with the lipophilic fluorescent dye DiI and observed the nodes of living embryos by confocal microscopy. Surprisingly, time-lapse imaging revealed that membranous parcels 0.3–5 μm in diameter flow leftward once every 5–15 s. These parcels are released one by one from the cell surface and dynamically protruding microvilli, flow down the stream of the nodal flow, and are finally fragmented by ciliated surfaces into several small pieces close to the left wall; this whole process takes about 30 s. Apparently, the transport results in a massive transfer of materials, because the lipophilic fluorescent-dye staining tended to be brighter on the left side of the node than on the right side. The transport is apparently triggered by FGF, because it is suppressed by the suppression of FGF by SU5402 or FGFR-DN. We named the parcels "nodal vesicular parcels" or NVPs. Typically, a parcel consists of multiple lipophilic granules sheathed by an outer membrane and is often associated with microvilli.

Does NVPs carry morphogens? To answer this we performed light- and electron-microscopic immunohistochemistry using an anti-SHH or an anti-RA antibody and found that SHH and RA are associated with NVPs.

Taken together, our data provide direct evidence that nodal flow transports NVP-associated morphogens toward the left, which is probably a critical phenomenon of symmetry breaking in mammalian embryos. The release of NVPs is triggered by FGF and NVPs carry SHH and RA – the morphogens positively involved in the left-determination.

5.4.6 Transport of N-Cadherin in Developing Neurons

Although the above studies clearly demonstrated the important roles of the KIF3 motor in left–right determination in vivo, it is likely that the KIF3 motor has other in vivo functions as well. Because KIF3 knockout mice have severe phenotypes and die at the embryonic stage, it is difficult to study other roles of KIFs using the knockout mice. Because of this, we decided to generate a conditional knockout mouse, and applied the Cre/*loxP* conditional knockout strategy to bypass the midgestation lethality using a *Cre* transgenic mouse drived by the promoter region of neurofilament-H (*Nefh*) gene [54]. We conditionally knocked out KAP3 gene. KAP3 binds to the tail end of KIF3 and could control the binding of the KIF3 motor and cargoes. Thus, the knockout of KAP3 could lead to the inhibition of KIF3 motor function.

In the conditional knockout mice, a tumorlike abnormal hypertrophy of the cerebral cortex was observed, and it was suggested that cells that lack KAP3 divide faster than control cells. We then examined whether the cadherin/catenin system, which controls cell-cell adhesion and proliferation, is affected by KAP3 deficiency. The immunoblotting of brain lysates showed no apparent changes in the expression levels of N-cadherin and β-catenin. However, when we probed brain sections against N-cadherin and β-catenin, we found that the cell peripheral level of N-cadherin and β-catenin markedly decreased. To further clarify the relationship between KAP3 and N-cadherin, we established an immortalized embryonic fibroblast cell line with the *kap3*-null genotype, and monitored the dynamics of GFP-tagged N-cadherin in vivo. It was revealed that the release of N-cadherin-GFP from the Golgi apparatus is significantly impaired, and the localization of N-cadherin-GFP at the cell-cell boundaries is significantly reduced in the knockout cells. We also observed the movements of individual post-Golgi organelles containing N-cadherin-GFP using time-lapse critical angle fluorescence microscopy (CAFM). Outward movements in the knockout cells are not as straight or continuous as those in control cells, and quantification showed that outward movements are significantly reduced in knockout cells. Immunoprecipitation showed the association of N-cadherin, β-catenin and the KIF3 motor, and immunostaining also showed the colocalization of N-cadherin and KAP3.

In summary, KIF3 transports newly synthesized N-cadherin with β-catenin from post-Golgi to the cell surface. This study suggests that KIF3 can be a potential tumor-suppressing factor by transporting N-cadherin with β-catenin from the cytoplasm to the plasma membrane. β-catenin in the nucleus can work as a transcriptional factor, a downstream of Wnt canonical pathway together with T cell factor (TCF), and control the expression of many genes including cyclin D1. Therefore, N-cadherin could function to control cell proliferation by recruiting cytoplasmic and nuclear β-catenin to the plasma membrane for cell adhesion machinery.

5.5 Monomeric Motor – How Can it Move?

Here, we discuss about the mechanism underlying motor motility. As we discussed above, we identified KIF1A motor, which exists as a monomer in solution [13]. It was an important finding because all the motors that had been identified before KIF1A are dimeric two-headed motors; for example, myosins that move along actin filaments, dyneins that generate the movements of cilia or flagella, or conventional kinesin. Therefore, the "walking" model or a "hand-over-hand" model is a widely accepted model of the processive movement, that is the movement of motor molecules along the rail without detaching for some distance (Fig. 5.5) [55]. In other words, it had been widely accepted that a motor needs two legs, just like when we walk. When one foot steps forward, the other foot needs to be firmly attached to the rail for a motor to

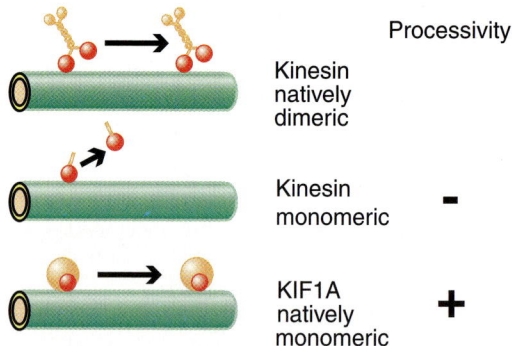

Fig. 5.5. Processivity of movement of monomeric motor along microtubules. Conventional kinesin is dimeric in its native form. It moves processively along microtubules in the dimeric forms, but not in the monomeric form. In contrast, KIF1A is monomeric in its native form and moves processively in the monomeric form

be able to "walk" without detaching from the rail. With this theory, a single-headed motor molecule is not expected to move processively along the rail, because a single-headed motor would dissociate from the rail when it steps forward [56, 57].

5.5.1 Processive Biased Brownian Movement of KIF1A

To determine whether KIF1A is a processive motor, we produced the shortest motor domain construct of the KIF1A molecule, C351, and labeled it with a fluorescent Alexa dye. A similar construct was produced for conventional kinesin [23]. We constructed a low-background video-intensified microscope and then visualized the behavior of single molecules, that is, their dynamics. It was revealed that a single KIF1A molecule moves along a microtubule processively over a distance of more than 1 μm (Figs. 5.6a and 6b-c shaded bar). The movement of single C351 molecules on the microtubule is not smooth, and actually appears to be oscillatory (Figs. 5.6a and 6b-d). Although on average they move unidirectionally, a single C351 molecule sometimes pauses for a while, or moves backward for a short distance, and then forward. To quantitatively analyze this apparently oscillatory movement, the distribution of displacement was plotted (Fig. 5.6b-f). The plot fitted well with a normal distribution whose mean and variance increased linearly against time. The linear increase of mean indicates a unidirectional constant-velocity movement on average. The linear increase of variance indicates a linear accumulation of random noise, as observed in Brownian movement [58]. Thus, the distribution of the displacement of C351 suggests that it was a biased Brownian movement. In contrast, a dimeric conventional kinesin molecule, K381, moves smoothly unidirectionally (Figs. 5.6b-e and 6b-c open bar). The distribution of its displacement fitted well with a normal distribution with linearly increasing mean and a constant variance (Fig. 5.6b-g). The latter can be explained by the error in position measurement.

We further analyzed the movement of C351 and K381 by plotting mean square displacement ([MSD, $\rho(t)$]) against time, which is a convenient quantitative measure of stochastic movement (Fig. 5.6b-h). The MSD plot of C351 (Fig. 5.6b-h open circle) fitted well with biased Brownian movement,

$$\rho(t) = 2Dt + \nu^2 t^2, \tag{5.1}$$

where D is the diffusion coefficient and ν is the mean velocity. The MSD plot of C351 fitted well with

$$D = 44{,}000 \pm 1200\,\mathrm{nm}^2/\mathrm{s} \tag{5.2}$$

and

$$\nu = 140 \pm 10\,\mathrm{nm/s}\,. \tag{5.3}$$

The estimated value of D is much larger than the value expected from fluctuation in ATPase activity (<2400 nm^2/s), but it is in good agreement

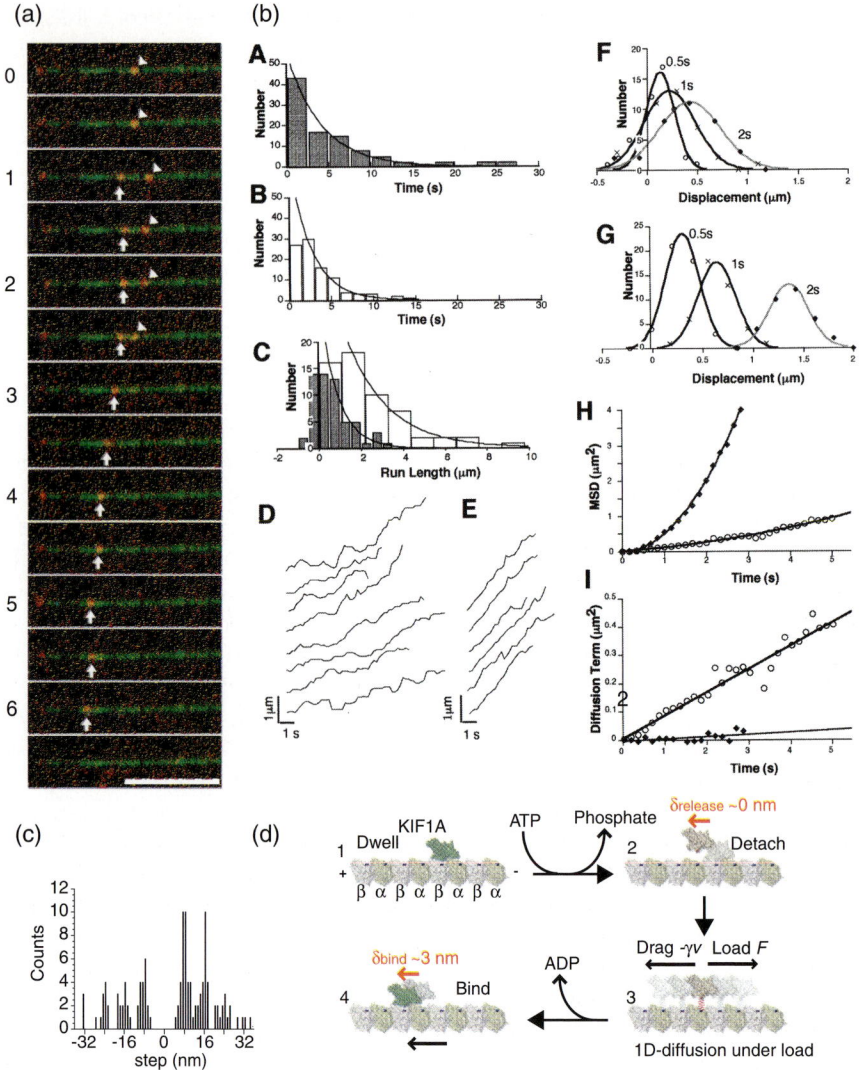

Fig. 5.6. Processive movement of monomeric motor KIF1A (**a**) Movement of fluorescently labeled single C351 molecules (*red*) along microtubule (*green*). Sometimes C351 moves backward (*arrowhead*), but it usually moves in one direction (*arrow*). Scale bar, 2 μm; frame interval, 0.5 s. (**b**) Analysis of movement. Distribution of duration of movement of C351 (A) and K381 (B). Distribution of run length of C351 (*shaded bar*) and K381 (*open bar*) (C). Typical traces of displacement of C351 (D) and K381 (E). Distribution of displacement of C351 (F) and K381 (G). MSD of C351 (○) and K381 (■) plotted against time (H). Diffusion term of C351 (○) and K381 (■) plotted against time (I). (**c**) Distribution of step size of KIF1A beads measured by optical trapping. (**d**) Flush ratchet model of KIF1A movement along microtubule. Panels (a and b) reproduced with permission from [23]. Panels (c and d) reproduced with permission from [40]

with the value previously reported for one-dimensional Brownian movement in microtubule-motor protein systems. In contrast, the MSD plot of K381 (Fig. 5.6b-h closed square) fitted well with

$$D = 2200 \pm 1000 \, \mathrm{nm}^2/\mathrm{s} \tag{5.4}$$

and

$$\nu = 710 \pm 10 \, \mathrm{nm/s} \,, \tag{5.5}$$

and the fluctuation in K381 movement can be explained by the fluctuation in ATPase activity.

The difference in the degree of fluctuation of movement between C351 and K381 is illustrated by the plot of the diffusion term,

$$\rho(t) - \nu^2 t^2 \,. \tag{5.6}$$

The movement of C351 is about 20 times as diffusive as K381 (Fig. 5.6b-I). This heavily stochastic nature of the movement cannot be explained by the error of position measurement or the Brownian noise, because these cannot accumulate and only contribute as constant terms to the diffusion term of MSD or the variance of displacement. Furthermore, our single-motor assay is free from the Brownian noise because the position of a fluorescent spot directly reflects the position of a motor molecule. These results are the first clear experimental demonstration that a motor could move by a biased Brownian movement.

5.5.2 "K-loop" is Essential for Processivity

The next question is why a one-headed monomeric motor can move processively along microtubules without detaching. It was suggested that the binding between motor and microtubule is mediated by electrostatic interactions. The microtubule surface is mainly negatively charged; therefore, we thought that KIF1A might have a highly positively charged region and we attempted to identify this region. We found a region with five lysines in tandem, and named this region the "K-loop" [23]. We considered that this highly positively charged region may play important roles and we confirmed it to be so. When mutants that have fewer lysines in the K-loop were produced, the processivity of the motor decreases, as determined by single-motor motility assay [59]. In other words, the duration of movement along a microtubule becomes shorter for the mutant KIF1A. The turnover rates (k_{cat}) of these mutants are the same level as that of the original construct, but $K_{MT(ATPase)}^{0.5}$ (microtubule concentration for half-saturation of ATPase), which is a measure of the average affinity to microtubule throughout the ATPase cycle, is significantly affected.

The time constant of the mechanical processivity (τ_{mec}) of C351 determined by single-molecule-motor assay is 6.1 ± 0.8 s (Fig. 5.6b-A), which agrees well with the time constant of kinetic processivity (τ_{kin}) of 6.3 s determined by microtubule-activated ATPase assay. In comparison, the time constant of the mechanical processivity of K381 (τ_{mec}) is 2.6 s and that of kinetic processivity (τ_{kin}) is 2.0 s (Fig. 5.6b-B), that is, about one-third of those of C351, indicating that monomeric KIF1A has a higher processivity than dimeric conventional kinesin.

The kinetic index of processivity, which indicates the number of ATPase cycles per motor-microtubule encounter, is 690 for C351. On average, C351 hydrolyzes nearly 700 ATPs before detaching from the microtubule. This is more than five times the number of ATPs hydrolyzed by the dimeric kinesin K381, and more than 40 times that hydrolyzed by the monomeric kinesin K351.

5.5.3 Biased Stepwise Movement During ATP Hydrolysis Cycle

We then analyzed the detail of movement by optical trapping nanometry [40]. For this, we attached a 0.2-μm-diameter bead to one end of a monomeric KIF1A construct, which contains almost solely the motor head domain and cannot dimerize. We then placed the beads on a polarity-marked microtubule fixed to a coverglass and performed optical trapping, that is, by pulling a bead with a laser beam, the optical tweezer, the movement of KIF1A is amplified as the movement of the bead.

A region called the "neck linker" at the C-terminus of the motor domain is proposed to play important roles in the motility of conventional kinesin [55,60]. To determine whether this is the case for the biased Brownian movement of KIF1A, a bead was attached to the N-terminus or C-terminus of the KIF1A motor domain. We also attached KIF1A to beads using methods that prevent dimerization. Binding beads at the N-terminus or C-terminus of KIF1A does not make any difference in the results of optical trapping nanometry; therefore, this indicates that the neck linker does not significantly contribute to at least motility based on biased Brownian movement.

The KIF1A attached beads showed a clear bidirectional stepwise movement. In addition, it was shown that a single ATP hydrolysis produces a single mechanical step. Furthermore, it was shown that the step size is not fixed, but distributed approximately in multiples of 8 nm (Fig. 5.6c). The histogram shows that the step size is distributed in both the directions of plus and minus ends, but step sizes are all in multiples of 8 nm. Therefore, the beads moved backward and forward in step sizes of multiples of 8 nm, but the net direction of the bead movement is towards the plus end, indicating biased Brownian movement. Because the size of the tubulin monomer is 4 nm and that of the heterodimer is 8 nm, the microtubule is predicted to have high affinity sites every 8 nm.

Next, to analyze at which steps the movement is biased towards the plus end, the movement was observed in the presence of ATP at a low concentration. The analysis showed that in the ADP-bound state, KIF1A was in Brownian motion, and when it released ADP and had no nucleotides, it bound tightly to the microtubule. At that moment, KIF1A moved towards the plus end by about 3 nm on average. This average distance of 3 nm is explained as follows. If a motor moves twice towards the plus-end over a distance of 8 nm and once towards the minus end over a distance of 8 nm once, the net distance it moved is 8 nm towards the plus end. However, it took three steps to make that movement. If 8 nm is divided by 3, the average movement of 2.66 nm per step is obtained. Therefore, KIF1A always moves at step sizes of multiples of 8 nm, but it moves towards the plus end over a distance of 3 nm on average. It was also demonstrated that when ADP is released, KIF1A makes a biased movement towards the plus end. In summary, the optical trapping showed that KIF1A is in Brownian movement in the ADP-bound state, and when ADP is released, it moves towards the plus end over a distance of 3 nm on average (Fig. 5.6d). This model also agrees well with the results of X-ray crystallography analysis, but we will discuss it later.

5.5.4 K-loop/E-hook Interaction Sustain Brownian Movement in Weak-Binding State – Tarzan Model

We also studied the mechanism of biased Brownian movement by cryoelectron microscopy [25]. The motor domain of KIF1A, C351, was bound to the microtubule in an ATP-like strong binding state and its structure was studied by cryoelectron microscopy at 15 Å resolution. In a state bound to a non-hydrolyzable analog of ATP, AMP-PNP, the KIF1A motor domain bound to microtubule protofilament like koala with three binding sites, namely MB1, MB2, and MB3 (Figs. 5.7a and c). When this was compared with conventional kinesin, a unique feature of KIF1A was found to be the high prominence of MB3. The next step was to determine which primary sequences, that is, amino acid sequence, form the MB1, MB2, and MB3 binding sites.

The X-ray crystallographic atomic structure of KIF1A was not available at the time, therefore, the available conventional kinesin atomic model was docked. Because the amino acid sequence of the motor domain of KIF1A have 60% similarity to that of conventional kinesin, certain parts would fit and others would not, revealing the differences. The docking showed that MB3 corresponded to the loop L12, that is, the K-loop region, and MB1 and MB2 corresponded mainly to L8 and L11.

When the tubulin atomic model was also docked, another important feature was revealed, that is the flexible C-terminus of tubulin can interact with the K-loop (Fig. 5.7b). This was confirmed by the cleavage of the C-terminus of tubulin [59]. Because this C-terminal region contains many glutamates (E in one-letter code), we named this region the "E-hook" [59]. It was also suggested that the interaction between the E hook and K loop is specifically important in the ADP-bound state. Furthermore, to directly show that the K-loop corresponds to MB3, we labeled the K loop with colloidal gold cluster by introducing reactive cysteine and examined it by cryoelectron microscopy. The gold-cluster labeling unambiguously proved that MB3 contains the K loop.

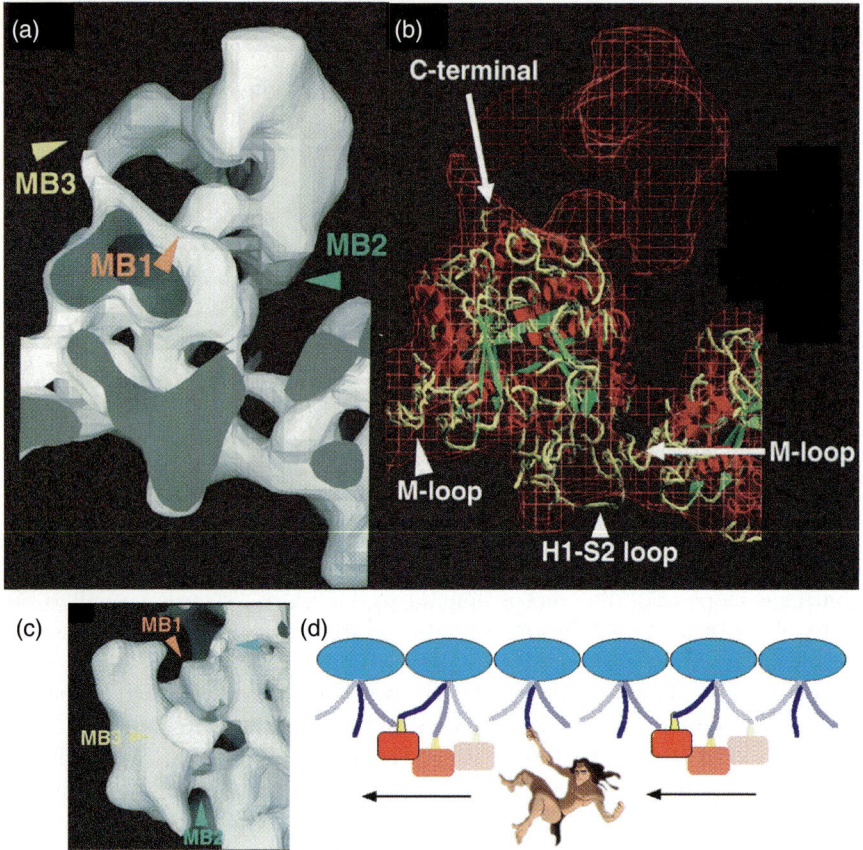

Fig. 5.7. (**a–c**) Docking of atomic model of KIF1A and tubulin. Surface representation from outside (c) and *top* view from plus end (a) and superposition of image on atomic model (b). (**d**) Tarzan model of biased Brownian movement of KIF1A. Panels (a–c) reproduced with permission from [25]

From these studies, we would like to propose a model for the biased Brownian movement of KIF1A called the Tarzan model (Fig. 5.7d). Tarzan is the KIF1A motor domain, and it is stretching its arm, the K-loop, upwards and grabbing a tether that hangs from the microtubule protofilaments composed of α- and β-tubulin. The tether is the flexible C-terminal E hook of tubulin. In the ADP-bound state, Tarzan does the Brownian motion via the interaction between the K loop and E-hook. Drag force, diffusion, might reflect the fluctuation of the flexible tether between KIF1A's K loop and tubulin's E hook. Upon the hydrolysis of ATP, Tarzan releases the tether and grabs another tether, but in doing so, movement has bias towards the plus end.

5.5.5 Conformational Changes in Switch I and II During ATP Hydrolysis

We then studied how the bias works by examining ATP hydrolysis steps by X-ray crystallography and cryoelectron microscopy (Fig. 5.8) [14]. First, the ADP-bound state and AMP-PCP-bound state of KIF1A were studied. The AMP-PCP state is considered to represent one of the ATP-like-states called preisomerization. X-ray crystallography revealed that switch I and II regions, which form a nucleotide-binding pocket interposed with the bridging water molecule and hydrogen bonds, show marked conformational changes. First, the switch I region forms a short β-hairpin structure in the ATP-like state, but it changes to a helix flanked by two short loops in the ADP-state. In addition, switch I region is closer to switch II region in the ATP-like state, resulting in the closure of the nucleotide-binding pocket. Second, in the switch II region, which is composed of the $\alpha 4$ helix and loop L11, the $\alpha 4$ tilts about 20 degrees and partly melts and L11 extends in the ATP-like state.

We then docked the atomic structures to cryoelectron microscopic images. Cryoelectron microscopy revealed that the "long axis" of the KIF1A motor domain in the ATP-state rotates ~ 20 degrees compared with that in the ADP-state. What roles this rotation plays is still a speculation, but this suggests that although the $\alpha 4$ helix tilts ~ 20 degrees in the crystal structure, when the motor is bound to the microtubule it remains fixed to the microtubule and the motor itself rotates relative to the microtubule.

We then analyzed additional states of ATP hydrolysis [43]. For this purpose, we produced a very short motor domain construct, C340. There are several ATP analogs that mimic the different states of the chemomechanical cycle of kinesin motors. First, three analogs mimic the strong-binding states: AMP-PCP, a nonhydrolyzable analog, mimics the preisomerization state; AMP-PNP, also a nonhydrolyzable analog, mimics the next prehydrolysis state; and ADP-AlFx mimics the early ADP-Pi state, in which ATP hydrolysis begins. ADP-Vi mimics the late ADP-Pi state, in which the association with the microtubule is the weakest, and is considered to be an actively detaching state (Fig. 5.8c–f). After that is the ADP state, a weak-binding state. As discussed above, we solved the AMP-PCP and ADP states by X-ray crystallography in a previous study. Therefore, we decided to determine three additional states, the AMP-PNP, ACP-AlFx, and ADP-Vi states; we have solved five states in total (Figs. 5.8b–e).

Comparing the five different states, the most significant conformational changes occur in switch I and II regions (Figs. 5.8c–f). In the preisomerization (AMP-PCP) state, γ-phosphate does not interact with any regions (Fig. 5.8c). In the prehydrolysis (AMP-PNP) state, γ-phosphate interacts with Ser[215] and Gly[251] (Fig. 5.8d). The switch I region elongates and changes its shape from a β-hairpin to a loop. Concomitantly, loop L11 in the switch II region extends downwards. In the early ADP-Pi (ADP-AlFx) state, a hydrogen bond between γ-phosphate and Gly[251] in switch II is broken and L11 is raised (Fig. 5.8e).

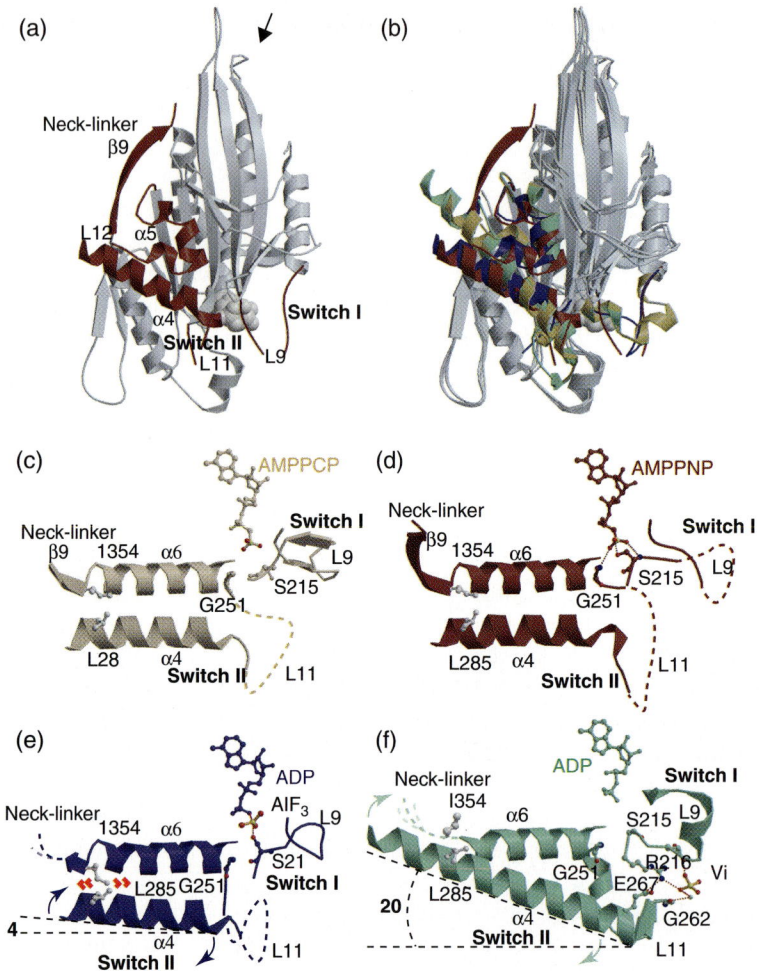

Fig. 5.8. (**a** and **b**) Crystal structures of KIF1A. (**a**) The AMP-PNP form. The switch I, switch II and neck linker regions are highlighted in *red*. (**b**) Superposition of AMP-PNP (*red*), ADP-AlFx (*blue*), ADP-Vi (*green*) and ADP (*yellow*) forms. (**c–e**) Conformational changes in two switch regions during ATP hydrolysis. AMP-PCP (**c**), AMP-PNP (**d**), ADP-AlFx (**e**) and ADP-Vi (**f**) forms. Reproduced with permission from [43]

The raised conformation of L11 triggers a slight tilt of the $\alpha4$ helix. In the late ADP-Pi (ADP-Vi) state, γ-phosphate is released from the nucleotide-binding pocket through the back door (Fig. 5.8f). This induces different interactions: loop L11 shortens and the elongated $\alpha4$ helix tilts further.

5.5.6 KIF1A Alternately Uses Two Loops to Bind Microtubules

We docked the atomic structures of KIF1A and tubulin (Fig. 5.9) [43]. In tubulin, helices H11 and H12 mainly serve as the rail for KIF1A movement. In the
strong-binding prehydrolysis (AMP-PNP) state, L11 is elongated and extends
downwards binding to H11', which is between H11 and H12 (Fig. 5.9a). In the
next early strong-binding ADP-Pi (ADP-AlFx) state, an early stage of hydrolysis, the extended L11 begins to rise (Fig. 5.9b). In the next late ADP-Pi
state (ADP-Vi) state, the loop L11 is completely raised, but the loop L12 is
also raised; therefore, the interaction of KIF1A with the microtubule rail is
the weakest (Fig. 5.9c). This is an actively dissociating state. After the release of Pi in the ADP state, L12 extends downwards and binds to the E-hook,
which is the C-terminus of tubulin (Fig. 5.9d). At this stage, the interaction

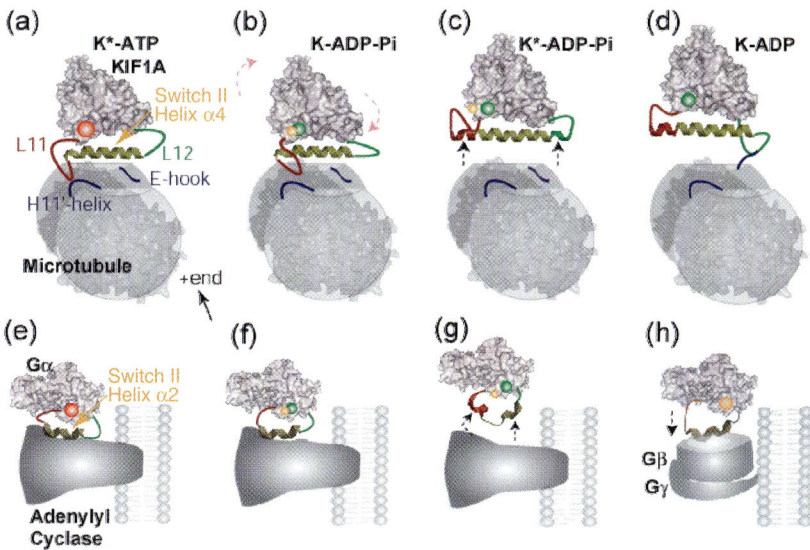

Fig. 5.9. Structural models of active detachment of KIF1A and G proteins. (**a–d**)
KIF1A (*pink*)-microtubule (*gray*) complex observed from minus end of microtubule
in different states of ATP hydrolysis. (a) Prehydrolysis state. (b) Early ADP-Pi
state. (c) Late ADP-Pi state, that is, detaching state (indicated by *arrows*). (d)
ADP state. L11, $\alpha 4$ and L12 are shown as colored ribbon model (*red, yellow* and
green). H11' and E-hook of tubulin are shown in *purple*. ATP, ADP and Pi are
shown as *red, green* and *orange spheres*, respectively. *Asterisks* refer to the second
ATP or ADP-Pi state of KIF1A. (**e–h**) The α subunit of G proteins and region
corresponding to switch II are color-coded as in (a–d). Adenylyl cyclase and β and
γ subunits of G proteins are shown in *gray*. (e) to (h) represent states during GTP
hydrolysis corresponding to those described in (a) to (d). The energy derived from
GTP hydrolysis is used for dissociation of Gα from adenylyl cyclase as indicated by
arrows in (g)

between K-loop (L12) and E-hook, similar to the interaction between Tarzan's arm and the tether, causes the Brownian movement (Fig. 5.7d). Because of this, the motor does not dissociate from the rail in spite of the fact that it is in the weak-binding state. Therefore, Brownian movement is the underlying force for the motor movement, and when ADP dissociates, KIF1A binds to the microtubule with bias towards the plus end, probably due to the presence of higher affinity sites on the plus end of tubulin.

We tested this hypothesis of the alternate use of two loops by introducing mutations to L11 and L12 and observed changes in the binding between KIF1A and microtubules [43]. In the AMP-PNP state, introducing mutations in L11 resulted in a decreased strength of binding. In contrast, in the ADP state, introducing mutations in L12 resulted in a decreased strength of binding. These results confirmed that L11 is used for binding at the AMP-PNP state and L12 is used for that at the ADP state. In addition, when the strength of binding between KIF1A and microtubules was compared at different ATP hydrolysis steps in wild-type constructs, ADP-Vi (late ADP-Pi) state was the weakest binding state and KIF1A actively dissociates from microtubules in this state.

In summary, KIF1A uses L11 and L12 alternately for binding to tubulin (Figs. 5.9a–d). In the ATP-like prehydrolysis state, KIF1A extends L11 and binds to the H11' helix of tubulin. Then L11 is raised, and KIF1A dissociates from tubulin in the late ADP-Pi state. In the ADP state, KIF1A extends L12, binding to the E-hook, and makes a Brownian movement. Therefore, energy derived from hydrolysis of ATP is used to dissociate from the microtubule at the late ADP-Pi state. This strategy is also used when kinases phosphorylate their substrates or when heterotrimeric guanine nucleotide-binding proteins (G proteins) interact with their target molecules (Figs. 5.9e–h). In both cases, energy derived from the hydrolysis of nucleotides is used for dissociation from the substrates or target molecules. Therefore, it was suggested that this is an evolutionarily conserved mechanism for nucleotide-binding proteins including kinases, G proteins, and motor proteins. The movement itself of motor proteins depends on the Brownian movement.

We would like to add some detailed discussion here. One of them is if neck linker regions play any roles in the movement of motor proteins by biased Brownian movement. The neck linker regions, which are located at the C-terminus of the motor domain of N-kinesins, are proposed to play important roles in dimeric motor movement. The neck linker of KIF1A is docked on the motor domain when in the ATP-bound state, but undocked when in the ADP-bound state, making conformational changes [14]. However, it is not likely that the neck linker contributes to biased Brownian movement, because results of optical trapping experiments were unaffected when beads were attached either to the N-terminus or C-terminus of the motor domain [40]. Another issue that needs to be discussed is the velocity achieved by biased Brownian movement. When a single molecule is bound to a bead and moves through the mechanism of Brownian movement, the bead moves at $\sim$0.2 µm/s, which is slower than the velocity observed in microtubule gliding assays. However, with the clustering of a few KIF1A molecules on the bead surface, but without dimerization, the bead moves much faster. This enhancement can be explained from the nature of biased Brownian movement. The increase in velocity is caused by the suppression of the backward movement, which is caused by the presence of multiple motors; this results in an increase in the efficiency of forward movement, that is, the plus-end-directed movement.

5.6 KIF2 – Microtubule Depolymerizing Motor

We have discussed translocators, that is, motor proteins that participate in transport processes. We would like to turn to another unique type of motor, KIF2 [12,39,42]. In the mammalian, the KIF2 gene family has three members, KIF2A, KIF2B, and KIF2C. They are the middle motor domain type, M-kinesins [32].

The first member we identified was KIF2A, which is expressed abundantly in neurons, particularly in the juvenile stage and in growth cones [12]. To examine its role in vivo, we generated a knockout mouse [39].

5.6.1 KIF2A Functions in Suppression of Collateral Branch Formation

The mutant mice were born alive, but all died within one day. Their brains showed laminary defects in the cerebral cortex, hippocampus, and cerebellum. During normal development, neurons are first born and differentiate in the ventricular zone and then they migrate outwardly towards the surface along radial glias in an orderly manner thus properly forming normal laminary structure. To examine whether there is any deficiency in neuronal migration in the mutant mice, we performed birth-date analysis by labeling cortical neurons with bromodeoxyuridin (BrdU). The E14-born neurons were labeled, and their distribution was analyzed at birth. When the mutant and wild-type mice were compared, the mutant mice showed a much slower migration of neurons. Then, to determine the events occurring at the cerebral cortex, the axonal morphology was observed by diI crystal staining. In the wild type, axons run straight longitudinally, but in the mutant mice, many horizontally running neurites were observed. This suggests that axonal branching might be abnormal in the mutants. When individual neurons were closely observed, wild-type neurons had several apical dendrites and a single long axon, but long axonal branches are rarely observed. In contrast, the mutant neurons had axons with many long axonal collateral branches. The results were also confirmed in cultured hippocampal neurons (Figs. 5.10c and 5.10d). In addition, it was observed that the frequencies of axonal collateral branch formation were the same, but the branches are short in the wild type, but long and overextend in the mutant neurons.

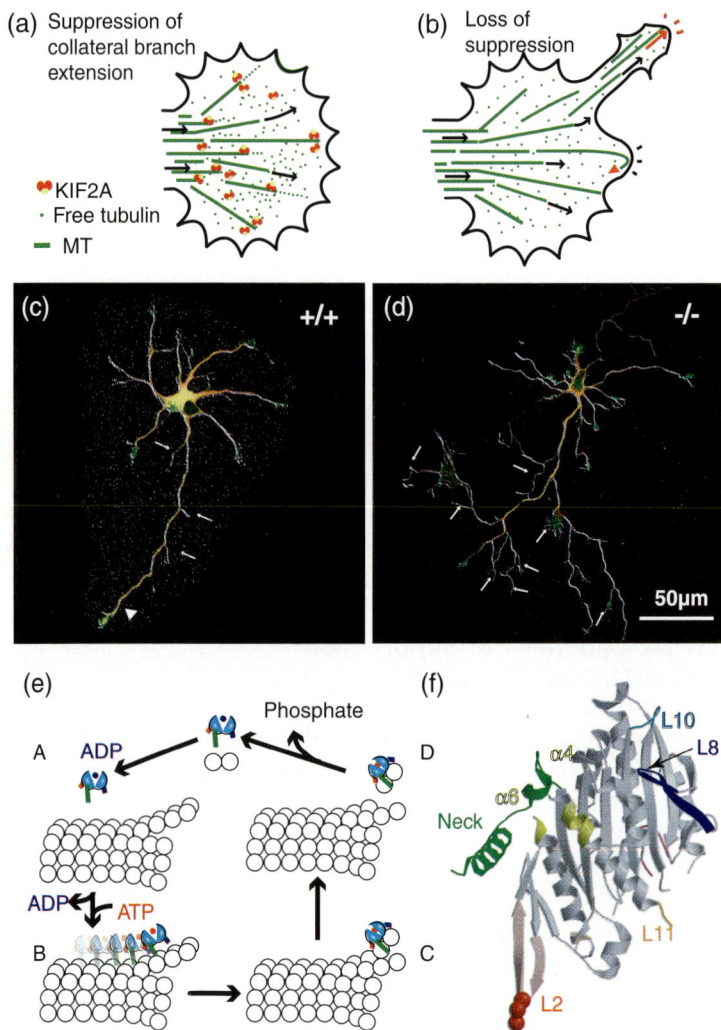

Fig. 5.10. (a and b) Suppression of axonal collateral branch extension by KIF2A. In the presence of KIF2A (a), growth cone extension is suppressed because microtubules maintain a dynamic equilibrium between polymerization and depolymerization, particularly at the edge of growth cones. In the absence of KIF2A (b), microtubule tips at the cell edge do not show controlled dynamics and hit the membrane. Microtubule tips turn back (*red arrowhead*) or push the membrane forward (*red arrow*). (c and d). Hippocampal neurons double-labeled for F-actin (*green*) and tubulin (*red*). (e) Structural model of the mechanism of microtubule depolymerization by M-kinesin (KIF-M). Neck region, KVD finger and L8 are shown in *green*, *red* and *blue* respectively. See texts for detail. (f) Structure of KIF2C (AMP-PNP-bound form). View from the microtubule-binding side. The residues corresponding to KVD motif are shown as *red spheres*. Panels (a–d) reproduced with permission from [39]. Panels (e and f) reproduced with permission from [42]

We then observed the migration of cultured neurons by time-lapse analysis. In the wild type, an axon extends from the cell body, and the cell body migrates forward following the extending axon. In contrast, in the mutants, many axonal collateral branches overextend forward, which hinder the forward migration of the cell body. Therefore, the results suggest that an abnormal axonal collateral branch elongation is one of the major cause of an abnormal cell migration in the absence of KIF2A.

To elucidate further the abnormal axonal collateral elongation, we examined KIF2A function in growth cones. Previous studies in other laboratories showed that KIF2C depolymerizes microtubules in an ATP-dependent manner [61]. Therefore, we examined whether KIF2A has the same function, and found that KIF2A also depolymerizes microtubules in an ATP-dependent manner. We then examined microtubule behavior in neuronal growth cones on the basis of fluorescence loss in photobleaching (FLIP) experiment. Microtubule dynamics in growth cones is blocked in the mutants compared with that in the wild type. When individual microtubule behavior in the cell periphery of KIF2A-expressing cells was observed, the elongating microtubules began to depolymerize when it reached the cell periphery. In contrast, in the absence of KIF2A, microtubules continued to elongate after reaching the cell periphery.

In conclusion, the extension of axon collaterals is suppressed in the wild type due to microtubule depolymerization at growth cones, which is regulated by KIF2A (Figs. 5.10a and 5.10b). In contrast, in the absence of KIF2A, microtubules continue to elongate and the extension of axonal collaterals is not controlled. Therefore, KIF2A is very important for controlling brain wiring by depolymerizing microtubules.

5.6.2 Common Mechanism for Microtubule Destabilizers

Lastly, we studied how KIF2 depolymerizes microtubules by X-ray crystallography [42]. For this purpose, we used the minimal functional domain necessary for depolymerizing microtubules, which is composed of the motor domain and the neck, which is located at the N-terminal side of the motor domain in M-kinesins. We also used KIF2C, which has a microtubule depolymerizing activity higher than that of KIF2A. Both the ADP-bound and AMP-PNP-bound forms were examined.

It was revealed that KIF2C has structural uniqueness primarily in three regions (Fig. 5.10f). The first is the neck region, which is located at the N-terminus of the motor domain and forms an α-helix-rich structure extending vertically downward to microtubules. The second, region called L2, which is a loop in other KIFs, is actually antiparallel two-stranded β sheets in KIF2C. There is a unique KVD motif at the tip of where two sheets folds back and it extends rigidly like a finger. Therefore, we named this region the "KVD finger". Third, the conformation of nucleotide-binding pocket does not alter significantly in the ATP-bound and ADP-bound states because of the unique position of L8. In other translocator KIFs such as KIF1A, the distance between the switch I and II regions in nucleotide-binding pocket becomes smaller in the ATP-bound state, and this conformational change triggers ATP hydrolysis. In contrast, in KIF2C, the nucleotide-binding pocket remains open in the ATP-bound state. However, if L8 is pulled up, the $\alpha3$ helix rotates and the nucleotide-binding pocket will close and trigger ATP hydrolysis.

We then fitted the atomic model of KIF2C to the microtubule protofilament [42]. The microtubule protofilament is straight in the shaft, but curved and frayed at its ends. KIF2C does not fit well with the straight protofilament on the shaft, but it fits well on the curved protofilament, in good agreement with the data that KIF2C preferentially binds to the plus end of microtubules [62]. Because the N-terminal helix extends vertically, KIF2C does not bind well to the straight portion of microtubules. However, at the curved ends, the N-terminal neck extends into the side groove of the protofilament and destabilizes the lateral interaction of adjacent protofilaments. At the same time, the KVD finger inserts into the interdimer interface and stabilizes the curved conformation of the protofilament. The stabilization of a curved conformation by itself enhances tubulin dissociation. For KIF2C to bind to tubulin tightly, L8 will be pushed up and this results in the rotation of the $\alpha3$ helix. The nucleotide-binding pocket will then close and ATP hydrolysis begins.

In conclusion, the mechanism of microtubule depolymerization by KIF2 can be represented by a "packman" model (Fig. 5.10e). In the ADP-state, the nucleotide-binding pocket is open and cannot bind to microtubules. When ADP is released and ATP is bound, KIF2 diffuses along the microtubule protofilament. When KIF2 reaches the microtubule end, it firmly attaches to the microtubule. The N-terminal neck destabilizes the lateral interaction of adjacent protofilaments and the KVD finger stabilizes the curved heterodimer and enhances its dissociation. L8 is pushed up and the $\alpha3$ helix rotates and the nucleotide-binding pocket will close. Then ATP hydrolysis begins. The tubulin heterodimer dissociates and phosphate is released. ATP becomes ADP and the next cycle begins. Thus, we have elucidated how the KIF2 family depolymerizes microtubules.

5.7 Conclusions and Future Perspectives

As we described, all types of cell including neurons use various motors, including the monomeric type, dimeric type, heterodimeric type, N-terminal motor domain type, middle motor domain type, and C-terminal motor domain type, and transport various molecules in the form of membrane vesicles, protein complexes and mRNA protein complexes to the proper destination at a proper velocity. These transport processes support basic functions of cells and play critical roles as bases for higher brain functions such as learning and memory or brain wiring. At the same time, the transport controls important events in development such as left-right axis determination and suppresses tumor formation.

In addition to KIFs, myosin superfamily proteins and dynein superfamily proteins are involved in the intracellular transport [3, 63]. Myosin superfamily proteins use actin filaments as rails and dynein superfamily proteins use microtubules as rails. KIFs are mostly used for long-distance transportation, whereas myosins are mostly used for short-distance transport near the plasma membrane. Dynein superfamily proteins move towards the minus end of microtubules. Because most of KIFs are classified as N-kinesins and move towards the plus end of microtubule, dyneins are involved in the transport in the direction opposite to most KIFs. It is likely that KIFs, myosins, and dyneins coordinate closely to accomplish complex intracellular transport, but the detail is a subject of future study.

There are still many KIFs whose functions are as yet unclarified. We need to clarify their functions using techniques in molecular cell biology and molecular genetics. We also need to clarify how motor molecules recognize and bind to their cargoes or how the binding is regulated. The transport is also regulated in terms of directional transport such as axonal vs. dendritic transport. As we demonstrated, the monomeric motors are the simplest motors, and studying further mechanisms underlying their motility mechanism is important. Because dimeric motors lose their processivity when a single molecule is studied, they need to be studied as dimers. However, in analyses using dimeric motors, the existence of a second head always hinders detailed analyses. Because of this, monomeric motors are best suited for the analyses of their mechanisms underlying motility.

Acknowledgments

We thank all members of the Hirokawa Laboratory and Okinaka Memorial Institute for Medical Research. This work was supported by a Center of Excellence grant to N. H. from the Ministry of Education, Culture, Sports, Science and Technology.

References

1. N. Hirokawa, R. Takemura (2005). *Nat. Rev. Neurosci.*, **6**, pp. 201–214
2. N. Hirokawa (1982). *J. Cell Biol.*, **94**, pp. 129–142
3. N. Hirokawa (1998). *Science*, **279**, pp. 519–526
4. N. Hirokawa, K.K. Pfister, H. Yorifuji, M.C. Wagner, S.T. Brady, G.S. Bloom (1989). *Cell*, **56**, pp. 867–878
5. N. Hirokawa (1996). *Trends Cell Biol.*, **6**, pp. 135–141
6. H. Aizawa, Y. Sekine, R. Takemura, Z. Zhang, M. Nangaku, N. Hirokawa (1992). *J. Cell Biol.*, **119**, pp. 1287–1296
7. N. Hirokawa, R. Takemura (2004). *Exp. Cell Res.*, **301**, pp. 50–59

8. N. Hirokawa, R. Sato-Yoshitake, N. Kobayashi, K.K. Pfister, G.S. Bloom, S.T. Brady (1991). *J. Cell Biol.*, **114**, pp. 295–302

9. S. Kondo, R. Sato-Yoshitake, Y. Noda, H. Aizawa, T. Nakata, Y. Matsuura, N. Hirokawa (1994). *J. Cell Biol.*, **125**, pp. 1095–1107

10. Y. Sekine, Y. Okada, Y. Noda, S. Kondo, H. Aizawa, R. Takemura, N. Hirokawa (1994). *J. Cell Biol.*, **127**, pp. 187–201

11. M. Nangaku, R. Sato-Yoshitake, Y. Okada, Y. Noda, R. Takemura, H. Yamazaki, N. Hirokawa (1994). *Cell*, **79**, pp. 1209–1220

12. Y. Noda, R. Sato-Yoshitake, S. Kondo, M. Nangaku, N. Hirokawa (1995). *J. Cell Biol.*, **129**, pp. 157–167

13. Y. Okada, H. Yamazaki, Y. Sekine-Aizawa, N. Hirokawa (1995). *Cell*, **81**, pp. 769–780

14. M. Kikkawa, E.P. Sablin, Y. Okada, H. Yajima, R.J. Fletterick, N. Hirokawa (2001). *Nature*, **411**, pp. 439–445

15. H. Yamazaki, T. Nakata, Y. Okada, N. Hirokawa (1995). *J. Cell Biol.*, **130**, pp. 1387–1399

16. T. Nakata, N. Hirokawa (1995). *J. Cell Biol.*, **131**, pp. 1039–1053

17. H. Yamazaki, T. Nakata, Y. Okada, N. Hirokawa (1996). *Proc. Natl. Acad. Sci. U.S.A.*, **93**, pp. 8443–8448

18. N. Saito, Y. Okada, Y. Noda, Y. Kinoshita, S. Kondo, N. Hirokawa (1997). *Neuron*, **18**, pp. 425–438

19. T. Nakagawa, Y. Tanaka, E. Matsuoka, S. Kondo, Y. Okada, Y. Noda, Y. Kanai, N. Hirokawa (1997). *Proc. Natl. Acad. Sci. U.S.A.*, **94**, pp. 9654–9659

20. Y. Yonekawa, A. Harada, Y. Okada, T. Funakoshi, Y. Kanai, Y. Takei, S. Terada, T. Noda, N. Hirokawa (1998). *J. Cell Biol.*, **141**, pp. 431–441

21. Y. Tanaka, Y. Kanai, Y. Okada, S. Nonaka, S. Takeda, A. Harada, N. Hirokawa (1998). *Cell*, **93**, pp. 1147–1158

22. S. Nonaka, Y. Tanaka, Y. Okada, S. Takeda, A. Harada, Y. Kanai, M. Kido, N. Hirokawa (1998). *Cell*, **95**, pp. 829–837

23. Y. Okada, N. Hirokawa (1999). *Science*, **283**, pp. 1152–1157

24. S. Takeda, Y. Yonekawa, Y. Tanaka, Y. Okada, S. Nonaka, N. Hirokawa (1999). *J. Cell Biol.*, **145**, pp. 825–836

25. M. Kikkawa, Y. Okada, N. Hirokawa (2000). *Cell*, **100**, pp. 241–252

26. S. Takeda, H. Yamazaki, D.H. Seog, Y. Kanai, S. Terada, N. Hirokawa (2000). *J. Cell Biol.*, **148**, pp. 1255–1265

27. M. Setou, T. Nakagawa, D.H. Seog, N. Hirokawa (2000). *Science*, **288**, pp. 1796–1802

28. Y. Kanai, N. Dohmae, N. Hirokawa (2004). *Neuron*, **43**, pp. 513–525

29. S. Terada, M. Kinjo, N. Hirokawa (2000). *Cell*, **103**, pp. 141–155

30. T. Nakagawa, M. Setou, D. Seog, K. Ogasawara, N. Dohmae, K. Takio, N. Hirokawa (2000). *Cell*, **103**, pp. 569–581

31. C. Zhao, J. Takita, Y. Tanaka, M. Setou, T. Nakagawa, S. Takeda, H.W. Yang, S. Terada, T. Nakata, Y. Takei, M. Saito, S. Tsuji, Y. Hayashi, N. Hirokawa (2001). *Cell*, **105**, pp. 587–597

32. H. Miki, M. Setou, K. Kaneshiro, N. Hirokawa (2001). *Proc. Natl. Acad. Sci. U.S.A.*, **98**, pp. 7004–7011

33. Y. Noda, Y. Okada, N. Saito, M. Setou, Y. Xu, Z. Zhang, N. Hirokawa (2001). *J. Cell Biol.*, **155**, pp. 77–88

34. M. Setou, D.H. Seog, Y. Tanaka, Y. Kanai, Y. Takei, M. Kawagishi, N. Hirokawa (2002). *Nature*, **417**, pp. 83–87

35. Y. Xu, S. Takeda, T. Nakata, Y. Noda, Y. Tanaka, N. Hirokawa (2002). *J. Cell Biol.*, **158**, pp. 293–303
36. R.W.-C. Wong, M. Setou, J. Teng, Y. Takei, N. Hirokawa (2002). *Proc. Natl. Acad. Sci. U.S.A.*, **99**, pp. 14500–14505
37. B. Macho, S. Brancorsini, G.M. Fimia, M. Setou, N. Hirokawa, P. Sassone-Corsi (2002). *Science*, **298**, pp. 2388–2390
38. L. Guillaud, M. Setou, N. Hirokawa (2003). *J. Neurosci.*, **23**, pp. 131–140
39. N. Homma, Y. Takei, Y. Tanaka, T. Nakata, S. Terada, M. Kikkawa, Y. Noda, N. Hirokawa (2003). *Cell*, **114**, pp. 229–239
40. Y. Okada, H. Higuchi, N. Hirokawa (2003). *Nature*, **424**, pp. 574–577
41. T. Nakata, N. Hirokawa (2003). *J. Cell Biol.*, **162**, 1045–1055
42. T. Ogawa, R. Nitta, Y. Okada, N. Hirokawa (2004). *Cell*, pp. **116**, 591–602
43. R. Nitta, M. Kikkawa, Y. Okada, N. Hirokawa (2004). *Science*, **305**, pp. 678–683
44. Y. Kanai, Y. Okada, Y. Tanaka, A. Harada, S. Terada, N. Hirokawa (2000). *J. Neurosci.*, **20**, pp. 6374–6384
45. Y. Tanaka, N. Hirokawa (2002). *Trends Genet.*, **18**, pp. S39–44
46. K. Jo, R. Derin, M. Li, D.S. Bredt (1999). *J. Neurosci.*, **19**, pp. 4189–4199
47. N. Hirokawa (2000). *Traffic*, **1**, pp. 29–34
48. N. Hirokawa, R. Takemura (2003). *Trends Biochem. Sci.*, **28**, pp. 558–565
49. Y. Okada, S. Takeda, Y. Tanaka, J.-C.I. Belmonte, N. Hirokawa (2005). *Cell*, **121**, pp. 633–644
50. Y. Okada, S. Nonaka, Y. Tanaka, Y. Saijoh, H. Hamada, N. Hirokawa (1999). *Mol. Cell*, **4**, pp. 459–468
51. Y. Tanaka, Y. Okada, N. Hirokawa (2005). *Nature*, **435**, pp. 172–177
52. E.N. Meyers, G.R. Martin (1999). *Science*, **285**, pp. 403–406
53. J. McGrath, S. Somlo, S. Makova, X. Tian, M. Brueckner (2003). *Cell*, **114**, pp. 61–73
54. J. Teng, T. Rai, Y. Tanaka, Y. Takei, T. Nakata, M. Hirasawa, A.B. Kulkarni, N. Hirokawa (2005). *Nat. Cell Biol.*, **7**, pp. 474–482
55. A. Yildiz, P.R. Selvin (2005). *Trends Cell Biol.*, **15**, pp. 112–120
56. J. Howard, A.J. Hudspeth, R.D. Vale (1989). *Nature*, **342**, pp. 154–158
57. B.J. Schnapp, B. Crise, M.P. Sheetz, T.S. Reese, S. Khan (1990). *Proc. Natl. Acad. Sci. U.S.A.*, **87**, pp. 10053–10057
58. R.D. Astumian (1997). *Science*, **276**, pp. 917–922
59. Y. Okada, N. Hirokawa (2000). *Proc. Natl. Acad. Sci. U.S.A.*, **97**, pp. 640–645
60. S. Rice, A.W. Lin, D. Safer, C.L. Hart, N. Naber, B.O. Carragher, S.M. Cain, E. Pechatnikova, E.M. Wilson-Kubalek, M. Whittaker, E. Pate, R. Cooke, E.W. Taylor, R.A. Milligan, R.D. Vale (1999). *Nature*, **402**, pp. 778–784
61. A. Desai, T.J. Mitchison (1997). *Annu. Rev. Cell Dev. Biol.*, **13**, pp. 83–117
62. A.W. Hunter, M. Caplow, D.L. Coy, W.O. Hancock, S. Diez, L. Wordeman, J. Howard (2003). *Mol. Cell*, **11**, pp. 445–457
63. R.D. Vale (2003). *Cell*, **112**, pp. 467–480
64. C.J. Lawrence, R.K. Dawe, K.R. Christie, D.W. Cleveland, S.A. Endow, L.S. Goldstein, H.V. Goodson, N. Hirokawa et al. (2004). *J. Cell Biol.*, **167**, pp. 19–22

6

Studies of DNA-Protein Interactions at the Single Molecule Level with Magnetic Tweezers

J.-F. Allemand[1], D. Bensimon[1], G. Charvin[1], V. Croquette[1], G. Lia[2], T. Lionnet[1], K.C. Neuman[1], O.A. Saleh[3], and H. Yokota[4]

[1] Laboratoire de Physique Statistique and Department of Biologie, Ecole Normale Superieure, UMR 8550 CNRS, 24 rue Lhomond, 75231 Paris Cedex 05, France
allemand@lps.ens.fr
david.bensimon@lps.ens.fr
croquette@lps.ens.fr
gilles.charvin@lps.ens.fr

[2] Harvard University Chemistry and Chemical Biology, 12, Oxford street, Cambridge, MA 02139, USA
lia@fas.harvard.edu

[3] Materials Department and Biomolecular Science and Engineering Program, University of California, Materials Dept, UCSB Bldg 503, Rm 1355 Santa Barbara, CA 93106-5050, USA
saleh@engineering.ucsb.edu

[4] Department of Molecular Physiology, The Tokyo Metropolition Institute of Medical Science, Tokyo, Japan
hiroaki_yokota@rinshoken.or.jp

Abstract. The development of tools to manipulate and study single biomolecules (DNA, RNA, proteins) has opened a new vista on the study of their mechanical properties and their joint interactions. In this short review we will focus on (single and double stranded) DNA and its interactions with various classes of proteins: structural DNA binding proteins such as gene repressors (e.g., the Galactose Repressor, GalR) and mechano-chemical enzymes that alter the DNA's topology (topoisomerases), unwind it (helicases) or translocate it (FtsK). We will show how the new tools at our disposal can be used to gain an unprecedented description of the binding properties (on and off-times) and the enzymes' kinetic constants that are often out of reach of more classical, bulk techniques.

6.1 Introduction

Biophysics is currently undergoing an important transformation due to the development of tools for manipulating, visualizing and studying single molecules and their interactions. New tools such as optical or magnetic tweez-

J.-F. Allemand et al.: *Studies of DNA-Protein Interactions at the Single Molecule Level with Magnetic Tweezers*, Lect. Notes Phys. **711**, 123–140 (2007)
DOI 10.1007/3-540-49522-3_6

ers, have allowed the manipulation of single DNA molecules and a detailed characterization of their elastic response (see [1] for review). These experiments have renewed theoretical interest in the mechanical properties of biomolecules. Consequently, we currently have a very good understanding of the response to tension and torsion of DNA over a large range of forces and torques [2]. Modification of these elastic properties induced by DNA binding proteins provides precious information about interaction and enzymatic kinetics. With these tools any protein that alters the DNA's extension at a given force can be studied at the single molecule level. Moreover, if the time resolution of the measuring device is sufficient, which is often the case, these studies permit real time monitoring of the DNA/protein interaction. As with all single molecule approaches, these investigations avoid the ensemble averaging inherent in bulk measurements. Such averaging may hinder the observation of some dynamical properties of the enzyme or obscure the existence of a heterogeneous population.

Micromanipulation experiments offer even more specific features. First they introduce force as a new thermodynamic parameter in in vitro experiments. Force can be used to alter the equilibrium of a reaction or modify its activation barriers, in addition to the temperature or buffer conditions, which are often the only control parameters in a conventional bulk assay [3–5]. Another interesting aspect of the single molecule assays is that they only measure active enzymatic complexes. Indeed, in any conventional biochemical experiment the intrinsic enzymatic activity is estimated by dividing the total activity by the concentration of enzymes, an estimate that often underestimates the real activity if a large portion of the enzymes are inactive or if the active complex is multimeric.

In the present review, after a short introduction to magnetic tweezers and a summary of the elastic properties of bare DNA molecules, we will present some of the results obtained in our group as an illustration of the previous considerations. First we will describe a loop forming protein, the Gal Repressor. We will then discuss the action of enzymes that modify the DNA topology, the so called topoisomerases. We will end by discussing the action of two DNA translocases: UvrD helicase and the very fast FtsK.

6.2 Magnetic Tweezers

Most of the motors that act on DNA convert the energy from NTP (or dNTP) hydrolysis into mechanical work. The hydrolysis of a single molecule of ATP, under physiological conditions, generates about 20 $k_B T$ (i.e. $8 \cdot 10^{-20} J$) of energy (k_B is Boltzmann constant and $T \sim 300°K$ room temperature). Since the characteristic displacement of a biological motor is on the order of a few nanometers, the relevant forces for micromanipulation of biological molecules are of order: $8 \cdot 10^{-20} J/10^{-8} m \sim 8 \cdot 10^{-12}$ N. To apply and detect such forces different techniques have been developed: Atomic Force Microscopy, Bio-

a

b

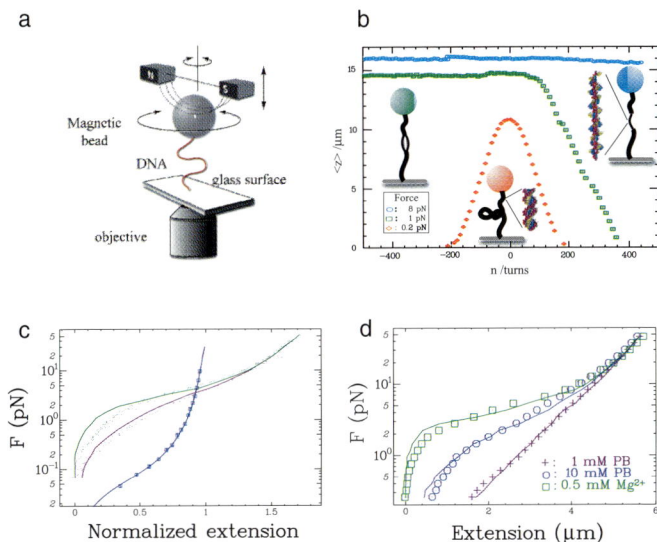

c

d

Normalized extension

Extension (μm)

Fig. 6.1. (a) Magnetic tweezers set-up. A few micron long DNA molecule is anchored at one end to a micron sized magnetic bead and at the other to the bottom surface of a square capillary tube, placed on top of a microscope objective. Small (few cm) magnets placed a few mm above the sample can be used to pull and twist the DNA molecule. **(b)** Twisting by $\Delta Lk = n$ turns a λ−DNA ($Lk_0 = 4620$) at various forces. At low forces (*red*), as the molecule is over or underwound its extension decreases identically by forming left or right-handed plectonemes. At intermediate forces (*green*), negatively supercoiled DNA denatures, while left-handed plectonemes can still be observed at positive supercoilings. At high forces (*blue*), positively supercoiled DNA undergoes a transition to an inside-out structure called P-DNA [6]. **(c)** Force extension curves of single stranded (green and magenta points) and double stranded (*blue points*) DNA, normalized by the crystallographic length of the dsDNA. The dsDNA data was fitted to a WLC model (*blue line*) with persistence length $\xi = 52$ nm. The two ssDNA data show that their elastic response depends on their nucleotide content: green points correspond to Charomid DNA (50% GC content), magenta points correspond to pXΔ2 DNA (30% GC content) (both in 10 mM Tris, 5 mM MgCl$_2$). Plain lines: results of MC simulations of an elastic model of ssDNA [7]. Note that above 5 pN, ssDNA is longer than dsDNA. **(d)** The elastic behavior of ssDNA strongly depends on the ionic conditions. Dots: experimental points in the indicated buffers: 1 mM and 10 mM Phosphate buffer (PB) and 10 mM PB buffer with 0.5 mM Mg$^+$+ ions. Plain line: results of MC simulations of an elastic model of ssDNA with a single fit parameter (the same for all curves) [7]

membrane Force Probe, glass micropipette manipulation, flow induced force, optical and magnetic tweezers (see [2] and references therein). For the purpose of this review, we provide only a brief description of magnetic tweezers.

Magnetic tweezers are used to manipulate magnetic beads tethered to a surface by a DNA molecule (but potentially by any polymer). When placed in the magnetic field of permanent (or electro-) magnets (see Fig. 6.1a), their magnetic dipole experiences both a force F (parallel to the magnetic field gradient) and a torque Γ (that tends to align its dipole with the magnetic field). The beads are subject to Brownian fluctuations and exhibit random displacements transverse to the direction of the pulling force that are inversely proportional to F: the stronger the force the less the bead fluctuates. In fact, the bead tethered by a DNA molecule of extension l and pulled by the magnetic field is similar to a damped pendulum. Applying the dissipation-fluctuation theorem to this system, yields a relation between the force and the amplitude of the transverse fluctuations $\langle \delta x^2 \rangle$: $F = k_B T l / \langle \delta x^2 \rangle$. Video microscopy allows measurement of the three dimensional position of the bead [8] and thus of l and $\langle \delta x^2 \rangle$. Forces have been measured from a few femtoNewtons to a hundred picoNewtons [8], with a precision that depends only on the quality of the statistical estimates of l and $\langle \delta x^2 \rangle$. In practice, we gather data for a time long enough to ensure a relative precision $\delta F / F < 10\%$. Notice that in a series of experiments the force has to be measured only once, since it is associated with a given distance of the magnets from the sample.

A distinct advantage of magnetic tweezers is that they are intrinsic force clamps. As the magnetic field gradient varies on a scale (mm) much larger than the size of the tethering DNA molecules (microns), the force applied to the bead is fixed and the resulting extension of the molecule measured. Other micromanipulation (e.g., optical tweezers) systems are natural extension clamps, distances are imposed and the force is measured. In those systems the application of a constant force requires implementation of a feedback loop [9]. As in all micromanipulation set-ups, the spatial and temporal resolutions of the system are limited by the dissipation, i.e., by the friction of the molecules of the solvent on the system. For a given dissipation, a high spatial resolution (obtain by averaging or filtering the time signal) comes at the expense of a lower temporal resolution, and vice-versa. To increase both the spatial and temporal resolution one needs to reduce the dissipation, i.e. the size of the bead and DNA. However as the maximal force one can apply on a bead depends on its volume, there is a trade-off between the force one can apply on a DNA molecule and the ensuing dissipation: the larger the maximal force, the larger the bead and the larger the dissipation. In practice, some of these numbers are set by the enzymatic system studied. Thus the typical forces one needs to apply on enzymatic systems are a few pN, which with our present set-up can be achieved with Dynal (Myone) beads of 1 μm diameter. The typical time scales associated with enzymatic reactions are between 0.01 and 1 sec, which on a 1 μm DNA pulled at $F = 1$ pN result in a typical spatial resolution of $\delta x \sim 10$ nm.

A unique feature of magnetic tweezers is that they offer a non-invasive way of twisting DNA by simply rotating the magnets. In this case, as long as the angular rotation is less that about 10 turns/sec, the magnetic bead follows the rotation of the field like the compass of a needle.

6.3 Stretching and Twisting DNA

The double stranded DNA (dsDNA) molecule is a long, double helical polymer. This particular secondary structure confers interesting mechanical properties to the molecule. Understanding these parameters is crucial before studying DNA bound motors as their action might induce changes in the DNA's
extension. Furthermore their activity might be controlled by the torque or the
force on the molecule.

6.3.1 Stretching dsDNA

Like any polymer in solution, dsDNA adopts a random coil geometry that
maximizes the number of accessible conformations and thus the entropy. As
one starts to stretch dsDNA, the molecule tends to align along the force axis,
which decreases the number of accessible conformations. This decrease in entropy dictates the magnitude of the stretching force. This so-called entropic
regime of forces lasts until the molecule is fully aligned (with a single conformation allowed) at a force $F \sim 10$ pN. In that regime the elastic response of
dsDNA is well described as a flexible tube with bending rigidity B (Worm-
Like Chain (WLC) model). At any given force, the extension of the molecule l
is determined by minimizing its free energy (i.e. weighting the loss in bending
energy against the gain in entropy [10, 11]).

The fit of the experimental force-extension data yields a persistence length
(i.e. the typical length over which thermal fluctuations can bend the DNA)
$\xi = B/k_B T = 52$ nm [11] under physiological conditions, see Fig. 6.1b. This
value is high compared to common man-made polymers (with $\xi \sim 1 - 2$ nm)
because the base stacking at the molecule's core stiffens it. Using this value
of ξ, we can estimate the mean radius R_0 of the $E.\ coli$ chromosome: $R_0 = \sqrt{2L\xi} \simeq 10$ μm, (L being the length of the DNA, $\simeq 1$ mm). The comparison
of this value to the typical $E.\ coli$ size of 1 μm suggests that the cell must
have more efficient ways to compact its DNA. This is indeed achieved by
specific DNA-compacting proteins (e.g., condensins) and by supercoiling the
molecule.

6.3.2 Twisting dsDNA

The behavior of DNA under torsion can be explored with magnetic tweezers
by simply rotating the magnets and thereby the bead tethered by a dsDNA.
At low twist, the DNA's extension is not affected significantly by the torsion.
However when the number of turns imposed on the molecule is large enough
it buckles under the torsional load. Additional turns applied to the molecule
make it writhe and form plectonemic loops, Fig. 6.1a or b. Thus its extension decreases linearly with the applied number of turns, n. The partitioning
of the excess linking number ΔLk (a topological constant here equal to n)
between writhe (plectonemes) and twist (change of the helical pitch) plays a

key role in major cellular processes, such as DNA compaction, replication or transcription. The formation of plectonemes is observed symmetrically under positive and negative torsion, as long as the stretching force is held below a (salt dependent) critical force, typically $F \sim 0.5$ pN. However above this force, the critical torque for denaturation is smaller than the critical torque for buckling Γ_c (which increases as $\Gamma_c \sim F^{1/2}$). In this case, a negatively supercoiled DNA will respond to a large unwinding by denaturing rather than by forming plectonemic loops [12].

6.3.3 Stretching ssDNA

The elastic behavior of ssDNA, unlike that of dsDNA, strongly depends on the salt conditions and the chain's nucleotide content. It cannot be fitted by a simple elastic polymer model such as the WLC model. First ssDNA is much more flexible than dsDNA: its persistence length ($\xi_{ssDNA} \sim 0.8$ nm) is about 60 times smaller. Therefore, the electrostatic repulsion between the charged phosphates on its backbone cannot be neglected (as they are for dsDNA), since they are screened over a Debye length that is similar to ξ_{ssDNA}. Second, ssDNA is a somewhat peculiar polymer: because of the possibility of pairing between the bases along its backbone, it is able to form hairpin structures at low forces, which are highly sequence- and salt-dependent, see Fig. 6.1b,c. All these effects can be incorporated (with some approximations) in a Monte Carlo (MC) simulation of a chain under tension. These simulations turn out to nicely describe the behavior of ssDNA over a large range of forces and ionic strengths [7]. Even though a complete theoretical understanding of the elastic behavior of ssDNA is still lacking, the experimental evidence clearly shows that (except near $F \sim 5$ pN) ssDNA and dsDNA have different extensions. As we shall see below this difference can be used to monitor the action of enzymes such as helicases (or DNA-polymerases) that transform dsDNA into ssDNA (or vice-versa).

6.4 Protein Induced DNA Looping

Looping occurs when a protein (or a protein-complex) binds simultaneously two different sites along the DNA, and bends the DNA molecule located between these two protein binding sites. This architectural modification is a very common mechanism utilized by all organisms to regulate transcription. For instance loop formation can physically block the binding of RNA polymerase (RNA-pol) [13,14], or it can block (or promote) the transition from the closed to the open promoter complex during isomerization of the RNA-pol in which the DNA helix is opened [15,16]. Loop formation also plays an important role in other cellular processes including DNA recombination, replication and repair [17–20]. To study loop formation, the force stretching the DNA molecule acts as a convenient thermodynamic parameter [21–23].

We chose here as a model the galactose (gal) operon system, a gene cluster composed of four genes involved in the metabolism of galactose in *E. coli*. It is regulated by a repressor loop encompassing the promoter region, which prevents binding of RNA-pol. The loop is formed by the interaction between a protein (the gal repressor or GalR) and two DNA binding sites for this protein (the operators). The repression induced by this loop also requires the presence of the DNA binding protein HU and a negatively supercoiled DNA [14].

We used magnetic tweezers to finely tune the tension and torsion applied on a single DNA molecule containing the promoter region of the gal operon. At a moderate force ($F = 0.9$ pN), in the presence of the proteins GalR and HU and a DNA molecule negatively supercoiled by at least 3%, we observed that the length of the molecule switched intermittently between two values, see Fig. 6.2, corresponding to the DNA length in the presence and absence of a loop [3].

As expected, higher forces shifted the thermodynamic equilibrium in favor of the unlooped form. From experimental traces like those shown in Fig. 6.2, we determined the mean lifetimes (i.e. inverse of the kinetic constants) as a function of force of the looped (τ_{on}) and unlooped (τ_{off}) DNA conformations [23]. The free energy change ΔG associated with formation of a loop at a given force F and torque Γ is obtained using the relation [23]: $\Delta G = k_B T \log \frac{\tau_{off}}{\tau_{on}} = \Delta G_0 + F\Delta l + \Gamma\Delta\theta$, where ΔG_0 is the free energy of binding at zero force and zero torque, Δl the size of the loop, i.e. the separa-

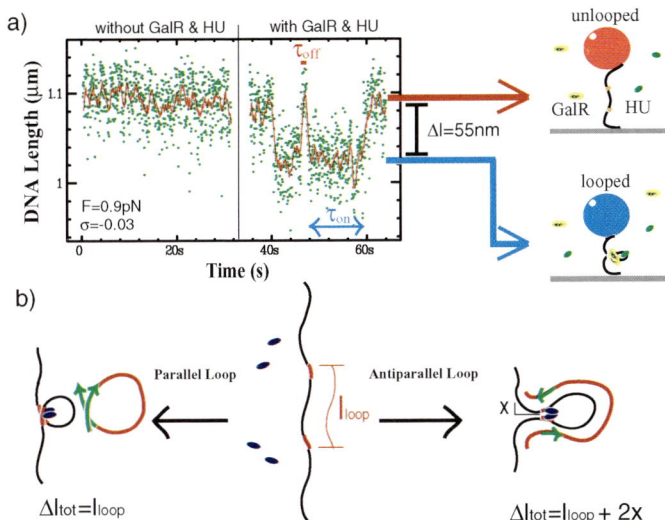

Fig. 6.2. (a) A typical signal observed without proteins and in presence of GalR and HU. The measurement was made by stretching a negatively supercoiled DNA ($\Omega = -3\%$) with a force $F = 0.9$ pN. (b) Parallel and antiparallel loops respond differently to a stretching force

tion between the two free energy minima and $\Delta\theta$ the change in twist between these minima. On a supercoiled molecule that change of twist in the DNA loop must be compensated by an opposite change in the rest of the molecule, i.e. by a change in the number of plectonemes on the DNA. On a supercoiled DNA molecule the last term can thus be written as: $\Gamma\Delta\theta = \pm F l_p \Delta\theta / \pi$, where $l_p \sim 50$ nm is the size of a plectoneme obtained by a change of linking number: $\Delta Lk = 1$ (the $\pm$ signs accounts for the opposite effects expected on $(+)$ and $(-)$scDNA). The observed total change in the DNA's extension is: $\Delta l_{tot} = \Delta l \pm l_p \Delta\theta / \pi$ arising in part from the size of the loop and in part from the compensating extra (positive or negative) writhe in the molecule.

Indeed, the observed decrease in the DNA's extension in presence of GalR and HU ($\Delta l_{tot} \sim 55$ nm) is larger than the distance $l_{rr} = 39$ nm that separates the two GalR binding sites (see Fig. 6.2). This result cannot be explained by the formation of a parallel loop of DNA (see Fig. 6.2a) for which $\Delta\theta = \pi$ (and $\Delta l_{tot} = l_{rr} + l_p > 55$ nm). It is however consistent with an antiparallel loop of DNA for which the DNA segment at the loop entrance is antiparallel to the segment at its exit, see Fig. 6.2b. The fact that $\Delta l_{tot} > l_{rr}$ could be due either to a slight twist of that loop (by $\Delta\theta < 1$ radian) and/or to the need for the two segments to bend gradually in order to align in the direction of stretching. The existence of this antiparallel loop was confirmed in recent work from the Adhya lab [24].

The force $F_0 \sim .88$ pN for which the system has a probability $1/2$ of being in either state (and thus $\Delta G = 0$) allows us to estimate: $\Delta G_0 = -F_0 \Delta l_{tot} \sim -12 k_B T$. It will be interesting to change the distance between the GalR binding sites l_{rr} by a few bps, since this will affect the change in twist $\Delta\theta$ in the loop formed by the operon. That change in twist will result in a variation of Δl_{tot} with the DNA pitch similar to the variation of the J-factor measured in DNA cyclization experiments [25].

6.5 Type II Topoisomerases

Type II topoisomerases (topos) are enzymes responsible for the regulation of DNA topology in the cell [26,27]. They catalyze the ATP-dependent passage of one double-stranded DNA segment (the transport, or (T), segment) through another (the gate, or (G), segment) [26, 28] by creating a transient break in the (G) segment (see Fig. 6.3a). Using this mechanism, they decatenate sister chromosomes during DNA replication but are also known to regulate the level of supercoiling during replication, transcription and recombination (the double-strand passage allows for a change in linking number by two units) [29]. As a consequence, their malfunction at mitosis or meiosis ultimately causes cell death. A particular feature of these enzymes is their ability to decatenate DNA molecules well below the thermodynamic entanglement equilibrium [30]: type II topos systematically shift the partition between knotted and unknotted DNAs towards unknotting. The mechanism by which this local action of type II topos on DNA results in overall unknotting is not fully understood, but from thermodynamic considerations (they are no "Maxwell demon"), it must be linked to the fact that to perform their feat these enzymes consume two ATP molecules per cycle [31]. Coupling between DNA transport and ATP hydrolysis has been investigated for years [26, 28, 32–34], but the detailed mechanochemistry of these fascinating enzymes remains still obscure.

Single Molecule Uncoiling Assay

In this context, single DNA micromanipulation provides a powerful way to track the activity of a single topoisomerase on a stretched supercoiled DNA. Figure 6.3b shows a typical experiment done at a saturating ATP concentration
(1mM) and a stretching force of 1 pN in the presence of the *S. cerevisiae* type II topo (as described in [35]). Mechanical coiling of the DNA leads to a noticeable reduction in the DNA's extension, due to the formation of plectonemic loops (see Fig. 6.1b) and sketch in Fig. 6.3b. At a sufficiently low enzyme concentration ($\simeq 100$ pM), there is a long waiting time between DNA winding and the binding and uncoiling of the molecule, leading to an increase in extension. This waiting time (longer than the relaxation time) ensures that plectoneme removal is catalyzed by a single enzyme. Interestingly, the enzyme acts processively so that all the plectonemes are removed within a single burst of activity. From this experiment, one can extract the enzyme turnover rate v, knowing that 2 links (i.e. turns) are removed per enzymatic cycle: $v \simeq 2.5$ cycles/s. By reducing the ATP concentration, Strick et al. could slow down the enzyme and thus resolve each cycle of activity and observe a single enzymatic cycle by a single enzyme [35]!

Investigating the Mechanism of Type II Topoisomerases

The versatility of the micromanipulation setup allows one to generate (+) supercoils as well as (−) supercoils [36] and thus to test the activity of type II topos on different substrates. It has been shown that Topo IV, one of the two type II topos found in prokaryotes, relaxes (+) supercoils much more efficiently than (−) ones [37], which raises the question: what is the mechanism of topological discrimination? Recent single molecule studies have shown that Topo IV displays similar topological selectivity when unlinking two DNA molecules, lending support to a model in which discrimination is achieved through recognition of the crossing angle between DNA segments [38, 39].

6.6 Study of Helicases

DNA helicases use the energy of ATP hydrolysis to catalyze the unwinding of a dsDNA substrate to produce two ssDNA strands. They play an essential role in a number of processes such as DNA replication, recombination and

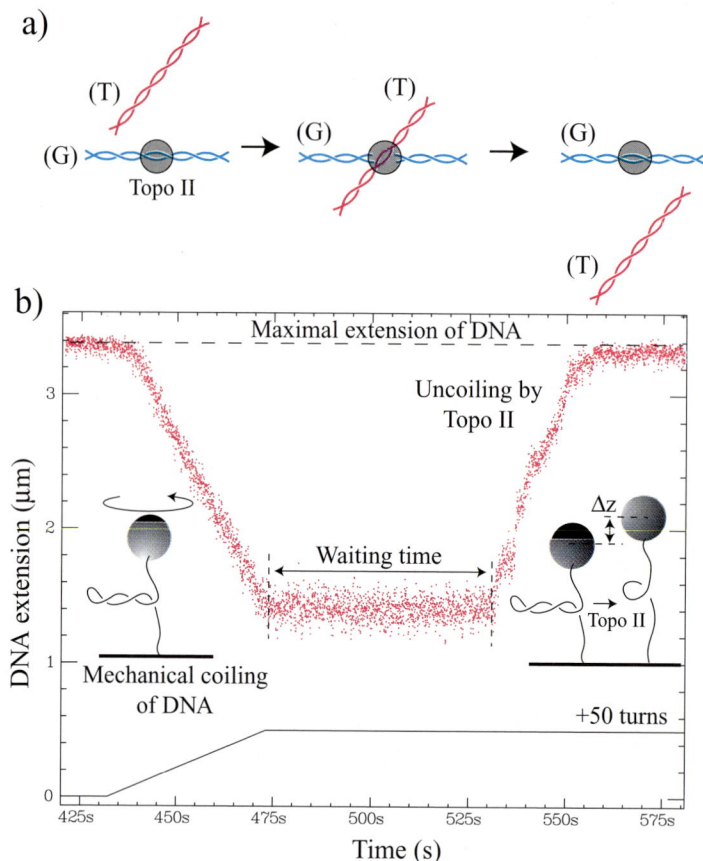

Fig. 6.3. Tracking the activity of Type II topoisomerases using DNA micromanipulation. (**a**) Type II topoisomerases are known to pass a double stranded DNA segment (the transport, T-segment) through another one (the gate, G-segment) by operating a transient break in the latter. Thus, it can decatenate entangled DNA molecules but also relax the supercoiling arising in twisted DNA. (**b**) Typical uncoiling assay at saturating ATP concentrations in the presence of *S. cerevisiae* topo II at $F = 1$ pN. Red points: extension of a twisted DNA molecule as a function of time. By mechanically rotating the magnets (the black straight line indicates the rotation of the magnets), we induce the formation of plectonemic structures in the DNA (its extension drops in the left part of the plot). After a random waiting time, an enzyme binds the DNA and processively uncoils it, thus leading to an increase in the DNA's extension

repair [40]. UvrD is one of the 12 different helicases in *E. coli*. It plays a central role in the mechanisms of DNA repair following UV damage [41]. It initiates DNA unwinding from a nick and then translocates along one strand with a 3′ to 5′ polarity [42,43].

To follow DNA unwinding (see Fig. 6.4) we pull on a nicked dsDNA molecule. As a helicase binds to the DNA and unwinds it, it creates a ssDNA region. As already mentioned, if the stretching force is above 5 pN, transformation of dsDNA into ssDNA results in a lengthening of the molecule. The dissociation of the enzyme from the DNA substrate is followed by rapid rehybridization of the two strands (an event labeled H), leading to the recycling of the DNA substrate. Thus, the time trace of the DNA extension displays separated sawtooth shaped events, each corresponding to the action of a single helicase (Fig. 6.4b). From these bursts it is possible to extract statistical distributions of many relevant parameters, such as the unwinding rate, the duration of unwinding and the number of base pairs unwound (respectively the slope of the rising part of a burst, its duration and its height). At saturating ATP concentrations, UvrD unwinds DNA at a constant, normally distributed, rate of 250 bp/s (see Fig. 6.4c). The helicase activity does not display any pauses. The mean lifetime of the enzymatic complex follows a Poisson distribution, with mean lifetime of ~ 1s.

Unexpectedly, we observed a second type of unwinding event. The ascending part (U for unwinding) of these events is the same as for the events already described: the enzyme binds to its substrate and unwinds it. However, the enzyme then switches strands (instead of dissociating) and translocates backwards on the opposite strand, with the fork closing in its wake (a slowly decreasing event labeled Z for reZipping, see Fig. 6.4d). Strand switching events are unlikely to be sequence-dependent, as they are continuously, Poisson distributed. The UvrD translocation rate for a reZipping event, i.e. when pushed by the ds/ssDNA fork, is only slightly higher than its unwinding rate ($v_Z \sim 300$ bp/s as compared to $v_U \sim 250$ bp/s). This indicates that the enzyme unwinds DNA by translocating along it like a "snow-plough" with most of the energy required for the translocation rather than for the opening of the strands. This observation does not support a passive model in which the enzyme would wait for the junction to breathe, and then trap the opening fluctuation by translocating forwards.

From an analysis of the noise during a U (or Z) event we can deduce the enzymatic step size, i.e. the number of base-pairs opened per enzymatic cycle (about 5 bps) [44]. This analysis uses the fact that the noise during an unwinding (or rezipping) event comes from two sources. First the experimental noise spectrum associated with the Brownian fluctuations of the bead which at low frequencies is white, i.e. frequency independent. Second, the noise spectrum associated with the stochastic stepping (DNA unwinding) of the enzyme which increases at low frequencies as $1/f^2$ and is larger the greater the step size (the larger the number of base-pairs unwound). The different spectral behavior of these two sources of noise permits their deconvolution and the extraction of a value for the enzymatic step size [45].

A remarkable feature of these single molecule data on the activity of UvrD is that they allow us to extract all the kinetic constants that define the transitions between the various U, Z or H states. These rate constants are simply

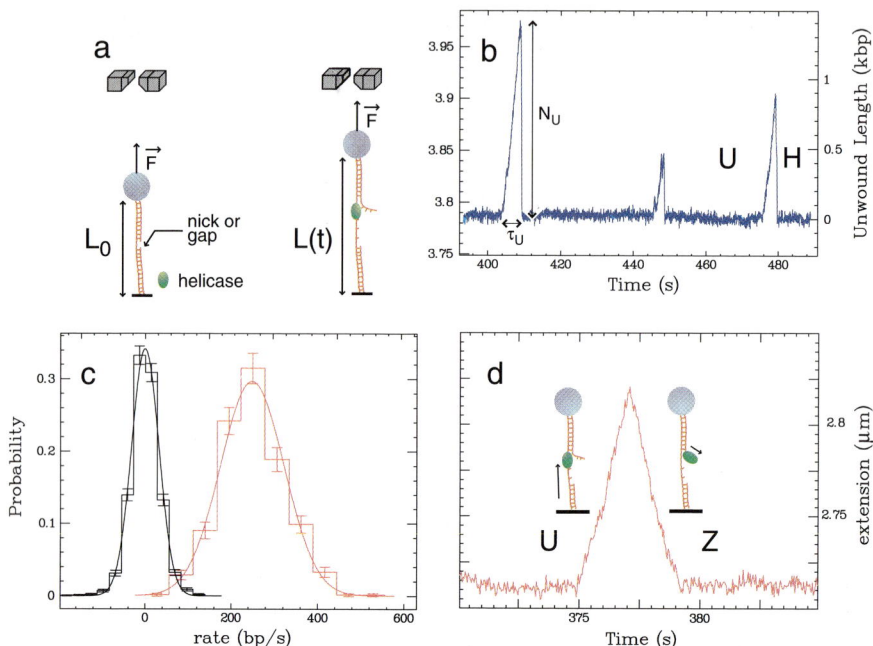

Fig. 6.4. (a) helicase assay setup. Magnets (*gray*) exert an adjustable force F on the DNA (*red*). Transformation of dsDNA into ssDNA as a result of helicase activity induces a lengthening of the molecule ($F = 35$ pN, [ATP] $= 500\,\mu$M for all data). (**b**) Typical sawtooth helicase events. From each burst it is possible to extract the rate V_U, the lifetime τ_U and the number of bp unwound N_U (see text). (**c**) Distribution of the unwinding rate (V_U) of UvrD (red, mean $\langle V_U \rangle = 248 \pm 3$ bp/sand S.D. $\sigma(V_U) = 74$ bp/s) and experimental noise (black, mean $\langle V_N \rangle = -0.6 \pm 0.7$ bp/s and S.D. $\sigma(V_N) = 32$ bp/s). The increased width of the unwinding rate distribution corresponds to additional noise arising from the stochastic stepping of the helicase. (**d**) Typical UvrD strand switching event

related to the probability of going to one state (say Z) from a given state (say U). These probabilities can be determined directly from the data, by simply counting the number of specific events (say a Z-event following a U-event) out of the total number of observed events [44]. This extremely powerful attribute of single molecule experiments, totally absent from bulk assays, has not yet been fully exploited.

6.7 The Fastest Known DNA Translocase: FtsK

The efficient transmission of genetic material from one cell to her daughters is essential for the survival of an organism. Genetic transmission requires both faithful replication of the genome and physical transfer of the replicated chromosomes to the daughter cells. In *E. coli*, the protein FtsK acts

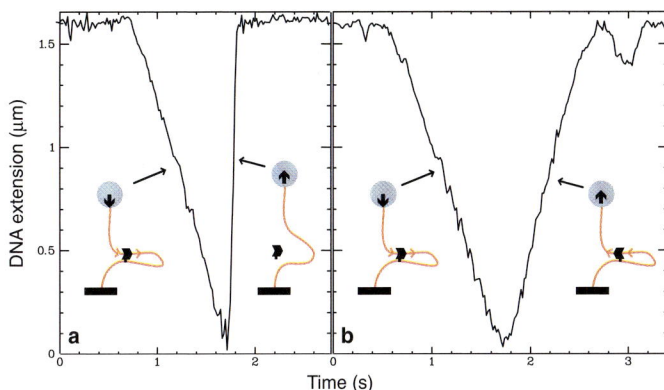

Fig. 6.5. Measurements of DNA extension vs. time in the presence of FtsK exhibits many transient decreases in extension; two such events are highlighted here. Both events were taken in the presence of 1 mM ATP, at $F = 3.6$ pN. (**a**) typical event showing a fast, constant-velocity decrease, corresponding to translocation by the protein and loop formation, followed by protein unbinding and a sudden recovery of the initial extension. (**b**) After causing the initial decrease, FtsK reverses and translocates DNA in the opposite direction, causing a much slower recovery of the initial extension than in (A). In such events, FtsK moves the same section of DNA in both directions. We conclude therefore that DNA sequence is not a strong determinant of the protein's translocation direction

in both chromosome transfer and in cell division. Each activity is associated with a different protein domain. The N-terminal domain of FtsK is bound to the cell membrane at the septum, and is essential for cell division. The C-terminal domain is an active DNA translocase: it hydrolyzes ATP in order to transfer DNA within the cell. It has been shown that DNA pumping by the C-terminal domain is important in the final stages of chromosome segregation. We have studied FtsK's action both on DNA molecules that are not torsionaly constrained (i.e. nicked) and those that are constrained (unnicked). In order to avoid the complications of supercoiling, our initial experiments utilized nicked molecules. With those molecules, immediately upon protein addition, we observed large transient decreases in the DNA's extension, as shown in Fig. 6.5. Analysis of these events revealed an initial ATP-dependent constant-velocity decrease in extension, followed by a fast (ATP-independent) recovery back to the initial extension; the speed of this recovery is limited only by the drag on the bead. We have identified the velocity of decrease in extension as the velocity of translocation by FtsK as it extrudes a loop of DNA (see Fig. 6.5). FtsK is an incredibly fast motor: it can translocate DNA at speeds up to 2µm/s, making it the fastest DNA motor protein yet discovered. Strikingly, it can also suddenly change its translocation direction. This can be seen in Fig. 6.5, where recovery of the initial DNA extension occurs at an ATP-dependent rate similar to the initial decrease, indicating that the motor is translocating but in the opposite direction, since the bead is rising. When utilizing unnicked DNA molecules, we are able to record the twist induced by FtsK into a DNA molecule; this property that can be easily quantified with magnetic tweezers. This measurement gives insight into the mechanism of FtsKs interaction with DNA, particularly the amount of relative rotation between FtsK and DNA during translocation. We have found that, unlike all previously studied DNA translocases, FtsK does not track DNAs helical pitch; that is, it does not induce a rotation of one turn for every DNA helical pitch (10.5 bases pairs) translocated, but instead a surprisingly low value of one turn for every 143 bp [46]. This value is almost exactly the opposite of the torsion already present in the *E. coli* genome. We proposed that this is not a coincidence but rather the most efficient way for FtsK to not disturb the torsional state of the genome during chromosome transfer. Another characteristic of this motor is the long distances it can travel: ∼3.4 μm (i.e. about 10 kbp) at $F = 0$ pN. FtsK's processivity decreases quickly with force; based on this, we proposed [47] a mechanism by which force might be used to direct FtsK's activity onto misplaced sections of the genome: in vivo, misplaced loops of the genome will not be bound by any protein, while the correctly-placed sections, in the center of the daughter cell, will be subject to the activity of next generation polymerases, condensins, and other proteins. If FtsK attempts to transport such a section, the bound protein will resist, creating a tension in the DNA. Since this tension is analogous to the force we apply in the measurement, existing bound protein on a section of chromosome will decrease FtsK's processivity, disallowing the transport of that section. Therefore, only those sections of the genome that are free of protein, i.e. those that are misplaced, will be efficiently dispatched to their correct location. This model may explain the in vivo role of FtsK in segregation of normal chromosomes. However, FtsK is also involved in resolving chromosome dimers that sometimes appear in the *E. coli* cell cycle. Genetic studies had demonstrated that an interaction of FtsK with short, polar DNA sequences is involved in the repair of dimers. In our first measurements a sequence having a polarizing effect in vivo showed no special effect on FtsK translocation [47]. Nevertheless, in a separate experiment Pease et al., [48] by directly observing the motion of an aggregate of FtsK, showed a direct effect of DNA sequence on the direction of translocation. A possible explanation for these apparently contradictory results was proposed recently [49]. Recent genetic studies led to the identification of the polarising sequences as an 8 bp motif named KOPS (FtsK Orienting Polar sequences). The in vitro use of KOPS in single molecule and bulk measurements showed a strong effect on FtsK translocation only when 3 consecutive KOPS were contained in the DNA molecule. Such a need for consecutive sequences was already demonstrated in the case of another translocase RecBCD [50]. It was then proposed that the reading of a single KOPS sequence is stochastic [47, 49] and does not have an all or none effect. The strong sequence effect observed by Pease et al. might be due to the presence of an aggregate of protein: multiple motors in the aggregate could simultaneously read the sequence, and thus have a much higher probability of stopping even on a single KOPS.

6.8 Conclusion

In this paper we have seen how one of the new tools of single molecule manipulation (magnetic tweezers) can be applied to the study of a broad range of DNA binding proteins and enzymes. The level of precision and detail of these experiments is often unmatched by bulk assays. These techniques allow one to generalize to a large class of enzymes the approach applied successfully and for many years to the study of ion channels.

We have hereby only described one aspect of recent single molecule studies, those involving their manipulation. Another equally exciting and powerful approach is using single molecule fluorescence and fluorescence resonant energy transfer (FRET) to monitor the motion of a single enzyme on its substrate [51] and its internal conformational changes [52] with nanometer accuracy and millisecond time resolution [53].

Combining both approaches [54, 55] will provide new opportunities for single molecule enzymology, where all (or most) of the parameters characteristic of a single molecular motor will be measured simultaneously (its rate, processivity, step-size, work done and number of ATP molecules consumed per cycle [56]). The characterization of the dynamic feature of an enzyme together with its static crystallographic data should provide an (almost) complete picture of its mechanism.

Acknowledgments

We thank our collaborators on the presented works: T.R. Strick, M.-N. Dessinges, B. Maier, X.-G. Xi, F.-X. Barre, L. Finzi, D. Dunlapp, S. Adhya, D. Lewis, N. Cozzarelli, N. Crisona, M. Peliti and Y. Zhang.

This work was supported by grants from CNRS, "ACI jeune chercheur", ARC and the EEC under the "MolSwitch" program. KCN was supported by a fellowship from the Human Fontier Science Program.

References

1. C. Bustamante, S.B. Smith, J. Liphardt, and D. Smith (2000). *Curr. Op. Structural Biology*, **10**, pp. 279–285.
2. T.R. Strick, M.-N. Dessinges, G. Charvin, N.H. Dekker, J.-F. Allemand, D. Bensimon, and V. Croquette (2003). Stretching of macromolecules and proteins. *Rep. Prog. Phys.*, **66**, pp. 1–45.

3. G. Lia, D. Bensimon, V. Croquette, J.F. Allemand, D. Dunlap, D.E.A. Lewis, S. Adhya, and L. Finzi (2003). Supercoiling and denaturation in gal repressor-heat unstable nucleoid protein (hu)-mediated dna looping. *Proc. Natl. Acad. Sci. (USA)*, **100**, pp. 11373–11377.

4. J. Liphardt, B. Onoa, S.B. Smith, I. Tinoco Jr., and C. Bustamante (2001). Reversible unfolding of single RNA molecules by mechanical force. *Science*, **292**, pp. 733–737.

5. J.F. Marko and E.D. Siggia. Driving proteins off DNA using applied tension (1997). *Biophys. J.*, **73**, pp. 2173–2178.

6. J.-F. Allemand, D. Bensimon, R. Lavery, and V. Croquette. Stretched and overwound DNA form a Pauling-like structure with exposed bases (1998). *Proc. Natl. Acad. Sci. USA*, **95**, pp. 14152–14157.

7. M.-N. Dessinges, B. Maier, Y. Zhang, M. Peliti, D. Bensimon, and V. Croquette (2002). Stretching ssdna, a model polyelectrolyte. *Phys. Rev. Lett.*, **89**, pp. 248102.

8. J.-F. Allemand, T. Strick, V. Croquette, and D. Bensimon (2000). Twisting and stretching single dna molecules. *Prog. Biophys. Molec. Biol.*, **74**, pp. 115–140.

9. F. Gittes and C.F. Schmidt CF (1998). Signals and noise in micromechanical measurements. *Methods in Cell Biology*, **55**, pp. 129–156.

10. J.F. Marko and E. Siggia (1995). Statistical mechanics of supercoiled DNA. *Phys. Rev. E*, **52**(3), pp. 2912–2938.

11. C. Bouchiat, M.D. Wang, S.M. Block, J.-F. Allemand, T.R. Strick, and V. Croquette (1999). Estimating the persitence length of a worm-like chain molecule from force-extension measurements. *Biophys. J.*, **76**, pp. 409–413.

12. T. Strick, J.-F. Allemand, D. Bensimon, and V. Croquette (1998). The behavior of supercoiled DNA. *Biophys. J.*, **74**, pp. 2016–2028.

13. H. Kramer, M. Niemoller, M. Amouyal, B. Revet, B. von Wilcken-Bergmann, and B. Muller-Hill (1987). Lac repressor forms loops with linear dna carrying two suitably spaced lac operators. *EMBO Journal*, **6**, pp. 1481–1491.

14. N. Mandal, W. Su, R. Haber, S. Adhya, and H. Echols (1990). Dna looping in cellular repression of transcription of the galactose operon. *Genes and Development*, **4**, pp. 410–418.

15. A.K. Vershon, S.M. Liao, W.R. McClure, and R.T. Sauer (1987). Interaction of the bacteriophage p22 arc repressor with operator dna. *J. Mol. Biol.*, **195**, pp. 323–331.

16. J.P. Hunt and B. Magasanik (1985). Transcription of glna by purified escherichia coli components: Core rna polymerase and products of glnf, glng and glnl. *Proc. Natl. Acad. Sci. (USA)*, **85**, pp. 8453–8457.

17. T. Schlick and W.K. Olson (1992). Supercoiled DNA energetics and dynamics by computer simulation. *J. Mol. Biol.*, **223**, pp. 1089–1119.

18. R. Tjian and T. Maniatis (1994). Transcriptional activation: a complex puzzle with few easy pieces. *Cell*, **77**, pp. 5–8.

19. D. Ristic, C. Wyman, C. Paulusma, and R. Kanaar (2001). The architecture of the human rad54-dna complex provides evidence for protein translocation along dna. *Proc. Natl. Acad. Sci. (USA)*, **98**, pp. 8454–8460.

20. S. Pathania, M. Jayaram, and R.M. Harshey (2002). Path of dna within the mu transposome. transposase interactions bridging two mu ends and the enhancer trap five dna supercoils. *Cell*, **109**, pp. 425–436.

21. G.I. Bell (1978). *Science*, **200**, pp. 618–627.

22. D. Bensimon (1996). Force: a new structural control parameter? *Structure*, **4**, pp. 885–889.
23. I. Tinoco Jr. and C. Bustamante (2002). *Biophys. Chem.*, **101-102**, pp. 513–533.
24. K. Virnik, Y.L. Lyubchenko, M.A. Karymov, P. Dahlgren, M.Y. Tolstorukov, S. Semsey, V.B. Zhurkin, and S. Adhya (2003). "antiparallel" dna loop in gal repressosome visualized by atomic force microscopy. *J. Mol. Biol.*, **334**, pp. 53–63.
25. D. Shore and R.L. Baldwin (1983). *J. Mol. Biol.*, **170**, pp. 957–981.
26. J. Roca and J.C. Wang (1994). DNA transport by a type II DNA topoisomerase: evidence in favor of a two-gate model. *Cell*, **77**, pp. 609–616.
27. J. J. Champoux (2001). DNA topoisomerases: structure, function, and mechanism. *Annu. Rev. Biochem.*, **70**, pp. 369–413.
28. J. Roca and J.C. Wang (1992). The capture of a DNA double helix by an ATP-dependent protein clamp: a key step in DNA transport by type II DNA topoisomerase. *Cell.*, **71**, pp. 833–840.
29. J. C. Wang (2002). Cellular roles of DNA topoisomerases: a molecular perspective. *Nat. Rev. Mol. Cell. Biol.*, **3**(6), pp. 430–40.
30. V.V. Rybenkov, C. Ullsperger, A.V. Vologodskii, and N.R. Cozzarelli (1997). Simplification of DNA topology below equilibrium values by type II topoisomerases. *Science*, **277**, pp. 690–693.
31. J.E. Lindsley and J.C. Wang (1993). On the coupling betzeen ATP usage and DNA transport by yeast DNA topoisomerase II. *J. Biol. Chem.*, **268**, pp. 8096–8104.
32. T.T. Harkins and J.E. Lindsley (1998). Pre-steady-state analysis of ATP hydrolysis by *saccharomyces cerevisiae* DNA topoisomerase II. 1. A DNA-dependent burst in ATP hydrolysis. *Biochemistry*, **37**, pp. 7292–7298.
33. T.T. Harkins, T.J. Lewis, and J.E. Lindsley (1998). Pre-steady-state analysis of ATP hydrolysis by *saccharomyces cerevisiae* DNA topoisomerase II. 2. Kinetic mechanism for the sequential hydrolysis of two ATP. *Biochemistry*, **37**, pp. 7299–7312.
34. C.L. Baird, T.T. Harkins, S.K. Morris, and J.E. Lindsley (1999). Topoisomerase II drives DNA transport by hydrolyzing one ATP. *Proc. Natl. Acad. Sci. USA*, **96**(24), pp. 13685–90.
35. T.R. Strick, V. Croquette, and D. Bensimon (2000). Single-molecule analysis of DNA uncoiling by a type II topoisomerase. *Nature*, **404**, pp. 901–904.
36. T. Strick, J.F. Allemand, D. Bensimon, A. Bensimon, and V. Croquette (1996). The elasticity of a single supercoiled DNA molecule. *Science*, **271**, pp. 1835–1837.
37. N. Crisona, T.R. Strick, D. Bensimon, V. Croquette, and N. Cozzarelli (2000). Preferential relaxation of positively supercoiled DNA by E.coli topoisomerase VI in single-molecule and ensemble measurements. *Genes & Developement*, **14**, pp. 2881–2892.
38. M.D. Stone, Z. Bryant, N.J. Crisona, S.B. Smith, A. Vologodskii, C. Bustamante, and N. R. Cozzarelli (2003). Chirality sensing by escherichia coli topoisomerase IV and the mechanism of type II topoisomerases. *Proc. Natl. Acad. Sci. USA*, **100**(15), pp. 8654–9.
39. G. Charvin, V. Croquette, and D. Bensimon (2003). Single molecule study of dna unlinking by eukaryotic and prokaryotic type II topoisomerases. *PNAS*, **100**(17), pp. 9820–9825.

40. T.M. Lohman and K.P. Bjornson (1996). Mechanisms of helicase-catalysed unwinding. *Annu. Rev. Biochem.*, **65**, pp. 169–214.
41. A. Sancar (1994). Mechanisms of dna excision-repair. *Science*, **266**, pp. 1954–1956.
42. G.T. Runyon and T.M. Lohman (1989). *Escherichia Coli* helicase ii (uvrd) protein can completely unwind fully duplex linear and nicked circular dna. *J. Biol. Chem.*, **264**, pp. 17502–17512.
43. S.W. Matson (1986). *Escherichia Coli* helicase ii (uvrd gene product) translocates unidirectionnaly in a 3' to 5' direction. *J. Biol. Chem.*, **261**, pp. 10169–10175.
44. M.-N. Dessinges, T. Lionnet, X. Xi, D. Bensimon, and V. Croquette (2004). Single molecule assay reveals strand switching and enhanced processivity of uvrd. *Proc. Nat. Acad. USA*, **101**, pp. 6439–6444.
45. G. Charvin, V. Croquette, and D. Bensimon (2002). On the relation betwen noise spectra and the distribution of time between steps for single molecular motors. *Single Molecule*, **3**(1), pp. 43–48.
46. O.A. Saleh, S. Bigot, F.-X. Barre, and J.-F. Allemand (2005). Analysis of DNA supercoil induction by FtsK indicates translocation without groove-tracking *Nat. Struct. Mol. Biol.*, **12**, 436–440.
47. O.A. Saleh, C. Perals, F.-X. Barre, and J.-F. Allemand (2004). Fast, DNA-sequence independent translocation by ftsk in a single-molecule experiment. *EMBO J.*, **23**, pp. 2430–2439.
48. P.J. Pease, O. Levy, G.J. Cost, J. Gore, J.L. Ptacin, D. Sherratt, C. Bustamante, and N.R. Cozzarelli (2005). Sequence-directed DNA translocation by purified FtsK. *Science*, **307**, pp. 586–590.
49. S. Bigot, O.A. Saleh, C. Lesterlin, C. Pages, M.El Karoui, C. Dennis, M. Grigoriev, J.-F. Allemand, F.-X. Barre, and F. Cornet. KOPS: DNA motifs that control *E. coli* chromosome segregation by orienting the FtsK translocase. *EMBO J., in press*
50. M. Spies, P.R. Bianco, M.S. Dillingham, N. Handa, R.J. Baskin, S.C. Kowalczykowski (2003). A molecular throttle: the recombination hotspot chi controls DNA translocation by the RecBCD helicase. *Cell*, **114**, pp. 647–654.
51. A. Yildiz, J.N. Forkey, S.A. McKinney, T. Ha, Y.E. Goldman, and P.R. Selvin (2003). Myosin v walks hand-over-hand: single fluorophore imaging with 1.5 nm localization. *Science*, **300**, pp. 2061–2065.
52. X. Zhuang, L.E. Bartley, H.P. Babcock, R. Russel, T. Ha, D. Herschlag, and S. Chu (2000). *Science*, **288**, pp. 2048–2051.
53. S. Weiss (1999). Fluorescence spectroscopy of single biomolecules. *Science*, **283**, pp. 1676–1683.
54. T. Funatsu, Y. Harada, M. Tokunaga, K. Saito, and T. Yanagida (1995). Imaging of single fluorescent molecules and individual ATP turnovers by single myosin molecules in aqueous solution. *Nature*, **374**, pp. 555–559.
55. M.J. Lang, P.M. Fordyce, A.M. Engh, K.C. Neuman, and S.M. Block (2004). Simultaneous, coincident optical trapping and single-molecule fluorescence. *Nat. Methods*, **1**(2), pp. 133–139.
56. A. Ishijima, H. Kojima, T. Funatsu, M. Tokunaga, H. Higuchi, H. Tanaka, and T. Yanagida (1998). Simultaneous observation of individual ATPase and mechanical events by a single myosin molecule during interaction with actin. *Cell*, **92**, pp. 161–171.

7

Membrane Nanotubes

I. Derényi[1], G. Koster[2,3], M.M. van Duijn[4], A. Czövek[1], and M. Dogterom[3], and J. Prost[2,5]

[1] Department of Biological Physics, Eötvös University, Pázmány P. stny. 1A,
H-1117 Budapest, Hungary
derenyi@angel.elte.hu
[2] Institut Curie, UMR 168, 26 rue d'Ulm, 75248 Paris Cédex 05, France
gerbrand.koster@curie.fr
[3] FOM Institute for Atomic and Molecular Physics (AMOLF), Kruislaan 407,
1098 SJ Amsterdam, The Netherlands
dogterom@amolf.nl
[4] Department of Bioengineering, University of California Berkeley, Berkeley, CA
94720-1762
vanduijn@berkeley.edu
[5] ESPCI, 10 rue Vauquelin, 75231 Paris Cédex 05, France
jacques.prost@curie.fr

Abstract. There is a growing pool of evidence showing the biological importance of membrane nanotubes (with diameter of a few tens of nanometers and length upto tens of microns) in various intra- and intercellular transport processes. These ubiquitous structures are often formed from flat membranes by highly localized forces generated by either the pulling of motor proteins or the pushing of polymerizing cytoskeletal filaments. In this chapter we give an overview of the theory of membrane nanotubes, their biological relevance, and the most recent experiments designed for the study of their formation and dynamics. We also discuss the effect of membrane proteins or lipid composition on the shape of the tubes, and the effect of antagonistic motor proteins on tube formation.

7.1 Introduction

Eukaryotic cells are typically a few microns to a few tens of microns in size. Even at these small scales, there is a clear organization of spatially and functionally separated compartments (organelles) for different cellular functions: the nucleus for coding and storing the genetic information, the endoplasmic reticulum (ER) for the synthesis of proteins and lipids, the Golgi apparatus for the sorting of proteins according to their destination, the mitochondria for ATP production, or the chloroplasts for photosynthesis. A much-simplified sketch of a eukaryotic cell is presented in Fig. 7.1. The cells, and the organelles within them, are bounded and separated from the rest of the world

I. Derényi et al.: *Membrane Nanotubes*, Lect. Notes Phys. **711**, 141–159 (2007)
DOI 10.1007/3-540-49522-3_7

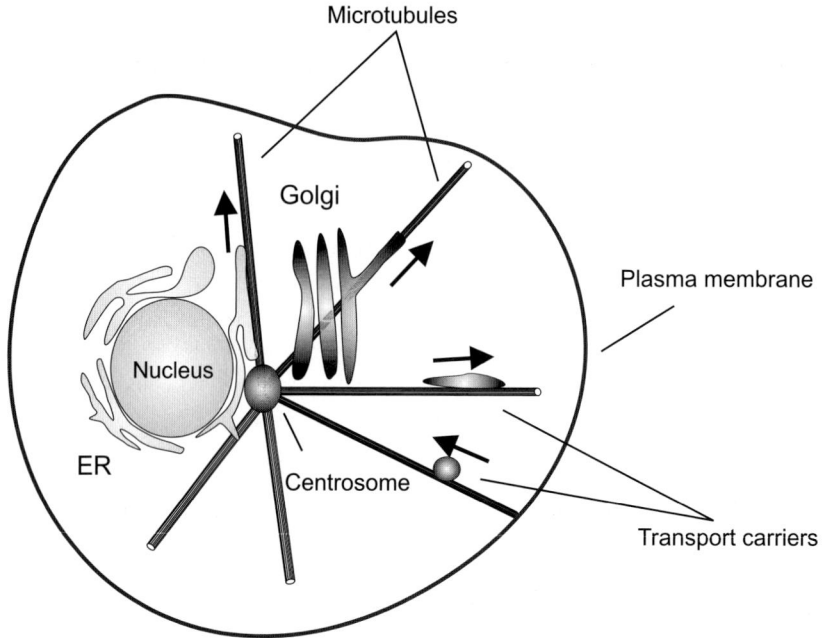

Fig. 7.1. Sketch of the internal organization of a eukaryotic cell. The *arrows* indicate the movement of membrane compartments along the microtubule cytoskeleton

by membranes. These membranes are made up of a mixture of different types of lipids that form a bilayer, and also contain a large number of membrane proteins.

7.1.1 In vivo Occurrences of Membrane Tubes

One important mechanism through which different compartments are shaped and spatially distributed is the action of motor proteins and the cytoskeleton. The cytoskeleton forms a dense network of tracks throughout the cell, and functions as an infrastructure for the movement of motor proteins that pull on the membrane compartments. This results either in the movement of the compartments through the cell or, if opposing forces are present, in their deformation, such as the formation and elongation of membrane tubes (also known as tethers), which are only a few tens of nanometers in diameter, but can reach tens of microns in length.

Motor proteins convert the chemical energy of ATP or GTP into mechanical work, and can typically generate forces up to 6 pN (the stall force [1]). Based on sequence homology, there are three main families of motors: kinesins, dyneins, and myosins [2]. Kinesin and dynein move along microtubules (MTs), whereas myosin along actin filaments. MTs grow by the polymerization of 8-nm-long tubulin dimers, which gives them a natural periodic and asymmetric structure. Motors recognize this asymmetry: kinesin moves towards the plus-end, while dyneins move towards the minus-end of the MT in 8-nm steps [3]. There are, however, exceptions: the kinesin-family protein ncd [4] moves in the minus-end direction. While the details of the movement of individual motors begin to be unraveled [2, 5–8], it is clear that dimerization (or oligomerization) of motor proteins is crucial to their processivity (i.e., the number of steps taken before dissociation from the MT).

Conventional kinesin e.g., which is a dimer, takes on average ∼100 steps of 8 nm, and it can move at speeds up to 1 µm/s [2]. When an external force is applied to kinesin (e.g., with optical tweezers or through cargo), the speed of the motor decreases until it stops moving at the stall force and, at the same time, the average number of steps before detachment decreases. Dynein, the dominant motor for minus-end directed movement along MTs, consists of a complex of proteins [9], making it difficult to work with for in vitro studies (genetic modification, expression, and purification). Ncd has been shown to be a non-processive motor: upon each contact with a MT it moves ∼9 nm and subsequently detaches [10]. The speed at which ncd can move MTs in gliding assay [2] has been measured to be 0.1–0.15 µm/s [11].

The different organelles in cells have characteristic shapes, which are dynamic in the sense that they are constantly being remodeled and deformed. From our point of view, in this review, the most interesting ones are the endoplasmic reticulum (ER) and the Golgi apparatus. The ER is often described as consisting of two different parts: the rough ER and the smooth ER. The rough part consists of flat sacs covered with ribosomes, whereas the smooth part consists of a network of interconnected membrane tubes. These tubes give the ER its characteristic appearance of a net-like labyrinth, which colocalizes with the MT cytoskeleton (see Fig. 7.2, [12, 13]).

In the smooth ER, new tubes are continuously being formed and existing ones disappear [14] by the action of motor proteins that move along MTs [15]. The importance of motors and the cytoskeleton is demonstrated by experiments in which the expression of kinesin is suppressed [16] or the MTs are depolymerized [17]. In both cases the tubular membrane network retracts towards the cell center and no tubes are being formed anymore. Tubular networks can also be observed in cell extracts, providing more insight into the relevant processes involved [18–20]. Such experiments have recently allowed for the determination of the forces required to form tubes from Golgi and ER membranes [21].

The Golgi apparatus is often characterized as a stack of flattened membrane sacs. Like the ER, the Golgi apparatus is also a dynamic organelle. On the cis side membranous cargo carriers that arrive from the ER fuse with the Golgi membrane, while on the trans side tubulovesicular membrane compartments form [22] and pinch off for further transport [23]. Motor proteins that move along MTs have been suggested to form and extend these tubes.

In addition to the shaping of larger organelles, motor proteins and the cytoskeleton are essential for intracellular transport as well. The compartmen-

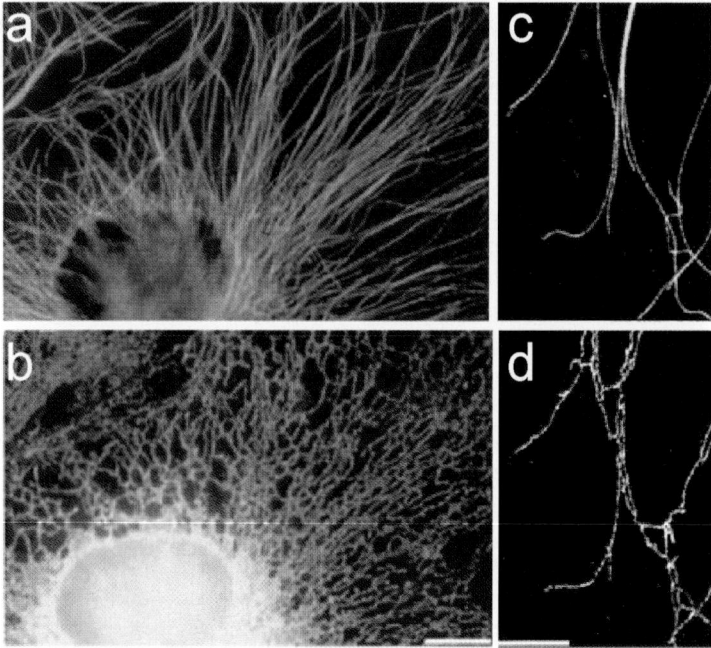

Fig. 7.2. Fluorescently labeled microtubules (**a**) and endoplasmic reticulum (**b**) are distributed in vivo throughout the cell, and show a close colocalization (**c** and **d**). Adapted from [12]. The bar in (b) is 10 µm and in (d) is 5 µm

talization of the cell requires the movement of material between the different organelles. Cargo carriers for intracellular transport are small membrane compartments. Historically, it was thought that they had a spherical shape, and were around 100 nm in size. Recent advances in microscopy, especially the specific fluorescent labeling of proteins (GFP technology [24]) have led to the observation that transport carriers in fact have many different shapes. For example, large parts of tubes formed from the Golgi apparatus are cleaved off at once, and subsequently transported [23, 25]. This process of cleavage, the correct movement to the target organelle and the subsequent fusion are intricate processes themselves that require the activity and assembly of protein complexes and cofactors on the membrane [26, 27].

Motor proteins are not the only molecules that can generate localized forces and pull out tubes from membranes. When polymerizing cytoskeletal filaments hit a flat membrane, then they can also generate tubes by pushing the membrane further. Polymerization forces can potentially even be larger (∼50 pN for MTs [28]) than the maximal force that a single motor protein can exert. Cilia and flagella [29] are such MT-based protrusions of the plasma membrane (cell membrane). They are responsible for the movement of a variety of eukaryotic cells. In addition, filopodia [29], which are exploratory motile

structures that form and retract with great speed, are generated by rapid local growth of actin filaments that push out the plasma membrane. Adhesive
fingers between cells can also be formed by actin polymerization [30]. Recent experiments show that intercellular nanotube networks can be generated
via actin polymerization (in rat neural and kidney cells [31], and also in human immune cells [32]) or via transient cell-cell contact (in human immune
cells [33]), and that these networks provide a novel mechanism for intercellular
communication.

Even though the important role of motor proteins and the cytoskeleton for
membrane tube formation is well-established, it should be noted that there are
other mechanisms through which curvature may be imposed on membranes
that result in shape changes. One may for example think of the assembly of
a protein coat with an intrinsic curvature on the membrane, or proteins or
lipids that change the local composition of one of the monolayers [34,35].

7.1.2 In Vitro Experiments

For a better understanding of the relevant physics involved in tube formation
and to measure the elastic properties of membranes, several experimental
techniques have been developed for pulling nanotubes from cell membranes
and synthetic vesicles. Historically, membrane nanotubes (tethers) were first
observed to be formed from red blood cells attached to glass surfaces and
subjected to hydrodynamic flows [36]. This was later followed by other hydrodynamic flow experiments [37,38], and the application of small beads that
are attached to the membranes and manipulated mechanically [39–41] or via
optical and magnetic tweezers [21,42–48]. Very recently, motor proteins and
polymerizing MTs have been used to form membrane tubes from artificial
vesicles [47,49–51]. Sheetz and co-workers have also shown that tubes can be
extracted from neuronal growth cones and other cells with optical tweezers
and were able to measure the extrusion force as a function of length [52–54].
In all cases, tethers were shown to be mainly membranous, i.e., devoid of
cytoskeleton [55,56].

In addition to the studies of individual nanotubes, networks of tubes between membrane vesicles have also been built for biotechnological applications [57–61]. Fluid in such tubes can be transported via surface tension difference: Marangoni flow drives the membrane and hence the fluid inside the
connecting tubes towards vesicles of higher tension, while Poiseuille flow (induced by Laplace pressure) occurs in the opposite direction [58,62]. For giant
vesicles the Marangoni flow dominates. When more than one tubes is pulled
from the same vesicle, then they tend to attract each other and coalesce [63].
Such coalescence has been observed experimentally [57,61] and also used to
measure the elastic properties of membranes [64].

7.2 Theory of Membrane Tubes

The basic question concerning membrane tubes is: why they form in the first place. When a largely flat piece of membrane is grabbed at a point (by a bead, motor protein, or a cytoskeletal filament) and pulled away, one would naturally expect the formation of some cone-like object rather than a narrow tube. An illustrative answer is that because membranes are usually under tension, they try to reduce their surface. The minimum surface area configuration is reached when the membrane is retracted to its original flat conformation and becomes connected to the point of pulling by an infinitesimally narrow tether (having practically zero surface area). As the membrane shrinks towards this tether, however, its curvature increases. And because membranes do not favor large curvatures, the bending rigidity will eventually prevent the membrane from collapsing entirely into such an infinitesimally narrow tether. The result is a narrow tube, the radius of which is set by the balance between the surface tension and the bending rigidity.

7.2.1 Free Energy of Membranes

To calculate the radius of membrane tubes and also to study tube formation, let us turn to the elastic theory of two-dimensional liquid bilayers. A general theoretical framework has been developed for the last three decades [65, 66]. In the earliest description [67, 68], known as the spontaneous curvature (SC) model, the membrane is treated as a thin sheet and locally characterized by its mean curvature

$$H = \frac{1}{2} \left(\frac{1}{R_1} + \frac{1}{R_2} \right), \tag{7.1}$$

where R_1 and R_2 are the two principal radii of curvature, as illustrated in Fig. 7.3a. The energy of the membrane, the so-called Helfrich-Canham free energy, is defined as

$$\mathcal{F}_{\text{H-C}} = \int \left[\frac{\kappa}{2} (2H)^2 - \kappa 2 H C_0 \right] \mathrm{d}A, \tag{7.2}$$

where κ denotes the bending rigidity of the membrane, C_0 is the spontaneous curvature (characterizing the asymmetry of the membrane if either the lipid composition or the surrounding medium is different on the two sides), and the integral goes over the entire surface of the membrane. The equilibrium shape of the membrane can then be found by minimizing this free energy while obeying system-specific constraints and boundary conditions.

In case of vesicles, e.g., the surface area A and the enclosed volume V are often considered constants (A_0 and V_0, respectively). These constraints can be taken into account by complementing the free energy with the terms $\sigma A - pV$, where σ and p are Lagrange multipliers. If, in addition to this, the distance L between the two poles of the vesicle along the z direction are also

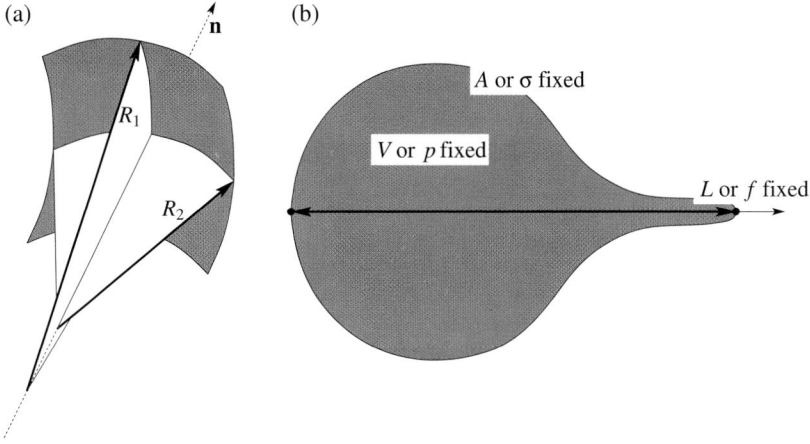

(a)

n

R_1

R_2

(b)

A or σ fixed

V or p fixed

L or f fixed

Fig. 7.3. (**a**) Sketch of a piece of membrane, with the two principal radii of curvature (R_1 and R_2) and the normal vector of the surface (**n**) indicated. (**b**) Either the extensive variables (such as A, V, and L) or their intensive conjugate variables (σ, p, and f) are kept fixed during the minimization of the free energy of the membrane

kept fixed (L_0), then the term $-fL$ should also be added to the free energy, where f is again a Lagrange multiplier (for illustration see Fig. 7.3b). Thus, the function to be minimized becomes:

$$\mathcal{F} = \mathcal{F}_{\mathrm{H-C}} + \sigma A - pV - fL . \tag{7.3}$$

The values of these Lagrange multipliers will have to be chosen such that the corresponding constraints ($A = A_0$, $V = V_0$, and $L = L_0$) be fulfilled. Then the physical meaning of σ, p, and f will become the surface tension of the membrane, the pressure difference between the inside and the outside of the vesicle, and the pulling force at the poles, respectively. This can be easily demonstrated through, e.g., the pulling force: at the minimum of $\mathcal{F}$, any of its first derivatives must be zero, thus, for fixed values of A and V the derivative $\partial \mathcal{F}/\partial L = \partial \mathcal{F}_{\mathrm{H-C}}/\partial L - f$ (representing an infinitesimal length change) is also zero, i.e., $-f = -\partial \mathcal{F}_{\mathrm{H-C}}/\partial L$, which is indeed, by definition, the force exerted by the membrane (or conversely, f is the external force pulling on the membrane).

Here we note that a term proportional to the integral of the Gaussian curvature (which is also quadratic in the curvature, because it is the product of the two principal curvatures) could also be included in the Helfrich-Canham free energy. But since such an integral is a topological invariant, it can be ignored in the energy minimization for vesicles.

If the surface area of the membrane is not constant, either because it is in contact with a lipid reservoir at a fixed chemical potential (as is often the case for biological membranes [44]), or because we are interested in only a small part of the vesicle (as in case of tube formation), then the constraint $A = A_0$

is released, the rest of the world at the boundaries of the membrane can be replaced by a surface tension σ and, thus, the term σA will become a real free energy contribution. Similarly, if instead of volume and length constraints, the conjugate variables (i.e., the pressure difference p and the pulling force f) are controlled, then the $-pV$ and $-fL$ terms will also become real contributions to the free energy (for illustration see also Fig. 7.3b).

Since a membrane is a bilayer (consisting of two monolayers of lipids), the description of a thin sheet is not always adequate. An additional term needs to be incorporated that takes into account the coupling between the two monolayers. When a bilayer is bent, the outer layer is stretched, while the inner monolayer is compressed. This differential stretching is the essence of the area difference elasticity (ADE) model [69], also known as the generalized bilayer-couple model [70] and it contributes an additional term (quadratic in the deviation of the area difference between the two layers ΔA from the preferred value ΔA_0) to the free energy. Although not all experimentally observed shapes and shape transitions can be explained by this model, for most cases the predicted behavior agrees well with the observations [71].

Since for short membrane tubes the free energy contribution of the differential stretching is much smaller than the other terms [72], we will omit it in the rest of this chapter. Similarly, in case of nanotubes the pressure term is also negligible [63, 70, 72, 73], and will be omitted. Thus, for the description of membrane tubes we will consider the following free energy:

$$\mathcal{F} = \int \left[\frac{\kappa}{2} (2H)^2 - \kappa 2HC_0 + \sigma \right] \mathrm{d}A - fL \,. \tag{7.4}$$

7.2.2 Effects of Membrane Proteins and Lipid Composition

So far we have considered the membranes to be spatially homogeneous. While this is indeed the case in most experiments with artificial vesicles, real biological membranes are usually made up of various different kinds of lipids and they also contain membrane proteins. If the different lipids and proteins have different elastic properties (e.g., contribute differently to membrane rigidity) then, as their distributions couple to the local membrane geometry, they can become inhomogeneously distributed or even cause shape instabilities (budding, pearling, fission, etc.) [74–80]. On top of this, the different molecular species, due to the different interactions between them, can even phase separate into distinct domains [81–83]. Such inhomogeneities and domain formations have often been observed experimentally (for examples in relation to membrane tubes see [35, 84–87]), and might be highly relevant in biological processes (such as protein and lipid sorting, or vesiculation).

As the simplest case, let us consider a two-component membrane and expand the free energy contribution of its local composition upto quadratic order in the mean curvature of the membrane ($2H$) and the concentration Φ of one of the two constituents:

$$\mathcal{F}_{\text{comp}} = \int \left[-\lambda 2H\Phi - \mu\Phi + \frac{\chi}{2}\Phi^2 \right] dA \, , \tag{7.5}$$

where λ is the coupling constant between the curvature and the concentration, μ can be regarded as a chemical potential, and χ as a susceptibility coefficient [74–76, 88]. The first experimental measurements of these parameters in case of a transmembrane protein have been reported very recently in [89].

Now the total free energy, which is the sum of Eqs. (7.4) and (7.5), has to be minimized with respect to both the geometry of the membrane and the concentration distribution Φ. However, since only $\mathcal{F}_{\text{comp}}$ depends on Φ, and only in a quadratic fashion, it alone can be minimized very easily with respect to Φ. The obtained minimum can then be added to Eq. (7.4) to get the effective total free energy, which thus depends only on the membrane geometry. It turns out that this effective free energy has the same functional form as Eq. (7.4), in other words, the presence of a second molecular species in the membrane simply rescales the elastic parameters of the membrane [90]. Therefore, in the following we will restrict ourself to the free energy as defined in Eq. (7.4).

7.2.3 Formation of Membrane Tubes

The radius of a long tube can easily be derived from Eq. (7.4). For a cylinder of radius R (yielding $2H = 1/R$) and length L the free energy simplifies to:

$$\mathcal{F} = \left(\frac{\kappa}{2} \frac{1}{R} - \kappa C_0 + \sigma R \right) 2\pi L - fL \, . \tag{7.6}$$

Minimizing this with respect to R results in

$$R_0 = \sqrt{\frac{\kappa}{2\sigma}} \tag{7.7}$$

for the tube radius. Plugging this back into Eq. (7.6), the free energy can be written as

$$\mathcal{F} = 2\pi\kappa \left(\frac{1}{R_0} - C_0 \right) L - fL \, . \tag{7.8}$$

Since this function is linear in L, the tube force f_0 (i.e., the force necessary to hold the tube) can be calculated by taking this free energy equal to zero:

$$f_0 = 2\pi\kappa \left(\frac{1}{R_0} - C_0 \right) \, . \tag{7.9}$$

If a pulling force f larger than f_0 is applied, then the free energy becomes negative and the tube grows to infinity, whereas if f is smaller than f_0, then the free energy is positive and, therefore, the tube retracts.

Note, that the tube force f_0 vanishes at $C_0 = 1/R_0$. For even larger spontaneous curvatures the tube force becomes negative, signaling that the membrane is unstable, and tubes will grow spontaneously even without pulling.

Interestingly, the tubes themselves will also become unstable against pearling above this critical spontaneous curvatures [72,90,91], thus, the growing objects will look like necklaces rather than tubes [86].

Although the calculation of the tube radius R_0 and tube force f_0 has been an easy exercise, the initial formation of the tubes is a non-trivial process. For $C_0 = 0$ the force starts to increase linearly as the flat membrane gets deformed, and then reaches a maximum (which is about 13% higher than f_0) before it converges to f_0 with an exponentially damped oscillation [63,70,73,92], see Fig. 7.4. For larger C_0, but still below $1/R_0$ (an illustration for $C_0 = 0.9/R_0$ can also be seen in Fig. 7.4) the oscillations of the force become much more pronounced, pearls appear initially, and even the long and well developed tubes get connected to the flat part of the membrane by extremely narrow necks, which can be prone to fission [90]. Thus, the curvature sensitivity of membrane proteins and the pulling of motor proteins together can easily lead to vesiculation.

The shapes and forces just described and exhibited in Fig. 7.4 have been calculated by numerically solving the so-called shape equations (for more details see [63]), which can be derived from the free energy of the membrane by variational methods [93–96].

In most experiments with artificial vesicles the spontaneous curvature can be considered zero. In this case the tube force simplifies to $f_0 = 2\pi\sqrt{2\sigma\kappa}$. This theoretical prediction has been verified experimentally [41,45,97,98], and also used for determining the bending rigidity κ of membranes.

For biological membranes the value of κ is usually in the $10^{-20} - 10^{-19}$ J range and the surface tension σ varies between 10^{-3} and 10^{-6} N/m [65]. Choosing some typical values ($\kappa \approx 4 \times 10^{-20}$ J and $\sigma \approx 5 \times 10^{-5}$ N/m) and ignoring the spontaneous curvature one finds that $R_0 \approx 20$ nm and $f_0 \approx 12.6$ pN. Thus, in agreement with the multitude of experimental observations, membrane nanotubes are indeed a few tens of nanometers wide, and can be formed by forces around ten piconewtons, which can be easily generated by either the pushing of polymerizing filaments or the pulling of a couple of motor proteins.

Because at the tip of the tubes the mean curvature (and thus the free energy density) of the membrane diverges [63,92,99], for biological systems it seems reasonable to protect the tips and distribute the pulling forces at larger areas (e.g., by utilizing cap proteins or lipid rafts). Recent in vitro experiments have demonstrated that if the pulling area greatly exceeds the tube radius, then a significant force barrier has to be overcome during tube formation, which depends linearly on the radius of the pulling area [46, 48, 100]. This observation is in good agreement with theoretical predictions [48].

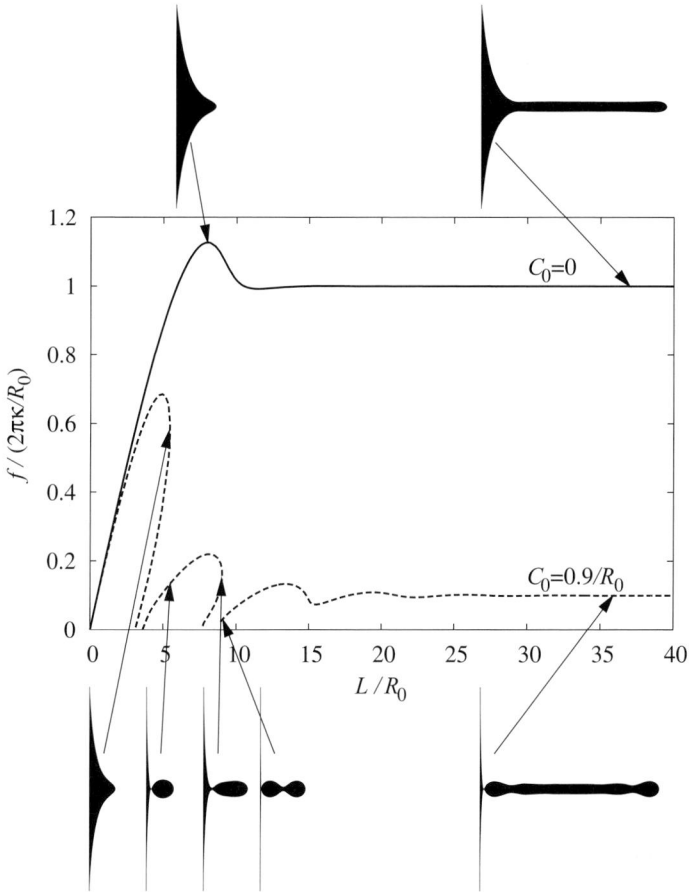

Fig. 7.4. Force vs. length curves for the formation of a membrane tube pulled out of a flat membrane that spans a ring of radius $20R_0$ for two different values of the spontaneous curvature ($C_0 = 0$ and $C_0 = 0.9/R_0$). The shape of the emerging tube is also depicted at certain stages

7.3 Membrane Tube Formation by Cytoskeletal Motor Proteins

The properties of individual motor proteins are being unraveled at a rapid pace. Much less is known about how and whether multiple motors *cooperate*, *interact*, and *coordinate* their activities. Although individual motors could in principle move objects (such as vesicles, DNA, proteins, membrane tubes, etc.), in vivo they often function together in groups. Many organelles are observed to move over larger distances than one motor can move [101], suggesting that multiple motors are working together. Cooperative functioning is also thought to be important for the spatial distribution and

morphology of organelles: multiple motors of opposite directionality are
present on organelles [102, 103], and it is yet unclear how their interaction is
orchestrated [104, 105]. Here, we will discuss recent in vitro studies that shed
some light on how motor proteins cooperate to generate enough force to form
and maintain membrane tubes. Next, we will discuss some promising future
studies to tackle the problem of oppositely directed motors, and we will give
some first results.

7.3.1 In vitro Studies Demonstrate that Molecular Motors Cooperatively Pull Membrane Tubes

Recent in vitro experiments with purified motors and synthetic vesicles have
shed some light on the cooperative activity of motor proteins. When purified motors were linked to beads, which were subsequently attached to a giant
unilamellar vesicle (GUV), and then these bead-motor-complex coated vesicles
were brought into contact with a network of MTs, the formation of membrane
tubes was observed (see Fig. 7.5a [49]). The fact that the force required to
form a tube is usually higher than what a single motor protein can exert suggests that each bead was pulled by several motors simultaneously. In contrast,
in other experiments with no beads (see Fig. 7.5b [51, 106]), it was found that
motors do not need to be cross-linked into multi-motor complexes for tube
formation. It was demonstrated that the extent of tube formation depends
on the concentration of motor proteins on the vesicle and the force that is
required to form a tube (set by global parameters like the membrane tension
and the bending rigidity). This led to the proposed mechanism of dynamic
association of motor proteins [51]: a steady-state cluster of motor proteins is
dynamically maintained at the tip of a membrane tube, taking into account
a force dependent departure rate of motor proteins [107, 108] and a concentration dependent arrival rate into the cluster. By simultaneously fluorescent
and biotin labeling of lipids, Leduc and Campàs et al. [50] succeeded in the
experimental demonstration of clusters at the tip of a tube. Their accompa-
nying theoretical description, which takes into account on and off rates of
motors and the active flux of motors along the tube as well, allowed for the
quantitative determination of parameters like the motor binding rate and the
minimal number of motors required for the extraction of a tube.

All the above-mentioned in vitro experiments were conducted with a biotinylated kinesin motor [109], which is a processive dimer. To determine
whether tube formation by dynamic clusters of motor proteins is a robust
and general mechanism, we recently studied the tube-formation potential of a
biotinylated ncd motor. This motor was attached to GUVs, and interestingly,
after getting into contact with a network of MTs, the first results indicate that
ncd motors are also able to pull membrane tubes [110]. This process seems to
occur in a similar manner as tube formation by kinesin. Some differences are
that tube formation occurs at a low velocity (in agreement with the velocity of ncd-mediated bead movement [10]) and a higher concentration of ncd

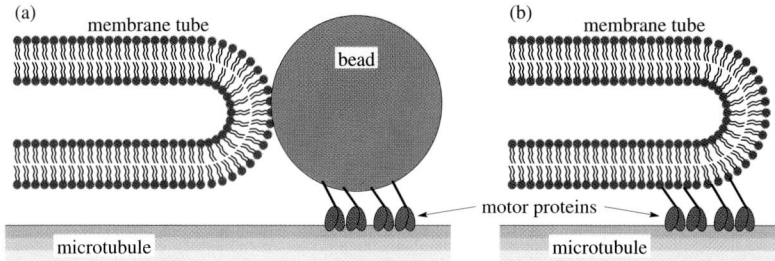

Fig. 7.5. Sketches of the in vitro experiments, where membrane nanotubes are formed by biotinylated motor proteins that pull on biotinylated lipids either (**a**) indirectly with the help of streptavidin coated vesicles or (**b**) directly through individual streptavidin molecules

is required for the same tube-formation potential as kinesin. As individual ncd motors are not processive, this result reveals an interesting cooperativity-effect: even if individual motors are non-processive, their joined activity can result in long-distance (processive) movement and drive tube formation and extension under an opposing load.

Interestingly, a recent finding indicates that the binding of motors is enhanced close to where other motors are bound [111]. This could stimulate the (dynamic) formation and stabilization of clusters of motors, which in turn would facilitate long distance movement and tube formation.

7.3.2 Interaction between Clusters of Antagonistic Motors?

There is substantial evidence that oppositely directed (antagonistic) motors are simultaneously attached to membrane organelles in the cell. This leads to bi-directional movement of organelles [102–105], and the morphology of membrane structures may also be dependent on such antagonistic motors [102]. In addition to a sufficiently high force, for membrane tube formation another requirement is that a sufficiently high counterforce is exerted on a membrane vesicle, as in the absence of a counterforce the vesicle would be dragged along. This counterforce may be generated by static linkers that immobilize a membrane vesicle, or can be developed by motors that exert force in the opposite direction.

To study the interaction between antagonistic clusters of motors, (plus-end directed) kinesin and (minus-end directed) ncd were simultaneously attached to the same vesicle. These motor-coated vesicles were brought into contact with a diluted concentration of immobilized MTs (in order to have only one MT as an interaction partner). Subsequently, a vesicle was grabbed with optical tweezers, and placed on top of the MT. In Fig. 7.6 we present a first result of such an experiment. After placing the vesicle on top of a MT, it started moving. This movement continued for several tens of micrometers,

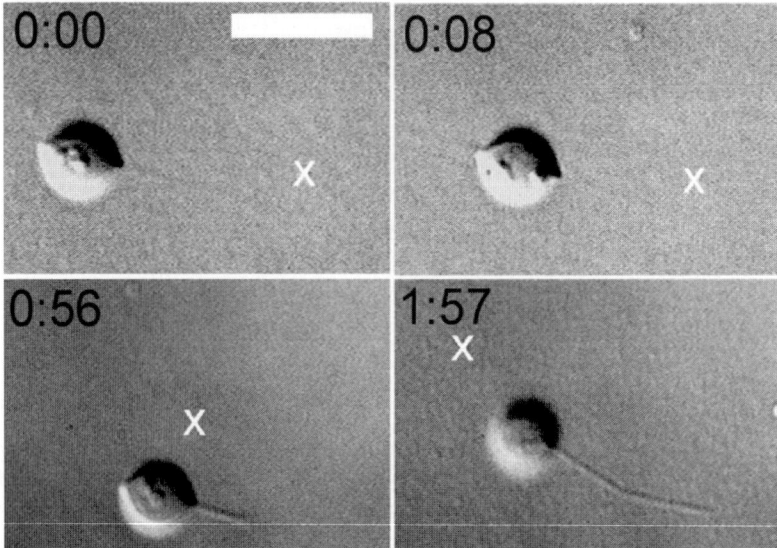

Fig. 7.6. VE-DIC microscopy of a membrane tube formed from a vesicle coated with kinesin and ncd. Microtubules are hardly visible. The mark X indicates a reference point on the surface of the coverglass. Time is in minutes and seconds. The bar is 10 μm

after which it stopped, while (at the same time) a membrane tube was formed and extended along the MT at about 0.4 micron/second.

Clearly more experiments are required, but one intriguing interpretation of this result is the following: first, the vesicle is moved along a MT by a cluster of motors (presumably kinesin, since it moves at ∼0.4 μm/s), Next, a cluster of ncd motors is nucleated at the trailing end of the vesicle, supplying sufficient counterforce to stop the movement of the vesicle by kinesin. Finally, as the cluster of kinesin motors can generate a high enough force the tug-of-war results in the formation of a tube.

An interesting point here is that naively one would expect two tubes to be formed, since motors are pulling in two opposing directions. It is, however, not energetically favorable to form two tubes, since in this case two neck regions with a high curvature need to be created [70], making it more favorable to extend an already existing tube. Also, the force (energy) barrier required to form a tube is larger than the force required to extend an existing tube [63,73], depending on the size of the area on which the force is exerted [48]. Thus, two tubes will only be formed if the vesicle is held at its position by an additional force.

Our preliminary data suggest that the competition between antagonistic clusters of motor proteins can lead to membrane tube formation. This tube formation is an additional degree of freedom, and it would prevent a tug-of-war in which case neither group of motors can move. Interestingly, the tubes effectively function as a sorting tool, since only motor proteins that move in one direction will move into and be enriched on a tube. If the tube were subsequently pinched off from the main vesicle body, the plus- and minus-end directed motors and their attached cargo would be segregated.

One important note is that, in the experiments where just kinesin is present on the vesicles and tubes are formed, kinesin motors supply the counterforce that holds the GUV by moving in all directions on the high-density network of randomly oriented MTs. It cannot be excluded that the interaction of the vesicle with the surface, or the presence of rigor motors may also play a role in the immobilization of the vesicle. In order to have interpretable experiments, in the antagonistic motor assay, the density of MTs on the surface was diluted to prevent the interaction of motor-coated vesicles with multiple MTs. However, this method does not prevent the interaction of the vesicle with the coverslip's surface. For a proper study of the interactions between antagonistic motor proteins, it will be important to immobilize the MTs, and at the same time prevent the vesicles from adhering. This is a big challenge since the underlying reason that MTs attach to a surface (charge), also makes the vesicles attach. A possible work-around for this issue is to use vesicles of a different charge. Alternatively, an elegant method may be to construct 3D structures like pillars [112] where MTs reach from one side to the other, and the middle part is not in contact with any surface. Vesicles can then be brought into contact with the elevated MT using the tweezers.

A subject connected with the antagonistic motors is the bi-directional movement of organelles that is observed in cells. When oppositely directed motors are present on a smaller vesicle, instead of tube formation their activity may lead to bi-directional movement (as tube formation from smaller vesicles will require a higher force because there is less excess surface area). Generally it is suggested that "coordination machinery" exists that controls whether one kind of motor or the other is active [104, 105], however, the physical interaction of antagonistic clusters of motors (transmitted through the vesicle) might play a role as well. One can imagine that at low motor concentrations, stochastic fluctuations in the number of motors may lead to alternating stable clusters of plus-end or minus-end directed motors, in turn leading to bi-directional movement. In that case the dynamical transition of collections of motors would act as number fluctuation amplifier [113–115]. The details of such "metastable" bi-directionality should depend on the rate of arrival into and the rate of departure from a cluster of motors [51, 110]. A further discussion of this bi-directionality is outside the scope of this work, but experiments are on the way to determine this system in more detail, and establish the regulatory role of physical parameters on antagonistic clusters of motors.

It is now well established that force-generating motor proteins are sufficient and enough for membrane tube formation. Resolving how (physical) parameters may regulate the dynamic interplay between motors and membranes is a promising line of research.

Acknowledgments

I. D. acknowledges support from the Hungarian Science Foundation (Grant No. OTKA F043756). The contribution by G. K. and M. D. is part of the research program of the "Stichting voor Fundamenteel Onderzoek der Materie (FOM)", which is financially supported by the "Nederlandse organisatie voor Wetenschappelijk Onderzoek (NWO)".

References

1. K. Visscher, M. J. Schnitzer, S. M. Block (1999). *Nature*, **400**, pp. 184–189
2. J. Howard. *Mechanics of motor proteins and the cytoskeleton*, (Sinauer Associates, Sunderland 2001)
3. K. Svoboda, C. F. Schmidt, B. J. Schnapp, S. M. Block (1993). *Nature*, **365**, pp. 721–727
4. H. B. McDonald, L. S. B. Goldstein (1990). *Cell*, **61**, pp. 991–1000
5. W. Hua, J. Chung, J. Gelles (2002). *Science*, **295**, pp. 844–848
6. C. L. Asbury, A. N. Fehr, S. M. Block (2003). *Science*, **302**, pp. 2130–2134
7. R. D. Vale (2003). *J. Cell Biol.*, **163**, pp. 445–450
8. A. Yildiz, M. Tomishige, R. D. Vale, P. R. Selvin (2004). *Science*, **303**, pp. 676–678
9. R. D. Vale (2003). *Cell*, **112**, pp. 467–480
10. M. W. Allersma, F. Gittes, M. J. deCastro, R. J. Stewart, C. F. Schmidt (1998). *Biophys. J.*, **74**, pp. 1074–1085
11. R. Chandra, E. D. Salmon, H. P. Erickson, A. Lockhart, S. A. Endow (1993). *J. Biol. Chem.*, **268**, pp. 9005–9013
12. J. Lane, V. Allan (1999). *Mol. Biol. Cell*, **10**, pp. 1909–1922
13. M. Terasaki, L. B. Chen, K. Fujiwara (1986). *J. Cell Biol.*, **103**, pp. 1557–1568
14. C. Lee, L. B. Chen (1988). *Cell*, **54**, pp. 37–46
15. C. M. Waterman-Storer, E. D. Salmon (1998). *Curr. Biol.*, **8**, pp. 798–806
16. F. Feiguin, A. Ferreira, K. S. Kosik, A. Caceres (1994). *J. Cell Biol.*, **127**, pp. 1021–1039
17. C. H. Lee, M. Ferguson, L. B. Chen (1989). *J. Cell Biol.*, **109**, pp. 2045–2055
18. S. L. Dabora, M. P. Sheetz (1988). *Cell*, **54**, pp. 27–35
19. R. D. Vale, H. Hotani (1988). *J. Cell Biol.*, **107**, pp. 2233–2241
20. V. Allan, R. D. Vale (1994). *J. Cell Sci.*, **107**, pp. 1885–1897
21. A. Upadhyaya, M. P. Sheetz (2004). *Biophys. J.*, **86**, pp. 2923–2928
22. H. H. Mollenhauer, D. J. Morré (1998). *Histochem. Cell Biol.*, **109**, pp. 533–543
23. N. Sciaky, J. Presley, C. Smith, K. J. M. Zaal, N. Cole, J. E. Moreira, M. Terasaki, E. Siggia, J. Lippincott-Schwartz (1997). *J. Cell Biol.*, **139**, pp. 1137–1155
24. J. Lippincott-Schwartz, E. Snapp, A. Kenworthy (2001). *Nat. Rev. Mol. Cell Biol.*, **2**, pp. 444–456
25. E. V. Polishchuk, A. Di Pentima, A. Luini, R. S. Polischuk (2003). *Mol. Biol. Cell*, **14**, pp. 4470–4485
26. T. Kirchhausen (2000). *Nat. Rev. Mol. Cell Biol.*, **1**, pp. 187–198
27. J. S. Bonifacino, B. S. Glick (2004). *Cell*, **116**, pp. 153–166

28. M. Dogterom, J. W. J. Kerssemakers, G. Romet-Lemonne, M. E. Janson (2005). *Curr. Opin. Cell Biol.*, **17**, pp. 67–74
29. B. Alberts, A. Johnson, J. Lewis, M. Raff, K. Roberts, P. Walter: *Molecular Biology of the Cell*, 4th edn (Garland Science, New York 2002)
30. H. Delanoë-Ayari, P. Lenz, J. Brevier, M. Weidenhaupt, M. Vallade, D. Gulino, J. F. Joanny, D. Riveline (2004). *Phys. Rev. Lett.*, **93**, pp. 108102
31. A. Rustom, R. Saffrich, I. Markovic, P. Walther, H.-H. Gerdes (2004). *Science*, **303**, pp. 1007–1010
32. S. C. Watkins, R. D. Salter (2005). *Immunity*, **23**, pp. 309–318
33. B. Önfelt, S. Nedvetzki, K. Yanagi, D. M. Davis (2004). *J. Immunol.*, **173**, pp. 1511–1513
34. K. N. J. Burger (2000). *Traffic*, **1**, pp. 605–613
35. K. Farsad and P. De Camilli (2003). *Curr. Opin. Cell Biol.*, **15**, pp. 372–381
36. R. M. Hochmuth, N. Mohandas, P. L. Blackshear (1973). *Biophys. J.*, **13**, pp. 747–762
37. R. E. Waugh (1982). *Biophys. J.*, **38**, pp. 29–37
38. O. Rossier, D. Cuvelier, N. Borghi, P. H. Puech, I. Derényi, A. Buguin, P. Nassoy, and F. Brochard-Wyart (2003). *Langmuir*, **19**, pp. 575–584
39. R. M. Hochmuth, H. C. Wiles, E. A. Evans, J. T. McCown (1982). *Biophys. J.*, **39**, pp. 83–89
40. R. E. Waugh, J. Song, S. Svetina, B. Zeks (1992). *Biophys J.*, **61**, pp. 974–982
41. E. Evans, A. Yeung (1994). *Chem. Phys. Lipids*, **73**, pp. 39–56
42. Z. Li, B. Anvari, M. Takashima, P. Brecht, J. H. Torres, W. E. Brownell (2002). *Biophys. J.*, **82**, pp. 1386–1395
43. T. Roopa, G. V. Shivashankar (2003). *Appl. Phys. Lett.*, **82**, pp. 1631–1633
44. D. Raucher, M. P. Sheetz (1999). *Biophys. J.*, **77**, pp. 1992–2002
45. V. Heinrich, R. E. Waugh (1996). *Ann. Biomed. Eng.*, **24**, pp. 595–605
46. H. Hotani, T. Inaba, F. Nomura, S. Takeda, K. Takiguchi, T. J. Itoh, T. Umeda, A. Ishijima (2003). *Biosystems*, **71**, pp. 93–100
47. D. K. Fygenson, J. F. Marko, A. Libchaber (1997). *Phys. Rev. Lett.*, **79**, pp. 4497–4500
48. G. Koster, A. Cacciuto, I. Derényi, D. Frenkel, M. Dogterom (2005). *Phys. Rev. Lett.*, **94**, pp. 068101
49. A. Roux, G. Cappello, J. Cartaud, J. Prost, B. Goud, P. Bassereau (2002). *Proc. Natl. Acad. Sci. USA*, **99**, pp. 5394–5399
50. C. Leduc, O. Campàs, K. B. Zeldovich, A. Roux, P. Jolimaitre, L. Bourel-Bonnet, B. Goud, J.-F. Joanny, P. Bassereau, J. Prost (2004). *Proc. Natl. Acad. Sci. USA*, **101**, pp. 17096–17101
51. G. Koster, M. VanDuijn, B. Hofs, M. Dogterom (2003). *Proc. Natl. Acad. Sci. USA*, **100**, pp. 15583–15588
52. J. Dai, M. P. Sheetz (1995). *Biophys. J.*, **68**, pp. 988–996
53. J. Dai, M. P. Sheetz (1999). *Biophys. J.*, **77**, pp. 3363–3370
54. R. M. Hochmuth, J. Y. Shao, J. Dai, M. P. Sheetz (1996). *Biophys. J.*, **70**, pp. 358–369
55. R. E. Waugh, R. G. Bauserman (1995). *Ann. Biomed. Eng.*, **23**, pp. 308–321
56. M. P. Sheetz (2001). *Nat. Rev. Mol. Cell Biol.*, **2**, pp. 392–396
57. E. Evans, H. Bowman, A. Leung, D. Needham, D. Tirrell (1996). *Science*, **273**, pp. 933–935
58. A. Karlsson, R. Karlsson, M. Karlsson, A.-S. Cans, A. Strömberg, F. Ryttsén, O. Orwar (2001). *Nature*, **409**, pp. 150–152

59. M. Karlsson, K. Sott, M. Davidson, A.-S. Cans, P. Linderholm, D. Chiu, and O. Orwar (2002). *Proc. Natl. Acad. Sci. USA*, **99**, pp. 11573–11578
60. M. Karlsson, M. Davidson, R. Karlsson, A. Karlsson, J. Bergenholtz, Z. Konkoli, A. Jesorka, T. Lobovkina, J. Hurtig, M. Voinova, O. Orwar (2004). *Ann. Rev. Phys. Chem.*, **55**, pp. 613–649
61. T. Lobovkina, P. Dommersnes, J.-F. Joanny, P. Bassereau, M. Karlsson, O. Orwar (2004). *Proc. Natl. Acad. Sci. USA*, **101**, pp. 7949–7953
62. P. G. Dommersnes, O. Orwar, F. Brochard-Wyart, J. F. Joanny (2005). *Europhys Lett.*, **70**, pp. 271–277
63. I. Derényi, F. Jülicher, J. Prost (2002). *Phys. Rev. Lett.*, **88**, pp. 238101
64. D. Cuvelier, I. Derényi, P. Bassereau, P. Nassoy (2005). *Biophys. J.*, **88**, pp. 2714–2726
65. U. Seifert, R. Lipowsky. Morphology of Vesicles. In: *Structure and Dynamics of Membranes*, vol 1A, ed by R. Lipowsky, E. Sackmann (Elsevier Science, Amsterdam 1995) pp. 403–462
66. U. Seifert (1997). *Adv. Phys.*, **46**, pp. 13–137
67. W. Helfrich (1973). *Z. Naturforsch. C*, **28**, pp. 693–703
68. P. B. Canham (1970). *J. Theor. Biol.*, **26**, pp. 61–81
69. L. Miao, U. Seifert, M. Wortis, H. G. Dobereiner (1994). *Phys. Rev. E*, **49**, pp. 5389–5407
70. V. Heinrich, B. Bozic, S. Svetina, B. Zeks (1999). *Biophys. J.*, **76**, pp. 2056–2071
71. H. G. Döbereiner, E. Evans, M. Kraus, U. Seifert, M. Wortis (1997). *Phys. Rev. E*, **55**, pp. 4458–4474
72. D. J. Bukman, J. H. Yao, M. Wortis (1996). *Phys. Rev. E*, **54**, pp. 5463–5468
73. T. R. Powers, G. Huber, R. E. Goldstein (2002). *Phys. Rev. E*, **65**, pp. 041901
74. S. Leibler (1986). *J. Phys.*, **47**, pp. 507–516
75. S. Leibler, D. Andelman (1987). *J. Phys.*, **48**, pp. 2013–2018
76. T. Taniguchi, K. Kawasaki, D. Andelman, T. Kawakatsu (1994). *J. Phys. II*, **4**, pp. 1333–1362
77. M. Seul, D. Andelman (1995). *Science*, **267**, pp. 476–483
78. J. B. Fournier (1996). *Phys. Rev. Lett.*, **76**, pp. 4436–4439
79. S. Komura, H. Shirotori, P. D. Olmsted, D. Andelman (2004). *Europhys. Lett.*, **67**, pp. 321–327
80. C.-M. Chen, P. G. Higgs, F. C. MacKintosh (1997). *Phys. Rev. Lett.*, **79**, pp. 1579–1582
81. F. Julicher, R. Lipowsky (1996). *Phys. Rev. E*, **53**, pp. 2670–2683
82. T. Kawakatsu, D. Andelman, K. Kawasaki, T. Taniguchi (1993). *J. Phys. II*, **3**, pp. 971–997
83. J.-M. Allain, C. Storm, A. Roux, M. Ben Amar, J.-F. Joanny (2004). *Phys. Rev. Lett.*, **93**, p. 158104
84. V. Kralj-Iglic, A. Iglic, M. Bobrowska-Hagerstrand, H. Hagerstrand (2001). *Colloids Surf. A*, **179**, pp. 57–64
85. I. Tsafrir, D. Sagi, T. Arzi, M.-A. Guedeau-Boudeville, V. Frette, D. Kandel, J. Stavans (2001). *Phys. Rev. Lett.*, **86**, pp. 1138–1141
86. I. Tsafrir, Y. Caspi, M.-A. Guedeau-Boudeville, T. Arzi, J. Stavans (2003). *Phys. Rev. Lett.*, **91**, p. 138102
87. B. J. Peter, H. M. Kent, I. G. Mills, Y. Vallis, P. J. G. Butler, P. R. Evans, H. T. McMahon (2004). *Science*, **303**, pp. 495–499

88. S. Ramaswamy, J. Toner, J. Prost (2000). *Phys. Rev. Lett.*, **84**, pp. 3494–3497
89. P. Girard, J. Prost, P. Bassereau (2005). *Phys. Rev. Lett.*, **94**, pp. 088102
90. I. Derényi, A. Czövek, F. Jülicher, J. Prost: (to be published)
91. H. J. Deuling, W. Helfrich (1977). *Blood Cells*, **3**, pp. 713–720
92. B. Bozic, V. Heinrich, S. Svetina, B. Zeks (2001). *Eur. Phys. J. E*, **6**, pp. 91–98
93. F. Jülicher, U. Seifert (1994). *Phys. Rev. E*, **49**, pp. 4728–4731
94. H. Jian-Guo, O.-Y. Zhong-Can (1993). *Phys. Rev. E*, **47**, pp. 461–467
95. W.-M. Zheng, J. Liu (1993). *Phys. Rev. E*, **48**, pp. 2856–2860
96. B. Bozic, S. Svetina, B. Zeks (1997). *Phys. Rev. E*, **55**, pp. 5834–5842
97. R. E. Waugh, R. M. Hochmuth (1987). *Biophys. J.*, **52**, pp. 391–400
98. L. Bo, R. E. Waugh (1989). *Biophys. J.*, **55**, pp. 509–517
99. R. Podgornik, S. Svetina, B. Zeks (1995). *Phys. Rev. E*, **51**, pp. 544–547
100. T. Inaba, A. Ishijima, M. Honda, F. Nomura, K. Takiguchi, H. Hotani (2005). *J. Mol. Biol.*, **348**, pp. 325–333
101. D. B. Hill, M. J. Plaza, K. Bonin, G. Holzwarth (2004). *Eur. Biophys. J.*, **33**, pp. 623–632
102. V. J. Allan, H. M. Thompson, M. A. McNiven (2002). *Nat. Cell Biol.*, **4**, pp. E236–E242
103. C. Kural, H. Kim, S. Syed, G. Goshima, V. I. Gelfand, P. R. Selvin (2005). *Science*, **308**, pp. 1469–1472
104. S. P. Gross (2004). *Physical Biology*, **1**, pp. 1–11
105. M. A. Welte (2004). *Curr. Biol.*, **14**, pp. 525–537
106. C. Leduc (2005). Système biomimétique d'intermediaires de transport tubulaires: étude quantitative. PhD thesis, Université Paris 7, Paris
107. C. M. Coppin, D. W. Pierce, L. Hsu, R. D. Vale (1997). *Proc. Natl. Acad. Sci. USA*, **94**, pp. 8539–8544
108. A. Parmegianni, F. Jülicher, L. Peliti, J. Prost (2001). *Europhys. Lett.*, **56**, pp. 603–609
109. T. Surrey, M. B. Elowitz, P.-E. Wolf, F. Yang, F. Nedelec, K. Shokat, S. Leibler (1998). *Proc. Natl. Acad. Sci. USA*, **95**, pp. 4293–4298
110. G. Koster (2005). Membrane tube formation by motor proteins. PhD thesis, AMOLF, Amsterdam
111. E. Muto, H. Sakai, K. Kaseda (2005). *J. Cell Biol.*, **168**, pp. 691–696
112. W. Roos, J. Ulmer, S. Grater, T. Surrey, J. P. Spatz (2005). *Nano Lett.*, **5**, pp. 2630–2634
113. F. Jülicher, J. Prost (1995). *Phys. Rev. Lett.*, **75**, pp. 2618–2821
114. D. Riveline, A. Ott, F. Jülicher, A. Winkelmann, O. Cardoso, J. J. Lacapere, S. Magnusdottir, J. L. Viovy, L. Gorre-Talini, J. Prost (1998). *Eur. Biophys. J.*, **27**, pp. 403–408
115. M. Badoual, F. Jülicher, J. Prost (2002). *Proc. Natl. Acad. Sci. USA*, **99**, pp. 6696–6701

8

Macromolecular Motion at the Nanoscale of Enzymes Working on Polysaccharides

M. Sletmoen[1], G. S.-Bræk[2], and B.T. Stokke[1]

[1] Biophysics and Medical Technology, Department of Physics, The Norwegian University of Science and Technology, NTNU, 7491 Trondheim Norway
[2] Department of Biotechnology, The Norwegian University of Science and Technology, NTNU, 7491 Trondheim Norway

Abstract. Polysaccharide modifying enzymes may utilise enzyme-substrate motion to generate specific sequence patterns in polymers (e.g. epimerisation), or to produce oligomers as a result of depolymerisation. Enzyme catalysed polymerisation of polysaccharides may also create motion of entire species (bacteria). The energy balance show the clear role of biochemically accessible energy in most of such examples, whereas in others, conformational changes associated with the mutual relocation of the enzyme relative to the polymer substrate is the most likely source. A comparison between the mechanism underlying the nanoscale motion of enzymes working on polysaccharides, with various known types of energy catalysing the motion is discussed.

8.1 Introduction

Life implies transformations and movement, which in the living world are powered by proteins. The term "molecular device" designates single proteins or protein assemblies having this ability. They fall into two broad classes: Catalysts and machines [1]. Catalysts are characterised by their ability to enhance the rate of a chemical reaction, and proteins having this ability are called enzymes. Chemical reactions may, despite having a favourable free energy change, proceed very slowly because of a large activation energy barrier. Molecular machines differ from catalysts in that they actively reverse the natural direction of a chemical or mechanical process by coupling it to another one. Progress made in the field of molecular biology has enabled a molecular level understanding of these biological devices. This new insight has attracted attention not only from biologists but also from chemists and physicists.

Most macroscopic machines invented to produce movement from chemicals require an intermediate form of energy, usually heat or electricity. In this sense, the generation of movement in cells is an exception since it in some cases involves the conversion of chemical energy directly to mechanical energy. The chemical energy often drives a conformational change within the proteins

M. Sletmoen et al.: *Macromolecular Motion at the Nanoscale of Enzymes Working on Polysaccharides*, Lect. Notes Phys. **711**, 161–180 (2007)
DOI 10.1007/3-540-49522-3_8   © Springer-Verlag Berlin Heidelberg 2007

that mediate the movement. Whereas some molecular machines depend on an internal source of energy, others depend on an external source. Molecular machines falling into the last category can again be divided into motors, characterised by their ability to transduce free energy into motion, pumps, which create concentration differences across membranes, and synthases which drives a chemical reaction, i.e. the synthesis of some product.

The interest in molecular machines increased with the realisation that the ability of motor proteins to generate force and to undergo directed motion form the basis of much of cellular behaviour and architecture. In eukaryotes, there are two major motility systems. The first involves interactions between the motor molecules kinesin and dynein and microtubules. Microtubules are abundant in interphase cells and are used as highways for a variety of intracellular transport processes. A second system is based on interactions between actin microfilaments and members of the myosin family of motor molecules. The action of these motility systems is fundamental to a wide range of cellular functions such as directed transport of macromolecules, membranes, or chromosomes within the cell's cytoplasm, and muscle contraction underlying locomotion. Other well-known examples of molecular machines found in cells are the DNA and RNA polymerases, which function in replication or transcription of DNA, respectively.

Despite the growing interest in molecular devices, most of the focus has been on molecular motors working on protein polymers, DNA or RNA. The examples of molecular motors given in recent reviews on this topic illustrate this trend [2, 3]. In the following we will present some examples illustrating the growing body of evidence revealing important roles played by molecular devices working on polysaccharides. In some cases these molecular devices are processive enzymes. Processive enzymes are interesting because they have the abilities of both catalysts and molecular motors, i.e. they enhance the rate of chemical reactions, and they use energy to undergo directed motion along their polymer substrate.

Polysaccharide modifying processive enzymes may utilise enzyme-substrate motion to generate specific sequence patterns in polymers, or to produce oligomers formed as a result of depolymerisation. Enzyme catalysed polymerisation of polysaccharides may also create motion of entire species (bacteria). The energy balance show the clear role of biochemically accessible energy in most of such examples, whereas in others, conformational changes associated with the mutual relocation of the enzyme relative to the polymer substrate is the most likely source. This chapter presents a comparison between the mechanism underlying the nanoscale motion of enzymes working on polysaccharides, with the various known types of energy catalysing the motion. Here, we focus on three enzyme substrate systems where recent research has provided new insight. Common to the examples chosen is also that models have been proposed in order to explain and predict the behaviour of the enzymes and the structures formed as a result of their action. Before describing these systems some relevant background information concerning polysaccharides and the requirement met by processive enzymes working on these polymers are given.

8.2 Polysaccharide Modifying Enzymes

Carbohydrates are among the most abundant organic compounds found in natural sources; they are widespread in both plants and animals. They range from small molecules with molecular weights exceeding one hundred to polymers with molecular weights exceeding one million. In nature, polysaccharides serve as stores of energy (starch and glycogen), they are found in bacterial cell walls or extrusions (scleroglucan, xanthan), in algae (alginate, agar, carrageenan) and they form a major portion of the supporting tissue of plants (cellulose, pectin) and of some animals i.e. crustacea (chitin). In view of this widespread occurrence and important roles played by polysaccharides, it is not surprising that many enzymes have evolved to synthesise, modify, or degrade them. One example of an enzyme working on a polysaccharide chain is depicted on the AFM topograph in Fig. 8.1, showing C-5-mannuronan epimerases catalysing the last step in the biosynthesis of the polysaccharide alginate. The enzyme catalysed reactions require that the enzyme recognise the substrate and form an enzyme-substrate complex. This complex stabilises the transition state and greatly reduces the activation energy of the reaction in comparison to that without the enzyme. When catalysed by enzymes, reactions can therefore proceed under mild conditions. The relationship between substrate and enzyme has been compared to a "key and lock" system. This

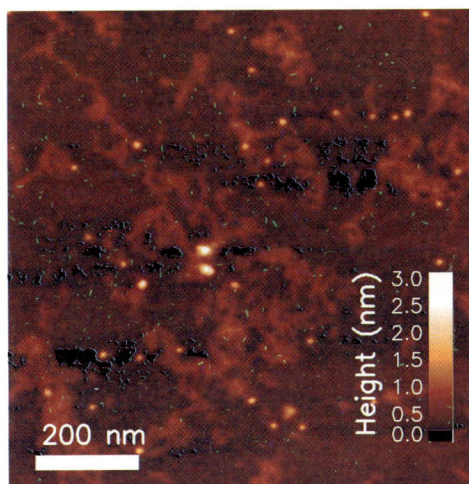

Fig. 8.1. Tapping mode AFM topograph showing flexible mannuronan polysaccharide with C5-mannuronan epimerase AlgE4 (*globular, bright dots*) bound to some of the polymer strands

analogy is illustrative of the properties of the complex formation, which is achieved through complementary interfaces on the two molecules, allowing specific non-covalent bonds to form.

8.3 The Action Patterns of Polymer Modifying Enzymes

The action patterns traditionally applied to describe enzymes that act on polysaccharide substrates are the following. *The single-chain mechanism* describes a situation where the enzyme forms an enzyme–substrate (ES) complex, perform its function along the chain and dissociate from the polymer at the end of the chain or when a non-convertible unit is reached. In a *multi-chain mechanism* the enzyme reacts randomly with a polymer unit and dissociates after each reaction. *The multiple-attack mechanism* represents an intermediate of the two extremes presented above by catalysing repeated reactions per effective encounter [4]. It has later been argued that the modes of action can be described more accurately in terms of enzyme processivity, i.e. the average number of times the reaction is repeated between association and dissociation of ES [5]. Processive enzymes differ greatly in the "extent" of their processivity. Any processive enzyme working on a substrate can be characterised by the mean number of subunits converted before enzyme-substrate dissociation. This number provides a quantitative measure of the processivity [6,7]. In the classical random attack model, as well as in the processive model, the enzyme is believed to randomly attack the polymer chains for each enzyme-substrate encounter (displays no preference for specific sequences). In the preferred attack model, on the other hand, the enzyme is supposed to have preference for specific monomer sequences within heteropolymeric substrates. Processivity requires a sliding motion of the enzyme relative to the polymer substrate. At least partial contact with the polymer lattice is therefore required at all times, since Brownian motions would otherwise rapidly separate detached enzymes from their track. In order to slide an enzyme must therefore strike the right energetic balance to remain associated with its polymeric substrate, while retaining the ability to move from site to site. Studies of different processive enzymes have shown that this problem of energetic balance is solved in different ways [8]. One strategy to prevent the protein from dissociating involves the creation of a physical barrier. A second strategy involves a "hand-over-hand" interaction with the polymer (also called "subunit switching" or "rolling"). This mechanism requires alternate binding of individual subunits of the motor to the lattice in a co-ordinated manner, so that at least one subunit is tightly bound to the lattice at any given time.

8.4 Polysaccharide Degrading Processive Enzymes

Polysaccharide modifying enzymes sometimes contain separate modules that afford the substrate binding capabilities of the enzyme. Non-catalytic polysaccharide-recognising modules of glycoside hydrolases are named carbo-hydrate-binding modules (CBM). Experimental studies have identified many CBMs, and a recent review summarised the current knowledge of their struc-ture and function [9]. By integrating different information the article discusses CBMs in the broader context of carbohydrate binding proteins and provides new insight into the mechanism of protein-carbohydrate recognition. Through their sugar-binding activity, CBMs maintain the enzyme in proximity with the substrate. In the case of polysaccharide degrading enzymes, this is thought to lead to more rapid degradation [10]. In some cases CBMs have evolved to become components of the substrate-binding sites of glycoside hydrolases, and sometimes even influence the substrate specificity and mode of action of the enzymes [11]. Comparative studies have revealed that cellobio-hydrolases belonging to different structural families all exhibit active sites with a tunnel-like topology [12–14]. The tunnels result from loops folding over the active site groove and allow the protein to progress unidirectionally along the polysac-charide chain without dissociating from its substrate between two hydrolysis events [12–15].

In the following we focus on family 18 chitinases which are reported to both have a CBM belonging to family 18 and a processive mode of action. Chitinases are found in bacteria, fungi, viruses, animals and in some plants. In these organisms they have both physiological and ecological roles. Inver-tebrates require chitinases for partial degradation of their old exoskeletons. Fungi produce chitinases to modify chitin, which is an important cell wall component. Bacteria produce chitinases to digest chitin and utilise it both as an energy and carbon source. The production of chitinases by higher plants has been suggested to be part of their defence mechanism against bacterial and fungal pathogens.

Chitosans are heteropolymers consisting of 2-acetamido-2-deoxy-D-glucose (GlcNAc, A) and 2-amino-2-deoxy-D-glucose (GlcN, D), which are formed by partial deacetylation of chitin, a $(1,4)$-β-linked linear polymer of GlcNAc. Biosynthesis or enzymatic degradation of chitin or chitosan produces chi-tooligosaccharides. These possess a number of biological activities, such as immune stimulation through activation of macrophages [16] and induction of chemotactic migration of polymorphonuclear cells [17]. It has been reported that compounds differing in degree of polymerisation (DP) and mole frac-tion of acetylated residues (F_A) cause different biological response [18]. Enzy-matic hydrolysis of polysaccharides produces oligosaccharides with specified lengths and, in the case of heteropolymeric substrates, specific sequences. The oligomer sequence and the mode of action of the enzyme determine the length and sequence characteristics of the products. A model based on knowl-edge about the enzymatic mode of action and substrate specificity would, if combined with structural information of the substrate, allow predicting and controlling the outcome of the enzymatic reaction. The development of such a model is therefore of significant scientific and industrial interest. In this

GlcNAc   GlcN   GlcNAc   GlcNAc   GlcNAc   GlcN

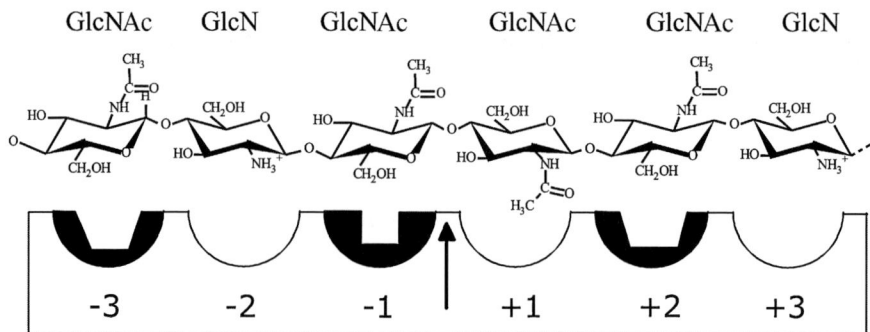

Fig. 8.2. Example of a chitosan oligomer sequence (ADAAAD) presented to a six-subsite model for hydrolysis of chitosans by ChiB chitinase. The glycosidic bond between the residues positioned in the −1 and +1 subsite is cleaved. The relative hydrolysis rates of a given oligomer sequence is determined by its composition. This can be accounted for by additive contributions depending on the residue-subsite pairwise contribution, to an overall apparent activation energy for the hydrolysis to occur [24]. In the processive mode of action, the chitosan polysaccharide undergoes a sliding motion of two sugar residues, about 1.03 nm, relative to the enzyme following each hydrolysis step

section we describe some recent studies aiming at providing such a model for a family 18 chitinase.

Hydrolyses of chitosans with various F_A by members of the family of processive enzymes known as 18 exochitinase ChiB yields mixtures of oligosaccharides, differing in sequence and DP [19]. ChiB is an elongated molecule. The active site is formed by a long, deep cleft (about 40 Å long, 15 Å wide, and 20 Å deep), which is characterised by a large number of exposed hydrophobic residues and harbours the catalytic residue [20]. It has six subsites, numbered from −3 to +3, and for all family 18 chitinases the cleavage occurs between site −1 and +1 [20] (Fig. 8.2); notation according to [21]. Hydrolysis of the glycosidic bond requires that the sugar bound in the −1 subsite be N-acetylated [22,23]. The ChiB enzyme also contains a chitin-binding domain, which presents an aromatic surface presumably involved in substrate binding. Chitosan and ChiB may form non-productive enzyme–substrate complexes. Binding of a D residue in the −1 subsite would represent one such non-productive complex. Experimental evidence suggests that non-productive binding do not necessarily result in enzyme substrate dissociation, but might instead result in displacement of the enzyme along the polymer chain allowing adjacent susceptible bonds to be reached and hydrolysed [19].

An approach for modelling chitosan degradation catalysed by ChiB was recently suggested [24]. Since the products obtained are the result of a complex interaction between the enzyme and the substrate, many parameters must be considered in order to predict their formation. The hydrolysis rate and the substrate binding energies will thus depend on the sequence of A and D units in the substrate. The model used experimental data obtained from the hydrolysis of well-defined and well-characterised polymeric substrates to estimate necessary parameters in the model. As the monomer sequence in the polymeric substrate is random [19, 25, 26], it was argued [24] that it was not possible to predict the reaction products using an analytical solution. A method of random sampling of the reaction space using Monte Carlo (MC) simulations was therefore chosen. The proposed model allowed predicting the outcome of hydrolysis experiments on chitosans with varying degrees of acetylation, as well as on oligomeric, fully acetylated substrates. Based on the Monte Carlo simulation the degree of processivity of the ChiB enzyme was determined from the fit to the experimental data. The results indicated an average number of residue moves per single attack equal to 16, corresponding to a total relocation of each enzyme along the polymer equal to 16.5 nm.

Biochemical and structural studies have revealed that the tunnel-forming loops observed in many processive enzymes often undergo conformational changes. Initiation of the reaction often demands an opening of the tunnel to allow an initial endo-attack. This is followed by a closing of the tunnel needed for the sliding motion of the enzyme, i.e. a processive mechanism [27–29]. The exo-chitinase ChiB is one example of a processive enzyme in which the tunnel-like active site undergoes conformational changes upon substrate binding [20, 23]. Conformational changes require energy, but the ChiB enzyme does not rely on any external source of energy. Instead, a likely source of energy for the conformational changes of the enzyme, underlying the directed movement of the enzyme along the polymeric substrate, is the energy contained in the chemical bonds of the substrate being cleaved. All molecules carry a definite amount of stored energy, residing in their chemical bonds. This chemical bond energy, just as any other form of stored energy, drives towards lower values, and it constitutes a potential source of the energy driving the conformational changes of enzymes.

The occurrence of tunnel-like active sites or CBMs in processive cellulases or chitinases, has been suggested to reflect the necessity to prevent isolated polysaccharide chains from re-adhering to the crystalline cellulose or chitin substrate [30]. In other systems this requirement does not exist, and other topologies which also promotes processivity are observed. The endo/exocellulase E4 from *Thermospora fusca* displays a catalytic domain with an open cleft topology, yet it degrades crystalline cellulose in a processive manner. This is due to the presence of an appended cellulose-binding domain, which remains loosely associated to the cellulose microfibril [11]. The processive character of α-amylases is proposed to result from a large interacting area in the open active cleft [31]. The high levels of multiple attacks observed in mammalian α-amylases may be facilitated by a movable loop, which interacts with subsites -2 and $+2$ upon substrate binding [32, 33]. In the case of more amorphous substrates, such as arabinans or pectins, processivity arises from the binding energies of individual subsites [30, 34]. In endopolygalacturonase a mutation at the -5 sugar binding site turn a non-processive enzyme into a processive one and vice versa. This observation can be explained by the residue found in this position being involved in sliding the substrate through the cleft at the active site [34, 35]. The examples presented above illustrate that in nature many alternative structural solutions which allow a protein to slide along its substrate has evolved. In the next section we shall focus on a class of processive enzymes displaying a catalytic domain with an open cleft topology. The experiments that first revealed the processive mode of action of these enzymes are also outlined.

8.5 Enzyme-Substrate Motion Can Explain the Formation of Specific Sequence Patterns in Polysaccharides

A number of different uronic acid-containing polysaccharides undergo post-polymerisation modifications to produce the final biologically active structure [36, 37]. Examples are found in the biosynthesis of heparin/heparan sulphates [38], dermatan sulphates [39] and alginates [40]. This modification has a profound effect on the secondary structure of these polysaccharides and therefore on their biological role. The apparatus allowing these modifications to occur are submitted to genetic control, and represent tools that allow tailoring the structure of the polysaccharides and thereby optimise functional properties fulfilling a specific biological role. In the following we focus on one of these systems: the alginates. Alginate consist of β-D-mannuronic acid ($^4C_1$ conformation) (ManA, M) and α-L guluronic acid ($^1C_4$ conformation) (GulA, G). Alginates are used commercially mainly as gelling agents, and the gelation properties depend both on the fraction of α-L-guluronic acid, and the sequence pattern of the G and M residues. The alginates available from various sources offer a certain, but limited selection of sequence patterns. Variations in sequence pattern found in alginates extracted from different parts of the algae (stem, blade etc) illustrate how the algae tailor the sequence pattern in order to afford its different parts with varying stiffness or strength [41]. When preparing alginate gels, this observed source dependence is exploited for tailoring the gel properties required for a given application. Another way of increasing the versatility in the properties of alginate gels involves controlled epimerisation of alginates. Such controlled epimerisation is expected to increase the application potential of these materials [42].

In alginates the conversion of β-D-mannuronic acid to α-L-guluronic acid at the polymer level is catalysed by enzymes known as mannuronan C-5 epimerases [37] (Fig. 8.3). The epimerase reaction is proposed to occur via a three-step reaction mechanism [43]. In the first step the negative charge on the carboxylate anion is neutralised, either by co-ordination to a metal ion or by the formation of a salt-bridge with an amino acid in the active site of the enzyme. The second step involves base-catalysed abstraction of the proton at C-5, which in the third step is replaced with a proton from the medium,

a

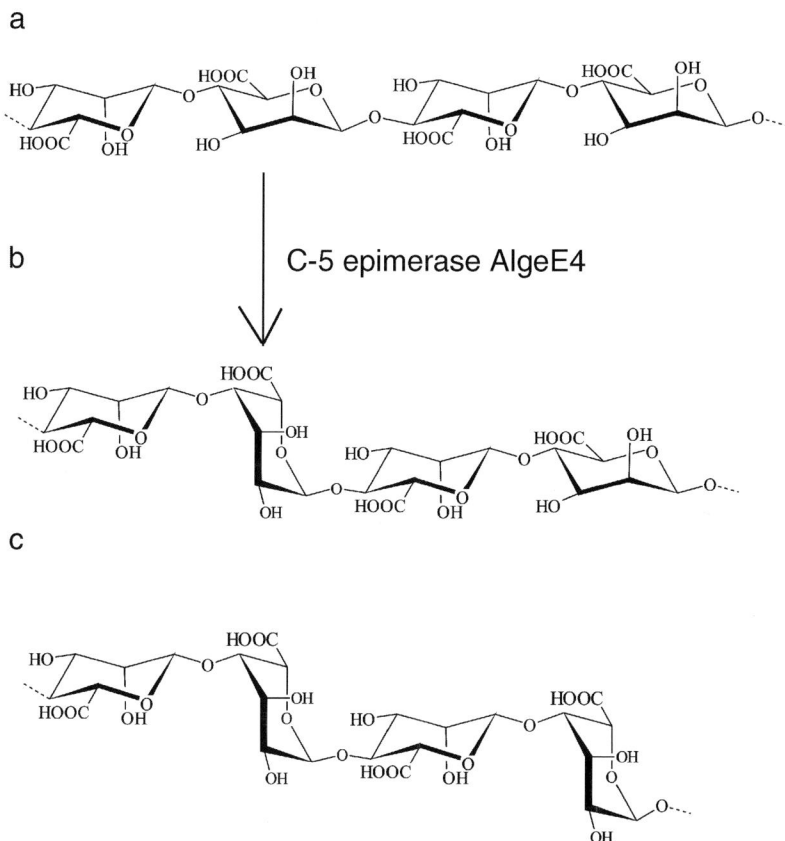

b

C-5 epimerase AlgeE4

c

Fig. 8.3. Epimerisation of mannuronan with the C-5 epimerase AlgE4. Analysis of the residue sequence developing as the epimerisation proceed [46] indicate that every second residue is converted yielding an overall polyalternating, or repeating dimeric, structure

which is added from the stereochemically opposite direction giving rise to the change in configuration around carbon 5 [43, 44]. Seven different natural mannuronan C-5-epimerases are known. Studies have revealed that each of them catalyse the production of alginates with a distinct monomer distribution pattern [45]. The mode of action of the different enzymes determines whether single guluronate units are introduced at random, or successively as blocks.

Monitoring the introduction of sugar sequence triads in the reaction product as G residues are being introduced by the activity of the epimerases AlgE2 and AlgE4 is accomplished by NMR spectroscopy [46]. The determined sequence distribution patterns revealed that the epimerase AlgE4 predominantly produced an alginate with long stretches of alternating MG sequences. The formation of such a pattern could be explained by a processive mode of action, and further studies therefore aimed at providing more

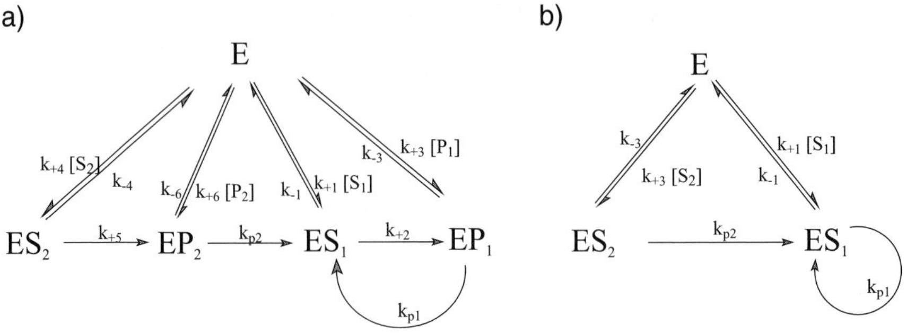

a)

b)

Substrate	S_2		P_2		S_1		P_1	
i)	-MM	MXX-	-MM	MXX-	-GM	MXX-	-GM	MXX-
ii)	-MM	G-			-MM	G-		

Fig. 8.4. Kinetic scheme (**a**) and simplified scheme (**b**) for modelling residue sequences introduced in alginates by C5-mannuronan epimerase AlgE4 [47]. Examples of residue sequences of M and G present in a hypothetical four subsite model (minimum number of required subsites) for productive modes (i) and non-productive modes (ii) for the AlgE4 enzyme are given. Mannuronan is the homopolymer of β-D-mannuronic acid (M), whereas naturally occuring alginates consist of various sequences of the two residues. The residue outlined in bold is the residue being epimerised. X depicts residues that do not yield nonproductive binding sequences. In the case of the productive mode, the epimerisation (Fig. 8.3) and associated prosessive movement (k_{p2} and k_{p1}) correspond to a sliding motion of two sugar residues, being displaced about 1.03 nm relative to the enzyme upon each epimerisation reaction

information about this reaction mechanism. The evolution of the sequence pattern in the alginate product, as determined by NMR, was investigated using Monte Carlo simulations [47]. Both a preferred attack mechanism and a processive mode of action could account for the non-random sequence arrangement of the residues. A model that allowed for preferred attack and/or processive mode of action was therefore used. Figure 8.4 shows the kinetic scheme used.

In this model, the preferred attack is obtained when both parameter k_{p1} and k_{p2} equal zero, whereas these differ from zero in a processive mode of action. The propagation probability, p, for the processive mode is given by $k_{p1}/(k_{p1} + k_{-1})$ (Fig. 8.4b), where k_{p1} and k_{-1} are the kinetic constants for propagation and dissociation, respectively, of the enzyme-product complex. The average number of converted residues per enzyme attack is equal to $1/(1 - p)$.

The analysis was carried out using an ensemble of 10,000 polymer chains. An experimentally determined degree of polymerisation and initial fraction of guluronic acid were used. For each iteration step, the concentrations of the enzyme–substrate complexes were updated based on knowledge about affinity constants between the enzymes and substrates of different size and sequence, and the relative abundance of the productive substrates. Additionally, in the processive mode model, a certain probability for propagation relative to dissociation was assumed at each step, limiting the length of the polyalternating MG-blocks introduced. Several factors potentially influence the length of the blocks formed by these processive enzymes. Bacterial alginates contain acetyl groups covalently linked to a fraction of the monomers. These groups function as termination signals, possibly due to steric hindrance introduced by the acetyl group. The concentration and type of ions present in the medium also influence the pattern observed in the end product. In the Monte Carlo simulation these factors were incorporated in the following way. After each cycle of epimerisation, the complexes were selected for either dissociation or propagation based on a propagation probability parameter p. The simulated residue sequences obtained were compared with the sequence information obtained using NMR. This comparison showed that for AlgE4 a situation without product inhibition, as well as a situation with product inhibition, could account for the experimental data, but demanded different substrate affinity constants. Additional experiments using ^{3}H-labelled alginate oligomers revealed that an octamer was the minimum size necessary to support enzyme activity of AlgE4 [47]. For AlgE2 the experimental data were more conclusive: the data suggested that AlgE2 function by a preferred attack mechanism.

The remaining questions concerning the mode of action of AlgE4 were readdressed in a later single molecular dissociation study between the epimerase and mannuronan, performed using an AFM. When measuring intermolecular forces with AFM, the AFM tip is functionalised with one of the interacting molecules. These molecules, exposed on the surface of the tip, are then allowed to interact with potential binding partners immobilised to a mica surface. Pulling experiments in which the tip-surface separation speed is changed in such a way that the force per unit time acting on the molecular bond under study is changed over many orders of magnitude, are referred to as dynamic force spectroscopy. In the study of the mannuronan-epimerase interaction, the polysaccharide-functionalised tip was brought in contact with the immobilised epimerase enzymes by controlling the z-piezo movement, allowed to form substrate-active cleft interactions, and withdrawn (Fig. 8.5a). Applying a continuous force ramp allowed determination of the most probable unbinding force at various force loading rates as well as determination of the ratio between the epimerase – mannuronan dissociation rate at zero force and the catalytic rate of epimerisation of single residues [48]. The results suggested a processive mode of action. Separate studies were performed on the two modular components, A and R, making up the AlgE enzymes. These studies revealed that the processive mode of action is associated with the A-module. The A-module alone was also found to bind more strongly to the alginate chain than the whole enzyme, and the R-module apparently did not interact directly with the polymer. These observations were taken to

a)

b)

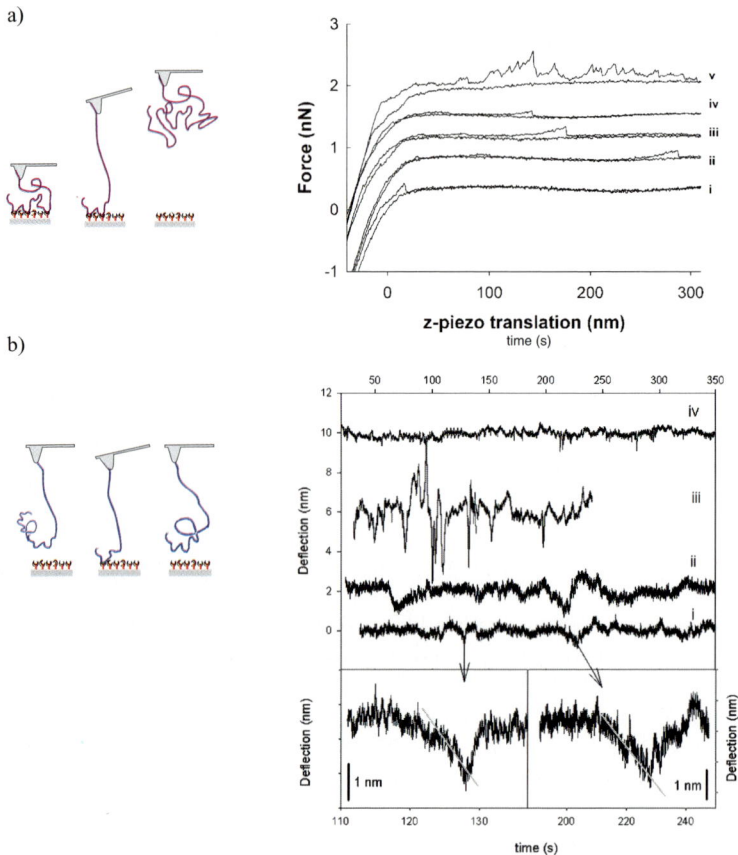

Fig. 8.5. (**a**) *Left*: Schematic illustration of the unbinding experiment between epimerases AlgE4 attached to the mica surface (*bottom*) and mannuronan attached to the AFM tip (*top*) retracted from the surface (*events towards right*). *Right*: Examples of force – z-piezo translation distance curves obtained for mannuronan functionalised AFM tips against C-5 mannuronan epimerase AlgE4 immobilised on mica. Curve i–iv: low mannuronan grafting densities on the AFM tip and low AlgeE4 density mica substrate. Curve v: effect of increasing the AlgeE4 density on the mica substrate. (**b**) *Left*: Schematic illustration of the unbinding experiment between epimerases AlgE4 attached to the mica surface (*bottom*) and mannuronan attached to the AFM tip (*top*) at constant separation 100–250 nm between the polymer functionalised tip and the enzyme covered surface. *Right*: Deflection of the cantilever holding the tip observed as a function of time. Curve i ii: low mannuronan grafting densities on the AFM tip and low AlgeE4 density on the mica substrate. Curve iii: effect of increasing the AlgeE4 density on the mica substrate. Curve iv: deflection of a tip having low mannuronan grafting densities positioned above a uncoated mica surface. The bottom panel shows magnifications of two of the force jumps observed in Curve i. The steepness of the cantilever deflection, indicated by the overlaid grey lines (*bottom panel*), contain information about the translocation speed of the enzyme

indicate that the R-module is involved in the regulation of the enzyme-substrate binding strength [48]. Another experimental setup which would give information concerning the mode of action of potentially processive enzymes is the one illustrated in Fig. 8.5b.

In this set-up, the polymer functionalised tip is held in a fixed position a certain distance above the enzyme-covered surface. The distance separating the tip and the surface should, in order to allow enzyme-substrate interactions, be shorter than the length of the polymers. If this requirement is fulfilled, rare binding events might be observed (Fig. 8.5b). This experimental set-up is interesting because the forces working on the enzyme-substrate complex, trying to bring the two components apart, are kept smaller than in the set-up illustrated in Fig. 8.5a, and the duration of the interaction can therefore be observed over a longer period of time.

Dynamic force spectroscopy studies do not provide detailed information about the length of the oligomer fully saturating the active cleft, and the position of the converted residues in that sequence. These questions were later addressed using NMR spectroscopy, electrospray ionisation mass spectrometry (ESI-MS) and capillary electrophoresis (CE). It was concluded that AlgE4 binds a minimum of 6 residues in the catalytic site, and for short chains the third residue from the non-reducing end is the first to be epimerised [49]. On average, the enzyme epimerises 10 units $((MG)_{10})$ in each reaction before leaving the chain [49]. This corresponds to a mutual translocation distance equal to 10.3 nm.

8.6 Possible Sources of Energy for the Epimerisation of Alginates at the Polymer Level

The studies cited above present convincing evidence that the AlgE4 enzyme is a processive enzyme, which upon each attack slides on average 10.3 nm relative to its substrate. Such a sliding movement of one macromolecule with respect to another one requires energy, but no external source of energy is supplied to the epimerase. It is therefore necessary to look for other possible sources of energy in order to explain this behaviour. As mentioned above, the atoms making up a molecule carry a definite amount of stored energy, and all spontaneous chemical reactions proceed in the direction that lowers the free energy. A negative ΔG value associated with the conformational change of the monomer units making up the polymer might therefore yield the driving force for the epimerisation reaction. This energy might also afford the net energy needed for conformational changes in the enzyme allowing it to relocate with respect to the polysaccharide. Alternatively, this relocation might be driven by the conformational change of the polysaccharide, relocating the polymer with respect to the binding sites in the enzyme.

A factor complicating the picture further is the Ca^{2+} dependence of the epimerases. All the epimerases are dependent on Ca^{2+} ions for activity, and all the AlgE epimerases are reported to have a calcium-binding domain. One possible function of Ca^{2+} is related to stabilisation of the active conformation of the enzyme. Alternatively, alternating binding and liberation of Ca^{2+} might produce repeated conformational changes of the enzyme, possibly underlying the translocation of the enzyme with respect to the polysaccharide. Identification of the molecular mechanism for the observed metal-ion dependence of the epimerisation reaction is complicated by the fact that the substrate also interacts with Ca^{2+} ions. The requirement for Ca^{2+} can therefore, at least in part, be explained by Ca^{2+} complexation of the carboxylate of a mannuronic acid residue about to undergo epimerisation. Homopolymers of mannuronic acid and guluronic acid both binds Ca^{2+} with dissociation constants equal to 2×10^{-4} M and 1×10^{-3} M, respectively [50]. It is conceivable that such complexation would withdraw electron density from the carbon atom α to the carboxylate (C-5), rendering the H-5 more acidic and hence more labile [50].

The 3-dimensional structure of the AlgE4 A-module was recently determined by X-ray crystallography at 2.1 Å [51]. Furthermore, it was recently shown that the interchange of certain amino acids from AlgE4, a processive enzyme, into AlgE2, a non-processive enzyme, converted the AlgE2 into a processive enzyme [52]. Hopefully, this structural information will allow further insight into the mechanism underlying the nanoscale motion of this enzyme.

8.7 High-Order Molecular Assembly of Cellulose

In the two systems described above, polysaccharide modifying processive enzymes utilised enzyme-substrate motion to degrade polysaccharides into oligosaccharides of well-defined size and structure, or to generate a specific sequence pattern. In the next example we focus on a system in which enzyme catalysed polymerisation of polysaccharides may create motion of entire bacteria.

Cellulose is a linear polysaccharide consisting of glucose units linked by β-(1,4)-glycosidic bonds. Structures responsible for cellulose synthesis have been identified by electron microscopy in freeze-fractured plasma membranes of many organisms [53]. Almost half a century ago it was proposed that the growth of a cellulose microfibril is due to the action of an enzyme complex at the tip of the microfibril [54]. These complexes were later given the name "terminal complexes" (TCs) because they are often observed at the end of microfibril imprints. The formation of cellulose microfibrils (CMF) involves a nearly simultaneous process of assembly, polymerisation, and crystallisation. In order to perform all these operations, the cellulose synthesising complexes contain, in addition to the polymerases, other components involved in providing the substrate, initiation or termination of the chain elongation or regulation of the activity of the complex. The spatial arrangement and the number of subunits in the terminal complexes to a large extent determine both the structure and morphology of the microfibrils (square or flat, ribbon-like type) and their width or diameter. Figure 8.6 is a schematic representation of different arrangements found in terminal complexes and the cross-sections of the corresponding cellulose microfibrils produced. That overview shows that the hexagonal structures with six-fold symmetry, rosettes, are observed in many organisms, i.e. mosses, ferns, algae and vascular plants. Each rosette is ∼25 nm in diameter, comprises six subunits and is believed to contain several, possibly six, synthetic units (CesA). The rosettes can be viewed as channels through which glucose from the interior of the cell is drawn, and subsequently polymerised into β-(1,4)-glucan chains. The glucan chains immediately associates with adjacent chains from the same rosette to form elementary microfibrils extruded on the outside of the plasma membrane.

Time-lapse studies have showed that cellulose synthesis in bacterial cells of the *Azetobacter* genera gives rise to a rotational movement of the cells along their longitudinal axes as they spin the cellulose ribbons [56]. This movement is thought to be due to the force of CMF crystallisation. The observed movement of the cell illustrates the strength of the forces of crystallisation. It also illustrates that in nature polysaccharides can, in much the same way as protein filaments, be involved in generating motion of entire cells. The assembly and disassembly of cytoskeletal protein filaments have for long been known to generate pushing and pulling forces that, together with motor proteins, contribute to the correct positioning of structures within cells [57]. The assembly (and disassembly) of these structures provide a source of energy for mechanical force generation.

Growing CMF produced by plant cells also give rise to forces. But, unlike the bacterial cells, these cells are parts of multicellular organisms and therefore do not share the ability of the *Azotobacter* cells to move around their axis. The strain formed by the crystallisation of the microfibrils as they are being deposited is therefore instead assumed to give rise to a displacement of the rosettes in the plasma membrane of the plant cells. If this is the case, the tracks followed by the rosettes would determine the orientation of the CMF in the cell membrane, and thus lead to the cell wall texture. This hypothesis has formed the basis of a model aiming at explaining the different cell wall structures observed in different plants [56, 58]. Previous models aiming at explaining the different patterns observed have claimed that CMF orientation is guided by microtubules [59]. However, this model seems to be ruled out for walls deposited in non-elongating cells, where CMF orientations are uncorrelated to the microtubule organisation [60]. Because these walls are also highly organized this mechanism cannot be considered a sufficient basis for the observed structures. The geometrical model for CMF ordering during their deposition is based on the observation that CMFs always appear approximately evenly spaced in close-packed lamellae and that their average distance apart does not depend on their orientation with respect to the cell axis. Based on these observations a geometrical close packing rule [56, 58] has been formulated. As shown in Equation (8.1), this model relates the CMF

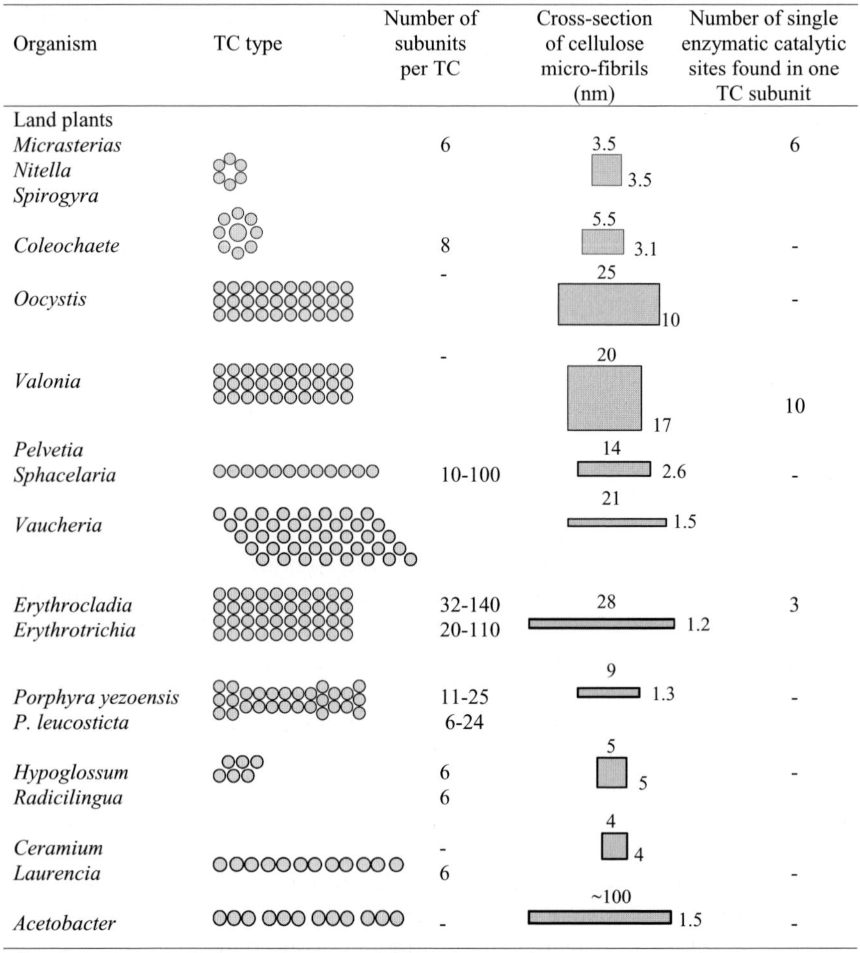

Organism	TC type	Number of subunits per TC	Cross-section of cellulose micro-fibrils (nm)	Number of single enzymatic catalytic sites found in one TC subunit
Land plants				
Micrasterias Nitella Spirogyra		6	3.5 / 3.5	6
Coleochaete		8	5.5 / 3.1	-
Oocystis		-	25 / 10	-
Valonia		-	20 / 17	10
Pelvetia Sphacelaria		10-100	14 / 2.6	-
Vaucheria			21 / 1.5	
Erythrocladia Erythrotrichia		32-140 20-110	28 / 1.2	3
Porphyra yezoensis P. leucosticta		11-25 6-24	9 / 1.3	-
Hypoglossum Radicilingua		6 6	5 / 5	-
Ceramium Laurencia		- 6	4 / 4	-
Acetobacter		-	~100 / 1.5	-

*Each single catalytic site synthesizes one glucan chain.

Fig. 8.6. Organisation of terminal complexes of various organisms and corresponding geometric cross-section of the extruded cellulose microfibril. Redrawn from Fig. 7 of Tsekos [55] with permission from the publisher to reproduce this graph

winding angle α with respect to the long axis of the cells to the number of CMFs being deposited (N), the distance d between them and the radius R of the cell.

$$\sin \alpha = \frac{Nd}{2\pi R} \tag{8.1}$$

The observed CMF ordering is both due to the restricted movement of the rosettes and to the limited area in which the fibrils are deposited as they are opposed during formation against already formed primary or secondary wall material. Observations have indicated that the available space for deposition of CMFs is optimally exploited i.e. the fibrils are packed as densely as possible. This results in a constant average distance between adjacent CMFs within a lamella, independent of their orientation. It also results in a coupling between the orientation of the newly deposited fibrils and the number of fibrils being deposited, which equals the number of active rosettes. This coupling provides the cell with a mechanism for manipulating the CMF orientations by controlling the number of active rosettes and/or the amount of wall matrix.

A diversity of cell wall texture patterns exists in nature. These include texture in which the angle of orientation between subsequent lamellae changes by a constant amount, texture with alternating lamellae with transverse and axial oriented CMFs, purely axial texture, and helical texture in which the CMFs have an almost constant winding angle α. Wood is a natural fibre composite. Each wood cell is typically built of cellulose fibrils spiralling around the macroscopic fibre direction. Experimental results have established a strong correlation between the CMF angle α of the cellulose fibrils in the cell wall and the modulus of elasticity on one hand, and the strain at maximum stress, characterising the extensibility of the cell wall, on the other hand. The variation of fibril angles has been found to provide an optimum combination of stiffness (small angles) and extensibility (large angles), in response to the actual mechanical demands in various parts of a tree. The large α found in juvenile wood confers low stiffness and gives the sapling the flexibility it needs to survive strong winds without breaking. It also means that timber from juvenile wood is unsuitable for use as high-grade structural timber.

The identification and the growing understanding of the structures involved in the CMF formation opens for a deeper understanding of the mechanism of cellulose synthesis as well as future improvements in the in vitro cellulose synthesis. The geometrical model proposes a mechanistic explanation of the formation of different patterns of cell wall textures found in nature. A combination of this information might enable future manipulations and modifications of the process of cellulose formation in plant cell walls for human benefit. The extent to which these techniques should be used to i.e. improve timber quality by reducing the CMF angle α in juvenile wood is, however, uncertain, since care must be taken to avoid compromising the safety of the tree.

8.8 Conclusions

The examples presented above show that in nature, the use of macromolecular motion in order to create specific structures or motion is not restricted to the field of protein filaments. The examples also illustrate the mutual benefit and close connection between applied and fundamental research on enzymes working on industrially important polymers. This connection was underlined in a recent review on enzymatic hydrolysis of cellulose [61], where important outstanding questions were divided into two groups: Fundamentally defined questions with applied implications and application-defined questions with fundamental implications. Progress in our understanding of molecular devices can be sought via either structural modelling, revealing protein function based on protein structure, or via functional modelling, aiming at an aggregated understanding based on knowledge about substrate features and protein function. Single molecule studies offer another interesting approach, which in the future may afford new insight into the function of single molecular devices. One motivation for seeking such understanding is related to the fact that biological machines have potential applications in the fabrication of movable nanodevices. In order to succeed in designing semibiological or fully synthetic molecules capable of performing directed motions in response to certain stimuli it is important to bring together the research areas of biological and artificial molecular machineries, and to initiate a new paradigm for the development of intelligent nanodevices [62]. We believe that in this context it is important to reveal the principles underlying not only the molecular motors playing the most important role in biological systems, but also to advance the knowledge about more rare systems. These systems are not only fascinating in themselves, but they may also reveal a diversity of solutions found in nature to the problems arising when trying to construct a nanodevice. They may therefore both lead to increased understanding of the possibilities and restrictions of such devices, and possibly inspire new solutions to problems met when designing artificial molecular machines.

Acknowledgement

The work was supported by the Norwegian Research Council Grant No. 145523/432.

References

1. P. Nelson (2004). *Biological Physics.* New York: Freeman.
2. R.D. Vale (1999). *Trends Genet,* **15**, pp. M38–M42
3. R. Mallik and S.P. Gross (2004). *Curr Biol,* **14**, pp. 971–982
4. J.F. Robyt and D. French (1967). *Arch Biochem Biophys,* **122**, pp. 8–16
5. P.H. Von Hippel, F.R. Fairfield, and M.K. Dolejsi (1994). *Ann N Y Acad Sci,* **726**, pp. 118–131
6. A. Kornberg and T.A. Baker (1992). DNA replication, 2nd edition. New York: W.H. Freeman.
7. S.P. Gilbert, M.R. Webb, M. Brune, and K.A. Johnson (1995). *Nature,* **373**, pp. 671–676
8. W.A. Breyer and B.W. Matthews (2001). *Protein Sci,* **10**, pp. 1699–1711
9. A.B. Boraston, D.N. Bolam, H.J. Gilbert, and G.J. Davies (2004). *Biochem J,* **382**, pp. 769–781
10. D.N. Bolam, A. Ciruela, S. McQueen-Mason, P. Simpson, M.P. Williamson, J.E. Rixon, A. Boraston, G.P. Hazlewood, and H.J. Gilbert (1998). *Biochem J,* **331**, pp. 775–781

11. J. Sakon, D. Irwin, D.B. Wilson, and P.A. Karplus (1997). *Nat Struct Biol*, **4**, pp. 810–818

12. J. Rouvinen, T. Bergfors, T. Teeri, K.C. Knowles, and T.A. Jones (1990). *Science*, **249**, pp. 380–386

13. C. Divne, J. Stahlberg, T.T. Teeri, and T.A. Jones (1998). *J Mol Biol*, **275**, pp. 309–325

14. G. Parsiegla, C. Reverbel-Leroy, C. Tardif, J.P. Belaich, H. Driguez, and R. Haser (2000). *Biochemistry*, **39**, pp. 11238–11246

15. G. Davies and B. Henrissat (1995). *Structure*, **3**, pp. 853–859

16. Y. Okamoto, A. Inoue, K. Miyatake, K. Ogihara, Y. Shigemasa, and S. Minami (2003). *Macromol Biosci*, **3**, pp. 587–590

17. Y. Usami, S. Minami, Y. Okamoto, A. Matsuhashi, and Y. Shigemasa (1997). *Carbohydr Polym*, **32**, pp. 115–122

18. P. Vander, K.M. Vårum, A. Domard, N.E. El Gueddari, and B.M. Moerschbacher (1998). *Plant Physiol*, **118**, pp. 1353–1359

19. A. Sørbotten, S.J. Horn, V.G.H. Eijsink, and K.M. Vårum (2005). *FEBS Lett*, **272**, pp. 538–549

20. D.M.F. van Aalten, B. Synstad, M.B. Brurberg, E. Hough, B.W. Riise, V.G.H. Eijsink, and R.K. Wierenga (2000). *Proc Natl Acad Sci USA*, **97**, pp. 5842–5847

21. G.J. Davies, K.S. Wilson, and B. Henrissat (1997). *Biochem J*, **321**, pp. 557–559

22. I. Tews, A.C.T. van Scheltinga, A. Perrakis, K.S. Wilson, and B.W. Dijkstra (1997). *J Am Chem Soc*, **119**, pp. 7954–7959

23. D.M.F. van Aalten, D. Komander, B. Synstad, S. Gåseidnes, M.G. Peter, and V.G.H. Eijsink (2001). *Proc Natl Acad Sci USA*, **98**, pp. 8979–8984

24. P. Sikorski, B.T. Stokke, A. Sørbotten, K.M. Vårum, S.J. Horn, and V.G.H. Eijsink (2005). *Biopolymers*, **77**, pp. 273–285

25. K.M. Vårum, M.W. Anthonsen, H. Grasdalen, and O. Smidsrød (1991). *Carbohydr Res*, **211**, pp. 17–23

26. K.M. Vårum, M.W. Anthonsen, H. Grasdalen, and O. Smidsrød (1991). *Carbohydr Res*, **217**, pp. 19–27

27. S. Armand, S. Drouillard, M. Schulein, B. Henrissat, and H. Driguez (1997). *J Biol Chem*, **272**, pp. 2709–2713

28. A. Varrot, M. Schulein, and G.J. Davies (1999). *Biochemistry*, **38**, pp. 8884–8891

29. J.Y. Zou, G.J. Kleywegt, J. Ståhlberg, H. Driguez, W. Nerinck, M. Claeyssens, A. Koivula, T.T. Teerii, and T.A. Jones (1999). *Structure Fold Des*, **7**, pp. 1035–1045

30. D. Nurizzo, J.P. Turkenburg, S.M. Roberts, E.J. Dodson, E.J. Taylor, H.J. Gilbert, and G.J. Davies (2002). *Nat Struct Biol*, **9**, pp. 665–668

31. E.A. MacGregor, S. Janecek, and B. Svensson (2001). *Biochimica et biophysica acta-Protein structure and molecular enzymology*, **1546**, pp. 1–20

32. M. Qian, R. Haser, G. Buisson, E. DuCe, and F. Payan (1994). *Biochemistry*, **33**, pp. 6284–6294

33. M. Machius, L. Vertesy, R. Huber, and G. Wiegand (1996). *J Mol Biol*, **260**, pp. 409–421

34. S. Pages, H.C.M. Kester, J. Visser, and J.A.E. Benen (2001). *J Biol Chem*, **276**, pp. 33652–33656

35. G. van Pouderoyen, H.J. Snijder, J.A.E. Benen, and B.W. Dijkstra (2003). *FEBS Lett*, **554**, pp. 462–466

36. M. Höök, U. Lindahl, G. Bäckström, A. Malmström, and L.-Å. Fransson (1974). *J Biol Chem*, **249**, pp. 3908–3915

37. B. Larsen and A. Haug (1971). *Carbohydr Res*, **20**, pp. 225–232
38. U. Lindahl, G. Bäckström, A. Malmström, and L.-Å. Fransson (1972). *Biochem Biophys Res Commun*, **46**, pp. 985–991
39. A. Malmström, L.Å. Fransson, M. Höök, and U. Lindahl (1975). *J Biol Chem*, **250**, pp. 3419–3425
40. S. Valla, J.-P. Li, H. Ertesvåg, T. Barbeyron, and U. Lindahl (2001). *Biochimie*, **83**, pp. 819–830
41. S.T. Moe, K.I. Draget, G. Skjåk-Bræk, and O. Smidsrød (1995). *Alginate*. In "Food Polysaccharides and their applications", A. Stephen, Ed., Marcel Dekker N.Y. Marcel Dekker N. 245–86.
42. K.I. Draget, B. Strand, M. Hartmann, S. Valla, O. Smidsrød, and G. Skjåk-Bræk (2000). *Int J Biol Macromol*, **27**, pp. 117–122
43. P. Gacesa (1987). *FEBS Lett*, **212**, pp. 199–202
44. D.S. Feingold and R. Bentley (1987). *FEBS Lett*, **223**, pp. 207–211
45. H. Ertesvåg, B. Doseth, B. Larsen, G. Skjåk-Bræk, and S. Valla (1994). *J Bacteriol*, **176**, pp. 2846–2853
46. M. Hartmann, A.S. Duun, S. Markussen, H. Grasdalen, S. Valla, and G. Skjåk-Bræk (2002). *Biochim Biophys Acta*, **1570**, pp. 104–112
47. M. Hartmann, O.B. Holm, G.A.B. Johansen, G. Skjåk-Bræk, and B.T. Stokke (2002). *Biopolymers*, **63**, pp. 77–88
48. M. Sletmoen, G. Skjåk-Bræk, and B.T. Stokke (2004). *Biomacromolecules*, **5**, pp. 1288–1295
49. C. Campa, S. Holtan, N. Nilsen, T.M. Bjerkan, B.T. Stokke, and G. Skjåk-Bræk (2004). *Biochem J*, **381**, pp. 155–164
50. C.A. Steginsky, J.M. Beale, H.G. Floss, and R.M. Mayer (1992). *Carbohydr Res*, **225**, pp. 11–26
51. H.J. Rozeboom, T.M. Bjerkan, K.H. Kalk, H. Ertesvåg, S. Holtan, F.L. Aachmann, S. Valla, and B.W. Dijkstra, in preparation.
52. T.M. Bjerkan, B.L. Lillehov, W.I. Strand, G. Skjåk-Bræk, S. Valla, and H. Ertesvåg (2004). *Biochem J*, **381**, pp. 813–821
53. S. Kimura, W. Laosinchai, T. Itoh, X. Cui, R.C. Linder, and R.M. Brown (1999). *Plant Cell*, **11**, pp. 2075–2085
54. P.A. Roelofsen (1958). *Acta Bot Neerl*, **7**, pp. 77–89
55. I. Tsekos (1999). *Journal of Phycology*, **35**, pp. 635–655
56. B.M. Mulder and A.M.C. Emons (2001). *J Math Biol*, **42**, pp. 261–289
57. M. Dogterom, J.W.J. Kerssemakers, G. Romet-Lemonne, and M.E. Janson (2005). *Curr Opin Cell Biol*, **17**, pp. 67–74
58. A.M.C. Emons and B.M. Mulder (2000). *Trends Plant Sci*, **5**, pp. 35–40
59. C. Wymer and C. Lloyd (1996). *Trends Plant Sci*, **1**, pp. 222–228
60. A.M.C. Emons (1986). *Ultramicroscopy*, **19**, p. 87
61. Y.H.P. Zhang and L.R. Lynd (2004). *Biotechnol Bioeng*, **88**, pp. 797–824
62. K. Kinbara and T. Aida (2005). *Chem Rev*, **105**, pp. 1377–1400

9

Brownian Motion after Einstein: Some New Applications and New Experiments

D. Selmeczi[1,2], S. Tolić-Nørrelykke[3], E. Schäffer[4], P.H. Hagedorn[5],
S. Mosler[1], K. Berg-Sørensen[6], N.B. Larsen[1,5], and H. Flyvbjerg[1,5,7]

[1] Danish Polymer Centre, Risø National Laboratory, 4000 Roskilde, Denmark
   niels.b.larsen@risoe.dk
[2] Department of Biological Physics, Eötvös Loránd University (ELTE), 1117
   Budapest, Hungary
   david.selmeczi@risoe.dk
[3] Max Planck Institute for the Physics of Complex Systems, Nöthnitzer Strasse
   38, 01187, Dresden, Germany
   tolic@nbi.dk
[4] Max Planck Institute for Molecular Cell Biology and Genetics, Pfotenhauer
   Strasse 108, 01307 Dresden, Germany
[5] Biosystems Department, Risø National Laboratory, 4000 Roskilde, Denmark
   peter.hagedorn@risoe.dk
[6] Department of Physics, Technical University of Denmark, Building 309, 2800
   Kgs. Lyngby, Denmark
[7] Corresponding author
   henrik.flyvbjerg@risoe.dk

9.1 Introduction

The first half of this chapter describes the development in mathematical models of Brownian motion after Einstein's seminal papers [1] and current applications to optical tweezers. This instrument of choice among single-molecule biophysicists is also an instrument of precision that requires an understanding of Brownian motion beyond Einstein's. This is illustrated with some applications, current and potential, and it is shown how addition of a controlled forced motion on the nano-scale of the tweezed object's thermal motion can improve the calibration of the instrument in general, and make it possible also in complex surroundings. The second half of the present chapter, starting with Sect. 9.1, describes the co-evolution of biological motility models with models of Brownian motion, including very recent results for how to derive cell-type-specific motility models from experimental cell trajectories.

D. Selmeczi et al.: *Brownian Motion after Einstein: Some New Applications and New Experiments*, Lect. Notes Phys. **711**, 181–199 (2007)
DOI 10.1007/3-540-49522-3_9   © Springer-Verlag Berlin Heidelberg 2007

9.2 Einstein's Theory

When Einstein in 1905 formulated the theory that quickly became known as his theory for Brownian motion, he did not know much about this motion[1]. He was looking for observable consequences of what was then called *the molecular-kinetic theory of heat*. So he was not concerned about the finer details of specific situations. In fact, apart from dated mathematical language, his papers on Brownian motion [1] remain paradigms for how to model the essence of a phenomenon with ease and transparency by leaving out everything that can possibly be left out.

The simplest version of his theory,

$$\dot{x}(t) = (2D)^{\frac{1}{2}} \eta(t) \ , \tag{9.1}$$

for the trajectory $x(t)$ of a Brownian particle, here in one dimension and in the language of Langevin [3,4], works so well also for real experimental situations that its extreme simplicity may be overlooked: No simplification of this theory is possible. The white noise $\eta(t)$ is the simplest possible:

$$\text{For all } t, t', \quad \langle \eta(t) \rangle = 0 \ \text{ and } \ \langle \eta(t)\eta(t') \rangle = \delta(t - t') \ . \tag{9.2}$$

When this noise is normalized as done here–as simple as possible–the dimensions of x and η require that a constant with dimension of diffusion coefficient appears where it does in (9.1). Equation (9.1) is mathematically equivalent to the diffusion equation, introduced by Fick in 1857, in which the diffusion coefficient D is already defined, and that determines the factor $2D$ in (9.1). The new physics was in Einstein's assumption that Brownian particles also diffuse, and in his famous relation, the fluctuation-dissipation theorem

$$D = k_{\mathrm{B}}T/\gamma_0 \ , \tag{9.3}$$

which relates their diffusion coefficient D and their Stokes' friction coefficient γ_0 via the Boltzmann energy $k_{\mathrm{B}}T$. It is derived by introducing a constant external force field in (9.1), and assuming Boltzmann statistics in equilibrium. For a spherical particle,

$$\gamma_0 = 6\pi\rho\nu R \ , \tag{9.4}$$

where ρ is the density of the fluid, ν its kinematic viscosity, and R is the sphere's radius.

9.3 The Einstein-Ornstein-Uhlenbeck Theory

Details left out in the model described in (9.1–9.4) will be found missing, of course, if one looks in the right places. For example, the length of the

---

[1] Just how much he knew seems an open question that may never be answered [2].

trajectory $x(t)$ is infinite for any finite time interval considered[2]. Ornstein and Uhlenbeck [5,6] showed that this mathematical absurdity does not appear in Langevin's equation [3],

$$m\ddot{x}(t) = -\gamma_0 \dot{x}(t) + F_{\text{thermal}}(t) \ , \tag{9.5}$$

where m is the inertial mass of the Brownian particle, and the force from the surrounding medium is written as a sum of two terms: Stokes friction, $-\gamma_0 \dot{x}$, and a random thermal force $F_{\text{thermal}} = (2k_B T \gamma_0)^{1/2} \eta(t)$ with "white noise" statistical properties following from (9.2). The random motion resulting from (9.5) is known as the *Ornstein-Uhlenbeck process* (OU-process). In the limit of vanishing m, Einstein's theory is recovered. Together, they make up the Einstein-Ornstein-Uhlenbeck theory of Brownian motion.

The OU-process improves Einstein's simple model for Brownian motion by taking the diffusing particle's inertial mass into account. As pointed out by Lorentz [7], however, this theory is physically correct only when the particle's density is much larger than the fluid's. When particle and fluid densities are comparable, as in the motion Brown observed, neither Einstein's theory nor the OU-process are consistent with hydrodynamics. This is seen from exact results by Stokes from 1851 and by Boussinesq from 1903 for the force on a sphere that moves with *non*-constant velocity, but vanishing Reynolds number, through an incompressible fluid. Hydrodynamical effects that the OU-process ignores, are more important than the inertial effect of the particle's mass. These effects are the frequency-dependence of friction and the inertia of entrained fluid. Stokes obtained the friction coefficient, (9.4), for motion with constant velocity [8]. Brownian motion is anything but that. Also, mass and momentum of the fluid entrained by a sphere doing rectilinear motion with constant velocity is *infinite* according to Stokes solution to Navier-Stokes equation [8,9]. This gives a clue that entrained fluid matters, and the pattern of motion too.

But since Einstein's theory explained experiments well, this hydrodynamical aspect of Brownian motion did not demand attention. Not until computers made it possible to simulate molecular dynamics.

---

[2] Consider an interval of duration t. Split it into N intervals of duration $\Delta t = t/N$. In each of these, the mean squared displacement of the Brownian particle is $2D\Delta t$. So on the average, the distance travelled in a time interval of duration Δt is proportional to $(\Delta t)^{1/2} \propto N^{-1/2}$. Consequently, the distance travelled in a time interval of duration t is proportional to $t^{1/2} \propto N^{1/2}$. Let $N \to \infty$, and the infinite trajectory has been demonstrated. The proof can be made mathematically rigorous in the formalism of Wiener processes, e.g., which is just the mathematical theory of Brownian motion.

9.4 Computer Simulations: More Realistic than Reality

In 1964–66 Rahman simulated liquid Argon as a system of spheres that interacted with each other through a Lennard-Jones potential [10,11]. He measured a number of properties of this simple liquid, including the velocity auto-correlation function $\phi(t) = \langle \boldsymbol{v}(t) \cdot \boldsymbol{v}(0) \rangle$, which showed an initial rapid decrease, followed by a slow approach to zero from below, i.e., there was a negative long-time tail. Several attempts were made to explain his results theoretically, with mixed success.

In the years 1967–1970 Alder and Wainwright simulated liquid Argon as a system of hard spheres and observed hydrodynamic patterns in the movement of spheres surrounding a given sphere, though all the spheres supposedly did Brownian motion [12, 13]. Using a simple hydrodynamical dimension argument, and supporting its validity with numerical solutions to Navier-Stokes equations, they argued that the velocity auto-correlation function has a positive power-law tail, $\phi(t) \propto t^{-3/2}$ in three-dimensional space. This result is in conflict with the velocity auto-correlation function for the OU-process, which decreases exponentially, with characteristic time m/γ. But the 3/2 power-law tail agrees also with Alder and Wainright's simulation results for a simple liquid of hard spheres doing Brownian motion.

This made theorists [14] remember Stokes' result from 1851 for the friction on a sphere that moves with *non*-constant velocity: There are actually *two* Stokes' laws, published in the same paper [8]. Einstein had used the simplest one, the one for movement with constant velocity, so the effect of accelerated motion is not accounted for in his theory. Nor is it in the Ornstein-Uhlenbeck theory. However, acceleration of a particle in a fluid also accelerates the fluid surrounding the particle, in a vortex ring (in three dimensions, and two vortices in two dimensions) that persists for long, disappearing only by broadening at a rate given by the kinematic viscosity [13]. In this way the fluid "remembers" past accelerations of the particle. This memory affects the friction on the particle at any given time in a manner that makes the dynamics of the particle depend on its past more than inertial mass can express. The result is an effective dynamical equation for the particle, Newton's Second Law with a memory kernel, as we shall see.

9.5 Stokes Friction for a Sphere in *Harmonic* Rectilinear Motion

The friction coefficient that is relevant for a more correct description of Brownian motion, differs from the friction coefficient that most often is associated with Stokes' name, (9.4), but it is actually the main subject of [8]. Stokes was not addressing the hydrodynamics of Brownian motion in 1851, but the hydrodynamics of an incompressible fluid surrounding a sphere that does rectilinear *harmonic* motion with no-slip boundary condition, at vanishing

Reynolds number, and with the fluid at rest at infinity. The equations describing this motion are linear, however, and *any* trajectory of a particle can be written as a linear superposition of harmonic trajectories, by virtue of Fourier analysis [15]. So the flow pattern around a sphere following any trajectory can be written as a superposition of flows around spheres in harmonic motion, as long as the condition of vanishing Reynolds number is satisfied by the arbitrary trajectory. It is for a Brownian particle's trajectory, so Stokes' result for harmonic motion is fundamental for the correct description of Brownian motion.

In general, the instantaneous friction experienced by a rigid body that moves through a dense fluid like water, depends on the body's past motion, since the past motion determines the fluid's present motion. For a sphere performing rectilinear harmonic motion $x(t; f)$ with cyclic frequency $\omega = 2\pi f$ in an incompressible fluid and at vanishing Reynolds number, Stokes found the "frictional" force [8], [9, §24, Problem 5],

$$F_{\text{friction}}(t; f) = -\gamma_0 \left(1 + \frac{R}{\delta(f)}\right) \dot{x}(t; f)$$

$$- \left(3\pi\rho R^2 \delta(f) + \frac{2}{3}\pi\rho R^3\right) \ddot{x}(t; f) \tag{9.6}$$

$$= -\gamma_{\text{Stokes}}(f)\, \dot{x}(t; f) \; ;$$

$$\gamma_{\text{Stokes}}(f) \equiv \gamma_0 \left(1 + (1 - i)\frac{R}{\delta(f)} - i\frac{2R^2}{9\delta(f)^2}\right) \;, \tag{9.7}$$

where only the term containing $\dot{x}(t; f) = -i2\pi f x(t; f)$ dissipates energy, while the term containing $\ddot{x}(t; f) = -(2\pi f)^2 x(t; f)$ is an inertial force from entrained fluid. The notation is the same as above: γ_0 is the friction coefficient of Stokes' law for rectilinear motion with constant velocity, (9.4). The *penetration depth* δ characterizes the exponential decrease of the fluid's velocity field as function of distance from the oscillating sphere. It is frequency dependent,

$$\delta(f) \equiv (\nu/\pi f)^{\frac{1}{2}} = R(f_\nu/f)^{\frac{1}{2}} \;, \tag{9.8}$$

and large compared to R for the frequencies we shall consider. For a sphere with diameter $2R = 1.0\,\mu\text{m}$ in water at room temperature where $\nu = 1.0\,\mu\text{m}^2/\mu\text{s}$, $f_\nu \equiv \nu/(\pi R^2) = 1.3\,\text{MHz}$.

Note that the mass of the entrained fluid, the coefficient to $\ddot{x}$ in (9.6), becomes infinite in the limit of vanishing frequency f, i.e., the flow pattern around a sphere moving with constant velocity has infinite momentum, according to Stokes' steady-state solution to Navier-Stokes' equations.

9.6 Beyond Einstein: Brownian Motion in a Fluid

The friction on a sphere that, without rotating, follows an arbitrary trajectory $x(t)$ with vanishing Reynolds number in an incompressible fluid that is at rest at infinity, is found by Fourier decomposition of $x(t)$ to a superposition of rectilinear oscillatory motions $\tilde{x}(f)$. Using (9.6) on these, gives

$$\tilde{F}_{\text{friction}}(f) = -\gamma_{\text{Stokes}}(f)(-i2\pi f)\,\tilde{x}(f) \ , \tag{9.9}$$

which Fourier transforms back to [15],

$$F_{\text{friction}}(t) = -\gamma_0\,\dot{x} \tag{9.10}$$
$$-\frac{2}{3}\pi\rho R^3\,\ddot{x}(t) - 6\pi\rho R^3 f_\nu^{1/2} \int_{-\infty}^{t} dt'(t-t')^{-1/2}\,\ddot{x}(t') \ .$$

So the Langevin equation (9.5) is replaced by [16, 17]

$$m\ddot{x}(t) = F_{\text{friction}}(t) + F_{\text{external}}(t) + F_{\text{thermal}}(t) \ , \tag{9.11}$$

where F_{external} denotes all external forces on the sphere, such as gravity or optical tweezers, and F_{thermal} denotes the random thermal force on the sphere from the surrounding fluid.

Several authors have derived expressions for the thermal force using different arguments and finding the same result

$$\tilde{F}_{\text{thermal}}(f) = (2k_{\text{B}}T\,\text{Re}\gamma_{\text{Stokes}}(f))^{\frac{1}{2}}\,\tilde{\eta}(f) \ ; \tag{9.12}$$

see overviews in [18, 19][3]. Briefly, Brownian motion in a fluid is the result of fluctuations in the fluid described by fluctuating hydrodynamics [9, Chap. XVII][4]. In this theory one assumes that the random currents split up into systematic and random parts, the former obeying (Navier-)Stokes equation, the latter obeying a fluctuation-dissipation theorem. From this theory one derives the expression of the thermal force on a sphere in the fluid.

Note that this description did *not* invoke a scenario of randomly moving molecules that bump into the micro-sphere and thus cause its Brownian motion. This scenario is correct for Brownian motion *in a dilute gas*. It is of great pedagogical value in undergraduate teaching. But it does not apply to fluids! The scientific literature shows that some undergraduates proceed to become scientists without realizing this limitation on the scenario's validity. However, the coarse-grained description that replaces a molecular description with a hydrodynamical one, is a very good approximation on the length- and time-scales of the thermal fluctuations that drive the Brownian motion of a micron-sized sphere *in a fluid*. This is why fluctuating hydrodynamics [9, Chap. XVII] is

---

[3] Here we have written the frequency-dependent noise amplitude explicitly, and to this end introduced $\tilde{\eta}(f)$, the Fourier transform of a white noise $\eta(t)$, normalized as in (9.2).

[4] Readers familiar with the Green-Kubo theory of linear response to perturbations may appreciate fluctuating hydrodynamics as a case where the order of *linearization* and *"stochastization"* [20, Sect. 4.6] is a non-issue by virtue of the Reynolds number for thermal fluctuations.

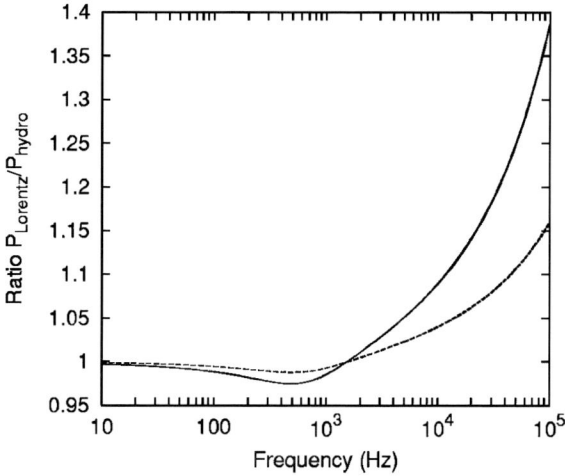

Fig. 9.1. The power spectrum of Brownian motion in an optical trap according to Einstein's theory, P_{Lorentz}, divided by the hydrodynamically correct power spectrum for the same motion, P_{Hydro}; see [21] for explicit expressions for the two spectra. *Fully drawn line:* Trap with Hooke's constant $3.8 \cdot 10^{-2}$ pN/nm for a micro-sphere with diameter 1 μm. *Dashed line:* Hooke's constant $1.9 \cdot 10^{-2}$ pN/nm for a micro-sphere with diameter 0.5 μm. At low time resolution, i.e., low frequency, the error vanishes. Einstein made an excellent approximation when he chose Stokes' law for *constant* velocity to characterize motion along a *fractal* trajectory

formulated by a "stochastization" [20, Sect. 4.6] of Navier-Stokes equation, and *not* by coarse-graining Langevin equations for individual molecules in the fluid. The correct physical scenario to bear in mind is one of molecules squeezed together "shoulder-to-shoulder" in a manner that allows only collective motion, similar to that observed in a tightly packed crowd of people.

Equations (9.10–9.12) constitute the accepted hydrodynamically correct theory for classical Brownian motion, i.e., in an incompressible fluid. It differs from Einstein's theory in a manner that matters in practise with the precision that optical tweezers have achieved recently [21, 22]; see Fig. 9.1.

Power spectra of micro-spheres in optical traps can be measured with stochastic errors below 1% [21]. So the differences in Fig. 9.1 between Einstein's simple theory and the hydrodynamically correct theory for Brownian motion in a fluid can be exposed experimentally [21, 23, 24]. The form of the thermal force in (9.12), on the other hand, remains a theoretical result. It is not a controversial result, it is not questioned. But because it is a small effect, it has not yet been demonstrated experimentally.

9.7 Power-Law Tails

In the absence of external forces, the position power spectrum of Brownian motion following from (9.10–9.12) is

$$P(f) \propto \langle |\tilde{x}^2| \rangle \propto \frac{2k_B T \mathrm{Re}\, \gamma_{\mathrm{Stokes}}(f)}{|m(2\pi f)^2 + i2\pi f \gamma_{\mathrm{Stokes}}(f)|^2} \ . \qquad (9.13)$$

Here, the frequency-dependent numerator is the power spectrum of the thermal force in (9.11), while the denominator is given by the other terms in (9.11). The frequency-dependent friction coefficient, $\gamma_{\mathrm{Stokes}}(f)$, appears both in numerator and denominator, and both appearances contribute, with opposite signs, to the $t^{-3/2}$ power-law tail in the velocity auto-correlation function.

By Wiener-Khintchine's theorem, the velocity auto-correlation function is

$$\phi(t) = \langle \dot{x}(t)\dot{x}(0) \rangle \propto \int_{-\infty}^{\infty} df e^{-i2\pi t f} (2\pi f)^2 P(f) \ . \qquad (9.14)$$

At asymptotically large values of t, $\phi(t)$ is given by $P(f)$'s behavior at small values of f,

$$(2\pi f)^2 P(f) = 2D(1 - (f/f_\nu)^{1/2} + \mathcal{O}(f/f_\nu)) \ . \qquad (9.15)$$

Hence

$$\phi(t) = \frac{D}{2\pi f_\nu^{1/2}} t^{-3/2} + \mathcal{O}(t^{-5/2}) \quad \text{for} \quad t \to \infty \ , \qquad (9.16)$$

quite different from the exponential decrease following from Einstein's simple theory, but not conceptually different from it [14, 16, 17].

Experimental evidence for this power-law tail remained sparse for years. Dynamic light scattering offered promise of its observation, but only Boon and Boullier [25, 26] reported an experimental result of the magnitude predicted theoretically, with statistical errors about half the size of the signal. Paul and Pursey used photon correlation dynamic laser light scattering to measure the time dependence of the mean squared displacement of polystyrene spheres with radius $R \sim 1.7\,\mu\mathrm{m}$ [27]. They found clear evidence for the expected $t^{-3/2}$-behavior ($t^{1/2}$ in the mean squared displacement), but with an amplitude of only $74 \pm 3\%$ of that predicted theoretically. They never found the reason why 26% of the theoretically expected amplitude was missing [28]. Ohbayashi, Kohno, and Utiyama [29] also used photon correlation spectroscopy, on a suspension of polystyrene spheres with radius $0.80\,\mu\mathrm{m}$, and found agreement between the theoretical amplitude of the $t^{-3/2}$ tail and their experimental results which has 9–10% error bars. Their results also agree with the predicted significant temperature dependence. This convincing experiment thus supports the validity of the theory (9.11). This is the current experimental status of the power-law tail of the velocity auto-correlation function of classical Brownian motion.

Or was, when this chapter was written. But before it went into print, [24] appeared. Strangely, the velocity auto-correlation function is not given in [24], though its authors have measured what it takes to display its power-law tail. Instead, they show the mean-squared-displacement of a diffusing micro-sphere. That quantity is essentially the velocity auto-correlation function integrated twice, and consequently contains the same power-law integrated twice.

The amplitude that was measured in all these experiments, albeit indirectly with photon correlation spectroscopy, is the first-order term in the expansion of $P(f)$ above, Eq. (9.15), in powers of $(f/f_\nu)^{1/2}$. This coefficient has two contributions: One from the denominator, from Stokes' frequency-dependent friction coefficient, and another from the numerator. The latter is half-as-large as the former, and with opposite sign. It stems from the noise term's frequency dependence.

Instead of measuring a photon correlation function for laser light scattered off a suspension of micro-spheres, developments in instrumentation [22] and data analysis [21] for optical tweezers have made it possible now to measure directly, with accuracy and precision, on a *single* micro-sphere [24]. Thus it just might be possible to observe directly the "color" of the thermal noise, the frequency dependence of the non-white power spectrum, in a very challenging single-particle experiment with optical tweezers [30].

9.8 In Situ Calibration of Optical Tweezers by Forced Nano-Scale Motion

There are many ways to calibrate an optical trap. Some ways are better than others if accuracy and precision is a concern. In that case, the best way is based on the motion's power spectrum [21]. Two aspects must be calibrated: The spring constant of the Hookean force exerted by the trap on a trapped micro-sphere (bead), and, to this end, the millivolt-to-nanometer calibration factor. The latter tells us which nanometer-displacement of the bead in the trap corresponds to a measured millivolt-change in output potential of a photo diode in the position detection system used with the tweezers. A common way to determine this calibration factor requires that one knows the radius of the bead, the temperature and dynamic viscosity of the fluid surrounding it, and its distance to the nearby surface of the microscope cover slip, if, as is usually the case, the experiment is done near this surface. One can then calculate the bead's diffusion coefficient in m^2/s using Stokes' law (9.4), Einstein's relation (9.3), and Faxén's formula [31,32] [21, Sect. XI]. By comparing the result with the same quantity measured experimentally in V^2/s, the calibration factor is determined.

A calibration of the photo diode that is much less dependent on a priori knowledge, can be achieved by moving the fluid cell with the bead harmonically relatively to the laboratory with the optical trap [33]. With a piezo-electric translation stage this can be done accurately with an amplitude of

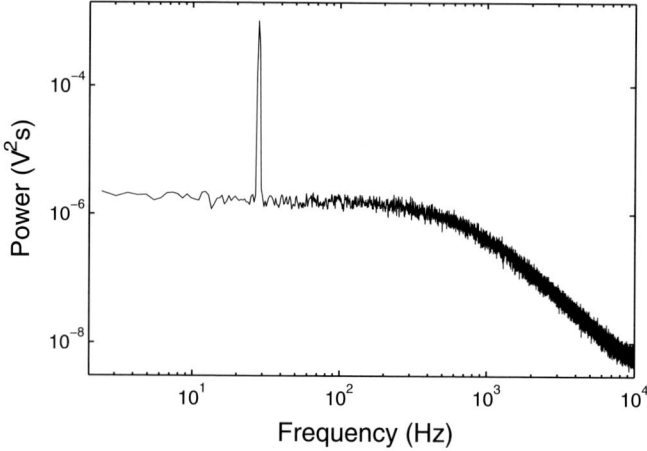

Fig. 9.2. Power spectrum of 1.54 μm diameter silica bead held in laser trap with corner frequency $f_c = 538$ Hz. The sample moves harmonically with amplitude $A = 208$ nm and frequency $f_{stage} = 28$ Hz. The power spectrum shown is the average of 48 independent power spectra, sampled at frequency $f_{sample} = 20$ kHz. The total sampling time was 79 s, which is six times more than we normally would need to calibrate. It was chosen for the sake of illustration, to reduce the relative amplitude of the Brownian motion, i.e., the scatter in the spectrum *away* from the spike at 28 Hz

order 100 nm and frequency of order 30 Hz. In the laboratory system of reference, the fluid flows back and forth through the stationary trap with harmonically changing velocity. This gives rise to an external force on the trapped bead in (9.11), a harmonically changing Stokes friction force,

$$F_{external}(t) = \gamma_0 v_{stage}(t) = \gamma_0 2\pi f_{stage} A \cos(2\pi f_{stage}(t - t_0)) , \qquad (9.17)$$

where A and f_{stage} are, respectively, the amplitude and frequency with which the stage is driven, and t_0 is its phase. The amplitude A can be chosen so small that the forced harmonic motion of the bead in the trap is masked by its Brownian motion, when observed in the time domain. Nevertheless, when observed long enough, the forced harmonic motion stands out in the power spectrum of the total motion as a dominating spike; see Fig. 9.2. This spike is the dynamic equivalent of the scale bar plotted in micrographs: The "power" contained in it is known in m^2 because the bead's motion in nanometers follows from its equation of motion and the known motion of the stage, measured in nanometers. The bead's motion is *measured* in Volts, however, by the photo-detection system, and the Volt-to-meter calibration factor depends on the chosen signal amplification, laser intensity, etc. So calibration is necessary. It is done by identifying the two values for the power in the spike: The measured value in V^2 with the known value in m^2 [33][5]. This method resembles an old method of calibration that moves the bead back and forth periodically with *constant* speed, but harmonic motion has a number of technical advantages. One is that the precision of power spectral analysis demonstrated in [21] can be maintained, while adding the advantage of not having to know the bead's radius, nor its distance to a nearby surface, nor the fluid's viscosity and temperature. On the contrary, the combination of these parameters that occurs in the expression (9.3) for the bead's diffusion coefficient, is determined experimentally from its Brownian motion, so, e.g., the bead's radius is measured to the extent the other parameter values are known. But also, this calibration method can be used in situ, where an experiment is to be done, by confining the bead's forced motion to this environment. This is useful for measurements taking place near a surface, in a gel, or inside a cell.

9.9 Biological Random Motion

Robert Brown did not discover Brownian motion, and he, a botanist, got his name associated with this physical phenomenon because he in 1827 carefully demonstrated what it is *not*, a manifestation of life, leaving the puzzle of its true origin for others to solve. Brownian motion has been known for as long as the microscope, and before the kinetic theory of heat it was natural to assume that "since it moves, it is alive." Brown killed that idea. But after Einstein in 1905 had published his theory for Brownian motion, Przibram in 1913 demonstrated that this theory describes also the self-propelled random motion of protozoa [35]. By tracking the trajectories $x(t)$ of individual protozoa, see Fig. 9.3, Przibram demonstrated that the net displacement $x(t)-x(0)$ averages to zero, while its square satisfies the relationship known for Brownian motion,

$$\langle d(t)^2 \rangle = 2n_{\text{dim}}Dt \ , \tag{9.18}$$

where n_{dim} is the dimension of the space in which the motion takes place.

In Einstein's theory D is the diffusion coefficient, and satisfies his famous relation (9.3). Przibram found a value for D which was much larger and much more sensitive to changes in temperature than Einstein's relation states. He used this as proof that it was not just Brownian motion that he had observed.

---

[5] A spike similar to the one shown here in Fig. 9.2 is seen in [34, Fig. 9.1b]. It was produced with a bead embedded in polyacrylamide, hence not moving thermally, and not optically trapped. It was used to demonstrate the high sensitivity of the authors' position detection system. It was also used for Volts-to-meters calibration of the detection system, and gave 10% agreement with the same calibration factor obtained from the power spectrum of Brownian motion. The optical properties of polyacrylamide differ from those of water, however, so it is an open question how accurate that calibration method can be made. Obviously, it is not an in situ calibration method.

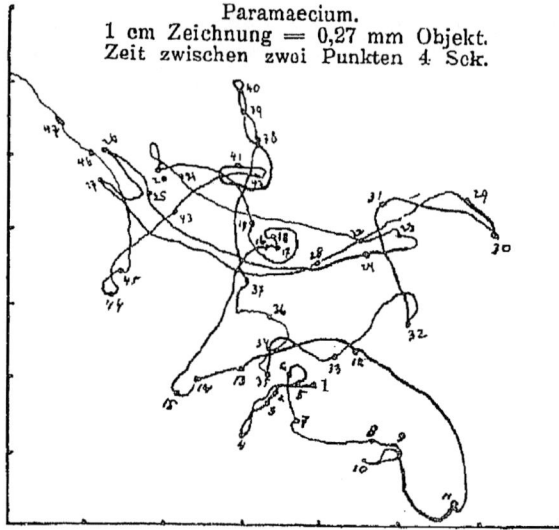

Fig. 9.3. Example of Przibram's motility data, a trajectory of a protozoon, hand-drawn with a mechanical tracking device operated in real time with a microscope. A metronome was used to mark time on the trajectory every four seconds [35]

If Przibram, a biologist, had used a better time resolution by marking out points in Fig. 9.3 more frequently than every four seconds, he might also have gotten ahead of the physicists in theoretical developments. But he was drawing by hand, marking time to a metronome, so marking points closer to 1 Hz must have been a challenge.

Fürth, a physicist at the German university in Prague where Einstein had been a professor for 16 months in 1911–12, also studied the motility of protozoa. First he repeated Przibram's results, apparently without knowing them [36]. Later he found that his data [37] were *not* described by (9.18). He consequently considered a random walker on a lattice, and gave the walker directional persistence in the form of a bias towards stepping in the direction of the step taken previously. By taking the continuum limit, he, independently of Ornstein [5,6], demonstrated that for random motion with persistence, (9.18) is replaced by

$$\langle \boldsymbol{d}(t)^2 \rangle = 2n_{\dim} D(t - P(1 - e^{-t/P})) \ , \tag{9.19}$$

where P is called the *persistence time*, and characterizes the time for which a given velocity is "remembered" by the system [37].

Ornstein solved (9.5), since known as the Ornstein-Uhlenbeck (OU) process. Its solution also gives (9.19), with $P \equiv m/\gamma$. The physical meaning of the three terms in the OU-process does not apply for cells: Their velocities are measured in *micrometers* per *hour*, so their inertial mass means absolutely nothing for their motion. Friction with the surrounding medium also is irrelevant – the cells are firmly attached to the substrate they move on – and it is not thermal forces that accelerate the cells. But as a mathematical model the OU-process is the simplest possible of its kind, like the harmonic oscillator, the Hydrogen atom, and the Ising model. It also agrees with the earliest data. Consequently, the OU-process became the standard model for motility. We can write it as

$$P\frac{d\boldsymbol{v}}{dt} = -\boldsymbol{v} + (2D)^{1/2}\boldsymbol{\eta} \ , \tag{9.20}$$

where each component of $\boldsymbol{\eta}$ is a white noise normalized as in (9.2) and uncorrelated with the other components.

$$\langle \boldsymbol{\eta}(t) \rangle = \mathbf{0} \ ; \quad \langle \eta_j(t')\eta_k(t'') \rangle = \delta_{j,k}\delta(t' - t'') \ . \tag{9.21}$$

Here $\delta(t)$ and $\delta_{j,k}$ are, respectively Dirac's and Kronecker's δ-functions, and $\boldsymbol{\eta}(t)$ is assumed uncorrelated with $\boldsymbol{v}(t')$ for $t \geq t'$. Fürth's formula (9.19) is a consequence of Equations (9.20) and (9.21), but follows also from other, similar theories. It was often the only aspect of the theory that was compared with experimental data, and with good reason, considering the limited quality of data.

Gail and Boone [38] seem to have been the first to model cell motility with (9.19). They did a time study of fibroblasts from mice by measuring the cells' positions every 2.5 hrs. Equation (9.19) fitted their results fairly well. Since then, cell motility data have routinely been fitted with (9.19). Its agreement with data can be impressive, and is usually satisfactory– sometimes helped by the size of experimental error bars and few points at times t that are comparable to P. Data with these properties cannot distinguish (9.19) from other functions that quickly approach $2n_{\dim}D(t - P)$.

Equation (9.19) is essentially a double integral of the velocity auto-correlation function $\phi(t)$ of the OU-process, where

$$\phi(t) = \langle \boldsymbol{v}(0) \cdot \boldsymbol{v}(t) \rangle = \frac{n_{\dim}D}{P}e^{-|t|/P} \ . \tag{9.22}$$

Experimental results for the velocity auto-correlation function are better suited for showing whether the OU-process is a reasonable model for given data. But experimental results for velocities are calculated as finite differences from time-lapse recordings of positions. If the time-lapse is short, precision is low on differences, hence on computed velocities. Yet, if the time-lapse is longer, the time resolution of the motion is poor. The solution is somewhere in between, compensating for lost precision with good statistics. Good statistics was not really achievable till computer-aided object-tracking became possible.

9.10 Enter Computers

We recently wanted to characterize the compatibility of human cells with various surfaces by describing the cells' motility on the various surfaces [39].

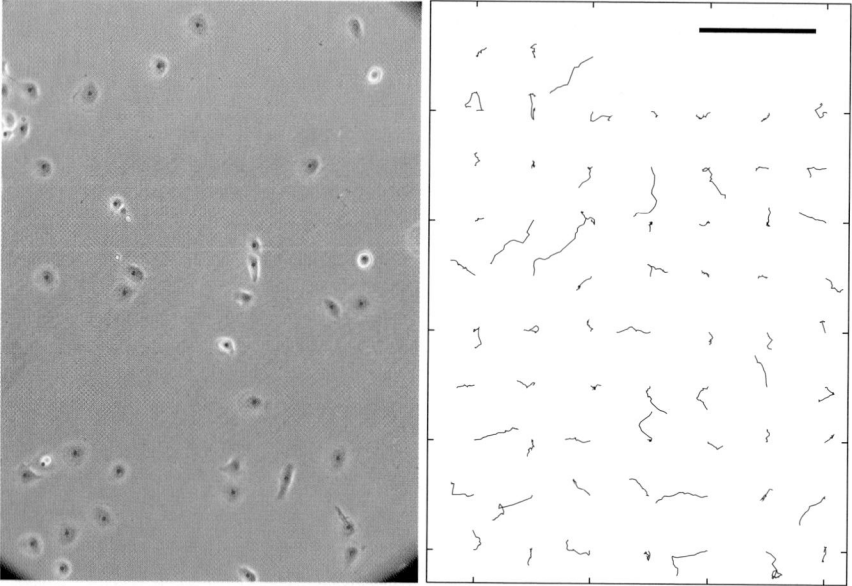

Fig. 9.4. Isolated human dermal keratinocytes are motile by nature. If not surrounded by other cells, they react as if in a wound: They search for other cells of the
same kind with which they can connect to form skin. Trajectories are formed from
15 min time-lapse photography. Trajectories as those shown here in the *right panel*
make up the raw data that are analyzed statistically to find a suitable stochastic
model for the motility of these keratinocytes. The black bar is 0.2 mm long

Computer-aided cell tracking – see Fig. 9.4 – quickly gave us so much data
that we found ourselves in a new situation with regards to modelling: We
were not limited to showing whether or not there is agreement between data
and a few consequences of a given model. We could investigate the model
itself experimentally, measure each term in its defining equation, check that
their assumed properties are satisfied, and whether together they satisfy the
equation of motion.

Furthermore, before we checked the equation of motion, we could check
whether the data are consistent with various assumptions of symmetry and
invariance on which the equation of motion is based. We found that the cells
behaved in a manner consistent with the assumptions that their surroundings
are isotropic, homogenous, and constant in time. This allowed us to average
data over all directions, places, and times. This in turn improved the statistics
of our investigation of the equation of motion [39].

9.11 Tailor-Made Theory Replaces "One Theory Fits All"

The theory in (9.20) states that for a given velocity $\boldsymbol{v}$ the acceleration is a stochastic variable with expectation value proportional to $\boldsymbol{v}$,

$$\left\langle \frac{d\boldsymbol{v}}{dt} \right\rangle_{\boldsymbol{v}} = -\boldsymbol{v}/P \ . \tag{9.23}$$

Figure 9.5a,b shows that this is also the case for experimental data.

The theory in (9.20) states also that

$$\frac{d\boldsymbol{v}}{dt} - \left\langle \frac{d\boldsymbol{v}}{dt} \right\rangle_{\boldsymbol{v}} = \frac{d\boldsymbol{v}}{dt} + \boldsymbol{v}/P = (2D)^{1/2}\boldsymbol{\eta}/P \ , \tag{9.24}$$

i.e., that this quantity in the OU-process is a white noise with the same speed-independent amplitude in both directions: parallel and orthogonal to the velocity.

Figure 9.5b shows that experimentally the amplitude of the two components of this noise are indeed indistinguishable in the two directions, but the two amplitudes are clearly *not* independent of the speed! Here we see the experimental data reject the OU-process as model. The distribution of experimentally measured values of the noise also reject the OU-process as model. Figure 9.5d shows clearly that it is *not* Gaussian, as it is in the OU-process. Apart from that, Fig. 9.5c shows that the noise is uncorrelated, like in (9.21), on the time scale where we have measured it. This result radically simplifies the mathematical task of constructing an alternative to the OU-process on the basis of experimentally determined properties of these cells' motility pattern.

The velocity auto-correlation function of the OU-process is a simple exponential, (9.22). Figure 9.5e shows the experimentally measured velocity auto-correlation function. It is fitted perfectly by the *sum* of two exponentials, so again the experimental data reject the OU-process as model.

The data shown in Fig. 9.5 are so rich in information that with a few assumptions favored by Occam's Razor one can deduce *from the data* which theory it takes to describe the data, and this theory is unambiguously defined by the data [39]. Results from this theory are shown as the fully drawn curves passing through the data points in Fig. 9.5. It is given by the stochastic integro-differential equation

$$\frac{d\boldsymbol{v}}{dt}(t) = -\beta\,\boldsymbol{v}(t) \tag{9.25}$$
$$+ \alpha^2 \int_{-\infty}^{t} dt' e^{-\gamma(t-t')} \boldsymbol{v}(t') + \sigma(v(t))\,\boldsymbol{\eta}(t) \ ,$$

where

$$\sigma(v) = \sigma_0 + \sigma_1 v \ . \tag{9.26}$$

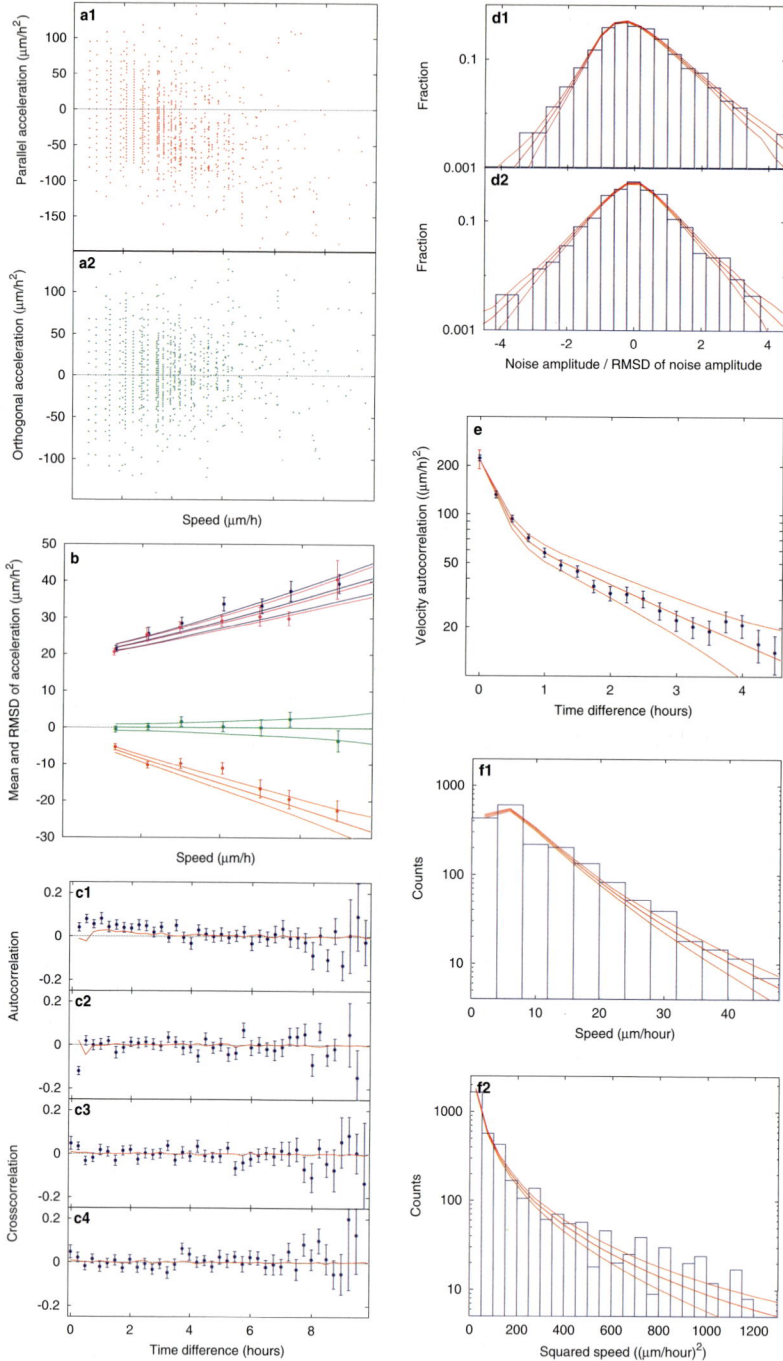

Fig. 9.5. Statistics accumulated from trajectories like those shown in Fig. 9.4. (**a**): The two components of the acceleration, as functions of speed. Panels A1 and A2 show

(*continued*)

The integral over past velocities in (9.25) is called a memory-kernel by mathematicians. It shows that these cells have memory. This is no surprise: The polarity of the cytoskeleton of a moving cell is a manifest memory of direction, and while its instantaneous velocity depends on the activity of transient pseudopodia, the fact that pseudopodia are active depends on states of the cell that last longer than the individual pseudopod, one would expect.

Note the similarity between (9.25) and the hydrodynamically correct theory for Brownian motion, (9.10, 9.11). Though both are more complex than Einstein's theory, applied by Przibram, and the OU-process, applied by Fürth, they still have much in common. This is so because they both are *linear* and both respect *causality* and the same *space-time symmetries*.

Linearity is simplicity, so wherever in modelling it is sufficient, one avoids going beyond it. This is why (9.10, 9.11) and (9.25) are both linear.

The *Principle of Causality* states that the future does not affect the present, including present rates of change of state variables. Only the past can do this. This principle is respected throughout physics, and we have of course built it into our motility models as well. This is why the rate of change of the velocity given in (9.10, 9.11), respectively (9.25), depends only on past

---

Fig. 9.5. (*continued*) the acceleration parallel with, respectively orthogonal to, the velocity. These scatter plots show that the two functions contain random parts, like the acceleration in (9.20). (**b**): Data points with error bars: Mean and standard deviation as function of speed for data shown in Panel A. *Curves* show the same quantities, plus/minus one standard deviation, calculated from the theory given in (9.25). (**c**): Correlation functions for scatter shown in Panel A. Panels C1 and C2 show the auto-correlations of the two components, C3 and C4 show the cross-correlation between the two, for both signs of the time difference. The many values shown are almost all indistinguishable from zero. This suggests that the scatter in data can be modelled with uncorrelated noise, as in (9.21). This is an experimental result for the theory we seek. The *curves* shown are not fits to the data shown, but results of (9.25) after it has been fitted to data in Panels B, E, and F. (**d**): Histograms of scatters shown in Panel A, measured relatively to the means shown in Panel B, and in units of the standard deviations shown in Panel B. The *curves* shown are not fits to the histograms shown, but results of (9.25) after it has been fitted to data in Panels B, E, and F. (**e**): Velocity auto-correlation function, calculated from trajectories like those shown in Fig. 9.4. It is not a simple exponential as in (9.22). But a sum of two exponentials fit data perfectly. So we assume that the theory we seek has a velocity auto-correlation function that is a sum of two exponentials. The *curves* through the data points are that correlation function, plus/minus one standard deviation, computed with the theory in (9.25), after it has been fitted to the data shown here, and simultaneously to the data in Panels B and F. (**f**): Histograms of speeds and (speed)^2 read off trajectories like those in Fig. 9.4. The *curves* shown are the same speed distributions calculated from the theory in (9.25), after it has been fitted to the data

and present velocities. The integral kernels occurring in both equations are *memory* kernels in order to respect this principle.

In a homogenous, isotropic environment that is constant in time, there is no absolute position, direction, nor time. A theory for a dynamical system in such an environment consequently cannot depend on the position variable x, nor can it depend explicitly on the time variable t, nor on explicit directions in space. The theory must be *translation invariant* in space, time, and with respect to direction. The last invariance is called *covariance under rotations*, because a theory for a vector variable like the velocity is not invariant under rotations of the coordinate system, it is covariant, i.e., transforms like the vector it describes. Because these space-time symmetries are shared by hydrodynamics and our cells, neither (9.10, 9.11) nor (9.25) depends on x, nor explicitly on t, and both models transform like a vector under rotations.

We conclude that with the rich data that one now can record and process, one should not be satisfied with the simplest possible model for persistent random motion, the OU-process. "One size fits all" is no longer true, if it ever was. Motility models can be made to measure. Here we have only presented the first phenomenological steps of that process: How to plot and read motility data in a manner that reveals mathematical properties of the theory sought. That done, it is another task to construct a model with the properties demanded. If that can be done, it is yet another task to decide whether the theory is unique or not. Two examples of such theories and their derivation are given in [39].

References

1. A. Einstein (1956). *Investigations on the Theory of the Brownian movement. Edited and annotated by R. Fürth. Translated by A. D. Cowper.* Dover Publications, Inc.
2. J. Renn (2005). *Ann. Phys. (Leipzig)*, 14(Suppl.), 23.
3. P. Langevin (1908). *C. R. Acad. Sci. (Paris)*, **146**, p. 530. Translated and commented in [4].
4. D. S. Lemons and A. Gythiel (1997). *Am. J. Phys.*, **65**, pp. 1079.
5. L. S. Ornstein (1918). *Proc. Amst.*, **21**, pp. 96–108.
6. G. E. Uhlenbeck and L. S. Ornstein (1930). *Phys. Rev.*, **36**, pp. 823–841.
7. H. A. Lorentz (1921). *Lessen over Theoretishe Natuurkunde.* E. J. Brill, Leiden.
8. G. G. Stokes (1851). On the effect of the internal friction of fluids on the motion of pendulums. *Transactions of the Cambridge Philosophical Society*, **IX**, pp. 8–106, Reprinted in Mathematical and Physical Papers, 2nd ed., vol. 3. New York: Johnson Reprint Corp., p. 1, 1966.
9. L. L. Landau and E. M. Lifshitz (1959). *Fluid Mechanics.* Addison-Wesley, Reading, MA.
10. A. Rahman (1964). *Phys. Rev.*, **136**, p. A405.
11. A. Rahman (1966). *J. Chem. Phys.*, **45**, p. 2585.
12. B. J. Alder and T. E. Wainwright (1967). *Phys. Rev. Lett.*, **18**, pp. 988–990.
13. B. J. Alder and T. E. Wainwright (1970). *Phys. Rev.*, **1**, pp. 18–21.

14. R. Zwanzig and M. Bixon (1970). *Phys. Rev. A*, **2**, pp. 2005–2012.
15. J. Boussinesq (1903). *Théorie Analytique de la Chaleur*, vol. II. Paris.
16. A. Widom (1971). *Phys. Rev. A*, **3**, pp. 1394–1396.
17. K. M. Case (1971). *Phys. Fluid*, **14**, pp. 2091–2095.
18. D. Bedeaux and P. Mazur (1974). *Physica*, **76**, pp. 247–258.
19. Y. Pomeau and P. Résibois (1975). *Phys. Rep.*, **19C**, pp. 63–139.
20. R. Kubo, M. Toda, and N. Hashitsume (1985). *Statistical Physics II. Nonequilibrium Statistical Mechanics*. Springer Verlag, Berlin, Heidelberg.
21. K. Berg-Sørensen and H. Flyvbjerg (2004). *Rev. Sci. Ins.*, **75**, pp. 594–612.
22. K. C. Neuman and S. M. Block (2004). *Rev. Sci. Instr.*, **75**, pp. 2782–2809.
23. E. J. G. Petermann, M. van Dijk, L. C. Kapitein, and C. F. Schmidt (2003). *Rev. Sci. Instr.*, **74**, pp. 3246–3249.
24. B. Lukić et al. (2005). *Phys. Rev. Lett.*, **95**, p. 160601.
25. J. P. Boon and A. Bouiller (1976). *Phys. Lett.*, **55A**, pp. 391–392.
26. A. Bouiller, J. P. Boon, and P. Deguent (1978). *J. Phys. (Paris)*, **39**, pp. 159–165.
27. G. L. Paul and P. N. Pusey (1981). *J. Phys. A: Math. Gen.*, **14**, pp. 3301–3327.
28. P. N. Pusey. Private communication.
29. K. Ohbayashi, T. Kohno, and H. Utiyama (1983). *Phys. Rev. A*, **27**, pp. 2632–2641.
30. K. Berg-Sørensen and H. Flyvbjerg (2005). *New J. Phys.*, **7**(38).
31. H. Faxén (1923). *Ark. Mat. Astron. Fys.*, **17**, p. 1.
32. J. Happel and H. Brenner. *Low Reynolds Number Hydrodynamics*. (Nijhoff, The Hague, 1983), p. 327.
33. S. F. Tolić-Nørrelykke, E. Schäffer, J. Howard, F. S. Pavone, F. Jülicher, and H. Flyvbjerg. arXiv: physics/0603037.
34. K. Svoboda, C. F. Schmidt, B. J. Schnapp, and S. M. Block (1993). *Nature*, **365**(6448), 721–727.
35. K. Przibram (1913). *Pflügers Arch. Physiol.*, **153**, pp. 401–405.
36. R. Fürth (1917). *Ann. Phys.*, **53**, p. 177.
37. R. Fürth (1920). *Z. Physik*, **2**, pp. 244–256.
38. M. H. Gail and C. W. Boone (1970). *Biophys. J.*, **10**, pp. 980–993.
39. D. Selmeczi, S. Mosler, P. H. Hagedorn, N. B. Larsen, and H. Flyvbjerg (2005). *Biophys. J.*, **89**, pp. 912–931.

10

Nonequilibrium Fluctuations
of a Single Biomolecule

C. Jarzynski

Theoretical Division, T-13, MS B213, Los Alamos National Laboratory,
Los Alamos, New Mexico 87545, LAUR-05-4612
cjarzyns@umd.edu

Abstract. In recent years it has been realized that equilibrium information is sub-
tly encoded in the fluctuations of a microscopic system that is driven away from
equilibrium, such as a biomolecule stretched irreversibly using optical tweezers. The
key to decoding this information resides in the external work, W, performed on
the system. I will give a brief summary of three theoretical predictions that relate
nonequilibrium statistical fluctuations in W to equilibrium properties of the system,
and which remain valid even in the far-from-equilibrium limit.

Over the past several decades, there has been a growing appreciation that
individual biomolecules and biomolecular complexes are capable of carrying
out remarkably sophisticated tasks. A prototypical example is the enzyme
RNA polymerase, which crawls along a segment of DNA, locally unzips the
two helical strands, and makes a messenger RNA copy of the gene. While this
behavior is in many ways reminiscent of a factory assembly line, it is not immediately clear that the laws of macroscopic thermodynamics, originally derived
with steam engines in mind, can be applied without modification to describe
that operation of a single RNA polymerase. In particular, fluctuations arising
from thermal noise introduce a degree of randomness into the behavior of this
enzyme – or for that matter the behavior of artificial nano-scale machines –
which we do not observe in the deterministic cycles of a properly functioning
macroscopic machine.

While the fluctuations experienced by a system in thermal equilibrium
are well described by traditional, Boltzmann-Gibbs statistical mechanics [1],
and linear response theory in turn reveals that these equilibrium fluctuations
govern the response of the system to small external perturbations [2], much
less is known about the fluctuations of a system *far* from thermal equilibrium.
Recently, however, it has been realized that such fluctuations are governed by
unexpectedly strong and general laws [3]. These results, which originated in
the theoretical community but have become the subject of active experimental
investigations, fall roughly into two classes. The first set describe fluctuations
in the entropy production of a system evolving in (or relaxing toward) a

C. Jarzynski: *Nonequilibrium Fluctuations of a Single Biomolecule*, Lect. Notes Phys. **711**,
201–216 (2007)
DOI 10.1007/3-540-49522-3_10

nonequilibrium steady state [4–13]. The second set pertain to fluctuations in the external work performed when driving a system away from an initial state of thermal equilibrium [14–32]. From the biomolecular perspective, we might expect the former set to reveal something interesting about the operation of a molecular machine in its natural environment (e.g. the RNA polymerase chugging along a piece of DNA can be viewed as a system in a nonequilibrium steady state), whereas the latter are more naturally applied to laboratory experiments in which properties of a system are investigated by observing its response to external perturbations.

Here I will focus on the second set of predictions mentioned above. My aim is to provide a brief, pedagogical introduction to three central results that have been derived in this context, Equations (10.4), (10.6), and (10.10) below. The emphasis is on theoretical, rather than experimental or computational, aspects. For details of experimental tests and applications of these predictions, see the work of Liphardt et al. [22] and Collin et al. [31] involving single-molecule manipulation, and that of Douarche et al. [30] using torsional oscillators. For a very nice discussion of issues related to computational applications of these theories, see Park and Schulten [29].

While the results I will discuss are quite general, they are most naturally presented in the context of single-molecule pulling experiments. Section 10.1 begins with a schematic depiction of such an experiment, then presents the theoretical predictions that constitute the focus of this chapter. These predictions relate the *work* performed on a system during a nonequilibrium thermodynamic process, to *free energy* differences between equilibrium states of the system. Section 10.2 gives a simple proof of one of these predictions, (10.4), for the special case of a thermally isolated system. Finally, Sect. 10.3 briefly discusses the relation between (10.4) and the second law of thermodynamics.

10.1 Setup and Statement of Theoretical Predictions

Figure 10.1 shows a schematic depiction of a single-molecule pulling experiment carried out by Liphardt and collaborators [22]. Two short DNA handles tether a strand of RNA between two microspheres, one of which is held in place by a micropipette, while the other is confined by an optical trap, shown as a harmonic potential. The entire system is immersed in a bath of water, perhaps including some cosolvents, at room temperature and pressure. Let λ denote the distance between the end of the pipette and the center of the optical trap, and imagine that this parameter can be manipulated externally, by varying the position of the trap.

When λ is held fixed, a state of thermal equilibrium is eventually achieved, in which the RNA, DNA, and beads – all subject to continual bombardment by the surrounding solvent molecules – jitter about randomly. These fluctuations

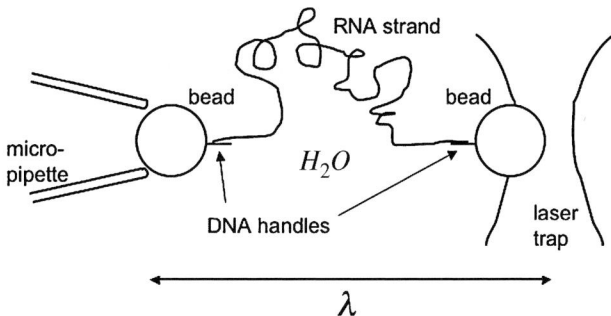

Fig. 10.1. A single strand of RNA, tethered between two beads (microspheres). The relative sizes are not drawn to scale

are described by the familiar formalism of classical statistical mechanics. In particular, the free energy of this equilibrium state is given by

$$G_\lambda = -k_B T \ln Z_\lambda \,, \tag{10.1}$$

where the partition function Z_λ is a sum – over microstates of the system – of the relative probabilities with which the system samples these microstates. (See the Appendix for details.)

The setup just described contains the elements that one ordinarily encounters in a textbook discussion of thermodynamic processes. The RNA strand, DNA handles, and the two beads constitute a *system of interest*; the aqueous bath provides a *thermal environment*; and the pipette-to-trap distance λ is a *work parameter*, akin to an externally manipulated piston pushing against a gas of particles.

Now imagine the following sequence of steps. First, (i) we prepare the system by holding the work parameter fixed at $\lambda = A$, allowing the biomolecule to relax to a state of thermal equilibrium. Next, (ii) we disturb the system, and we perform external *work* on it, by varying the work parameter from its initial value $\lambda_0 = A$ to some final value $\lambda_\tau = B$, according to an arbitrary but pre-determined schedule. The notation λ_t specifies the value of the work parameter at a time t during this interval, from $t = 0$ to $t = \tau$. Let W denote the total amount of work that we perform on the biomolecule during this interval of time. Finally, (iii) we hold the work parameter fixed at the value $\lambda = B$, and allow the system to relax to a new state of equilibrium. No work is performed during this final relaxation stage, although typically there is some systematic exchange of energy between the system and its thermal surroundings. These steps constitute an irreversible thermodynamic process, during which the system begins and ends in equilibrium states (corresponding to $\lambda = A$ and B), but is driven away from thermal equilibrium at intermediate times.

The external work performed on the system is defined as follows:[1]

$$W = \int_0^\tau dt \, \dot{\lambda} \, \frac{\partial H}{\partial \lambda}(\mathbf{x}_t, \lambda_t) \,. \tag{10.2}$$

Here, $\mathbf{x} = (\mathbf{r}_1, \mathbf{r}_2, \cdots ; \mathbf{p}_1, \mathbf{p}_2, \cdots)$ is a many-dimensional vector specifying the *microstate* of the system of interest, i.e. the position and momentum of every one of its microscopic degrees of freedom; and $\mathbf{x}_t$ is the microstate reached by the system at time t during a given realization of the process. The function $H(\mathbf{x}, \lambda)$ is the Hamiltonian of the system, which gives its internal energy in terms of both the microstate and the value of the work parameter. This definition of work (10.2) was introduced implicitly by Gibbs [33] (see Uhlenbeck and Ford [34] for a pedagogical motivation), but differs from the familiar notion of mechanical work as an integral of force versus displacement, $W_{\mathrm{mech}} = \int F \, dz$. Hummer and Szabo [20] discuss the relationship between the two quantities, W and W_{mech}, and when they can or cannot be used interchangeably (to a good approximation), in the context of single-molecule pulling experiments (see also [35, 36]).

As indicated by (10.2), the amount of work W performed during a single realization of this process depends both on how we act on the system (the schedule λ_t for varying the work parameter) and on how the system responds (the microscopic motions described by the trajectory $\mathbf{x}_t$). Since the system is continually jostled by the surrounding water, there is an element of randomness in its response to the perturbation: if we carry out the same process repeatedly – e.g. we stretch the RNA molecule repeatedly, always using the same schedule λ_t – we will typically observe different microscopic trajectories, $\mathbf{x}_t$, and therefore different values of W, from one realization to the next. In other words, the *thermal* fluctuations of the system give rise to *statistical* fluctuations in the amount of work performed on it. These fluctuations in W represent a specific experimental signature of the underlying nonequilibrium response of the system, and are the central focus of this chapter.

10.1.1 Nonequilibrium Work Theorem

Imagine that we repeatedly subject our system to the process described above, carefully noting down the work, W, performed during each realization of the process. After infinitely many realizations, we construct the empirical distribution of these work values, $\rho(W)$, depicted in Fig. 10.2, defined such that $\rho(W) \, dW$ is the fraction of realizations for which the work was observed to fall between W and $W + dW$. This distribution provides a quantitative description of the statistical fluctuations in W.

By the second law of thermodynamics (in the form given by the Clausius inequality [37]), we expect the mean of ρ to exceed the free energy difference between the initial and final equilibrium states:

---

[1] Since λ is presumed to be held fixed during both the initial and final relaxation stages, the integration limits can be extended from $-\infty$ to $+\infty$.

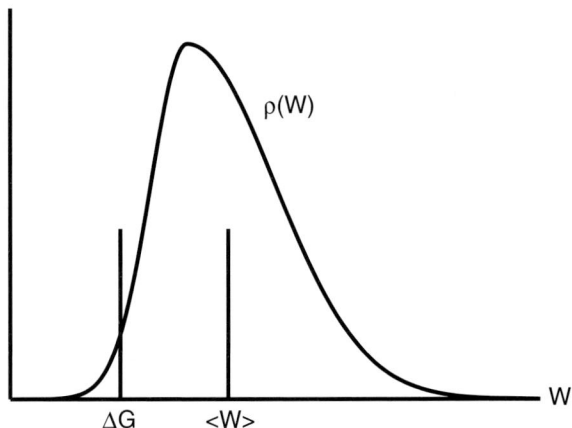

Fig. 10.2. Distribution of work values. According to the second law of thermodynamics, the mean of this distribution is greater than the free energy difference $\Delta G = G_B - G_A$

$$\langle W \rangle \geq \Delta G \ , \tag{10.3}$$

where $\langle W \rangle = \int dW \, \rho(W) \, W$ and $\Delta G = G_B - G_A$ is the free energy difference between the equilibrium states at $\lambda = A$ and B. However, as suggested by the "tail" of $\rho(W)$ extending to the left of ΔG in Fig. 10.2, we might occasionally observe a work value less than the free energy difference, i.e. $W < \Delta G$. Such realizations are fortuitous, occurring when the nearby water molecules jostle the system in a way that facilitates the stretching of the molecule. Events of this sort are sometimes described as "violations" of the second law [11], although it is understood that this pillar of science has in no way been toppled: when $W < \Delta G$, it is simply a case of a relatively large fluctuation in a very small system.

These expectations regarding $\rho(W)$ reflect our understanding of the second law of thermodynamics as a statistical statement. However, the *nonequilibrium work theorem* [14,15] (see also [32]) predicts that fluctuations in W are subject to a constraint that is considerably stronger than what we might reasonably have predicted by extrapolating down from macroscopic experience. Specifically, this theorem states that

$$\langle e^{-\beta W} \rangle = e^{-\beta \Delta G} \ , \tag{10.4}$$

where $\langle e^{-\beta W} \rangle \equiv \int dW \rho(W) e^{-\beta W}$, and $\beta \equiv 1/(k_B T)$, where k_B is Boltzmann's constant, and T denotes the bulk temperature of the surrounding thermal environment.[2]

---

[2] Since we assume that the system is prepared in equilibrium with its surroundings, we can also interpret T as the *initial* temperature of the trapped biomolecule itself. However, once we begin to stretch the biomolecule, its temperature might no longer be well defined, as it is not in thermal equilibrium.

The work distribution $\rho(W)$ depends strongly on the schedule for varying λ: if we change the work parameter very slowly, then ρ will be sharply peaked around ΔG, whereas for fast and violent perturbations we expect a broader distribution, with a mean substantially higher than ΔG. However, according to (10.4), the value of the integral $\int dW \rho(W)e^{-\beta W}$ does *not* depend on the schedule for varying λ, but only on the difference between initial and final equilibrium free energies, ΔG. In this respect the nonequilibrium work theorem extends, to *irreversible* processes, a familiar result from classical thermodynamics, namely that the work performed on a system during a *reversible*, isothermal process depends only on the initial and final equilibrium states [37].

10.1.2 Crooks Fluctuation Theorem

Equation (10.4) pertains to many repetitions of a single thermodynamic process. Let us now consider a comparison between two processes, designated *forward* and *reverse*. The forward process (F) is just the one considered above: the work parameter is varied from A to B according to a schedule λ_t^F. During the reverse process (R), λ is varied from B (at $t = 0$) back to A (at $t = \tau$), using a schedule that is the time-reverse of that used in the forward case:

$$\lambda_t^R = \lambda_{\tau-t}^F . \tag{10.5}$$

Thus if we move the trap away from the pipette at a speed u during the forward proces, then during the reverse process we move the trap toward the pipette at the same speed. Each process (F and R) also includes the initial and final relaxation stages, hence the system begins and ends in thermal equilibrium.

Imagine that we carry out each process infinitely many times, and let $\rho_F(W)$ and $\rho_R(W)$ denote the corresponding work distributions. The Crooks fluctuation theorem [17] then predicts that these distributions are related by the following simple formula:

$$\frac{\rho_F(W)}{\rho_R(-W)} = \exp[\beta(W - \Delta G)] , \tag{10.6}$$

where $\Delta G = G_B - G_A$, as earlier. This prediction applies even if the process is highly irreversible, that is, even if the work parameter is varied sufficiently violently to drive the biomolecule substantially far from equilibrium during the process.

The distributions $\rho_F(W)$ and $\rho_R(-W)$ are depicted schematically in Fig. 10.3. In agreement with the second law of thermodynamics, the means of these distributions are shown to lie to the right and to the left, respectively, of ΔG. [Thus the mean of $\rho_R(W)$ is greater than $-\Delta G = G_A - G_B$.] An immediate consequence of (10.6) is the nonequilibrium work theorem: multiplying both sides of (10.6) by $\rho_R(-W) \exp(-\beta W)$ and integrating over the W-axis, we obtain (10.4), for the forward process. [If we instead multiply both sides by

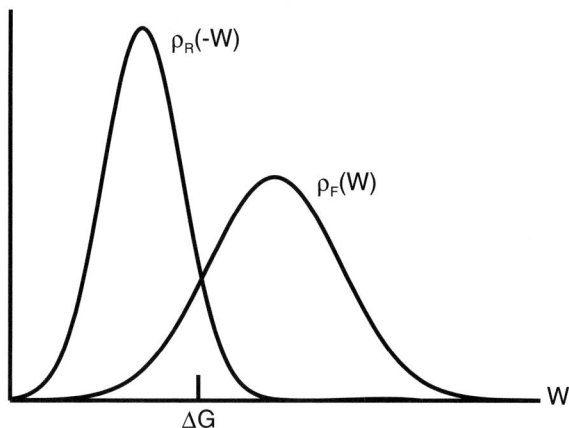

Fig. 10.3. Comparison between two distributions of work values, corresponding to the forward and reverse processes. Note the minus sign in $\rho_R(-W)$: in effect the distribution $\rho_R(W)$ has been reflected around the point $W = 0$

$\rho_R(-W)\exp(\beta\Delta G)$ and integrate, then we get the corresponding prediction for the reverse process.] Another consequence is that the crossing point of $\rho_F(W)$ and $\rho_R(-W)$ occurs precisely at $W = \Delta G$ [31], as seen by setting the numerator and denominator on the left side of (10.6) equal to one another.

10.1.3 Reweighting Theorem

Equations (10.4) and (10.6) are both relatively easy to state, and can be subjected to direct experimental verification. Moreover, these predictions reveal that it is possible to determine equilibrium free energy differences ΔG, from observations of systems that are driven away from equilibrium. A third prediction related to the response of a system driven far from equilibrium, (10.10) below, is somewhat more abstract, but reveals that equilibrium quantities other than ΔG can similarly be determined from nonequilibrium data.

Let us again consider infinitely many repetitions of the thermodynamic process during which the work parameter is varied from A to B (i.e. the *forward* process). Now, however, imagine that during each realization we record the entire trajectory $\mathbf{x}_t$ ($0 \le t \le \tau$) describing the evolution of every degree of freedom of the system of interest. Also, instead of simply noting down the total work, W, we keep track of the work as it is being performed: for a given realization, let w_t be the work that we perform from time 0 up to time t; thus $w_0 = 0$, and $w_\tau = W$.

If $\mathbf{x}_t$ denotes the microscopic history of the system during one realization of the process, then our ensemble of realizations consists of a set of such trajectories, $\{\mathbf{x}_t^{(1)}, \mathbf{x}_t^{(2)}, \cdots\}$, where $\mathbf{x}_t^{(n)}$ is the evolution observed during the n'th realization. Typically, we represent such an ensemble by a time-dependent phase space density,

$$f(\mathbf{x}, t) = \langle \delta(\mathbf{x} - \mathbf{x}_t) \rangle . \tag{10.7}$$

Here as in (10.3) and (10.4), angular brackets denote an average over a statistical ensemble of realizations of the process. If we picture our ensemble of trajectories as a swarm of fictitious particles evolving independently in many-dimensional phase space, then $f(\mathbf{x}, t)$ simply represents the time-dependent number density of these particles (normalized to unity).

For the same ensemble of trajectories, let us now consider an alternative statistical representation, $g(\mathbf{x}, t)$, in which each realization carries around a time-dependent statistical weight, $\exp(-\beta w_t)$:

$$g(\mathbf{x}, t) = \langle \delta(\mathbf{x} - \mathbf{x}_t) \exp(-\beta w_t) \rangle . \tag{10.8}$$

If we imagine that each particle in our swarm has a fictitious, time-dependent "mass", $\mu_t = \exp(-\beta w_t)$, then $g(\mathbf{x}, t)$ is the corresponding time-dependent mass density (normalized to the average particle mass).

The distributions $f(\mathbf{x}, t)$ and $g(\mathbf{x}, t)$ are two different ways of representing the same data, namely the set of microscopic histories $\{\mathbf{x}_t^{(1)}, \mathbf{x}_t^{(2)}, \cdots\}$ observed during repeated stretches of the biomolecule. The difference between them is that while f is "democratic", in the sense that each realization contributes equally to the number density, g is "undemocratic": each realization contributes an amount μ_t at time t. While the assignment of such statistical weights might appear to be an unnecessary complication, it leads to an elegant result, (10.10) below, with useful consequences.

Since we assume an initial state of thermal equilibrium, and since $w_0 = 0$ by definition, both f and g are initially given by the Boltzmann-Gibbs distribution corresponding to $\lambda = A$:

$$f(\mathbf{x}, 0) = g(\mathbf{x}, 0) = \frac{1}{Z_A} \exp[-\beta H^*(\mathbf{x}; A)] . \tag{10.9}$$

Here $\exp[-\beta H^*(\mathbf{x}; \lambda)]$ is the Boltzmann factor describing the equilibrium state at fixed control parameter value λ; the partition function Z_λ is obtained by integrating this factor over all of phase space (see Appendix). Once we drive the system away from equilibrium, $f(\mathbf{x}, t)$ generally evolves in a complicated way, all the more so if the system is driven far from equilibrium. The *work reweighting theorem*, however, asserts that the weighted density $g(\mathbf{x}, t)$ evolves as follows [15, 18, 20]:

$$g(\mathbf{x}, t) \equiv \langle \delta(\mathbf{x} - \mathbf{x}_t) \exp(-\beta w_t) \rangle = \frac{1}{Z_A} \exp[-\beta H^*(\mathbf{x}; \lambda_t)] . \tag{10.10}$$

Note the normalization mismatch: the Boltzmann factor for $\lambda = \lambda_t$ is divided by the partition function for $\lambda = A$. Apart from normalization, however, the right side of (10.10) is exactly the *equilibrium* distribution associated with the current value of the control parameter. In other words, (10.10) predicts that even though the system is out of equilibrium at time t, information about the $\lambda = \lambda_t$ equilibrium state is nevertheless encoded in the ensemble of trajectories $\{\mathbf{x}_t^{(1)}, \mathbf{x}_t^{(2)}, \cdots\}$, and the key to extracting this information is to reweight each realization by a time-dependent factor $\exp(-\beta w_t)$.

By integrating both sides of (10.10) over phase space ($\int d\mathbf{x} \cdots$), we are led immediately to the nonequilibrium work theorem (10.4). Moreover, since (10.10) gives a prescription for constructing equilibrium distributions from repeated observations of the system's behavior as it is driven away from thermal equilibrium, it in principle enables us to use nonequilibrium data to compute equilibrium averages of arbitrary observables. Hummer and Szabo [20] have made use of this property to develop a method for constructing potentials of mean force (equilibrium free energy profiles along a reaction coordinate, rather than as a function of an externally manipulated parameter), using the data from nonequilibrium pulling experiments.

10.2 Proof of Nonequilibrium Work Theorem for a Thermally Isolated System

To date, a number of derivations of the predictions discussed above have been presented in the literature [14–20, 23, 24, 26, 28, 29]. These differ from one another with respect to how one models the microscopic evolution of a system in contact with a heat reservoir. However, under fairly standard assumptions of classical (or quantum [19, 24]) statistical mechanics, these derivations lead to the same results, (10.4), (10.6), or (10.10), suggesting that these predictions are relatively robust.

In this section I present a simple proof of (10.4) for the special case of a system that is thermally isolated as the work parameter is varied. This situation is admittedly not the most physically interesting, or even plausible, in the context of single-molecule pulling experiments, which are inevitably carried out in solution. However, my aim here is simply to illustrate how one might derive this sort of result from first principles, while avoiding some of the technical details that necessarily arise when allowing for the possibility of heat exchange with a surrounding environment. For a derivation that is similar in spirit to the one given below, but which includes heat exchange with a reservoir (and unlike earlier treatments allows for strong interaction between the system and reservoir), see [28].

As above, $\mathbf{x}$ denotes a microstate of the system (a point in its phase space); λ a work parameter; and $H(\mathbf{x}; \lambda)$ the system Hamiltonian. Imagine that we prepare the system (prior to $t = 0$) by placing it in weak thermal contact with a heat reservoir at temperature T, holding the work parameter fixed at $\lambda = A$, and then removing the reservoir after a sufficiently long equilibrium time. At the end of such a procedure, the system is in a microstate $\mathbf{x}$, which can be viewed as a random variable sampled from the Boltzmann-Gibbs distribution

$$p^{\mathrm{eq}}(\mathbf{x}; A) = \frac{1}{Z_A} \exp[-\beta H(\mathbf{x}; A)] \,, \qquad (10.11)$$

with

$$Z_A = \int d\mathbf{x} \, \exp[-\beta H(\mathbf{x}; A)] \,. \qquad (10.12)$$

The free energy corresponding to this equilibrium state of the system of interest is given by $G_A = -\beta^{-1} \ln Z_A$ (10.1). (Both Z and G depend on the temperature at which the system is equilibrated, but I will suppress this dependence in the notation. As noted in the Appendix, for a system that is strongly coupled to a thermal environment, the Hamiltonian H must be replaced by an effective Hamiltonian $H^* = H + \phi$, to account for the free-energetic cost of solvation.)

After removing the heat reservoir, the time evolution of the (now isolated) system is determined by Hamilton's equations:

$$\dot{\mathbf{q}} = \frac{\partial H}{\partial \mathbf{p}} \,, \qquad \dot{\mathbf{p}} = -\frac{\partial H}{\partial \mathbf{q}} \,. \qquad (10.13)$$

Here $\mathbf{q}$ is a vector representing the configurational coordinates of the system, and $\mathbf{p}$ gives the associated momenta, hence $\mathbf{x} = (\mathbf{q}, \mathbf{p})$.

Now imagine that the system evolves in time as we vary the control parameter according to some protocol/schedule λ_t, from the initial value $\lambda_0 = A$ to a final value $\lambda_\tau = B$. Let $\mathbf{x}_t$ denote the trajectory describing the microscopic evolution of the system during this time interval. Since an isolated system does not exchange energy with thermal surroundings, any change in the energy of the system must be due to the work that is performed on it. The total amount of work performed on the system is thus equal to the net change in the value of H:[3]

$$W = H(\mathbf{x}_\tau; B) - H(\mathbf{x}_0; A) \,. \qquad (10.14)$$

Since evolution under Hamilton's equations is deterministic, the microstate reached by the system at the final time τ is determined uniquely by the initial microstate:

$$\mathbf{x}_\tau = \mathbf{x}_\tau(\mathbf{x}_0) \,. \qquad (10.15)$$

Now consider the Jacobian matrix $\partial \mathbf{x}_\tau / \partial \mathbf{x}_0$, whose (i, j) element is the derivative of the i'th component of the vector $\mathbf{x}_\tau$ with respect to the j'th component of $\mathbf{x}_0$. By Liouville's theorem [38], the determinant of this matrix is equal to unity,

$$\left| \frac{\partial \mathbf{x}_\tau}{\partial \mathbf{x}_0} \right| = 1 \,. \qquad (10.16)$$

---

[3] For a system evolving under Hamilton's equations, (10.2) and (10.14) are equivalent, as follows from the identity $dH/dt = \dot{\lambda} \, \partial H / \partial \lambda$. See [38], Equations (8–35).

This has a simple geometric interpretation: phase space volume is preserved under Hamiltonian evolution. That is, a cell of initial conditions in phase space, after propagation under Hamilton's equations, gets mapped onto a cell of final conditions of equal volume.

With these elements in place, the derivation of (10.4) is straightforward. By (10.14) and (10.15), the work performed on the system can be viewed as a function of the initial conditions:

$$W = W(\mathbf{x}_0) = H(\mathbf{x}_\tau(\mathbf{x}_0); B) - H(\mathbf{x}_0; A) . \tag{10.17}$$

The evaluation of $\langle e^{-\beta W} \rangle$ is now a matter of averaging the quantity $\exp[-\beta W(\mathbf{x}_0)]$ over an equilibrium distribution of initial conditions:

$$\langle e^{-\beta W} \rangle = \int d\mathbf{x}_0 \, p^{\mathrm{eq}}(\mathbf{x}_0; A) \, \exp[-\beta W(\mathbf{x}_0)] = \frac{1}{Z_A} \int d\mathbf{x}_0 \, \exp[-\beta H(\mathbf{x}_\tau(\mathbf{x}_0); B)] , \tag{10.18}$$

using (10.11) and (10.17). The last integrand above depends on $\mathbf{x}_0$ *implicitly*, through $\mathbf{x}_\tau(\mathbf{x}_0)$. Because there is a one-to-one correspondence between initial and final conditions, we can perform a change of variables in this integral, replacing the integral over $\mathbf{x}_0$ with an integral over $\mathbf{x}_\tau$. This change of variables involves the usual Jacobian factor, i.e.

$$\int d\mathbf{x}_0 \cdots \quad \rightarrow \quad \int d\mathbf{x}_\tau \left| \frac{\partial \mathbf{x}_\tau}{\partial \mathbf{x}_0} \right|^{-1} \cdots , \tag{10.19}$$

but by (10.16) this factor is equal to unity. We thus get

$$\langle e^{-\beta W} \rangle = \frac{1}{Z_A} \int d\mathbf{x}_\tau \, \exp[-\beta H(\mathbf{x}_\tau; B)] = \frac{Z_B}{Z_A} = e^{-\beta \Delta G} , \tag{10.20}$$

using (10.1). This concludes the derivation.

Note that the Boltzmann factor $\exp[-\beta H(\mathbf{x}_\tau; B)]$ appears in the last step of this derivation. At first glance this seems to suggest that we have somehow slipped in the assumption that at $t = \tau$, the system is in the equilibrium state corresponding to $\lambda = B$. This is not the case, however. The appearance of this Boltzmann factor in (10.20) is the result of a cancellation [in (10.18)] between factors associated with the assumption of an *initial* state of equilibrium, and the expression for W. In general the system is *not* in equilibrium at $t = \tau$.

10.3 Relation to Second Law

Equation (10.4) relates the work performed on a system during a nonequilibrium process (W), to the change in the value of a thermodynamic state function (ΔG). This *equality* immediately implies two *inequalities* that, not surprisingly, are closely related to the second law of thermodynamics. First,

Jensen's theorem [39] tells us that $\langle e^{-\beta W} \rangle \geq e^{-\beta \langle W \rangle}$. Combining this with (10.4) gives us

$$\langle W \rangle \geq \Delta G . \tag{10.21}$$

Thus, the nonequilibrium work theorem is consistent with our expectation that the mean of the distribution $\rho(W)$ is no less than ΔG (Fig. 10.2).

With just a bit more effort we can use (10.4) to derive a more stringent inequality. For a given schedule λ_t, let $P(W < X_n)$ denote the probability of observing a work value no greater than $X_n \equiv \Delta G - nk_BT$, where n is an arbitrary positive constant. We can think of this as the probability of observing an apparent violation of the second law, where the magnitude of the violation (the amount by which W falls short of ΔG) is at least n units of k_BT. This probability is equal to the area underneath the tail of $\rho(W)$, to the left of $W = X_n$, which leads us to the following inequality chain:

$$P(W < X_n) = \int_{-\infty}^{X_n} dW \, \rho(W) \tag{10.22}$$

$$\leq \int_{-\infty}^{X_n} dW \, \rho(W) \, e^{\beta(X_n - W)} \tag{10.23}$$

$$\leq e^{\beta X_n} \int_{-\infty}^{\infty} dW \, \rho(W) e^{-\beta W} = \exp(-n) , \tag{10.24}$$

using (10.4) in the last line. (The first inequality is the result of multiplying the integrand by a factor that is never greater than unity over the range of integration; the second is obtained by extending the upper limit of integration.) This result is independent of both the size of the system and the schedule for perturbing it. It reveals that the likelihood of an apparent violation of the second law diminishes exponentially (or faster!) with the degree of violation: no matter how we stretch the biomolecule, it is extremely unlikely that we will observe a violation whose magnitude is, say, at least $10\,k_BT$ ($P \leq e^{-10} \approx 4.5 \times 10^{-5}$).

10.4 Conclusion and Discussion

My central aim in this chapter has been a brief introduction to three closely related theoretical predictions – (10.4), (10.6), and (10.10) – pertaining to the statistical fluctuations of a microscopic system that is driven away from an initial state of thermal equilibrium. The take-home message is that such fluctuations are not quite as random as might be expected. Even when the system is driven far from equilibrium, these fluctuations satisfy strong constraints, which in principle make it possible to deduce equilibrium properties of a system, from observations of its non-equilibrium behavior.

In principle, the *reversible* unfolding or refolding of a nucleic acid or protein provides the most direct measurement of the free energy difference between the two end states. However, in an actual single-molecule pulling experiment it is often difficult or impossible to carry out the process reversibly. There are two time scales of relevance: the time over which the molecule is stretched (τ), and the intrinsic relaxation time of the molecule (t_{rel}), which can be taken as the transition time between the folded and unfolded configurations (when the externally applied forces are such that the free energies of these two states are equal). For a reversible process, we require $\tau > t_{\text{rel}}$, which might not be experimentally feasible if the free energy barrier between the configurations is high. Moreover, current single-molecule experiments are often subject to (poorly understood) apparatus drift [31], which introduces a systematic error that grows with the duration of the pulling process, τ. Thus, even if it is possible to pull the molecule nearly reversibly, the benefits might be outweighed by the error arising from drift. For these reasons, the nonequilibrium approaches described in this chapter, and validated in recent single-molecule pulling experiments [22, 31], represent a potentially useful route for the experimental determination of equilibrium free energy differences.

Any method based on the theoretical predictions described in this chapter inevitably involves a number of repetitions of the thermodynamic process in question (or of two processes, the forward and the reverse). An obvious practical question involves the number of realizations needed for the method to give an accurate estimate of ΔG. This issue remains to be fully investigated, especially for methods based on (10.6) and (10.10). However, assuming Gaussian work distributions, Gore et al. [40] have suggested that the number of realizations needed for the convergence of (10.4) grows exponentially with the average dissipated work, $\langle W_{\text{diss}} \rangle = \langle W \rangle - \Delta G$. This conclusion is supported by a more recent and general analysis [41].

Another issue of practical importance is the evaluation of the work performed during a given realization of the thermodynamic process. While the value of W is given by (10.2), in an experimental situation this integral must be estimated as a sum over discrete time steps of size δt:

$$W \approx \sum_n \delta \lambda_n \frac{\partial H}{\partial \lambda}(\mathbf{x}_{t_n}, \lambda_{t_n}) , \qquad (10.25)$$

where $\delta \lambda_n$ denotes the increment in the value of the work parameter during the n'th time step, from t_n to $t_{n+1} = t_n + \delta t$. For this approximation to be valid, the time step δt must be small enough to accurately capture the fluctuations in $\partial H/\partial \lambda$ as the system evolves in time. Specifically, the change in $\partial H/\partial \lambda$ between consecutive time steps ought to be small in comparison with the characteristic size of fluctuations in $\partial H/\partial \lambda$. Otherwise, (10.25) will give an overly coarse-grained estimate of (10.2), which will artificially suppress statistical fluctuations in W from one realization of the process to the next.

An explicit expression for $\partial H/\partial \lambda$ in terms of experimental data depends on the physical nature of the work parameter, and its coupling to the system of interest. For instance, if λ specifies the position of a potential well created by laser tweezers, and this trap acts on a bead attached to a polymer (Fig. 10.1), then $\partial H/\partial \lambda$ is simply the instantaneous force that the trap exerts on the bead, along the direction of motion of the trap.

Finally, it bears emphasis that the definition of work given by (10.2), and approximated by (10.25), differs from the more familiar expression for mechanical work as the integral of force over displacement. As mentioned in Sect. 10.1, the criteria for using W_{mech} in place of W were derived in [20]. In the single-molecule pulling experiments of [22,31], these criteria were found to be satisfied. By contrast, in the torsional oscillator experiment of [30], there is a substantial difference between these two definitions of work, hence W_{mech} cannot be used as a substitute for W. The difference between W and W_{mech} is related to the difference between using (10.4) for estimating a free energy difference ΔG, and using (10.10) for estimating a potential of mean force along a reaction coordinate [20]. While (10.4) and (10.6) have been confirmed experimentally [22,30,31], to date there has been no direct experimental test either of (10.10), or of the method proposed by Hummer and Szabo for using this result to reconstruct potentials of mean force [Equation (10.8) of [20]].

This work was supported by the United States Department of Energy, under contract W-7405-ENG-36.

Appendix

When the work parameter is held fixed at a particular value λ, and the system is allowed to exchange energy with its surroundings, then these thermal fluctuations drive the system to a state of equilibrium. Equation (10.1) expresses the free energy of this state, G_λ, in terms of the associated partition function, Z_λ. In textbook discussions of classical statistical mechanics, the partition function is ordinarily defined as an integral of the Boltzmann factor over all microstates of the system, $Z = \int d\mathbf{x}\, e^{-\beta H}$ (omitting prefactors that render Z dimensionless, and which do not contribute to free energy differences). This familiar expression is valid when the interaction energy between the system and its thermal surroundings is weak in comparison with the bare internal energy H of the system, as is generally the case when the system of interest is macroscopic. For a single solvated biomolecule, however, this is typically a poor approximation. Indeed, solvation effects such as thermodynamic sensitivity to the presence of cosolvents, provide direct evidence that the coupling between the biomolecule and its aqueous surroundings is not at all negligible. It is therefore important that the statistical-mechanical formalism properly account for a non-negligible interaction energy.

As discussed in [28] (see also references therein), when the system-environment coupling is substantial, then the equilibrium distribution is given by the usual Boltzmann-Gibbs formula, but with an effective Hamiltonian H^* in place of the bare Hamiltonian H:

$$p^{\mathrm{eq}}(\mathbf{x}; \lambda) = \frac{1}{Z_\lambda} \exp[-\beta H^*(\mathbf{x}; \lambda)] \tag{10.26}$$

$$Z_\lambda = \int d\mathbf{x} \, \exp[-\beta H^*(\mathbf{x}; \lambda)] \, . \tag{10.27}$$

This effective Hamiltonian has the form

$$H^*(\mathbf{x}; \lambda) = H(\mathbf{x}; \lambda) + \phi(\mathbf{x}) \, , \tag{10.28}$$

where $\phi(\mathbf{x})$ can be interpreted as the free-energetic cost of transferring the biomolecule from vacuum to the aqueous solution under consideration. [See Equation (10.20) of [28] for an explicit expression for ϕ]. In the limit of infinitely weak system-environment coupling, $\phi(x)$ vanishes and we recover the usual formula.

References

1. R.C. Tolman (1938). *The Principles of Statistical Mechanics*. Oxford.
2. L. Onsager (1931). *Phys. Rev.*, **37**, p. 405; *Ibid.* (1931). **38**, p. 2265. L. Onsager and S. Machlup (1953). *Phys. Rev.*, **91**, p. 1505.
3. C. Bustamante, J. Liphardt, and F. Ritort (2005). *Physics Today*, **58**, p. 43.
4. D.J. Evans, E.G.D. Cohen, and G.P. Morriss (1993). *Phys. Rev. Lett.*, **71**, p. 2401.
5. D.J. Evans and D.J. Searles (1994). *Phys. Rev. E*, **50**, p. 1645.
6. G. Gallavotti and E.G.D. Cohen (1995). *J. Stat. Phys.*, **80**, p. 931.
7. J. Kurchan (1998). *J. Phys. A*, **31**, p. 3719.
8. J.L. Lebowitz and H. Spohn (1999). *J. Stat. Phys.*, **95**, p. 333.
9. C. Maes (1999). *J. Stat. Phys.*, **95**, p. 367.
10. D.J. Evans and D. Searles (2002). *Adv. Phys.*, **51**, p. 1529.
11. G.M. Wang et al. (2002). *Phys. Rev. Lett.*, **89**, p. 050601.
12. D.M. Carberry et al. (2004). *Phys. Rev. Lett.*, **92**, p. 140601.
13. R. van Zon, S. Ciliberto, and E.G.D. Cohen (2004). *Phys. Rev. Lett.*, **92**, p. 130601.
14. C. Jarzynski (1997). *Phys. Rev. Lett.*, **78**, p. 2690.
15. C. Jarzynski (1997). *Phys. Rev. E*, **56**, p. 5018.
16. G.E. Crooks (1998). *J. Stat. Phys.*, **90**, p. 1481.
17. G.E. Crooks (1999). *Phys. Rev. E*, **60**, p. 2721.
18. G.E. Crooks (2000). *Phys. Rev. E*, **61**, p. 2361.
19. A. Yukawa (2000). *J. Phys. Soc. Japan*, **69**, p. 2367.
20. G. Hummer and A. Szabo (2001). *Proc. Natl. Acad. Sci. (USA)*, **98**, p. 3658.
21. C. Jarzynski (2002). In *Dynamics of Dissipation*, P. Garbaczewski and R. Olkiewicz, eds. Springer, Berlin.
22. J. Liphardt et al. (2002). *Science*, **296**, p. 1832.
23. D.J. Evans (2003). *Mol. Phys.*, **101**, p. 1551.
24. S. Mukamel (2003). *Phys. Rev. Lett.*, **90**, p. 170604.
25. F. Ritort (2003). *Séminaire Poincaré*, **2**, p. 193.
26. S.X. Sun (2003). *J. Chem. Phys.*, **118**, p. 5769.

27. E.G.D. Cohen and D. Mauzerall (2004). *J. Stat. Mech.: Theor. Exp.* P07006.
28. C. Jarzynski (2004). *J. Stat. Mech.: Theor. Exp.* P09005.
29. S. Park and K. Schulten (2004). *J. Chem. Phys.*, **120**, p. 5946.
30. F. Douarche, S. Ciliberto, A. Petrosyan, and I. Rabbiosi (2005). *Europhys. Lett.*, **70**, p. 593.
31. D. Collin et al. "Experimental test of the Crooks Fluctuation Theorem", *Nature* (accepted for publication).
32. For *cyclic* thermodynamic processes, in which the work parameter returns to its initial value ($A = B$, i.e. $\Delta G = 0$), (10.4) reduces to a result derived earlier by G.N. Bochkov and Yu. E. Kuzovlev, *Zh. Eksp. Teor. Fiz.*, **72**, p. 238 (1977) [*Sov. Phys. – JETP*, **45**, p. 125 (1977)]; *Physica*, **106A**, p. 443 (1981); *Physica*, **106A**, p. 480 (1981).
33. J.W. Gibbs (1902). *Elementary Principles in Statistical Mechanics*. Scribner's, New York, 42–44.
34. G.E. Uhlenbeck and G.W. Ford (1963). *Lectures in Statistical Mechanics*. Americal Mathematical Society, Providence. Chap. I, Sect. 7.
35. J.M. Schurr and B.S. Fujimoto (2003). *J. Phys. Chem. B*, **107**, p. 14007.
36. O. Narayan and A. Dhar (2004). *J. Phys. A*, **37**, p. 63.
37. C.B.P. Finn (1993). *Thermal Physics* (2nd Ed.). Chapman and Hall, London.
38. H. Goldstein (1980). *Classical Mechanics* (2nd Ed.). Addison-Wesley, Reading, Massachusetts.
39. D. Chandler (1987). *Introduction to Modern Statistical Mechanics*, Oxford University, New York, p. 137.
40. J. Gore, F. Ritort, and C. Bustamante (2003). *Proc. Natl. Acad. Sci. (USA)*, **100**, p. 12564.
41. C. Jarzynski (2006), *Phys. Rev. E.*, **73**, p. 046105.

11

When is a Distribution Not a Distribution, and Why Would You Care: Single-Molecule Measurements of Repressor Protein 1-D Diffusion on DNA

Y.M. Wang[1], H. Flyvbjerg[2], E.C. Cox[3], and R.H. Austin[1]

[1] Department of Physics, Princeton University, Princeton, NJ, 08544
   austin@princeton.edu
   ymwang@wuphys.wustl.edu
[2] Biosystems Department and Danish Polymer Centre, Risø National Laboratory,
   4000 Roskilde, Denmark
[3] Department of Molecular Biology, Princeton University, Princeton, 08544, USA
   ecox@princeton.edu

Abstract. We address the long-standing puzzle of why some proteins find their targets faster than allowed by 3D diffusion. To this end, we measured the one-dimensional diffusion of LacI repressor proteins along elongated Lambda DNA using single molecule imaging techniques. We find that (1) LacI diffuses along nonspecific sequences of DNA in the form of 1D Brownian motion; (2) the observed 1D diffusion coefficients D_{DNA} vary over an unexpectedly large range, from 2.3×10^{-12} cm^2/s to 1.3×10^{-9} cm^2/s; (3) the lengths of DNA covered by these 1D diffusions vary from 120 nm to 2920 nm; and (4) the mean values of D_{DNA} and the diffusional lengths indeed predict a LacI target binding rate 90 times faster than the 3D diffusion limit. The first half of this chapter is a tutorial on the models we use to think about the physics, the limited and noisy data, and how to squeeze the maximum amount of physics from these data. The second half is about our experiments and results.

11.1 Introduction

Single-molecule techniques have helped emphasize a fact of biological complexity that was for a long time not generally recognized: biomolecules, because of their complex structural elements, can exist in a vast number of different conformational states [1]. Single-molecule work has also helped emphasize another aspect of statistical physics, namely that thermal fluctuations are important at the single-molecule scale in biology. This then leads to a problem in single-molecule experiments, e.g. with a protein: When are you looking at an essentially time-invariant distribution of conformational states of a given protein – time-invariant because they are separated by free-energy-barriers that

Y.M. Wang et al.: *When is a Distribution Not a Distribution, and Why Would You Care: Single-Molecule Measurements of Repressor Protein 1-D Diffusion on DNA*, Lect. Notes Phys.
711, 217–240 (2007)
DOI 10.1007/3-540-49522-3_11   © Springer-Verlag Berlin Heidelberg 2007

are large compared to $k_B T$ – and when are you looking at thermal fluctuations around a single energy minimum?

This may seem like an academic question, but it is important in biology because much of biology is about specificity: proteins make conformational changes which greatly alter their reactivity. Prions are but an extreme example [2]. A less controversial, but biologically much more important, example is the specific recognition of gene control sites by protein complexes [3]. We will explore in this paper a related issue to conformational changes in proteins: the statistics of single molecule diffusion measurements, and the extraction of reliable values for diffusion coefficients in single molecule experiments, which involve protein-DNA complexes. The connection to conformational distributions should be clear (we hope) by the end of the chapter.

11.2 Random Walks, Random Motion, Diffusion

First, some fundamental statistical mechanics before we get to the biology[1]. In single molecule measurements the experimenter watches the motion of a single molecule in time. Let's consider the simplest of examples, a trajectory of a particle doing random Brownian motion in one dimension, i.e., simple diffusion. About names: A *random walk* consists of discrete steps, often of fixed length, and usually taken at fixed time intervals. So random walks typically take place in discrete time, and often in discrete space as well. The direction of each step is chosen at random, and in the simplest and most common type of random walk, each step is chosen independent of all previous steps, and with equal probability for all directions. *Random motion*, on the other hand, usually denotes motion in continuous space and time, such as diffusional motion of a particle. Thus *random motion* has a continuous trajectory $x(t)$, where x is a real-valued spatial coordinate and a continuous function of a real variable t that parameterizes continuous time – or x is *a set* of such coordinates in case of motion in two or more dimensions.

About geometry: The trajectory $x(t)$ of a random motion is a continuous variable, but in its simplest mathematical form, described by Langevin's equation from 1908 (See Eq. (11.1) in the chapter by Selmeczi et al. in this volume),

---

[1] Many textbooks give good introductions to this subject. Howard Berg's delightful *Random Walks in Biology* (Princeton University Press, 1993) is gentle *and* solid. *The Feynmann Lectures on Physics* (Benjamin Cummings, 2005) are always good for a good introduction. For this particular subject, with the added twist of something rare from Feynman: The opening sentence on Brownian motion is totally wrong! Robert Brown did *not* discover Brownian motion! Its discovery is, naturally, as old as the microscope [4, 5, Easily available thanks to Peter Hänggi, at www.physik.uni-augsburg.de/theo1/hanggi/History/BM-History.html]. Another pedagogical introduction is given in F. Reif, *Fundamentals of Statistical and Thermal Physics* (McGraw-Hill, 1965). Van Kampen's book is a standard graduate-level [6].

it is not a *smooth* motion: It has no tangent vector, but zigzags in such a crazy manner that the *length* of its trajectory is *infinite*, no matter for how short a time it is measured! The trajectory is actually a *fractal*! This obviously is not a realistic physical description of diffusion, because the actual path traced out in finite time must have a finite length. The infinite trajectory arises, because the inertial mass of the diffusing particle is ignored in Langevin's equation. This mathematical artifact is the price paid for the, in other ways, mathematical simplicity of Langevin's description: It has *no inherent length-scale*. So if you come across the statement that "diffusion has no inherent length-scale," this is what is referred to. Check the diffusion equation:

$$\frac{\partial}{\partial t}\rho(x,t) = D\frac{\partial^2}{\partial x^2}\rho(x,t) \ . \tag{11.1}$$

Do you see a length-scale anywhere? The density or concentration ρ is a function of space that depends on time, and may of course be created initially with a built-in length scale. But the equation itself has none. The dimension of the diffusion coefficient D is obviously $(\text{length})^2/(\text{time})$, and any length-scale built into $\rho(x,0)$ will change with time, and become increasingly difficult to discern in $\rho(x,t)$.

The diffusion equation is a *very* simple equation. It describes the deterministic time evolution of the density of a large number of identical, non-interacting particles that all behave the Langevin equation, so it can be deduced from the Langevin equation [6]. It consequently contains the same artifact of infinitely long transport in any finite time: Let all particles in an ensemble be located at $x = 0$ at $t = 0$. Then at any later time, however shortly after $t = 0$, there is finite probability for particles everywhere on the x-axis, according to the solution $\rho(x,t) = (4\pi Dt)^{-1/2}\exp(-x^2/(4Dt))$ to the diffusion equation with this particular initial distribution.

Mathematicians actually had to develop a whole new formalism in order to speak correctly about Brownian motion. They invented the *Wiener process* for the purpose, a new kind of function at its time of invention [6]. It is just Brownian motion. Except, what is meant by that? Well, this is what the mathematicians pinned down: You may think of the Wiener process as a limit of random walks with steps in space and time made infinitesimal. This limit is similar in spirit to approximating a *smooth* trajectory with a sequence of straight-line trajectories. The only difference is that the sequence of straight-line trajectories zigzags freely in the case of the Wiener process.

We mention all this in order (i) to teach you a few words that you may come across; (ii) to make sure that you *don't* think of random walks as smooth curves. Check Fig. 11.1, the curves are *not smooth*; (iii) to make you realize that all this goes on "under the hood," so to speak, of the innocent-looking diffusion equation when you look into the behavior of the individual diffusing particle; and, not the least, (iv) to explain to you how the computer program works that produced the data for this paper. It simulates Brownian motion, obviously, or we couldn't use it for illustration. But computers can only work

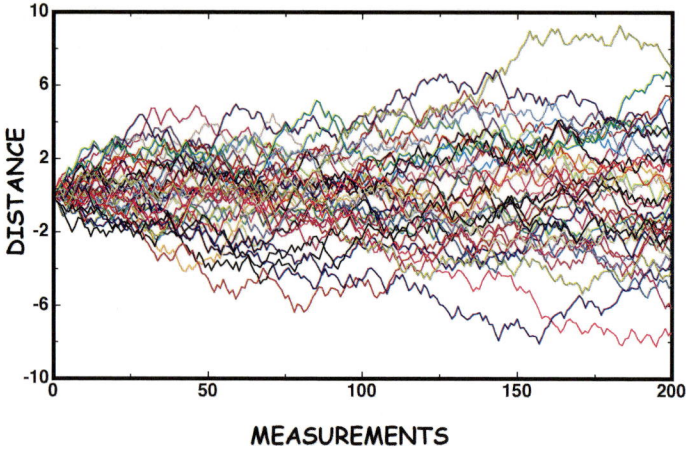

Fig. 11.1. Forty random paths, outlined as 200 points on each path. The position coordinate of each point is plotted against its time of recording. The plotted points are connected with *straight lines* to guide the eye. The true trajectories are fractals that zig-zag from one plotted point to the next. The length- and time-scales used are arbitrary. In a single-molecule measurement, each one of these paths can be traced in time. In an ensemble measurement, on the other hand, a large number of particles is simultaneously measured but are not individually distinguished from one another

with finite numbers, so we didn't go all the way to the limit, just close enough to be far away from all physical scales. Then physical results extracted from our simulation do not depend on our discretization of time – the usual procedure in numerical analysis.

11.3 Einstein's Theory for Brownian Motion

If by now you feel thrown into the pool at its deep end, you may find consolation in the fact that the 26-year-old patent clerk with a four-year university education who first described Brownian motion correctly, in 1905, knew none of this, as he was only about to prompt its development. He was a physicist, and knew that inertia would smooth the trajectory of the thermal motion of the micron-sized particles that he was thinking about, but would do this on a scale of space and time that could not be resolved in 1905. So he thought up an *effective* theory, valid at the scales that *could* be resolved with an optical microscope and was correspondingly simpler.

In this theory [7], the observer records the position $x_i = x(t_i)$ of the particle at discrete times $t_i = i\Delta t$ where Δt is so large that displacements $\Delta x_i = x_i - x_{i-1}$ corresponding to different i-values are effectively uncorrelated, i.e., $\langle \Delta x_i \Delta x_j \rangle = 0$ for $i \neq j$. Here the angular brackets denote the expectation value, a theoretical ideal that is approached by making many, say M, identical

experiments, and then calculate the average over these experiments of the
entity inside the brackets. Thus, if we record M trajectories of the *same*
particle, each trajectory consisting of N measurements, we can compute the
mean squared displacement achieved in N time-steps:

$$\langle (x_N - x_0)^2 \rangle = \left\langle \left(\sum_{i=1}^{N} \Delta x_i \right)^2 \right\rangle$$

$$= \sum_{i=1}^{N} \langle (\Delta x_i)^2 \rangle + \sum_{i \neq j=1}^{N} \langle \Delta x_i \, \Delta x_j \rangle$$

$$\sim N \langle (\Delta x)^2 \rangle = \left(\langle (\Delta x)^2 \rangle / \Delta t \right) t \ . \tag{11.2}$$

Here we have said that $\langle (\Delta x_i)^2 \rangle$ does not depend on i, and have called it
$\langle (\Delta x)^2 \rangle$. This is really an assumption: we assume that the random motion we
are considering, is a *stationary process*, i.e., everything about it is constant in
time, including the probability distribution from which it derives its random
nature.

The problem with this approach is that one has to be assured that one
always is measuring the same particle (different particles could have different
values for D!), and it requires $M \gg 1$ independent trajectory measurements:
one often doesn't have that luxury in optical measurements, where the measurement process can destroy or alter the sample.

The diffusion equation is much older than Einstein, and it has the property
that the second moment of a concentration $\rho(x,t)$ that at time $t = 0$ is
concentrated at $x = 0$, is $2Dt$. By assuming that colloid particles also diffuse,
Einstein thus had the prediction that the thermal motion of a colloid particles
should satisfy

$$\langle (x(t) - x(0))^2 \rangle = 2Dt \tag{11.3}$$

where the diffusion coefficient D is a so far unknown constant of proportionality, but *a property of the individual particle*.

Already Eq. (11.3) was a great step forward in understanding thermal
motion: It states that the mean *squared* displacement is proportional to time.
Experimenters at the time had tried to measure the *speed* of thermal motion,
and arrived at conflicting results. Equation (11.3) explains this. They got the
mean speed by looking at the mean *displacement*, as in $v = N^{-1} \sum_{i}^{N} |\Delta x_i| / \Delta t$.
In this formula, $\Delta x_i / \Delta t$ gives approximately the speed at time t_i if the trajectory is *smooth* and Δt small. But Einstein's theory states that trajectories
are *not* smooth, since the displacements Δx_i, $i = 1, \ldots, N$, are uncorrelated
random increments. Also, Eqs. (11.2) and (11.3) taken together give that
$\langle |\Delta x| \rangle \propto \sqrt{\Delta t}$ if D is a physical quantity, hence independent of our choice
of Δt. Consequently, $v \propto (\Delta t)^{-1/2}$. So you can see why experimenters got
conflicting results for v: their results depended strongly on their choice of Δt!
You can also see that the distance traveled along the trajectory in a time t is
$vt \propto t/(\Delta t)^{1/2}$, which diverges in the limit $\Delta t \to 0$ that defines the Wiener

process. So now you have seen that the trajectory has infinite length in finite time, in theory.

Einstein also considered diffusion in a gravitational field, of colloid particles experiencing a downward net force mg from gravitation minus buoyancy, and a Stokes friction force $-\gamma\dot{x}$ for particles sinking with velocity $\dot{x}$. He found that a steady-state-distribution would result, with density profile $\rho(x) \propto \exp(-mgx/(\gamma D))$. Boltzmann had already found that this distribution should be $\rho(x) \propto \exp(-mgx/(k_B T))$. In this way Einstein found his famous relation for the diffusion coefficient,

$$D = k_B T/\gamma \ , \tag{11.4}$$

and suggested that one could use Stokes result from 1851 for the friction coefficient,

$$\gamma = 6\pi\eta a \ , \tag{11.5}$$

where a is the radius of the colloid particle and η is the dynamic viscosity of the surrounding liquid. Stokes derived this result for a sphere moving with constant velocity and zero Reynolds number through a liquid at rest. Einstein used it for a particle moving with anything *but* constant velocity through a liquid in incessant thermal motion. Nevertheless, it is an extremely good approximation; see Fig. 11.1 in Selmeczi et al. in this volume. The relative importance of corrections to Einstein's theory coming from hydrodynamical and inertial effects is also shown in this figure, as a function of the time scale.

11.4 The Problem of Tracing Single Trajectories

With all this theory in place, you might also appreciate the following experimental problem: An experimenter follows a particle that does random motion, and records the position $x_i = x(t_i)$ at discrete times $t_i = i\Delta t$. The particle is unaffected by the fact that its position is recorded at these discrete times. The question now is: How to measure a particle's diffusion coefficient from its trajectory?

The biggest problem is the measurement process itself (such as fluorescence excitation): the exciting light can very rapidly destroy the chromophore that makes tracking possible. We may not be able to record M trajectories. We might be able to make just N measurements on one trajectory, in fact. Can't we just drop the ensemble average in Eq. (11.3), and gate our measurement light so that we only make two measurements, one at time 0 and one at time t? Is there a time t that is sufficiently long to make a good approximation of

$$(x_N - x_0)^2 = 2Dt \ ? \tag{11.6}$$

We know the left-hand side, we know t, so we can deduce D *if* this relation is OK. It is not at all OK, however. The problem is the size of the fluctuations

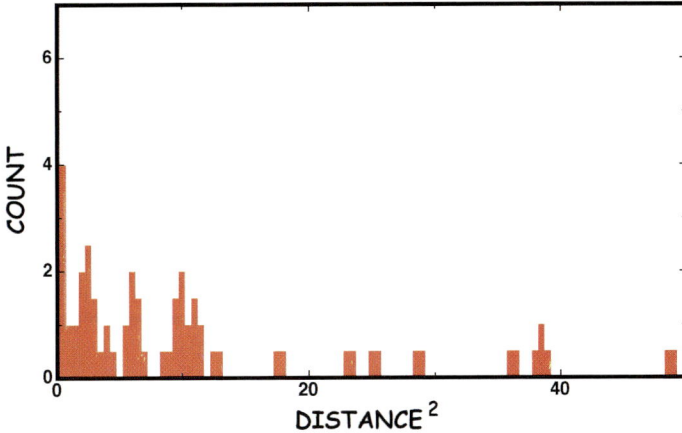

COUNT

DISTANCE^2

Fig. 11.2. A histogram of the distribution of net distances from the origin after 200 measurements

of x_N. Figure 11.2 gives a histogram of the quantity $(x_N - x_0)^2$ for our 40 computer simulated trajectories, and the distribution of diffusion coefficients that would come from that histogram if we used Eq. (11.6). That distribution of D-values is wrong, all the trajectories have the same parent diffusion coefficient.

The diffusion equation, Eq. (11.1), describes these. It describes how the probability distribution for the particle's position at time t evolves from a delta-function distribution at time zero: For $\rho(x,0) = \delta(x - x_0)$, Eq. (11.1) gives $\rho(x_N, t) = (4\pi D t)^{-1/2} \exp(-(x_N - x_0)^2/(4Dt))$. So $x_N - x_0$ has a *random value*, Gaussian distributed with variance $2Dt$. It consequently doesn't matter for how long we measure: $(x_N - x_0)^2$ is also a random number, and does not *self-average* as the technical term goes. On the contrary, it scatters more widely the larger t is. While its expectation value is $2Dt$, as we have already mentioned, a small calculation gives that its standard deviation (SD) is $2\sqrt{2}Dt$. So the signal-to-noise ratio in Eq. (11.6) is $2^{-1/2}$, i.e., less than one! That is why Eq. (11.6) is no good for estimations of D.

11.5 Time Average vs Ensemble Average

What was said above for the interval t, holds also for the interval Δt. Thus $x_i - x_{i-1}$ is Gaussian distributed with expectation value zero and variance given by Eq. (11.3),

$$\langle (x_i - x_{i-1})^2 \rangle = 2D\Delta t \ , \tag{11.7}$$

while $(x_i - x_{i-1})^2$ has the expectation value just given and standard deviation $2\sqrt{2}D\Delta t$. This SD exceeds the expectation value. But $(x_i - x_{i-1})^2$ we have measured N times, one for each value of $i = 1, \ldots, N$, and these measurements are statistically independent, because the N intervals do not overlap. So consider the time average

$$\overline{(\Delta x)^2} = N^{-1} \sum_{i=1}^{N} (\Delta x_i)^2 \ . \tag{11.8}$$

Its expectation value is $2D\Delta t$, but its SD is $2D\Delta t \sqrt{2/N}$, so its signal-to-noise ratio is $\sqrt{N/2}$. Consequently, the more positions x_i, $i = 0, \ldots, N$ we measure, the more precisely we determine D, when we estimate it as

$$D \sim \overline{(\Delta x)^2}/(2\Delta t) \ . \tag{11.9}$$

11.6 Check First, Interpret Later

Estimating values for parameters in a theory is one thing, proving data consistent with the theory is another. The former makes sense only after the latter has been done. Thus estimation of D, as just described, is a correct approach only if Einstein's simple theory describes all aspects of data. Not that there are that many aspects, according to his theory, because it assumes that the displacements Δx_i, $i = 1, \ldots, N$, are uncorrelated Gaussian variables.

So we must first demonstrate that the displacements we have measured, are not auto-correlated, i.e., we must demonstrate that

$$\overline{\Delta x_i \Delta x_{i+j}} = 0 \quad \text{for} \ j > 0 \tag{11.10}$$

is satisfied up to statistical errors on this quantity. Here the bar denotes time-averaging over the dummy variable $i = 0, \ldots, N - j$. Consequently, the statistical error to compare with, the standard deviation, is

$$\sigma(\overline{\Delta x_i \Delta x_{i+j}}) = \langle (\Delta x)^2 \rangle / \sqrt{N - j} \ , \tag{11.11}$$

a brief calculation shows. That done, we must show that a histogram of the recorded values for the N displacements Δx_i, $i = 1, \ldots, N$, is described by a Gaussian with zero mean and, obviously, variance $\overline{(\Delta x)^2}$.

11.7 And Now with Experimental Errors ...

In the two previous sections we found stochastic errors on our results which depended on our measurements as $N^{-1/2}$ and $(N - j)^{-1/2}$ with no mentioning of t, the duration of our measurement. Apparently, we can achieve *any degree of precision* on our estimates for D and the auto-correlation function for displacements merely by recording positions x_i more frequently in the same time interval! This seems so because there is no inherent time- nor length-scale in Einstein's theory for Brownian motion. In reality – and here we mean physical reality – he knew and we know that inertial motion and, more important, hydrodynamic effects not accounted for with Stokes simple friction coefficient, will provide a cut-off that is not there in Einstein's theory; see Selmeczi et al. in this volume. In practice – and here we mean experimental practice – you are likely to run into limits in your data acquisition system before you see inertial effects. In the literature on optical tweezers, you may come across plots of Brownian motion trajectories that are claimed to demonstrate ballistic motion, i.e., demonstrate the effect of inertial mass in Brownian motion. They look that way, with smooth trajectories. But such plots only demonstrate that the data acquisition system contains a low-pass filter [8]. Know thy filters! Or inoculate yourself against that mistake by calculating the characteristic time m/γ with which momentum is dissipated by friction according to Eq. (11.1) in Selmeczi et al., this volume.

If you are using fluorescent imaging, you are likely to have limited spatial resolution, set by the point-spread function of your optics in combination with a finite number of photons. The finite pixel width of your digital camera may add to this. So let's consider how a random recording error on x_i affects the precision with which we can determine D: Suppose the recorded position, $x_i^{(\mathrm{rec})}$, is the true position, $x_i^{(\mathrm{true})}$, *plus* an error s_i

$$x_i^{(\mathrm{rec})} = x_i^{(\mathrm{true})} + s_i \qquad (11.12)$$

where we imagine s_i drawn from a Gaussian distribution with zero mean, SD σ_s, and no correlation between s_i and s_j for $i \neq j$, i.e.,

$$\langle s_i s_j \rangle = \sigma_s^2 \delta_{i,j} \ . \qquad (11.13)$$

Then estimation of D must take into account the fact that Eq. (11.7) no longer holds for the positions we have recorded. Instead we have

$$\langle (\Delta x_i^{(\mathrm{rec})})^2 \rangle = 2D\Delta t + 2\sigma_s^2 \ . \qquad (11.14)$$

We can still estimate the left-hand side well with $\overline{(\Delta x^{(\mathrm{rec})})^2}$, but the right-hand side now contains two unknowns, D and σ_s^2, so how to determine D? If we simply ignore the second term on the right-hand side, we have a *biased* estimator for D in Eq. (11.9), the bias being $\sigma_s^2/\Delta t$. We systematically overestimate D by this amount, on the average.

Also, before D is estimated, we must check that our data are consistent with the theory as such, we have emphasized. To that end, we must now take into account that Eqs. (11.10) and (11.11) no longer hold for the positions we have recorded. Instead we have that consecutive displacements *are* correlated, because the same noise term s_i occurs both in $\Delta x_i^{(\mathrm{rec})}$ and in $\Delta x_{i+1}^{(\mathrm{rec})}$. A brief considerations shows that

$$\langle \Delta x_i^{(\text{rec})} \Delta x_{i+1}^{(\text{rec})} \rangle = -\sigma_s^2 \ . \tag{11.15}$$

So $\langle \Delta x_i^{(\text{rec})} \Delta x_j^{(\text{rec})} \rangle = 0$ holds only for $|i - j| \geq 2$.

Equations (11.14) and (11.15) can be combined to form an *unbiased* estimator for D. There is no free meal, however: By ridding ourselves of the *systematic* error caused by the bias, we increase the *stochastic* error. A simpler approach that uses σ_s^2 does not depend on the value of Δt. So by estimating the left-hand side of Eq. (11.14) with $\overline{(\Delta x^{(\text{rec})})^2}$ for several different values of Δt, one over-determines both D and σ_s^2. That is useful, since the over-determination provides a reality-check of our assumptions about the noise.

11.8 Over-sampling

We now discuss *over-sampling*, a powerful trick of improving statistics by using correlated data. It is a very useful technique in situations where it is much cheaper (or only possible!) to take data more frequently than it is to run the experiment the extra time it takes to produce more data that are less correlated.

We need not measure again to obtain time series with different values for Δt. Our original time series of $N + 1$ positions $x_i^{(\text{rec})}$ recorded at intervals Δt contains two overlapping subseries recorded at intervals $2\Delta t$, the one having $i = 0, 2, 4, \ldots$ and the one having $i = 1, 3, 5, \ldots$. Our original time series also contains three overlapping subseries recorded at intervals $3\Delta t$, etc. So for any integer $n \ll N$ we have n overlapping subseries recorded at intervals $n\Delta t$ on which we can test our theory and, if it works, from which we can estimate D, σ_s^2, and the stochastic errors on our estimates.

Each of these n subseries defines a series of displacements $\Delta x^{(\text{rec})}$ that, apart from their correlations through s_i, are uncorrelated, according to Einstein's theory. Each displacement in one series is, on the other hand, correlated with the two displacements that it overlaps with, in each of the other $n - 1$ subseries. So the n subseries are correlated with each other. It is easier to do statistics with uncorrelated numbers. But we get better statistics if we use all available information, here all n subseries. The mean-squared-displacement average MSD_n over *all* displacements in *all* the n subseries uses this information from a single trajectory:

$$MSD_{n,N} = \overline{(\Delta_n x^{(\text{rec})})^2} = \frac{1}{N - n + 1} \sum_{i=n}^{N} (x_i^{(\text{rec})} - x_{i-n}^{(\text{rec})})^2$$

$$\sim 2Dn\Delta t + 2\sigma_s^2 \ . \tag{11.16}$$

The great power of $MSD_{n,N}$ comes from the variance of $MSD_{n,N}$. In the absence of recording errors, i.e., for $\sigma_s^2 = 0$ or negligible, and for $2n - 1 \leq N$, the stochastic error on the estimate obtained with Eq. (11.16), its standard deviation, is [9],

$$\sigma\left(\overline{(\Delta_n x)^2}\right) = 2D\Delta t \sqrt{\frac{2n(2n^2 + 1)}{3(N - n + 1)} - \frac{n^2(n^2 - 1)}{3(N - n + 1)^2}}$$

$$\approx 2Dn\Delta t \sqrt{\frac{2}{N}\frac{2n^2 + 1}{3n}} \quad \text{for } N \gg n ; \qquad (11.17)$$

see also [10, Chap. 4 and Appendix D2]. For $n = 1$, we recover the SD $2D\Delta t\sqrt{2/N}$ already given above, i.e., a relative error of $\sqrt{2/N}$. For $n > 1$ the error is larger and grows faster with n than the estimated quantity, i.e., the relative error grows with n.

If $MSD_{n,N}$ is plotted against $n\Delta t$, this mean squared displacement falls on a straight line with slope $2D$ and intercept $2\sigma_s^2$ with the second axis. Such a plot looks like a nice way to test the theory and get an idea about the value of D and the relative importance of the noise term $2\sigma_s^2$. You should *not*, however, determine D by fitting a first-degree polynomial in $n\Delta t$ to the plotted points! At least not without realizing that the values they represent, are highly correlated! These values were obtained by applying the estimator in Eq. (11.16) repeatedly to the same time-series x_i, $i = 0, \ldots, N$, just with different values of n. You plug in the same data each time, only tweak n a bit, and out you get: highly related numbers, of course. So if your estimate happens to overshoot (undershoot) the true value at one n-value, you'll see that it also overshoots (undershoots) the true value for many nearby n-values. But the least-squares fitting routine that you use to fit a straight line to your points, will treat these points as statistically independent data! Your fitting routine yields a value for D which isn't necessarily wrong, but it may also output an error estimate for this value, and *that* value is unfounded, even if you feed the program the correct stochastic error for each data point, as given below. I don't have to know your program to know this, because least-squares fitting routines that can fit to correlated data are rare, and users who know how to turn on that facility are possibly rarer and do not read this far in the present kind of text.

Don't despair, however. The best estimate for D is really obtained in the case of $n = 1$, provided $\sigma_s^2 = 0$, and the long-winded story that you just read, is all about estimating whether $\sigma_s^2 = 0$ is a good approximation, and if not, what to do (see below).

11.9 Estimating D

Here's how to estimate D if σ_s^2 is not negligible: Plot the values you get with the estimator in Eq. (11.16) against n after having divided each value with $2n\Delta t$. This is a plot of estimates of D. These estimates are biased, you know, since σ_s^2 is not negligible, and the bias is $\sigma_s^2/(n\Delta t)$, i.e., decreases as $1/n$. Plot the error bar given in Eq. (11.17) on each data point after having divided it as well with $2n\Delta t$. This is not the true error, since it ignores contributions from the recording error. Never mind that for now. You need a value for D in Eq. (11.17) in order to get a value for the error bar. Use your estimate for D. Either the individual estimate you have for each n-value or a sensible-looking average of them, or, for that matter, the value found in the slope of estimates for $2Dn\Delta t$ plotted against $n\Delta t$. Any reasonable value will do, as we won't worry about small errors on error bars, and you can always redo this part of the procedure when you know D's value better. You now have a plot of estimates for D which have minimal stochastic error and maximal systematic error at lowest n-values, and vice versa at large n-values. Between this Scylla and Charybdis you must navigate.

If your data are consistent with the theory presented here and you have good statistics, navigation is unproblematic, there is a range of intermediate n-values for which your estimates for D are the same because n is large enough for the systematic error to be negligible, yet the stochastic error remains small. Any of these estimates is correct, and comes with the correct error bar.

If you don't have good statistics, but it's good enough to show that σ_s^2 is substantial, you'll need formulas we won't bother you with here – or you'll have to cut the usual corners.

11.10 "Give Me a Random Number Between 1 and 10!" "*Seven!*" . . . . "*Seven* Doesn't *Look* Random!"

The reader should be warned that a given "random path" needs not look obviously random. As an example, we extracted one "unusual" path and have plotted it in Fig. 11.3. If you didn't know better, this certainly looks like the

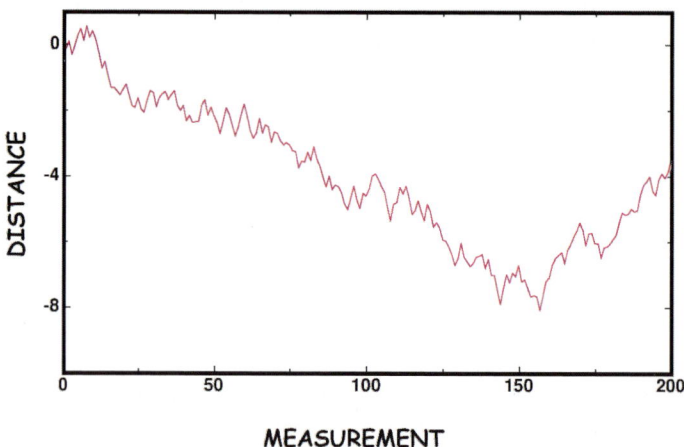

Fig. 11.3. The trajectory of a particular path in the set of 40, plotted over 200 time measurements. It is a random trajectory, though it does not look random

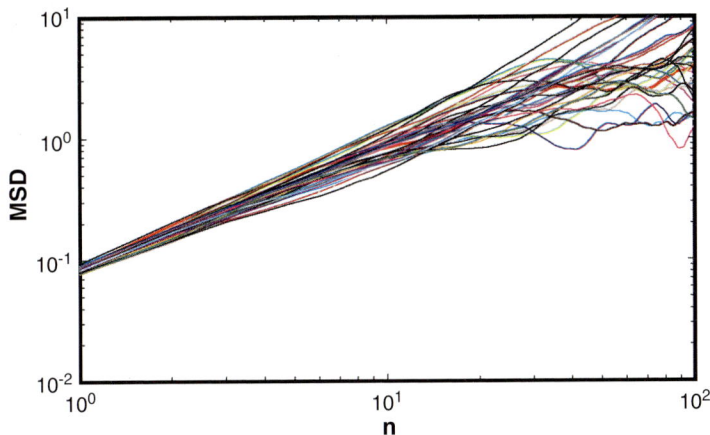

Fig. 11.4. $\log MSD_n$ vs. $\log n$ plot for all 40 trajectories. All 40 trajectories collapse onto a common line, with a fixed diffusion coefficient

path of a particle which moved with constant negative speed on top of some noise until about the 150th measurement and then reversed course. But, it is simply one random path out of a rather small set of 40 of them, and there is no deterministic motion in spite of appearances.

Figure 11.4 shows the same 40 trajectories we have used in the previous figures analyzed using Eq. (11.16). Unlike the huge, and incorrect, apparent distribution of diffusion coefficients that arise from doing the simple net displacement analysis presented in Fig. 11.2, Fig. 11.4 shows that indeed all 40 trajectories have the same diffusion coefficient D within a relative standard deviation of $\sqrt{2/200} \sim 10\%$, and that they are all random trajectories since they have the same unit slope in a log-log plot. Even the "bizarre" trajectory, which looked non-random, is no statistical outlier.

11.11 The Random Diffusion of Transcription Factors

Now we try to apply this analysis to the problem in biological physics, which, as we mentioned, motivated the lengthy theoretical introduction: the presumably random motion of transcription factors as they move over non-specific sequences in dsDNA. The expected bimolecular association rate k_{diff} by which molecules, such as transcription factors, find specific sequences on double-strand DNA (dsDNA) by a purely Brownian motion search in 3-D solution is $4 \pi D_{\text{soln}} \ell_{\text{seq}}$, where ℓ_{seq} is the protein-DNA interaction length which we set equal to the physical length of the specific sequence on the DNA, which the transcription factor binds to, and D_{soln} is the diffusion coefficient of the protein in solution [11–13]. As we mentioned above, the diffusion coefficient of a protein in solution satisfies Einstein's famous relation $D_{\text{soln}} = k_B T / \gamma$. Here

k_B is the Boltzmann constant, T is the absolute temperature, and $\gamma = 6\pi\eta a$ is Stokes friction coefficient, with η the viscosity of the solvent and $2a = 5$ nm the typical diameter of the protein. Thus $D_{\text{soln}} = 8.8 \times 10^{-7}\,\text{cm}^2/\text{s}$ at room temperature. If we assume that $\ell_{\text{seq}} \approx 3\,\text{bp} = 1\,\text{nm}$, then the DNA-binding rate k_{diff} should be approximately $10^8\,\text{M}^{-1}\text{s}^{-1}$. A $k_{\text{diff}} = 10^8\,\text{M}^{-1}\text{s}^{-1}$ can actually be a problematic rate in biology. The original in vitro study on LacI-*lacO* binding by Riggs et al. was with 45.5 kbp DNA of 15.5 μm length [11], and the *lacO* association rate k_a was measured to be $10^{10}\,\text{M}^{-1}\text{s}^{-1}$, 100 times higher than $k_a \approx 10^8/\text{M}/\text{s}$ of the diffusion limit [11] (the $10^{10}/\text{M}/\text{s}$ binding rate was also reported in [13–15]). It has been proposed that such high rates can be achieved if encounters between transcription factors and the specific sequence are facilitated by a combination of diffusion in 3-D solution, searching for a 1-D object, the DNA, followed by diffusion along the 1-D DNA, searching for the specific sequence. Since a 3-D search for a 1-D object is faster than a 3-D search for an effectively 0-D object, and a 1-D search for an effectively 0-D object is even faster, *facilitated diffusion*, as this scenario is called, is a proper name for it.

Thus, in the *facilitated diffusion* scenario, the key to faster target binding lies in the long non-specific DNA sequences that flank the target site. A protein that binds non-specifically to a flanking region and then diffuses along the DNA can search through the nonspecific stretches of DNA for a binding site with an effective rate that can be substantially higher than the simple 3-D rate. The facilitated 1-D protein-target association rate coefficient k_a has been derived by Halford and Marko [13] to be:

$$k_a = \frac{D_{\text{soln}}\ell_d}{1 + \left[\frac{D_{\text{soln}}}{D_{\text{DNA}}}\right]\ell_d^2 Lc} \tag{11.18}$$

where D_{DNA} is the 1-D diffusion coefficient of the protein bound non-specifically to dsDNA, L is the total length of the DNA molecule, ℓ_d is the total distance over which the protein diffuses while bound to the DNA non-specifically, c is the concentration of the specific target site.

In order to check the facilitated association rate scenario directly, it is necessary to know D_{DNA} for LacI bound to DNA, and $\ell_d = x_{\text{max}} - x_{\text{min}}$, the length of the interval covered by the protein during its 1-D random motion along the dsDNA before their dissociation. In fact, if these values do not fall within a certain range– if the time spent by the protein on the DNA is spent inefficiently by too slow diffusion, albeit in a lower dimension – then "facilitated" diffusion can actually slow the search times, so it is critical to experimentally determine D_{DNA} and ℓ_d.

In order to visualize the transcription factor proteins, in our case LacI, we used a LacI protein, which was fused to a fluorescent green fluorescent protein (GFP). GFP13 (S65T):LacI-I12 fusion (GFP-LacI) and DNA constructs with 256 tandem copies of *lacO* (*lacO*$_{256}$) were used. *lacO*$_{256}$-DNA was 42.06 kbp long with a contour length of 14.3 μm, and the 9.22 kbp *lacO*$_{256}$ insertion

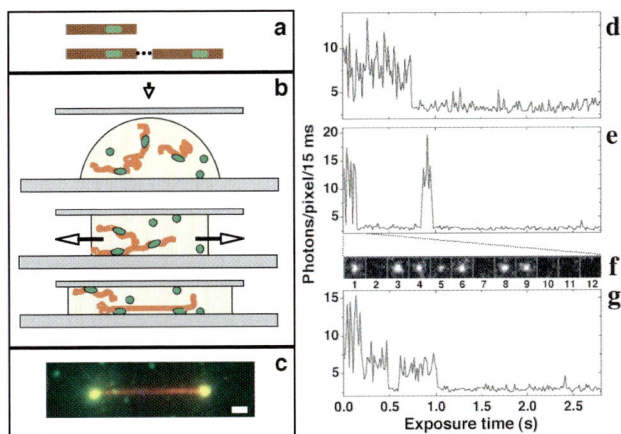

Fig. 11.5. (a) Schematics of a GFP-LacI (*green*) bound $lacO_{256}$-DNA monomer and dimer (*red*). (b) Elongation of the DNA. (c) Frame averaged superposed image of GFP-LacI bound to an elongated $lacO_{256}$-DNA dimer molecule. The scale bar is 1 µm. (d) A GFP-LacI monomer of frequent blinking and unitary bleaching. (e) A GFP-LacI monomer that blinked, recovered the first bleaching in 3 sec, and finally irreversibly bleached. (f) The GFP-LacI dots for the first 12 frames of (e), showing blinking at frames 2 and 7, and bleaching at frame 10. (g) A GFP-LacI dimer with two bleaching events. This dimer was made from one monomer of type (d) and one of type (e)

started at 24.02 kbp. The synthesis methods for the fusion protein and the $lacO_{256}$-DNA, and the sample preparation method are described in [16]. The final GFP-LacI concentration was 100 nM, and $lacO_{256}$-DNA concentration was 0.6 µg/ml. The dimeric cyanine dye BOBO-3 was used to fluorescently image the stretched dsDNA molecules. Cyanine dyes, when intercalated in dsDNA, are known to stretch dsDNA by 30% in length at 1 dye per 5 bp [17], so at 1 dye per 10 bp, our DNA molecules should have been stretched by 15% to 16 µm. There were $lacO_{256}$-DNA dimers as well as monomers in the solution; the dimers were formed by the sticky-end-hybridization of two $lacO_{256}$-DNA monomers. A catalytic oxygen scavenging solution was used to maximize dye lifetimes [16]. The dsDNA+ GFP-LacI solution was deposited onto a fused-silica chip, then a cover slip was used to flatten the solvent between the fused-silica chip and the cover slip, and the edges of the cover slip were then sealed with nail polish. As the cover slip flattened the droplet, hydrodynamic flow elongated the DNA dimers, and the two LacI-$lacO_{256}$ sites stuck to the surface, creating an anchored elongated DNA molecule (Fig. 11.5b and c). At pH 8, dsDNA doesn't stick to fused-silica surfaces, and the elongated DNA molecules were suspended, so unbound GFP-LacI molecules interacted only with free unattached and nonspecific DNA. The elongated DNA molecules were stretched up to 90% of their native contour length; the tension on DNA was a few nano-Newtons [18]. The transverse mean Brownian displacement of the suspended DNA was observed to be less than 50 nm, thus making the longitudinal Brownian fluctuations of the DNA-chain negligibly small and they do not influence the observed of order 100 nm longitudinal protein movement measurements along the DNA.

Near-field excitation techniques are critical for single-molecule imaging. Our technique was described in [16]. Briefly, a prism type Total Internal Reflection Fluorescence Microscopy (TIRFM) method was used. The laser excitation was synchronized to the 3.4 Hz data acquisition rate of the I-CCD camera. The emitted photons from BOBO-3 and GFP were collected using a 100X TIRF oil-immersion objective (N.A. = 1.45), went through a custom-designed dichroic mirror and emission filter set (Chroma Technology Corp, Rockingham, VT) and were recorded by an I-CCD camera (I-PentaMAX:HQ Gen III, Princeton Instruments, Trenton, NJ). The point spread function width of the optical system was measured to be approximately 280 nm. The pixel count of the camera was converted to a photon count using known conversion factors [16]. The mean 488 nm illumination intensity over the point spread function was 1000 W/cm^2. The centroid location of a GFP-LacI dot was determined by fitting its 1-D fluorescence intensity profile to a Gaussian profile.

Knowledge of the time-dependent fluorescence characteristics of free single GFP-LacI monomers and dimers attached to fused-silica surfaces are essential and place fundamental bounds on how well we can determine l_d and D_{DNA}. GFP-LacI monomers blink frequently (short fluorescence dips to near noise level), and bleach with no recovery (Fig. 11.5d). At our excitation intensity of 1000 W/cm^2, mean exposure time of 10 ms, and synchronized imaging frequency of 3.4 Hz, the mean net observation time of each GFP-LacI molecule was 5 s before it bleached (giving a total laser exposure time of 0.15 s). The mean number of photons emitted by the bound GFP-LacI molecules before bleaching was $\approx 4 \times 10^4$ photons. This 5 s observation time gave the *instrumental* limit to the maximum mean distance that we observed GFP-LacI motion on DNA in this experiment.

An image sequence of a single GFP-LacI molecule diffusing along DNA is shown in Fig. 11.6. This is 1 out of 70 trajectories that were observed, and chosen for its large displacement. Figure 11.6a shows the frame averaged superposed image of the anchored DNA, the diffusing GFP-LacI on DNA, and the two LacI-$lacO_{256}$ anchors. Time-lapse images of the diffusing protein show clear relative displacements (Fig. 11.6b), with one immobile anchoring site used as a reference point. We know that we are observing a GFP-LacI dimer from the fluorescence time trace in Fig. 11.6d, which clearly shows two bleaching steps. Both GFP-LacI monomers (80%) and dimers (20%) have been observed to diffuse on DNA. As is evident in Fig. 11.6d, fluorescence time traces of bound GFP-LacI molecules were identical to that of immobile GFP-LacI (Fig. 11.5d–g), with the same blinking rate and characteristic bleaching time of ≈ 0.15 s (5 s net observation time). We thus can infer that each observed diffusing GFP-LacI was one single protein that remained bound

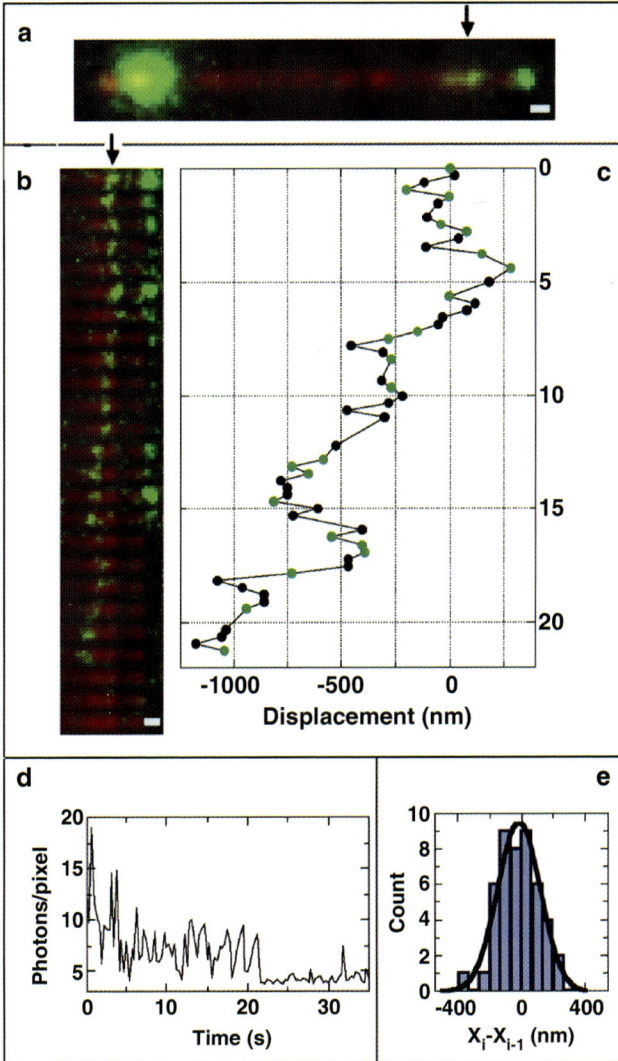

Fig. 11.6. A diffusing GFP-LacI molecule along DNA. (**a**) Frame averaged, super-posed image of a GFP-LacI molecule diffusing along DNA. The two large dots at the DNA ends are LacI-$lacO_{256}$ sites, and the green segment on the nonspecific DNA denoted by *arrow* is the single GFP-LacI protein. In this panel we have summed the frames in Panel (b), so that the net path of the molecule shows as a *green line*. (**b**) Image series of the diffusing protein (*arrow*) of selected clear relative displacements corresponding to green dots in (**c**), the displacement vs. t curve of the diffusing protein. (**d**) The fluorescence time trace of the diffusing GFP-LacI. It is a dimer. (**e**) Gaussian distribution of consecutive displacements $x_i - x_{i-1}$. The scale bar is $0.5\,\mu m$

to DNA until bleached, rather than different proteins that jumped on and off the DNA.

We are faced here with the same dilemma that we faced in the Introduction: how can we extract good diffusion coefficients from these time series data? Obviously the MSD technique is the one to apply. That the GFP-LacI molecular motion was a Brownian motion, can be established by plotting $\log \overline{(\Delta_n x)^2}$ against $\log(n)$, which, as we have shown, should yield a straight line with slope 1 on a log-log plot, and an intercept at $n = 1$ whose value is $2D\Delta t$ for that particular trajectory. Figure 11.7 shows an analysis of 15 trajectories in which no more than 5 contiguous GFP blinks occured. Figure 11.7a is a plot of simple displacement $x(t)$ vs. time, Fig. 11.7c is a plot of the $\overline{(\Delta_n x)^2}$ vs. n, while Fig. 11.7d is a plot of the $\log \overline{(\Delta_n x)^2}$ vs. $\log n$. It is clear that the log-log plot yields the most information concerning the nature of these diffusional motions, and makes clear that all the trajectories are basically Brownian in nature since $\overline{(\Delta_n x)^2}$ scales linearly with n. The dashed line in Fig. 11.7d is a weighted fit to Eq. (11.16) for a particular trajectory. Thus, while all the trajectories are Brownian in nature, the different intercepts at $n = 1$ indicate that there is a large distribution in diffusion coefficients and that there is not a unique, single value for the one-dimensional

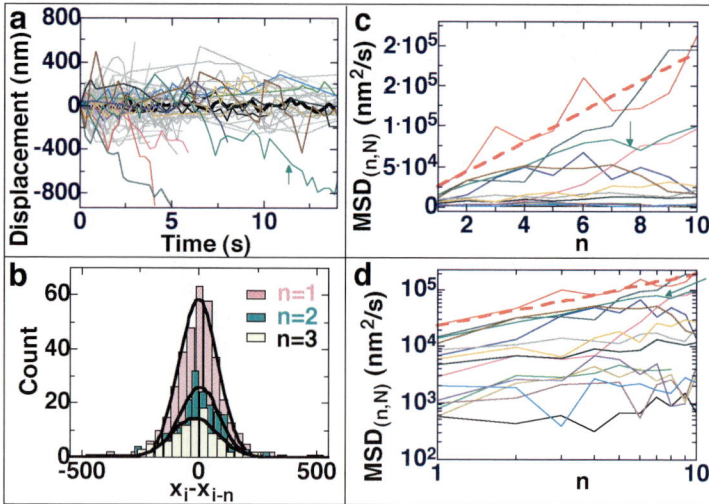

Fig. 11.7. (a) x vs. t for 70 diffusion events of GFP-LacI on DNA. The *black line* is a stationary protein stuck to fused-silica surface (not bound to DNA), and the *colored lines* are the 15 events for which we have obtained D_{DNA}. (b) Gaussian uncorrelated $x_i - x_{i-n}$ distributions for $n = 1$, 2, and 3 for the first 15 points of all 70 events. (c) $\overline{(\Delta_n x)^2}$, here called $MSD_{(n,N)}$, vs. n for the 15 colored trajectories in linear scale and (d) in log-log scale. The *arrows* in (a), (c), and (d) denote the trajectory in Fig. 11.2. The *dashed lines* in (c) and (d) are a fit of Eq. (11.16) to the uppermost random trajectory

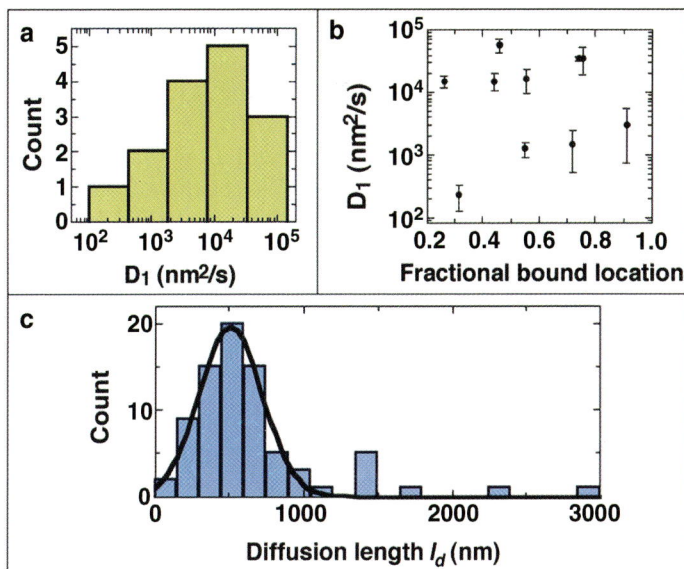

Fig. 11.8. (a)D_{DNA} distribution of the 15 single diffusion events. (b) D_{DNA} vs. fractional bound location on the nonspecific segment of the $lacO_{256}$-DNA dimer. The error bars were obtained from the fit of $MSD_{(n,N)}$ to n with weighted errors at each n given by Eq. (11.17). (c) Histogram of the 1D diffusion length $x_{max} - x_{min} = \ell_d$ for the 70 observed diffusing events. The *solid line* is a Gaussian fit with a mean of 500 ± 220 nm (mean $\pm$ SD). Values in (a) and (c) have been adjusted to DNA contour length

diffusion coefficient D_{DNA}. Note how the variance of $\sigma \overline{(\Delta_n x)^2}$ in Fig. 11.7c increases rapidly with n as predicted by Eq. (11.17). Random trajectories with more than $N = 15$ position measurements were used to obtain values for D_{DNA}. Figure 11.8a shows 15 single diffusion D_{DNA} values, which span a large range from 2.3×10^2 nm^2/s to 1.3×10^5 nm^2/s (or 2.3×10^{-12} cm^2/s to 1.3×10^{-9} cm^2/s). Clearly, each LacI molecule diffuses with dramatically different speeds. Figure 11.8b shows that the different D_{DNA} values are distributed randomly along the nonspecific DNA, apparently showing a lack of correlation between the diffusion coefficient and the position on the DNA where the protein lands.

We can use our data to examine the question of the extent to which facilitated diffusion can enhance the rate at which transcription factors find their specific binding sites, the biological purpose behind these measurements. Just as there is a distribution in the 1-D diffusion coefficients, there is also a distribution in the sliding lengths ℓ_d, which is further compromised by the mean observation-time-to-bleaching of the GFP chromophore, approximately 5 seconds. Figure 11.8c shows the measured distribution in the sliding lengths ℓ_d from our data. Since there is a large distribution in both $D_{\rm DNA}$ and ℓ_d, we use the mean value of $\langle \ell_d \rangle$ (probably a lower bound really due to bleaching) of 500 nm, and the mean diffusion coefficient of $\langle D_{\rm DNA} \rangle = 2.1 \times 10^{-10}$ cm^2/s in Eq. (11.18). Using $\langle \ell_d \rangle = 500$ nm, Riggs' concentration of 1 $lacO$ per 1670 μm^3, $D_{\rm soln} \approx 4 \times 10^{-7}$ cm^2/s for LacI tetramers ($a \approx 10$ nm), $\langle D_{\rm DNA} \rangle = 2.1 \times 10^{-10}$ cm^2/s, and $L = 15.5$ μm, the predicted acceleration factor from our data is 93 ± 20. We conclude from these measurements that facilitated diffusion does increase the LacI-$lacO$ binding rate well over the apparent diffusion limit. This result demonstrates that facilitated diffusion in the form of 1-D Brownian motion is the mechanism responsible for the faster-than-diffusion binding of LacI to $lacO$, and quite possibly, the reason also for the observed faster-than-diffusion binding in other protein-DNA interactions.

11.12 Real Distributions?

Hopefully we have convinced the reader that we have done a reasonably careful analysis of the data. As we discussed in the section on Brownian dynamics, it is easy to see distributions in single-molecule experiments where there are none due to the statistical nature of the data. In our case, we believe that the statistical analysis has been done right and that there is indeed a large distribution in the 1-D diffusion coefficients on DNA. What could be the origin of such a distribution?

One of the authors (RHA) began his scientific work with Professor Hans Frauenfelder. We found a simple problem related to the fairly simple problem discussed in this paper, namely the recombination of carbon monoxide (CO) after photolysis from the heme protein myoglobin [19].

The recombination of the CO with Fe^{2+} iron in the heme should be a bimolecular recombination with the (temperature dependent) rate $k(T)$ and irreversible binding:

$$\frac{d[Mb]}{dt} = -k(T)[Mb][CO] \qquad (11.19)$$

The standard treatment in biochemistry and chemistry is based on two equations: The time course of a reaction is usually fitted to one or possibly two exponentials,

$$N(t) = N(0)\exp[-k(T)t] \qquad (11.20)$$

and the temperature dependence of the rate coefficient $k(T)$ is usually fitted to an Arrhenius (Transition State) expression

$$k(T) = A\exp[-H/k_BT] \qquad (11.21)$$

where H is the activation energy for the chemical reaction.

We might have expected that the recombination curve would be a simple exponential decay law as the CO recombines with the iron atom and that there would be a single activation barrier H. However, that is not what was

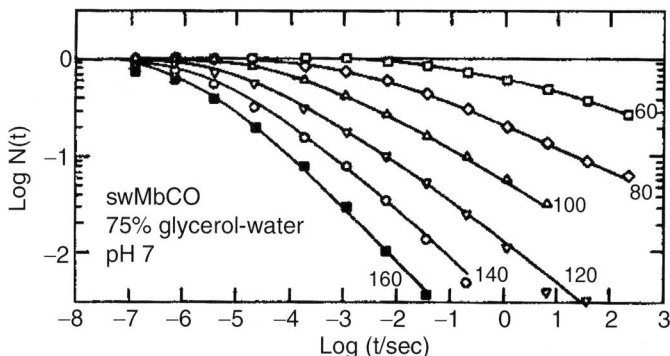

Fig. 11.9. Time dependence of the rebinding after photodissociation of CO to Mb. $N(t)$ is the survival probability, the fraction of Mb molecules that have not bound a CO at time t after the flash. The *solid lines* correspond to a theoretical fit with the activation enthalpy distribution g(H)

observed if the temperature is varied over a much bigger range than is typically down in biological experiments. At temperatures below about 200 K, the recombination is highly non-exponential. A typical result is shown in Fig. 11.9 for the rebinding of CO to sperm whale myoglobin. Note that the plot does not display $\log N(t)$ versus t, but $\log N(t)$ versus $\log t$. In a log-log plot, an exponential appears nearly like a step function, and a power law gives a straight line.

We postulated that proteins exist in different conformational substates, that these different conformational substates have different barrier heights, and denoted by $g(H_{BA})dH_{BA}$ the probability of a protein having a barrier height between H_{BA} and $H_{BA} + dH_{BA}$. Thus we got

$$N(t) = \int dH_{BA} G(H_{BA}) \exp[-k_{BA}(H_{BA}, T)t] \ , \qquad (11.22)$$

where the rate constant k_{BA} is assumed to be given by the Arrhenius relation:

$$k_{BA}(H_{AB}, T) = A_{BA} \exp(-H_{BA}/k_B T) \ . \qquad (11.23)$$

If $N(t)$ is measured over a wide range of temperatures and times, the pre-exponential A_{BA} and the activation enthalpy distribution $g(H_{AB})$ can be found by numerical Laplace transformation. This has to be done numerically, since the real exponential functions do not form an orthonormal set. The pre-exponential factors typically have values on the order of $10^9 \, s^{-1}$, and some distributions are shown in Fig. 11.10.

We are faced with a similar dilemma in this work. Instead of a single 1-dimensional diffusion coefficient, we measured a large range of diffusion coefficients. In the case of the Mb work, it was essential to cool the system to extract the distribution of rates because the protein Mb presumably rapidly

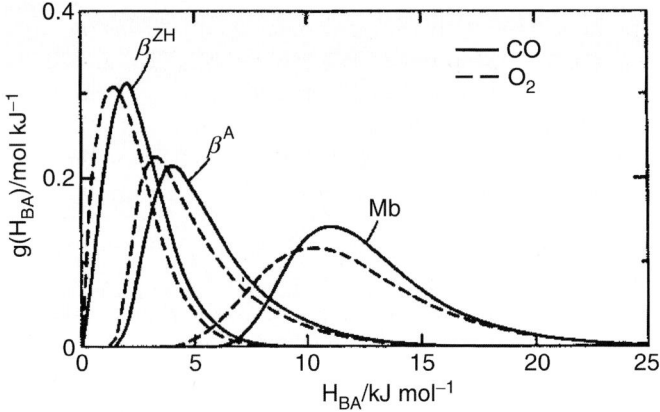

Fig. 11.10. Activation energy distribution $g(H)$, for rebinding of CO and O_2 to myoglobin and the separated beta chains of normal and mutant hemoglobin

samples (diffuses) over many states during the time it takes a CO molecule to find the protein and thus one gets an average rate at room temperature. Had we been able to do single-molecule recombination, the result would have been the same: only a lower temperature could have frozen out the rapid thermal sampling of states, and the full ensemble of states would not have been revealed through single-molecule techniques. In the present experiment, the single-molecule nature of the measurements does allow us to see directly that there is a distribution in diffusion coefficients, but once again it must be true that this distribution has a characteristic relaxation time much longer than the 10 seconds or so that we observe a given protein on the DNA.

We don't know at present the origin of the distribution in diffusion rates for the GFP-LacI/DNA system. There are several possibilities. For example, the protein interaction with nonspecific DNA may vary with local DNA sequence variations, such as AT and CG composition differences, given that the AT/CG ratio along Lambda DNA fluctuates from 30% (per 100bp unit) to 70%. The dependence of DNA stiffness with base pair composition [20] could then influence protein diffusion coefficients. It could also be that the conformation of LacI changes upon nonspecific binding to DNA and that also influences D_{DNA}: when LacI first lands on a sequence of non-specific DNA, it adapts to one of its many configurations that best suits the sequence [21]. Since this configuration lasts for a certain time, it affects the over all protein's motion on DNA. This suggestion is supported by the fact that for diffusion events of different D_{DNA}, the one-step displacements have $x_i - x_{i-1}$ have comparable sizes, rather than sparse large steps and many small steps. So LacI diffuses with the same diffusion constant for each diffusion event, on the average 1.5 kbp of DNA of various sequences, indicating that the LacI's characteristics upon binding to DNA may play a role in D_{DNA}.

However, the source of this large distribution in diffusion coefficients remains to be explained at present. It is true that there is large variance in sequence on Lambda DNA in the nonspecific region with $\pm 30\%$ different in AT and CG concentration, thus it is possible that the diffusion constants are a function of local sequence. It is also possible that the large distribution in D_{DNA} is caused by conformational distributions, either in the protein and/or the pinned DNA. NMR measurements have shown that LacI has at least 21 different conformations [21]. We discussed at the beginning of this article how early experiments showed that proteins exist in a free energy basin where there are many accessible discrete conformational states. If the conformational relaxation time [19] is greater than the mean diffusional search time, this conformational distribution could influence the effective 1-D diffusion rate. The DNA molecule also has many conformational states accessible to it, due to the 50 nm persistence length for both twisting and bending of the polymer [20]. The bending and twisting of the DNA molecule results in a thermally accessible set of topological states, which could get frozen in when the ends of the molecule get tacked down. Thus, there could also be a static distribution in DNA supercoiling numbers due to the non-equilibrium pinning of the two ends of the DNA molecule [22] and there could be changes in the diffusion rates with the supercoiling. Further experiments are needed to answer these questions about the role of conformational distributions in this biological system. But, we think you should care.

Acknowledgements

We thank Xiao-Juan Guan and Ling Guo for sample synthesis, Jonas Tegenfeldt for design of the prism TIRF, and Ido Golding, Walter Reisner, Robert Riehn and Eric Siggia for discussions. This document was written as part of a submission to Physical Review Letters and contains figures taken from that submission.

References

1. R. H. Austin and C. M. Chen (1992). *The Spin-Glass Analogy in Protein Dynamics*, pp. 179–223. World Scientific, Singapore.
2. S. B. Prusiner, M. R. Scott, S. J. DeArmond, and F. E. Cohen. (1998). Prion protein biology. *Cell*, **93**, pp. 337–348.
3. J. T. Kadonaga (1998). Eukaryotic transcription: An interlaced network of transcription factors and chromatin-modifying machines. *Cell*, **92**, pp. 307–313.
4. *Chapter entitled* Bemerkungen über den Gebrauch des Vergrösserungsglases *in J. Ingen-Housz, Verm. Schriften physisch-medicinischen Inhalts.* Christian Friederich Wappler, Wien, (1784).
5. J. Ingen-Housz (1789). *Nouvelles expériences et observations sur divers objets de physique.* Théophile Barrois le jeune, Paris.

6. N. G. Van Kampen (2001). *Stochastic Processes in Physics and Chemistry.* North-Holland.
7. A. Einstein (1905). On the movement of small particles suspended in a stationary liquid demanded by the molecular-kinetic theory of heat. *Ann. d. Phys.,* **17**, p. 549, In German. English translation published in [23].
8. K. Berg-Sørensen and H. Flyvbjerg (2004). Power spectrum analysis for optical tweezers. *Rev. Sci. Instrum.,* **875**, pp. 594–612.
9. Hong Qian, Michael P. Sheetz, and Elliot L. Elson (1991). Single particle tracking: Analysis of diffusion and flow in two-dimensional systems. *Biophysical Journal,* **60**, pp. 910–921.
10. D. Frenkel and B. Smit. *Understanding Molecular Simulation.* Elsevier, USA, 2nd edition.
11. A. D. Riggs, S. Bougeois, and M. Cohn (1970). The *lac* repressor-operator interaction. 3. Kinetic studies. *Journal of Molecular Biology,* **53**, pp. 401–417.
12. O. G. Berg and P. H. vonHippel (1985). Diffusion-controlled macromolecular interactions. *Annual Review of Biophysics and biophysical chemistry.* **14**, pp. 131–160.
13. Stephen E. Halford and John F. Marko (2004). How do site-specific DNA-binding proteins find their targets? *Nucleic Acids Research,* **32**, pp. 3040–3052.
14. M. D. Barkley (1981). Salt dependence of the kinetics of the *lac* repressor-operator interaction: role of nonoperator deoxyribonucleic acid (DNA) in the association reaction. *Biochemistry,* **20**, pp. 3833–3842.
15. M. Hsien and M. Brenowitz (1997). Comparison of the DNA association kinetics of the Lac repressor tetramer, its dimeric mutant Laci^{adj} and the native dimeric Gal repressor. *Journal of Biological Chemistry,* **272**, pp. 22092–22096.
16. Y. M. Wang, J. Tegenfeldt, W. Reisner, R. Riehn, Xiao-Juan Guan, Ling Guo, Ido Golding, Edward C. Cox, James Sturm, and Robert H. Austin (2005). Single-molecule studies of repressor-DNA interactions show long-range interactions. *Proceeding of the National Academy of Sciences.* **102**, pp. 9796–9801.
17. T. T. Perkins, D. E. Smith, R. G. Larson, and S. Chu (1995). Stretching of a single tethered polymer in a uniform flow. *Science,* **268**, pp. 83–87.
18. Steven B. Smith, Laura Finzi, and Carlos Bustamante (1992). Direct mechanical measurements of the elasticity of single DNA molecules by using magnetic beads. *Science,* **258**, pp. 1122–1126.
19. R. H. Austin, K. Beeson, L. Eisenstein, H. Frauenfelder, I. Gunsalus, and V. Marshall (1974). Activation energy spectrum of a biomolecule: Photodissociation of carbonmonoxy myoglobin at low temperatures. *Physics Review Letters,* **32**, pp. 403–405.
20. M. E. Hogan and R. H. Austin (1987). Importance of DNA stiffness in protein-DNA binding specificity. *Nature,* **329**, pp. 263–266.
21. Charalampos G. Kalodimos, Nikolaos Biris, Alexandre M. J. J. Bonvin, Marc M. Levandoski, Marc Guennuegues, Rolf Boelens, and Robert Kaptein (2004). Adaptation in nonspecific and specific protein-DNA complexes. *Science,* **305**, pp. 386–389.
22. N. R. Cozarelli, T. Boles, and J. White (1990). *Topology and its Biological Effects.* Cold Spring Harbor Press.
23. A. Einstein (1985). *Investigations on the Theory of the Brownian Movement.* Dover Publications, Inc. New York, edited with notes by R. Fürth, translated by A. D. Cowper. edition.

12

BioNEMS: Nanomechanical Systems
for Single-Molecule Biophysics

J.L. Arlett[2], M.R. Paul,[5] J.E. Solomon,[2] M.C. Cross,[2] S.E. Fraser,[1,3,4]
and M.L. Roukes[1,2,4]

[1] Kavli Nanoscience Institute;
[2] Division of Physics, Mathematics and Astronomy, California Institute of
   Technology Pasadena, CA 91125 USA
   arlett@cco.caltech.edu
[3] Division of Biology; and
[4] Division of Engineering and Applied Science, California Institute of Technology
   Pasadena, CA 91125 USA
   roukes@caltech.edu
[5] Department of Mechanical Engineering, Virginia Polytechnic Institute
   and State University Blacksburg, VA 24061
   mrp@vt.edu

Abstract. Techniques from nanoscience now enable the creation of ultrasmall electronic devices. Among these, nanoelectromechanical systems (NEMS) in particular offer unprecedented opportunities for sensitive chemical, biological, and physical measurements [1]. For vacuum-based applications NEMS provide extremely high force and mass sensitivity, ultimately below the attonewton and single-Dalton level respectively. In fluidic media, even though the high quality factors attainable in vacuum become precipitously damped due to fluid coupling, extremely small device size and high compliance still yield force sensitivity at the piconewton level – i.e., smaller than that, on average, required to break individual hydrogen bonds that are the fundamental structural elements underlying molecular recognition processes. A profound and unique new feature of nanoscale fluid-based mechanical sensors is that they offer the advantage of unprecedented signal bandwidth ($\gg$1 MHz), even at piconewton force levels. Their combined sensitivity and temporal resolution is destined to enable real-time observations of stochastic single-molecular biochemical processes down to the sub-microsecond regime [2].

12.1 Introduction: Mechanical Sensors for Biology

In 1992, Hoh et al. [3] pioneered the application of micron scale mechanical sensors to the study of biological molecules and their interaction forces. These UCSB researchers noted discrete steps in the adhesive interaction between a silicon nitride atomic force microscope [AFM] tip and a glass surface, steps that were interpreted as corresponding to an interaction force of

J.L. Arlett et al.: *BioNEMS: Nanomechanical Systems for Single-Molecule Biophysics*, Lect.
Notes Phys. **711**, 241–270 (2007)
DOI 10.1007/3-540-49522-3_12

10 pN – apparently the average strength of individual hydrogen bonds. Since this pioneering experiment, a growing literature of force spectroscopy, also known as chemical force microscopy, has shown that a chemically-modified AFM can indeed be tailored to measure the binding force of interactions for a wide range of affinity-based processes, such as single receptor-ligand interactions. In such experiments, the tip of an AFM cantilever is typically functionalized with a receptor of interest, a molecular complex with the analyte is formed and then pulled, and the interaction force applied is read out as proportional to the measured tip deflection. Early studies included probing the biotin-streptavidin interaction force by Lee et al. [4] and the study of biotin-avitin, desthiobiotin-avidin, and iminobiotin-avidin by Florin et al. [5]. Force spectroscopy experiments were extended to cells by Dammer et al. [6] who studied intercellular adhesion forces (of proteoglycan from a marine sponge) and Evans et al. [7] who also observed adhesion forces at the cell surface. A large body of literature has ensued. Force spectroscopy is not limited to studies of interaction forces but has also been applied to the study of conformational changes, e.g., of the enzyme lysozyme [8] and of the unfolding of the protein titin [9]. Force spectroscopy has also been used to study DNA hybridization processes [10].

Clearly, force spectroscopy has proven to be an extremely fruitful approach for studying biomolecular interactions. However, cantilevers employed for this work are primarily research-oriented probes and not, *per se*, directly applicable for biosensing in the conventional sense (at least not for large scale, real-time applications). However, a related cantilever-based biosensor technology has developed within the field of microelectromechanical systems (MEMS) and has been successfully applied to biological detection in liquids. These *surface-stress sensors* are based upon a direct measurement of the stress induced by the binding of a *layer* of ligands to an appropriately-prepared ("biofunctionalized") device surface [11–16]. Despite the impressive advances in this field, these MEMS-based surface-stress sensing devices have critical limitations. In order to achieve a notable change in surface stress, these sensors require the binding of many ligands (a significant fraction of a monolayer); consequently, surface-stress sensors, in general, suffer from the inability to respond to forces that vary rapidly in time. Especially important are variations on the few μs timescale, characteristic of both important classes of conformational changes in large biomolecules and of low-affinity or non-specific binding processes. As regards the latter, high frequency response may ultimately prove critical to following the stochastic nature of affinity-based interactions, such as the receptor-ligand interaction, especially in the presence of an overabundance of weakly-binding entities. Receptor-ligand pairs interact dynamically: binding, remaining engaged for times ranging from microseconds to seconds (depending on the exact binding affinities and concentrations), and then releasing. Temporal resolution would provide an additional new "dimension" for discrimination between processes.

Ostensibly, the utilization of MEMS or NEMS as sensor devices is simpler than the setup required for force spectroscopy – with sensor devices one waits for the molecule of interest to bind, whereas with force spectroscopy one actively searches for it. Hence, at first glance, biosensor systems may appear simpler in that they do not apparently require a scanning system, feedback, or even a sharp tip, which are essential elements of a force spectroscopy apparatus. However, advanced MEMS and NEMS sensors do indeed require additional system components: microfluidic delivery systems, differential biofunctionalization (to allow multiplex assays), and complex readout systems, which, in fact, may be feedback-based for optimal performance.

The current state-of-the art of mechanical biosensors and force spectroscopy devices labor with an addition limitation: the majority of force spectroscopy and surface-stress sensor devices employ optical displacement detection of the probe. This detection scheme has proven extremely fruitful for scientific studies of cantilevers with feature sizes of order 1 μm and greater. However, in this paper we show that greatly increased sensitivity can be attained if the cantilever dimensions are further reduced into the nanometer size regime. Here, however, diffraction limits one's ability to focus optical "interrogation" upon the nanocantilever tips (as required for optical detection); hence alternative detection schemes become essential. Moreover, a totally integrated approach to sensing, such as is provided by electrical schemes, is crucial for the large-scale integration required for multiplex sensing applications. In situ, that is *on chip*, electrical displacement transduction eliminates the need for external equipment typically employed for optical detection. This greatly increases the potential applications of electrically-transduced devices beyond those of the research laboratory setting.

In what follows we describe a new class of nanomechanical biosensors: BioNEMS. These sensors are distinctly different from the aforementioned surface stress sensors. The central point we emphasize in this review is that BioNEMS, as we define them, do not sense quasistatic surface-layer induced stress. Instead, BioNEMS sense analyte-induced changes that measurably alter *dynamical* device properties. Moreover, the mechanisms that underlie BioNEMS sensing scale to the realm of single molecules (a fact that is confirmed by force spectroscopy itself), whereas stress sensing devices do not. We shall describe in what follows the changes that arise from analyte binding to BioNEMS sensors. Among these are: alteration of the nanomechanical device compliance (force constant), changes to device damping, or direct imposition of additional forces or motional correlations to the device. The latter are effective in multiple-cantilever assays, which are described below. At the outset, we stress that for fluid-coupled nanomechanical devices the inertial mass is *not* among the parameters most strongly affected by target capture (analyte binding); hence BioNEMS as described herein are not mass sensing devices.

12.2 Motion Transduction via Piezoresistive Sensing

Our work utilizes electrical displacement sensing based upon integrated piezoresistive strain sensors. The piezoresistive effect in silicon was discovered in the 1950s [17] and its application has played an important role in early silicon sensor devices [18, 19] and more recently for sensitive detection of displacement within microelectromechanical devices [MEMS], both in vacuum [20] and air [21]. Piezoresistive sensors were first applied to AFM cantilevers by Tortonese et al. in 1993 [22] and have since been adopted by a number of researchers. The sensitivity attained is comparable to that of optical sensors, yet significant advantages for size scalability and integration both in the context of reducing device dimensions to sizes where optical detection would not be possible and for arrays where optical detection is more complicated (due to the extra hardware required). Piezoresistive sensors have recently been applied to surface-stress sensor devices by Rasmussen et al. [23] who have detected immobilization of ssDNA, and Wee et al. [24] who used piezoresistive surface-stress sensors to detect prostate specific antigen and C-reactive proteins (specific markers for prostate cancer and cardiac disease, respectively). Modeling of piezoresistive surface-stress sensors has been performed by Rasmussen et al. [25, 26] Kassegne et al. [27] and Yang et al. [28].

12.3 Nanoscale Mechanical Devices: BioNEMS

Previous techniques have all involved force sensors with active surface areas that are quite large compared to the molecular scale. This is the case since surface-stress sensors rely on interactions between a layer of molecules on the device surface; hence a significant fraction of the surface area must be coated with molecules to enable detection. It is unclear how surface-stress sensors will scale, if at all, to the few-to-single molecule regime. These are among the issues that motivate our investigation of alternative sensing techniques with BioNEMS. We are interested in both passive and active sensors for which the binding of an analyte molecule of interest leads to a direct change in the vibrational behavior of a cantilever. As mentioned, in the case of a single cantilever device this may originate from several mechanisms [2]. Increased damping will arise if an analyte with surface area comparable to the device dimensions binds to the device. A change in the device compliance (the inverse of force constant) arises for the case where the analyte is allowed to bind in such a fashion as to bridge a gap between the suspended device and a rigid support structure. In both cases, the signal arises from a change in the dynamical, fluid-coupled response function of the device as a direct consequence of the binding of the analytes under study. Clearly, with sufficient sensitivity the approach scales to the realm where single analytes may be resolvable. In what follows we demonstrate this possibility; more than a decade of aforementioned work in force spectroscopy confirms this.

A separate class of BioNEMS devices involves multiple-cantilever devices [2]. For these, the binding of an analyte between two cantilevers leads to a novel biomechanical coupling. This, in turn, is then reflected in motional correlations between the two devices as well as the ability to directly transmit forces between the two devices – both mediated by the soft "biological linker", i.e. the target analyte. These two consequences of coupling offer two modalities of biosensing; they correspond to passive and actively-driven sensing. We shall demonstrate that by reducing the dimensions of the device to the size regime of the cells, spores, and molecules of interest for biosensing, BioNEMS offer the prospect of high force sensitivity in liquid with fast response times. We now embark on a discussion of response functions in liquid for the devices of interest and then to an evaluation of their ensuing force sensitivities.

12.4 Overview: Realizable Force Sensitivity of Piezoresistive BioNEMS Devices

For nanometer-scale cantilevers, the most important dynamical regime is that of very low Reynolds number flow. Below we evaluate the realizable sensitivity of fluid-coupled systems that take into account noise of practical readouts; our results are based upon initial experiments, analytic calculations, and numerical modeling. Thermal noise is typically characterized by its power spectral density (PSD), $S(f)$, defined as the Fourier transform of the autocorrelation function for the process of interest (in this case the stochastic thermal noise). The "sensitivity" is then given by $\sqrt{S(f)}$. The total r.m.s. noise is the integral of this force sensitivity over the experimental frequency bandwidth. Limits to force sensitivity imposed by thermal fluctuations appears better than 10 fN/$\sqrt{\text{Hz}}$ for small but realizable device dimensions via advanced nanofabrication techniques. *Transducer-coupled* force sensitivity (referred to the input, i.e., force domain) – which includes all additional noise processes generated downstream from the mechanical probe by the displacement transducer and its essential electrical readout scheme – is still well below the 100 fN/$\sqrt{\text{Hz}}$ realm and permits bandwidths greater than 1 MHz. This opens a new range of possibilities for biological force measurements on extremely short time scales.

12.5 Fluid-Coupled Nanomechanical Devices: Analysis

Many important biological processes involve complicated fluidic interactions on the micro- and nano-scales [29]. The nondimensional equations governing the motion of an incompressible fluid are the well-known Navier-Stokes equations,

$$\Re_f \frac{\partial u}{\partial t} + \Re_u u \cdot \boldsymbol{\nabla} u = -\boldsymbol{\nabla} p + \nabla^2 u \quad \text{and} \qquad (12.1)$$

$$\nabla \cdot \boldsymbol{u} = 0 \,. \tag{12.2}$$

In these equations $\boldsymbol{u}$ represents the fluid velocity and p the pressure. The frequency- and velocity-based Reynolds numbers are $\Re_f$ and $\Re_u$, respectively. The velocity-based Reynolds number $\Re_u = uw/2\nu$ expresses the ratio between inertial convective forces and viscous forces where ν is the kinematic viscosity (defined as η/ρ where η is the dynamic or shear viscosity and ρ the density). The characteristic length scale is chosen as the cantilever half-width, $w/2$. For MEMS and NEMS devices both the characteristic velocity and length scale become quite small resulting in low Reynolds numbers where viscous effects dominate inertial effects. For the nanoscale devices under consideration $\Re_u \ll 1$ making the nonlinear convective inertial term negligible which greatly simplify analysis. The frequency-based parameter $\Re_f = fw^2/4\nu$ expresses the ratio between inertial acceleration forces and viscous forces where the inverse oscillation frequency, $1/f$, has been used for the characteristic time scale. $\Re_f$ can become important when oscillations are imposed externally. This is the case when micron and nanometer scale cantilevers immersed in fluid exhibit stochastic dynamics due to their constant buffeting by fluid particles, i.e. by their Brownian motion. For the nanoscale cantilevers under consideration here $\Re_f \sim 1$. As a result the local inertial term must be included making the resulting analysis more difficult. This has led to the development of an experimentally accurate numerical approach to calculate the stochastic dynamics of micron and submicron scale cantilevers (discussed below) [30]. Hereafter we shall simply refer to the frequency-based Reynolds number as $\Re$.

12.6 Analytical Calculations for Experimentally Relevant Conditions

The motion of a cantilever in a fluid vibrating in its fundamental mode can be described by the equation of a simple damped harmonic oscillator:

$$F(t) = M_{eff}\frac{\partial^2 x}{\partial t^2} + \gamma_{eff}\frac{\partial x}{\partial t} + Kx \,. \tag{12.3}$$

Here we take x to describe the motion of the free end of the cantilever and F(t) as the effective force acting at cantilever tip. For a cantilever of width, thickness and length w, t, l, respectively, vibrations in the fundamental mode involve a force constant [31] $K \cong 0.25\,Ew\,(t/l)^3$. In vacuum $M_{eff} \rightarrow M_0 \cong 0.243\rho_c wt\,l$ and $f_{eff} = \sqrt{K/M_{eff}}/2\pi \rightarrow f_0 \cong 0.507\sqrt{E/\rho_c}(t/l^2)/\pi$. Here, ρ_c is the cantilever's mass density and E is its Young's modulus. An oscillating device in a fluidic medium will have a characteristic boundary layer of fluid (dependent on the device geometry, frequency of oscillations, and fluid properties) which effectively oscillates with the device. This is referred to as fluid loading and adds an additional contribution to the effective mass of the device (described below). The constant γ_{eff} is the effective damping coefficient; as described below, in all cases of interest here the cantilever's coupling to the fluid completely dominates its internal materials-dependent loss processes.

Noise is often characterized by its power spectral density, S, defined as the Fourier transform of the autocorrelation function of the parameter of interest. The sensitivity is then given by $\sqrt{S}$. Finally, the total r.m.s. noise is given by the integral of the sensitivity over the measurement bandwidth. Although we are concerned here with quite small, i.e. nanoscale, systems, the mechanical structures themselves are still quite large compared with the size of solvent molecules. The thermal motion of a fluid-loaded nanocantilever may thus be modeled in terms of stochastic forces, which are Markovian (because the time scale of molecular collisions are short compared with frequencies of macroscopic motion) and Gaussian (because many molecular collisions combine to force macroscopic motion). Hence, the fluctuation-dissipation theorem may be used to analyze this motion. The force spectral density is given by the Nyquist formula, $S_F = 4k_B T \gamma_{eff}$ [32]. This is the fundamental limiting force sensitivity of a sensor in fluid. Hence, the force sensitivity can be optimized by minimizing this force noise while at the same time optimizing the sensing protocol to allow a force sensitivity very close to this fundamental limit to be achieved. One finds that a uniform reduction of *all* device dimensions accomplishes both goals. The damping of the cantilever arising from the fluid loading is most dependent upon its dimensions transverse to the motion, i.e. its width and length, and the reduction of these dimensions therefore leads to a direct reduction of the force spectral density from fluctuations. Reduction in device thickness, by contrast, leads to improvements in the device responsivity, which allows signals at forces comparable to the Brownian noise floor to be readily "transduced" to experimentally measurable voltage signals at levels above the noise floor set by transducer noise processes and readout preamplifier noise limitations. It will be shown that such a scaling downward of dimensions has the profound additional benefit of leading to a marked reduction in response time. This decrease arises primarily from the reduction in cantilever mass loading with the decreased dimensions.

A very rough estimate of the drag constant is possible by considering the drag on a sphere in low Reynolds number flow far from any surface. In this case, $\gamma_{eff} = 6\pi\eta a$, where $\eta = \rho\nu$ is the shear viscosity of the solution and a is the radius of the sphere. For $a = 1$ μm, in water, the Nyquist formula yields $S_F^{1/2} \sim 17$ fN/Hz$^{1/2}$.

12.7 BioNEMS Displacement Response Functions

The motion of the free end of a cantilever in fluid at frequency f is described by

$$S_x = \frac{S_F}{[(K - 4\pi^2 M_{eff} f^2)^2 + 4\pi^2 \gamma_{eff}^2 f^2]} . \qquad (12.4)$$

This equation represents the average squared magnitude of the Fourier transform of Eq. (12.3) [32]. S_x is the power spectrum density of the tip displacement (i.e. the Fourier transform of the autocorrelation function for the cantilever's tip displacement). The constant γ_{eff} is the effective damping coefficient. This provides a complete description of the cantilever's displacement response both to the externally applied forces and, through the fluctuation-dissipation theorem, to the stochastic forces imparted from the fluid. As shown below, in all cases of interest here the cantilever's coupling to the fluid completely dominates its internal materials-dependent loss processes.

Sader has presented a very useful analysis of the coupling of the fluid to long thin cantilevers in the context of the atomic force microscope [33]. Numerical evidence suggests that loading of a rectangular cantilever is well approximated by the loading of a circular cylinder of diameter equal to the width of the beam [34]. The fluid loading of an infinite cylinder, first calculated by Stokes, is well known [35] and can be written as an equivalent mass per unit length:

$$L(f) = \frac{\pi \rho_L w^2}{4} \Gamma(\Re). \qquad (12.5)$$

The prefactor is simply the volume displaced by the cylinder while the function Γ which depends solely upon Reynolds number must be calculated from the motion of the fluid. In this approximation, the fluidic forces at each frequency and on each section of the cantilever are proportional to the displacement at that point. For this case it can be shown that the structure of the cantilever modes is unchanged – only their frequency and damping is modified. The Stokes calculation for a cylinder yields

$$\Gamma(\Re) = 1 + \frac{4i K_1(-i\sqrt{i\Re})}{\sqrt{i\Re} K_0(-i\sqrt{i\Re})} \qquad (12.6)$$

where K_o and K_1 are modified Bessel functions. There are two important consequences of this relation; first, $2\pi f \, \text{Im}\{\Gamma\}$ gives an effective, frequency-dependent, viscous force per unit length, $\Re \, \text{Im}\{\Gamma(\Re)\} \, \pi \eta u$, where u is the velocity. The prefactor, $\Re \, \text{Im}\{\Gamma(\Re)\}$, is of order 4 at $\Re = 1$ and is only a slowly varying function of $\Re$. The similarity with the expression for the Stokes force $6\pi \eta a u$ acting upon a sphere of radius a is apparent. However, unlike the case for the sphere, the dissipative drag coefficient for a cylinder does not asymptotically approach a constant value at low Reynolds numbers – instead the prefactor decreases asymptotically as $8/\ln \Re$ at very small $\Re$. For the fundamental mode of a rectangular cantilever, the fluidic damping term can be written as

$$\gamma_{eff} \cong \alpha \, \ell \, [2\pi f \, \text{Im}\{L\}] . \qquad (12.7)$$

This is weakly frequency dependent, since the factor $2\pi f \, \mathrm{Im}\{\Gamma\}$ is not constant. The parameter α relates the mean square displacement along the beam to the displacement at its end. For the fundamental mode of a simple rectangular cantilever, $\alpha = 0.243$ [31]; for a cantilever that acts as a hinge, $\alpha = 0.333$.

The second consequence of mass loading is an increase in the effective mass per unit length given by $\mathrm{Re}\{\Gamma\}$. This term becomes quite large at small $\Re$. For the fundamental mode of a cantilever,

$$M_{eff} \cong \alpha \left(\rho_C V_c + \ell \mathrm{Re}\{L\} \right). \tag{12.8}$$

Here, V_c is the cantilever volume. Note that the fluid loading is determined by w^2 and not wt; hence thin beams experience relatively large fluid loading. The value of $\mathrm{Re}\{\Gamma\}$ is unity for large $\Re$, is around 4 at $\Re = 1$, and continues to increase as $\Re$ decreases. Hence, for a silicon cantilever in water at a value of $w/t = 2$, the mass loading factor (defined as the ratio of fluid loading to inertial mass) is approximately 3 at $\Re = 1$ and increases for proportionately thinner beams and lower Reynolds numbers.

If we assume that the fluid mass dominates, the quality factor Q of the oscillator can be estimated simply from the fluid properties as

$$Q \sim \frac{2\pi f \, M_{eff}}{\gamma_{eff}} \sim \frac{Re\{\Gamma(\Re)\}}{Im\{\Gamma(\Re)\}}. \tag{12.9}$$

This expression is rather independent of frequency, varying only over the range $0.2 < Q < 0.9$ as $\Re$ changes from 10^{-3} to 1. As expected, this is many orders of magnitude smaller value than the Q's obtained from semiconductor resonators in vacuum [1, 36]. Note that since M_{eff} and γ_{eff} are frequency-dependent, this notion of Q is only approximate.

The displacement response function is given by the Fourier transform of Eq. (12.3):

$$H(f) = \frac{\tilde{x}}{\tilde{F}} = \frac{1}{K - M_{eff}(f) + i2\pi f \gamma_{eff}(f)}. \tag{12.10}$$

(The average squared magnitude of $H(f)$ was given in Eq. (12.4)). We shall use this in the analysis below to relate effective sources of displacement noise back to the force domain (in electrical engineering parlance, "refer them to the input") to enable evaluation of the practical force sensitivity attainable. The resultant motion of the cantilever tip from an applied force, F, is consequently dependent on two parameters, the spring constant, K, which depends on the elastic properties of the material and device geometry and the normalized displacement response function which characterizes the frequency-dependent response of the beam in fluid. (This is analogous to the response function for a resonant device in vacuum, except that in the case of the latter the effective mass and damping are frequency-independent. In Fig. 12.1 we plot theoretical calculations of the response function, $H(f)$, for two different cantilever geometries (the dimensions and properties are delineated in Table 12.1). At

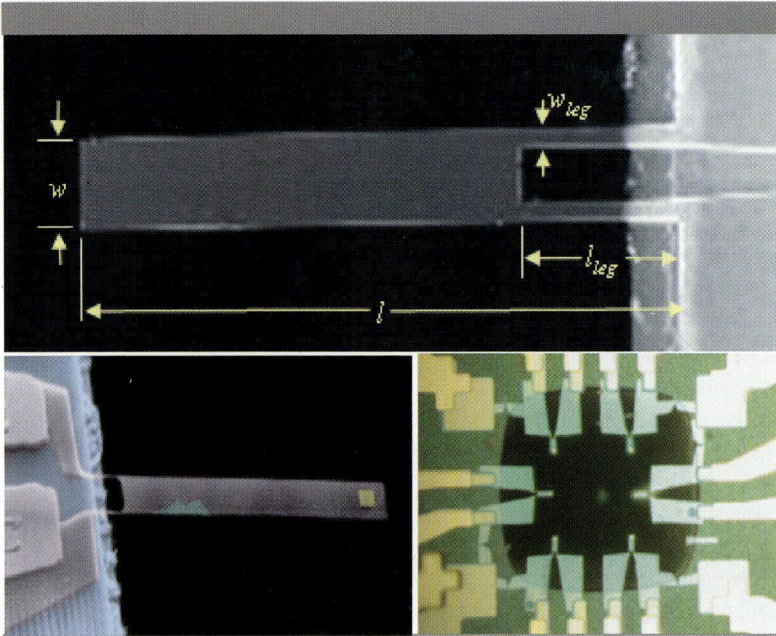

Fig. 12.1. Prototypical silicon nanocantilevers. The cantilevers extend over a fluidic via (*dark region*) formed by deep-etching the wafer through to its backside. The topmost electron micrograph shows the following geometrical parameters for this particular prototypical two-leg device: $\ell = 15\,\mu m$, $w = 2.5\,\mu m$, $w_{leg} = 0.58\,\mu m$, and $\ell_{leg} = 4\,\mu m$. The cantilever thickness is $t = 130\,nm$, of which the top $30\,nm$ forms the conducting layer (with a boron doping density of $4 \times 10^{19}/cm^3$). From this *top* layer the transducer and its leads are patterned. The two electrical terminals are visible on the *right*. For this cantilever, the current path is along the $\langle 110 \rangle$ direction for which $\pi_\ell \sim 4 \times 10^{-10}\,Pa^{-1}$ [37, 38]. The two *lower* colorized images show other nanocantilevers above their respective fluidic vias (*dark regions*). The small gold pad visible at the cantilever tip is used for thiol-based biofunctionalization protocols

high frequencies (greater than 10% of the vacuum resonance frequency) the roll-off in device response due to fluid induced effective stiffening (from fluid loading) is evident. At low frequencies (less than 1% of the vacuum resonant frequency) the effect of fluid on the cantilever response is slight. It is in the intermediate region that a sharp resonance would be observed for a vacuum based device. In fluid there is a peak in responsivity (at least for cantilever 1), but it is a very broad peak, greatly suppressed compared to the resonance in vacuum. The experiments of Viani et al. [39] involving silicon nitride micro-cantilevers in water confirm this; a peak intensity response of order of twice the low frequency response is found.

Table 12.1. Physical parameters for two prototype Si nanocantilevers. Parameters tabulated are thickness, t; width, w; length, ℓ; constriction (leg) width, W_{leg}, and length, ℓ_{leg}; fundamental mode resonant frequency in vacuum, f_0; force constant, K; and Reynolds number at the resonant frequency in vacuum, $\Re$. (For a description of the dimensions referred to here see Fig. 12.1)

#	t(nm)	W(μm)	ℓ(μm)	ℓ_{leg} (μm)	W_{leg} (nm)	f_0 (MHz)	K (mN/m)	$\Re$
1	130	2.5	15	4.0	600	0.51	34	0.8
2	30	0.1	3	0.6	33	3.4	2.2	0.01

12.8 Transducer Performance and Noise Analysis

We turn now to consider a concrete implementation of fluid-based mechanical force detection. Perhaps one of the most important engineering challenges is that of the readout system, which provides continuous interrogation of cantilever displacement. The devices under study will be referred to as "cantilevers" but in fact they are somewhat more complex than simple "diving board" cantilever geometry. Prototypical devices are shown in Fig. 12.2. By removing a region to create two "legs" near the anchor point at the end of the device, a higher degree of compliance is attained along with a slight reduction in cantilever mass. Because these "two-legged" geometries are rather non-standard, we provide the following approximate expression for the effective force constant, K, (for point loading at the end of the cantilever) [40]:

$$K = \frac{Et^3}{\frac{4\ell^3}{w} + \left(2\ell_{leg}^3 - 6\ell\ell_{leg}^2 + 6\ell^2\ell_{leg}\right)\left(\frac{1}{w_{leg}} - \frac{2}{w}\right)}, \tag{12.11}$$

The displacement transducer converts the motion of the cantilever into an electrical signal, in this case this occurs via the strain-induced change of the resistance of a conducting path patterned from a p+ doped Si epilayer on the topmost surface of the device [41]. We characterize the transducer's performance by its responsivity (with units volts/m), $\mathcal{R}_T = I_b \frac{\partial R_d}{\partial x}$, where I is the bias current, $G = \frac{\partial R_d}{\partial x} = \frac{3\beta\pi_\ell}{2w_{leg}t^2}(2\ell - \ell_{leg})KR_d$ is the resistance change per unit displacement of the cantilever tip and R_d is the two-terminal resistance of the transducer. $\mathcal{R}_T \delta x$ is the signal (in volts) that will be observed for a tip displacement δx. (Note that this *transducer* responsivity is distinct from the device's compliance, which is its *mechanical* responsivity.) Here, the parameter π_ℓ is the piezoresistive coefficient of the p+ transducer material. The parameter β was introduced by Harley and Kenny to account for the decrease in $\partial R_d/\partial x$ due to the finite thickness of the conducting epilayer and its overlap with the induced variation of the strain field along the thickness dimension of the device. For their devices, which are similar to those discussed here, $\beta \sim 0.7$ [20]; β approaches unity in the limit where the carriers are confined to a surface layer of infinitesimal thickness.

To account for the effect of degradation of force sensitivity by read-out process; we add three additional terms (arising from electrical noise) to the spectral density of fluid-induced displacement fluctuations, $S_x^\gamma = 4k_BT\gamma |H(f)|^2$. Again, to assess their role in limiting the practical force sensitivity of the transducer-coupled device, these must be referred back to the input, i.e. to the displacement domain using the factor $1/\mathcal{R}_T^2$. The first arises from the thermal voltage noise of the piezoresistive transducer, $S_V^J = 4k_BTR_d$, while the second arises from the readout amplifier's voltage and current noise, $S_V^A = S_V + S_I R_d^2$, where S_V and S_I are the spectral density of the amplifier's voltage and current noise, respectively. If the response extends down to low frequencies, we must also consider a third term, the flicker noise (often termed "$1/f$" noise) in the transducer, $S_V^{1/f}$. The sum of these fluctuations yields what we term the total *coupled* displacement noise, which is the actual displacement sensitivity of the entire system:

$$S_x^{(tot)} = S_x^\gamma + \frac{1}{\mathcal{R}_T^2}\left\{S_V^J + S_V^A + S_V^{1/f}\right\}. \tag{12.12}$$

From this we can determine the coupled force sensitivity of the electro-mechanical system:

$$S_F^{(tot)} = S_x^{(tot)}/|H(f)|^2 = 4k_BT\gamma_{eff}$$
$$+ \frac{1}{|H(f)|^2 \mathcal{R}_T^2}\left\{S_V^J + S_V^A + S_V^{1/f}\right\}. \tag{12.13}$$

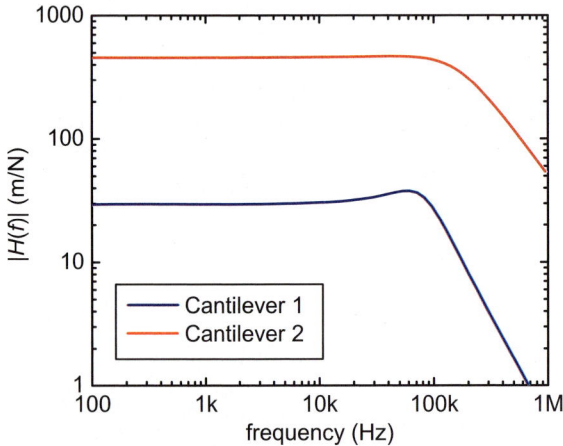

Fig. 12.2. Calculated Amplitude response functions (defined by Eq. (12.10) in the text) for two prototypical fluid-loaded nanocantilevers. The *curves* correspond to two examples whose properties are delineated in Table 12.1. The Reynolds number at the vacuum resonance frequency ranges from 0.8 to 0.01 (cf. Table 12.1) and the response is seen to evolve from nearly critically damped to strongly overdamped

12.9 BioNEMS: Practical Considerations Determining Realizable Sensitivity

12.9.1 Maximal Transducer Current Bias

We now investigate the constraints upon the level of current bias that can be applied. The force sensitivity attainable clearly hinges on the maximum current level that is tolerable, given that the responsivity is proportional to bias current, $\mathcal{R}_T = I_b \frac{\partial R_d}{\partial x}$. However, this applied bias current leads to self-heating of the device. Several considerations are important in determining the optimal (maximum) current level: first, the highest tolerable temperature rise within the device must be considered both at its biofunctionalized tip (to avoid damage to the SAM) and at the position of peak temperature rise within the device (to avoid device failure). Additionally, if the electrical elements of the device are unpassivated, it may prove important to limit the transducer voltage drop to below 0.5 V to prevent undesired electrochemical processes. Here we consider the self-heating of the device due to electrical power dissipation. The geometry of the prototype devices causes dissipation to occur predominantly within the constriction regions, treated as a beam of width, W_{leg}, length, ℓ_{leg}, and cross sectional area A (cf. Fig. 12.2) with a one dimensional heat sink at the supporting end. For $x > \ell_{leg}$, a rough estimate of the heat loss to the surrounding fluid may be obtained through the relationship κ_{Si} A $\frac{d^2 T}{dx^2} = \kappa_{H_2O} P \nabla_n T$, where P is the perimeter around cross-sectional area A of the beam. Estimating $\nabla_n T \sim T/w$, $\frac{d^2 T}{dx^2} \sim \frac{2(w+t)\kappa_{H_2O}}{\kappa_{Si} tw^2}$ where $\kappa_{Si} = 1.48 \times 10^2$ W/m K [42] is the thermal conductivity of silicon and $\kappa_{H_2O} = 0.607$ W/mK [43] the thermal conductivity of water. In the dissipative region $x < \ell_{leg}$ we have $2\kappa_{Si} tw_{leg} \frac{d^2 T}{dx^2} \sim -\frac{I_b^2 R_d}{\ell_{leg}} + 4\left(w_{leg} + t\right) \frac{T}{w_{leg}} \kappa_{H_2O}$. As boundary conditions we have that the temperature is continuous at ℓ_{leg}, as is the heat flux, and the temperature must monotonically decrease for $x > \ell_{leg}$. For a bias current of 30 μA, this calculation yields a temperature rise at the cantilever tip of 0.01 K. The maximal temperature rise of 0.1 K occurs within the constricted region, approximately 2.3 μm from the support. For this bias current, our prototype device yields a responsivity $\mathcal{R}_T = I_b \frac{\partial R_d}{\partial x} \sim 20 \mu V/nm$.

With knowledge of these parameters we now estimate the coupled force sensitivity of the prototype system. For cantilever 1, assuming that a 1 K rise at the tip is tolerable, the transducer-induced displacement sensitivity is found to be $\sqrt{S_V^J}/5\times10^{-13}$ m/$\sqrt{Hz}$. This number represents the PSD for displacement fluctuations of the cantilever tip due to thermal mechanical motion. For a typical low noise readout amplifier with voltage and current noise levels (referred to input) of ~4nV/$\sqrt{Hz}$ and ~5 fA/$\sqrt{Hz}$, respectively,[1] these same parameters yield an amplifier term $\sqrt{S_{VA}}/\mathcal{R}_T = 2\times10^{-13}$ m/$\sqrt{Hz}$ (this

---

[1] This is a typical value for JFET input low noise amplifiers, for frequencies beyond the $1/f$ noise knee.

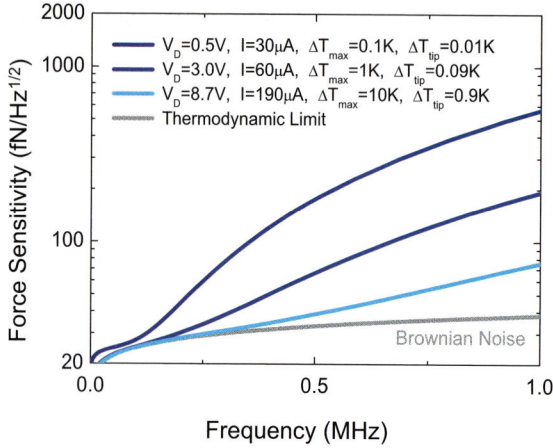

Fig. 12.3. Total, coupled force spectral density for cantilever 1. The analytic model, based upon fluid coupling for an infinite cylinder, is used for analysis of the device that has parameters shown in Table 1. As described in the text (Eq. (12.14)), the force noise includes noise contributions from the electrical domain, referred back to the input (force domain) – specifically, electrical fluctuations arising from Johnson noise in the piezoresistive transducer and noise from the subsequent read-out amplifier. The fluidic fluctuation limit is displayed for reference; at higher bias currents this is more closely approached

number represents the PSD for displacement fluctuations that would yield a noise equivalent to the PSD for the combined Johnson and preamplifier noise).

Figure 12.3 shows the coupled force sensitivity per unit bandwidth for this device, as given by Eq. (12.13). This includes the combined noise from fluidic, transducer, and readout amplifier sources referred to input (i.e. the force domain) using the above parameters. For comparison, separate traces representing only the fluidic noise are displayed; these are calculated from the Nyquist formula with Eq. (12.7) for the damping term in water.

Figure 12.3 shows that even the largest device (cantilever 1) yields a remarkably low *coupled* force sensitivity, $[S_F^{(tot)}]^{1/2} \leq 30$ fN/$\sqrt{\text{Hz}}$ for frequencies below 100 KHz, at a bias voltage of 0.5 V. This means dynamical measurements on the $\sim 10\,\mu$s scale are possible for absolute forces on the level of < 8 pN without averaging. This is comparable to the average force of a single hydrogen bond (~ 10 pN) [45, 46, 47], whereas typical antibody-antigen binding forces are in the 50–300 pN range [48, 49]. Two higher bias voltages are also shown in Fig. 12.3. The heating at the cantilever tip remains modest at these voltages. For a bias voltage of 4.3 V across the device, the expected temperature rise at the tip is only 0.9 K, and a coupled force sensitivity $[S_F^{(C)}]^{1/2} \leq 33$ fN/$\sqrt{\text{Hz}}$ for frequencies below 500 KHz can be attained. This is only 20% above the thermomechanically determined thermal noise floor and means that dynamical measurements on the $\sim 2\,\mu s$ scale are possible for absolute forces on the level of <22 pN without averaging.

In Fig. 12.3 the transducer noise is clearly dominant. At fixed voltage, V, the Johnson noise contribution to the force sensitivity is given by

$$\sqrt{S_F^J} = \frac{4t^2\sqrt{2w_{leg}\ell_{leg}\rho_{epi}k_B T}}{3\beta\pi_\ell V\,(2\ell - \ell_{leg})\sqrt{t_{epi}}K\,|H\,(f)|}. \qquad (12.14)$$

where the product $K\,|H\,(f)|$ is a dimensionless response function (Fig. 12.1). It is clear from Eq. (12.14) that to minimize the Johnson noise it is desirable to maximize β, π_ℓ and V. Since π_ℓ decreases with increasing doping (boron concentration), [37, 44] in the Johnson noise limited regime it is desirable to work at low doping concentrations. (Below 1×10^{17} cm$^{-3}\pi_\ell$ is essentially independent of doping concentration.) While modest improvements can be made in these areas, there is more potential in improved device design through the reduction of device dimensions, particularly w_{leg}, t, and ℓ_{leg}. It is actually also beneficial to increase the total device length ℓ although this increased sensitivity comes at the expense of bandwidth. However, by far the greatest improvement is achieved through a reduction in device thickness.

With these considerations in mind we evaluate an additional device; *cantilever 2*, for which the dimensions have been reduced. We stress that all of these dimensions, as tabulated in Table 12.1, are *practical*; they are readily achievable by top-down nanofabrication processes in our laboratory. For cantilever 2, with dimensions tabulated in Table 12.1, we obtain $R_d = 270$ kΩ and $\partial R_d/\partial x = 9.9 \times 10^9$ Ω/m. We again first assume that a maximum voltage drop across the device of 0.5 V is tolerable. The resulting force sensitivity is shown in Fig. 12.4. This smaller device provides very impressive force sensitivity: $[S_F^{(tot)}]^{1/2} \leq 6$ fN/$\sqrt{Hz}$ for frequencies below 0.5 MHz. This permits dynamical measurements on the 2 μs scale with force resolution at the 4 pN level without averaging. For a larger bias voltage of 4.1 V across the device, the bandwidth can be increased to 2 MHz while maintaining a coupled force sensitivity $[S_F^{(tot)}]^{1/2} \leq 8$ fN/$\sqrt{Hz}$, allowing dynamical measurements on the 500 ns time scale for absolute forces at the 10 pN level. Again, resulting temperature rises for these higher biases are negligible on biochemical scales.

12.9.2 Low Frequency Displacement Transducer Noise

An important consideration is the low-frequency "flicker" noise that arises within the transducer under conditions of current bias from its intrinsic resistance fluctuations, $S_{1/f} = I_b^2 S_R$. An empirical model first proposed by Hooge [50] and applicable where the number of carriers is small, relates the spectral density of the transducer's resistance fluctuations to the number of carriers involved in conduction, $S_R = \varsigma R_d^2/Nf$. Here, ς is a sample-specific materials parameter and N is the number of carriers within the sample. This formula assumes uniform conduction; based upon the work of Harley and

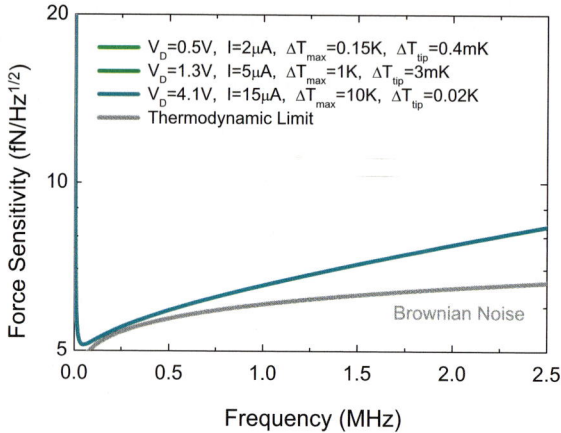

Fig. 12.4. Total coupled force spectral density for cantilever 2. The noise performance obtained is again qualitatively similar to that of the two larger devices considered, but substantially higher sensitivity is obtained for this, the smallest device analyzed. Note that at frequencies <1 MHz, the transducer is quite capable of matching the thermodynamic limit imposed by Brownian fluctuations, and the coupled sensitivity remains within about 20% of the fluidic noise floor at a bias voltage of 0.5 V. Below 0.25 MHz, the spectral density is only of order 5 fN/$\sqrt{\text{Hz}}$ (for reasonable bias voltages). The low frequency noise rise in these spectra includes the expected contributions from low frequency resistance fluctuations arising within the piezoresistors (see text), which is more clearly displayed in Fig. 12.5

Kenny [44] this is a reasonable approximation if N is taken to represent the combined number of carriers in the two legs, and from their work we expect $\varsigma \sim 10^{-5}$ for our devices. (This assumes a 3 hour anneal at 700°C.) Based on numerical calculations of the depletion length, the number of carriers in the doped region is estimated to be 1.2×10^4 for cantilever 2.[2] In the plots of Fig. 12.5 we have included the low frequency flicker noise for this device at the current bias levels used in Fig. 12.4. Note that above about 10 kHz the total noise floor for the coupled device is dominated by either Brownian fluidic fluctuations or the combined Johnson noise of the transducer and its readout amplifier.

---

[2] In addition to concerns over the depletion length, one might be concerned as to whether a total thickness of 30 nm (comprised of 7 nm doped and 23 nm intrinsic silicon) is feasible due to migration of carriers into the intrinsic region. Numerical calculations show that for this dopant concentration, 2.3 nm into the intrinsic region, the concentration of holes has dropped by about an order of magnitude, and by 7.3 nm it has dropped by 2 orders of magnitude.

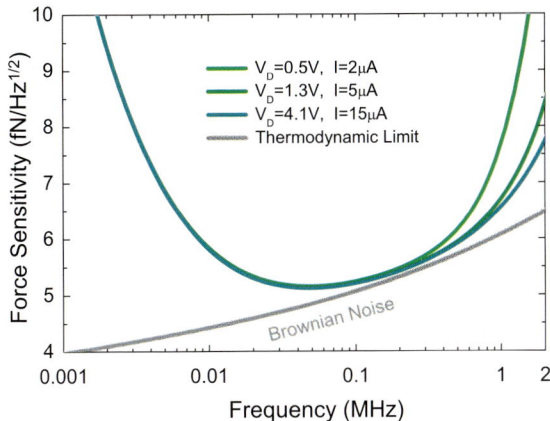

Fig. 12.5. Total, low frequency, transducer-coupled force spectral density for cantilever 3. The increase in the low frequency spectral densities arises from flicker noise processes in the displacement transducers (see text). This unavoidable voltage noise contribution from the semiconducting piezoresistors of cantilever 3 (cf. Table 12.1) is referred back to the input (force domain) and contributes to the overall noise spectra for the transducer-coupled devices. As can be seen, despite this contribution, above a few kHz the low frequency noise is not significant for these cantilever dimensions

12.10 Simulations of the Stochastic Dynamics of Fluid-Coupled Nanocantilevers

To realize the goal of making single molecule biophysical measurements with NEMS it is important to have a baseline of understanding concerning their stochastic motion in the absence of any immobilized biomolecules. In our previous discussion we have modeled the stochastic dynamics of a long and slender cantilever (as is common in atomic force microscopy) by coupling the classic equations of beam elasticity with the two dimensional flow field around an oscillating circular cylinder [33, 52]. However, the nanoscale biosensors we are developing raise new concerns that require a more precise and quantitative approach. Four of these concerns are discussed below.

(1) The cantilevers under consideration are an order of magnitude smaller than those considered in the past. As a result, the fluidic damping is much stronger and the fluid coupling with the cantilevers is large. When there is less dissipation (i.e. for devices with higher Q) it is possible to take advantage of the fact that the mode shape of the cantilever remains unchanged by developing an analytical solution based upon an expansion of these modes. However for the BioNEMS cantilevers under consideration here the complex and highly dissipative flow field affects the beam mode shapes, making this approach quite difficult. (2) The geometries under consideration are actually *not* simple cantilevers but are instead complex structures designed to improve experimental measurement (cf. Fig. 12.2). The flow field around these

complex geometries is also more complicated and, for example, the flow field off the tip of the cantilever is now also of importance. As a result, the approximation of an oscillating cylinder of infinite length is no longer appropriate. (3) Advanced mechanical bioassays involve multiple cantilevers – for example, cantilevers biofunctionalized for different target biomolecules and placed in close proximity to one another. Their proximity results in mutual coupling of the cantilevers through the intervening fluid. Understanding, and ideally, *exploiting* this fluidic coupling will be an important and essential component in advanced BioNEMS designs. This mutual fluidic coupling imposes important, ultimate limits to the performance of complex BioNEMS assays. First, it determines the maximum density at which BioNEMS arrays can be assembled. Second, for force transmission assays (§15.2 and Fig. 12.11), it also determines the threshold force magnitude that is detectable, i.e. it sets the biomolecular coupling required to exceed this unavoidable background fluidic coupling. There is currently no analytical theory available to quantitatively describe the fluid coupled motion of arrays of elastic structures. (4) For many lab-on-a-chip applications arrays of biosensors will be placed in small volumes, comparable in size to the sensors themselves. The fluid disturbance caused by an oscillating cantilever is long range compared to its dimensions. As the cantilevers are scaled in size downward the range over which fluidic effects are significant increases. For the BioNEMS devices this can be on the order of microns.

12.11 Stochastic Dynamics of Fluid-Coupled Nanocantilevers: Theoretical Approach

A possible approach to the analysis of fluid-coupled nanomechanical devices would be to perform a stochastic simulation allowing resolution of the Brownian dynamics of the surrounding fluid and the cantilever. This approach, however, would be both extremely difficult and computationally intensive. On the other hand, we can exploit the fact that the system is always at, or near, thermal equilibrium. This consideration allows use of powerful concepts of statistical mechanics such as the fluctuation-dissipation theorem. Utilizing this has led to a deterministic numerical approach that allows the stochastic dynamics of BioNEMS cantilevers to be calculated with sufficient accuracy to model the experimental devices and their performance. Below we discuss in more detail how this is accomplished; the logic follows the development given by Paul and Cross [30].

For the case of fluid loaded cantilevers, the fluctuations are caused by the collisions of fluid molecules with the cantilever and the dissipation is due to the fluid viscosity. However, dissipation from other sources such as the internal elastic dissipation of the cantilever could be included if desired.

The approach can be clearly described using a classical formulation of the fluctuation-dissipation theorem [53, 54]. Consider a dynamical variable

A which is a function of the microscopic phase space variables consisting of $3N$ coordinates and conjugate momenta of the cantilever, where N is the number of particles in the cantilever. For now we will leave A general but, in what follows, it will represent displacement of a nanomechanical device, e.g. deflection of a cantilever tip. Next, consider the situation where a force $F(t)$ that couples to A is imposed. In this case the Hamiltonian of the system is,

$$H = H_o - AF(t), \qquad (12.15)$$

where H_0 is the Hamiltonian of the system in the absence of the force $F(t)$. We are interested in the linear response, or in other words, we are interested in the case where $F(t)$ is very small. For the special case of a step function force applied in the distant past and turned off at $t = 0$,

$$F(t) = \begin{cases} F_o & \text{for } t < 0 \\ 0 & \text{for } t \geq 0 \end{cases}. \qquad (12.16)$$

In the linear response regime, the change in the average value of a second dynamical quantity B from its equilibrium value in the absence of F is given by,

$$\Delta \langle B(t) \rangle = \beta F_o \langle \delta A(0) \, \delta B(t) \rangle_o. \qquad (12.17)$$

Here $\beta = (k_B T)^{-1}$, k_B is Boltzmann's constant, T is the absolute temperature, $\delta A = A - \langle A \rangle_o$, $\delta B = B - \langle B \rangle_o$ and the subscript zero on the average $\langle \rangle$ denotes the equilibrium average in the absence of the force, F. (Here, as above, we are considering B as a general parameter, but in what follows we shall specify it as displacement of a second, adjacent nanomechanical device.) Thus, by rearranging the above expression, a relation for the general equilibrium cross-correlation function is found in terms of the linear response as

$$\langle \delta A(0) \, \delta B(t) \rangle_o = \frac{k_B T}{F_o} \Delta \langle B(t) \rangle. \qquad (12.18)$$

This result is exact in the approximation of classical mechanics and linear behavior. If, in addition, the dynamical variables are sufficiently macroscopic such that the mean $\langle B(t) \rangle$ can be calculated using deterministic, macroscopic equations, we have our desired result. In other words, if $\Delta \langle B(t) \rangle$ can be calculated deterministically from simplified analytical models, experiment, or numerical simulations then Eq. (12.18) will yield the desired result describing the precise stochastic dynamics that would be measured in experiment. The frequency response is given by the noise spectrum which is the cosine Fourier transform of Eq. (12.18).

An important advantage of this approach is that the full correlation function is given by a single deterministic calculation. This becomes particularly straightforward for the response of the system to the removal a step force. The approach is advantageous for the low-quality factor (Q) dynamics characteristic of fluid-damped nanoscale cantilevers, since their spectral response is very broad and a large number of fixed-frequency simulations would otherwise be needed to characterize this response. Finally, Eq. (12.18) allows the calculation of the correlation function and noise spectrum of *precisely* the quantity that is measured in experiment. This is accomplished by first tailoring the applied force to couple to one physical variable measured in the experiment (A), and then determining the effect on the second physical variable (B). For the case of the BioNEMS sensors discussed in the previous section, the displacement of the cantilever is measured through the strain of a piezoresistive layer near the pivot point of the cantilever. It is possible to tailor the force F to couple to this distortion, and so determine the "strain-strain" correlation function of one or more cantilevers.

12.12 Stochastic Dynamics of Fluid-Coupled Nanocantilevers: Implementation and Results

The numerical problem is reduced to the deterministic calculation of the response of a fluid-immersed elastic body to the removal of step force. Numerical methods for such fluid-solid interaction problems are quite well developed, both in the research literature as well as in industry (see for example [55, 56]). We have validated this approach to calculate the stochastic dynamics of an AFM cantilever in fluid [30]. We now demonstrate how this approach can be used to calculate the auto- and cross-correlation functions for the displacements $x_1(t)$ and $x_2(t)$ of the tips of one or two nanoscale cantilevers with experimentally-realistic geometries, as shown in Fig. 12.6. For this case neither analytical nor, at present, experimental results are available; this makes the numerical approach a valuable tool for developing further insight. These

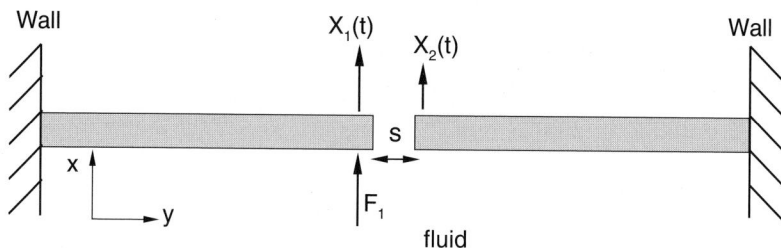

Fig. 12.6. Schematic of two adjacent cantilevers immersed in fluid. The step force F_1 is released at $t = 0$ resulting in the deterministic cantilever motion referred to by $X_1(t)$. The deterministic motion of the adjacent cantilever is, $X_2(t)$, and is driven through the response of the fluid. The cantilever separation is given by s. The numerical simulations are three-dimensional (the figure shows only a side view) and the cantilevers are placed in a large box (only the side-walls are shown in the figure) with the no-slip boundary condition on all surfaces

calculations are accomplished by first calculating the deterministic response of the displacement of each tip, denoted as $X_1(t)$ and $X_2(t)$, after switching off at $t = 0$ a small force applied to the tip of the first cantilever, F_1, given by Eq. (12.16). For this case the equilibrium auto- and cross-correlation functions for the fluctuations x_1 and x_2 are precisely,

$$\langle x_1(t)\, x_2(0)\rangle = \frac{k_B T}{F_1} X_1(t),\qquad\qquad (12.19)$$

$$\langle x_2(t)\, x_1(0)\rangle = \frac{k_B T}{F_1} X_2(t).\qquad\qquad (12.20)$$

The cantilever autocorrelation function and the two cantilever cross-correlation functions are shown in Fig. 12.7 (a) and (b), respectively. The quantities actually measured in experiment, the noise spectra $G_{11}(f)$ and $G_{12}(f)$, are simply the cosine Fourier transforms of the auto and cross-correlation functions, respectively, and are shown in Fig. 12.8 (a) and (b).

It is informative to explore these results further. The autocorrelation at time $t = 0$ is $\langle x_1(0)\, x_1(0)\rangle \approx 0.471$ nm^2. The square root of this gives an indication of the cantilever deflection due to Brownian motion that would be measured in an experiment; this yields a deflection of 0.686 nm or about 2.3% of the thickness of the cantilever. The cross-correlation of the Brownian fluctuations of two cantilevers is small compared with the individual fluctuations. The largest magnitude of the cross-correlation is –0.012 nm^2 for s $=$ h and –0.0029 nm^2 for s $=$ 5h. Notice that tuning the separation could be used to reduce the correlated noise in some chosen frequency band.

The variation in the cross-correlation behavior with cantilever separation as shown in Fig. 12.7 can be understood as an inertial effect resulting from the non-zero Reynolds number of the fluid flow. The flow around the cantilever can be separated into two components. The first is a potential component, which is long range and propagates instantaneously in the incompressible fluid approximation. The second is a vorticity-containing component which propagates diffusively. For small cantilever separations the viscous component dominates for nearly all times, and the result of this is an anticorrelated response of the adjacent cantilever. This behavior is in agreement with an experimental study of the hydrodynamic coupling between two micron scale beads in fluid [57]. However, with increasing cantilever separation the amount of time during which the adjacent cantilever is subject solely to the potential flow field increases resulting in the initially correlated behavior.

12.13 Implementation of Practical Biosensing Protocols

In this section we outline the translation of conceptual device design into realistic biosensing protocols. Possible BioNEMS sensors can be differentiated into two main classes that we describe in turn below: single cantilever

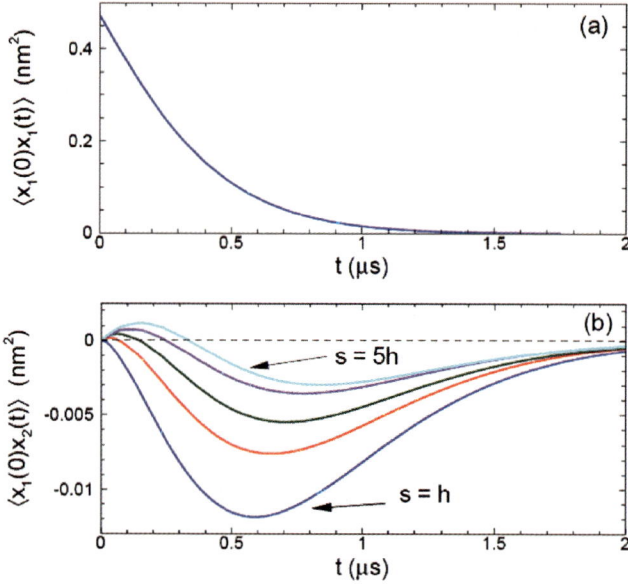

Fig. 12.7. Absolute predictions of the auto- and cross-correlation functions of the equilibrium fluctuations of a fluid-coupled pair of BioNEMS cantilevers. The calculation assumes the cantilevers are configured as shown in Fig. 12.6. Panel (**a**) shows the autocorrelation of the fluctuations; panel (**b**) shows their cross-correlations. In panel (b) 5 different cantilever separations (from 5 independent numerical simulations) are shown for s = h, 2h, 3h, 4h, 5h. (Only s = h, 5h are labeled; the remaining *curves* lie between these values in sequential order.) The initial conditions are a step force applied to the tip of the first cantilever, $F_1 = 0.75$ pN at $t = 0$. The *dashed line* indicates where the cross-correlation is equal to zero, and is included here to highlight the novel bipolar evolution of the cross-correlations as separation is increased

and multiple-cantilever devices. Single cantilever devices can be actuated in a number of ways including: *piezoelectric actuation* – e.g., via a piezoelectric actuation element patterned locally upon the cantilever; *fluidic actuation* – e.g., via pulsatory fluidic drive delivered through the microfluidic environment in which BioNEMS devices are housed [58]; *magnetic actuation* – e.g., via inclusion of nanomagnets upon the cantilever tip and use of external or on-chip drive coils [40]; or by "*passive actuation*" – i.e., *stochastic sensing*, which employs Brownian fluctuations as the "drive" force [2,40]. Independent of the method of actuation employed, device response in fluid is governed by the response function discussed previously.

Biosensing with NEMS is based upon effecting a change in the dynamical properties of the device upon capture of a target analyte. To realize this, the device must be biofunctionalized in a manner such that the immobilization event induces a change in device response – either through change in the device compliance, a change in its damping, or by the imposition of new

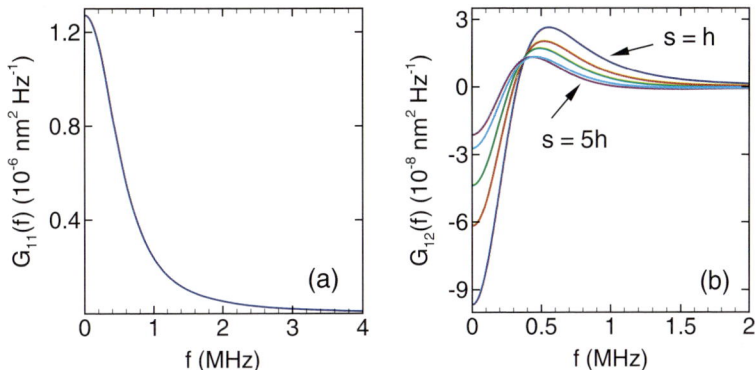

Fig. 12.8. Noise spectra and cross-noise spectra for a fluid-coupled pair of nanocantilevers. Panel (**a**): The noise spectrum, $G_{11}(f)$. Panel (**b**): The noise spectrum, $G_{12}(f)$, as a function of cantilever separation, s, for two adjacent experimentally realistic cantilevers. Five separations are shown $s = $ h, 2h, 3h, 4h, 5h where only $s = $ h, 5h are labeled and the remaining *curves* lie between these values in sequential order

correlations or actuation forces as a result of target immobilization. We now consider concrete realizations for each of these types of devices.

12.13.1 BioNEMS Detection Based on a Change in Device Compliance

The binding of an analyte will lead to a change in the cantilever's effective spring constant if its immobilization forms a structural bridge between the motional element and some other body. This is most effective if the other body is a rigid structure; Fig. 12.9 shows such a device in which a cantilever becomes tethered to a rigid dock upon target capture. In this example, the fine gold lines at the cantilever tip and along the dock in the vicinity of the tip are functionalized with a receptor that is specific to the analyte of interest. This illustrates the general protocol involving the binding of an analyte between two functionalized regions to form a biomolecular bridge, which increases the effective device spring constant.

12.13.2 BioNEMS Detection Based on a Change in Device Damping

The analysis of Sects. 12.9 and 12.10 makes it clear that fluid coupling depends upon the cross-sectional surface presented by the device, normal to the direction of its motion. To effect a measurable change in this coupling, the cross section presented must be increased by some fraction of the total effective surface area; the absolute fractional change required depends upon

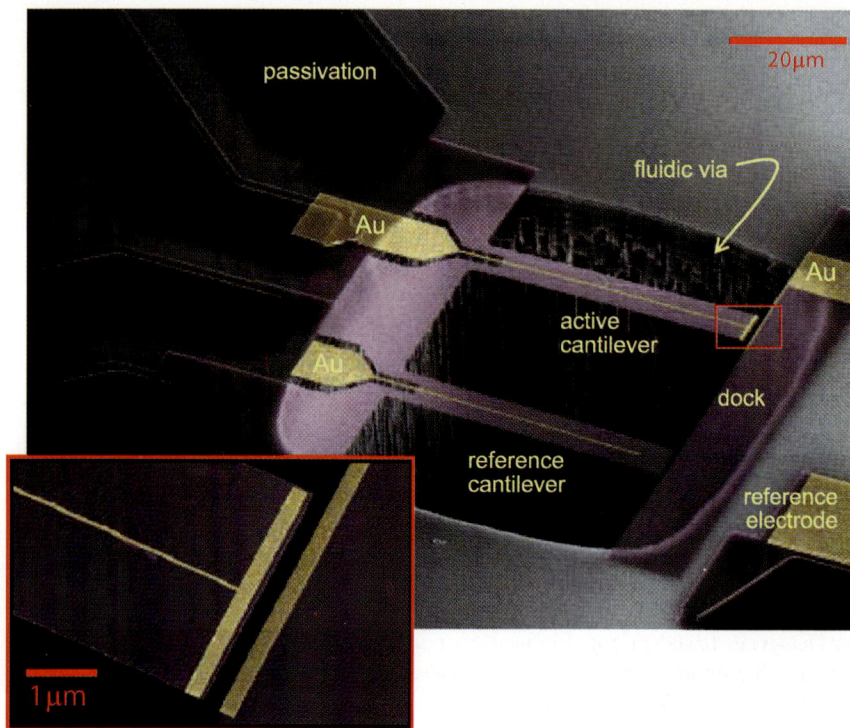

Fig. 12.9. BioNEMS device employing single-analyte detection via change in device compliance. In these devices analytes bind across the gap between the active cantilever and the dock, and their immobilization is reflected as a measurable change in the overall device compliance [58]. This particular family of devices is configured for detection of bacterial pathogens; the fine *Au lines* at the active device tip and on the dock are biofunctionalized with antibody-terminated alkanethiol SAMs [59]. Electrical connections to these lines permit electrically-activated biofunctionalization protocols in situ. Upon specific binding, the target pathogens subtend the submicron gap between these biofunctionalized regions (see magnified view in the inset). Both the gap width and the two biofunctionalized regions are specifically tailored to capture an individual biological target of interest

the measurement resolution. These considerations make it clear that the effective surface area of an analyte to be detected must be sufficient as to lead to an appreciable change in the fluid loading of the device. If this is the case, then the binding of an analyte can lead to a direct change in the device response through an increase in damping. This, in turn, leads to a change in the response function of the device which can also lead to a change in the requisite driving force (the latter would be the case for both fluidic drive and drive based on Brownian fluctuations). This "damping assay" is conceptually depicted in Fig. 12.10.

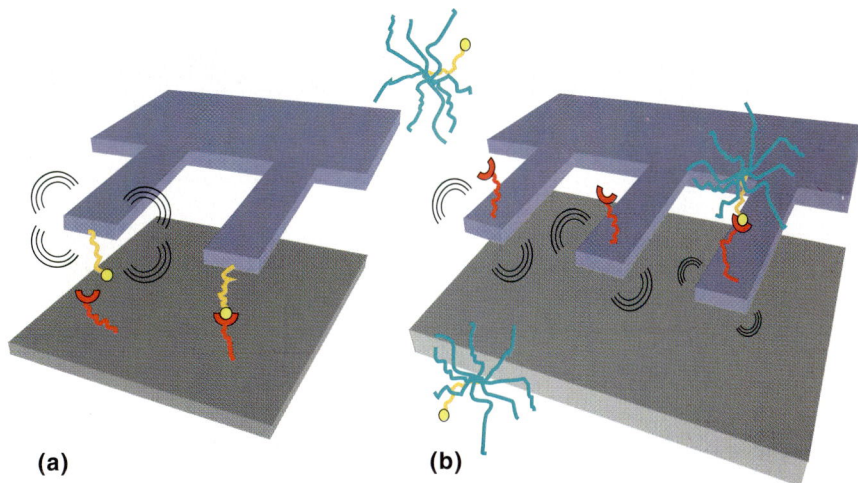

Fig. 12.10. Schematic of analyte detection based on single-cantilever devices. These images detect two single-molecule sensing modalities that remain effective for heavily-fluid damped nanomechanical sensors, even when the associated molecular mass-loading is undetectable [2]. Hence, these are not mass sensing protocols. **(a)** Detection based upon a change in device compliance. The presence of a single biomolecular bond changes the force constant of the device sufficiently to alter its vibrational properties in fluid, as described in the text. **(b)** Detection based upon a change in device damping. The immobilization of a target analyte causes increased damping which is detectable by the resulting change in the device response function. The impact of an individual analyte binding event can be amplified by attaching a "label", e.g., a star dendrimer, which causes enhanced fluid coupling (compared to that which is induced by immobilization of an unlabelled analyte)

12.13.3 Coupled, Multiple-Cantilever Devices

More sensitive detection schemes can be based upon the use of two or more coupled cantilevers. These multiple cantilever devices fall into two general classes schematically depicted in Fig. 12.11, namely stochastic sensors and actively-driven sensors.

Stochastic sensors make use of Brownian fluctuations of the fluid for their "actuation." As discussed previously, adjacent cantilevers are coupled by local fluidic correlations. If placed sufficiently far apart, these background correlations can be very small, and the immobilization of a biomolecular target between the two devices, as depicted in Fig. 12.11(a), can induce new, measurable correlations between their motion in the Brownian drive "field."

Actively-driven "force transmission" devices involve, at minimum, two cantilevers, one of which is actuated while the second serves as a sensor. This is shown schematically in Fig. 12.11(b). Ideally, for such a detection scheme there would be no background coupling between the actuator and sensor in the absence of analyte binding. Again, in reality there is a background

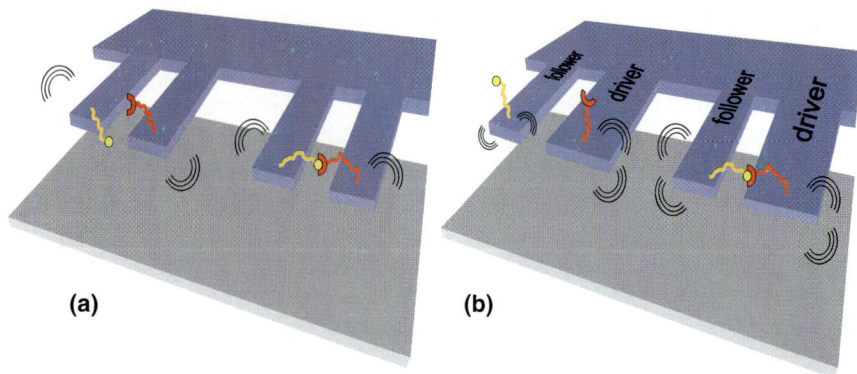

(a)   **(b)**

Fig. 12.11. Schematic of two-cantilever, analyte-coupled devices. In these devices coupling between the adjacent cantilevers is induced by specific interactions between immobilized biomolecules [2]. **(a)** Passive stochastic correlation sensing. In this modality the biomolecular linkage induces additional motional correlations between two cantilevers – correlations that exceed the "parasitic" background present due to mutual fluidic coupling (see text). **(b)** Active driver/follower lock-in force sensing. This two-cantilever sensing protocol involves an actuator cantilever that is actively-driven (the "driver"), which imparts forces of specific frequency and phase to a second detector sensor (the "follower"). The resulting "lock-in" force detection provides increased resolution of the binding of an analyte linking two cantilevers. Both of these concepts can be translated into realistic two-cantilever biosensor protocols in multiple ways. For example, a pre-programmed linkage (such as pictured above) can be designed to become destabilized upon the arrival of a target analyte that displaces an initial ligand to bind with the immobilized receptor with higher affinity. Alternatively, a high specificity, two-cantilever sandwich assay may be formed by functionalizing the adjacent cantilever pair with two different immobilized antibodies, which then permit specific binding to two different epitopes upon the same individual target antigen

coupling imposed by the fluidic coupling of the devices discussed previously. This imposes a minimum threshold for the strength of the analyte-induced coupling [60]. With a three-cantilever design (using two actuators, driven with a specific phase difference, and one follower) it is possible to place the sensor in a "node" of the fluidic coupling from the two actuators so that a null baseline can be achieved [2].

12.13.4 Practical Considerations

Embedding the BioNEMS device in a microfluidic assembly is essential in order to complete the goals of simplification and miniaturization for which the integrated sensors were chosen [40]. Single-molecule biosensing protocols, in some sense, become practically useful only within microfluidic assembly where total fluidic volumes is minimized, recirculation protocols can be realized to create enhanced binding probabilities, and the environment to which the device is exposed can be actively controlled. The silicon devices discussed here are readily compatible with most of the common microfluidic designs, including surface-micromachined [61] and polymer- [62, 63] based protocols.

12.14 Specificity and the Stochastic Nature of Single-Analyte Binding Events

The biomolecular specificity of the devices discussed herein is governed entirely by their biofunctionalization; this is the biosensors true "front end." Hence, surface biochemistry of the analyte-receptor interaction is critical – high affinity capture probes will ensure a highly selective response with good rejection of non-specific background interactions. The design of high affinity capture probes is clearly absolutely essential to future progress and real applications in all areas of biosensing, and this includes BioNEMS. However, discussion of this important topic is outside the scope of the present review.

The stochastic nature of single molecular events certainly makes for interesting science, but its also opens very significant new challenges for biosensing. This is true both in terms of device design and interpretation of data that is acquired [64]. For biosensing, it is usually "go/no-go" sensing, rather than the ensemble-averaging of events, that is the goal. Hence, protocols must be formulated that deal directly with such considerations, for example, as the fact that the residence time and binding constant for individual events – that is successive events involving the same biochemical species – may vary [65]. One possible avenue to minimize this intrinsic "uncertainty," i.e., provide biosensing protocols with high confidence levels even for single-shot measurements, is to make use of extremely high-affinity capture probes. This, again, puts the onus upon obtaining or developing optimal biofunctionalization; with extremely high affinity capture probes, false positives will be minimal and residence times will be sufficiently long as to permit temporal rejection of the ubiquitous background of non-specific binding. A second approach, which we have involved in our work, involves electrically-actuated analyte release [59]. This utilizes the fact that, although *capture* is a stochastic process, *release after capture* can be programmed deterministically, and the resultant response used to provide good rejection of false positives [2].

Acknowledgements

We gratefully acknowledge interactions and work with the Caltech BioNEMS collective, spanning the Roukes, Fraser, Cross, and Phillips research groups. We are also grateful for support from DARPA/DSO SIMBIOSYS, via AFOSR grant number F49620-02-1-0085. M.P. thanks the Wellcome Foundation "Interfaces" program for additional support.

References

1. M.L. Roukes, Technical Digest of the 2000 Solid-State Sensor and Actuator Workshop, Hilton Head, NC, (Cleveland: Transducer Research Foundation, 2000); downloadable at http://arxiv.org/pdf/cond-mat/0008187
2. M.L. Roukes, S.E. Fraser, J.E. Solomon, and M.C. Cross (2001). *"Active NEMS arrays for biochemical analyses"*, United States Patent Application 20020166962, Filed August 9, referencing U.S. application Ser. No. 60/224, **109**, Filed Aug. 9, 2000.
3. J.H. Hoh, J.P. Cleveland, C.B. Prater, J.-P. Revel, and P.K. Hansma, (1992). *J. Am. Chem. Soc.*, **114**, pp. 4917–4918.
4. G.U. Lee, D.A. Kidwell, and R.J. Colton (1994). *Langmuir*, **10**, pp. 354–357.
5. E.L. Florin, V.T. Moy, and H.E. Gaub (1994). *Science*, **264**, pp. 415–417.
6. U. Dammer, O. Popescu, P. Wagner, D. Anselmetti, H.-J. Güntherodt, and G.N. Misevic (1995). *Science*, **267**, pp. 1173–1175.
7. E. Evans, K. Ritchie, and R. Merkel (1995). *Biophys. J.*, **68**, pp. 2580–2587.
8. M. Radmacher, M. Fritz, H.G. Hansma, and P.K. Hansma (1994). *Science*, **265**, pp. 1577–1579.
9. M. Rief, M. Gautel, F. Oesterhelt, J.M. Fernandez, and H.E. Gaub (1997). *Science*, **276**, pp. 1109–1112.
10. G.U. Lee, L.A. Chirsley, and R.J. Colton (1994). *Science*, **266**, pp. 771–773.
11. H.-J. Butt (1995). *Journal of Colloid and Interface Science*, **180**, pp. 251–259.
12. R. Raiteri, G. Nelles, H.-J. Butt, W. Knoll, and P. Skládal (1999). *Sensor and Actuator B-Chem.*, **61**, pp. 213–217.
13. R. Raiteri, M. Grattarola, H-J. Butt, and P. Skládal (2001). *Sensor and Actuator B- Chem.*, **79**, pp. 115–126.
14. G. Wu, R.H. Datar, K.M. Hansen, T. Thundat, R.J. Cote, and A. Majumdar (2001). *Nature Biotechnology*, **19**, pp. 856–860.
15. J. Fritz, M.K. Baller, H.P. Lang, H. Rothuizen, P. Vettiger, E. Meyer, H.J. Güntherodt, Ch. Gerber, and J.K. Gimzewski (2000). *Science*, **288**, pp. 316–318.
16. G. Wu, H.F. Ji, K. Hansen, T. Thundat, R. Datar, R. Cote, M.F. Hagan, A.K. Chakraborty, and A. Majumdar (2001). *PNAS*, **98**, pp. 1560–1564.
17. C.S. Smith (1954). *Phys. Rev.*, **94**, pp. 42–49.
18. O.N. Tufte, P.W. Chapman, and D. Long (1962). *J. Appl. Phys.*, **33**, p. 3322.
19. A.C.M. Gieles (1969). *IEEE Int. Sol. St.*, pp. 108–109.
20. J.A. Harley and T.W. Kenny (1999). *Appl. Phys. Lett.*, **75**, pp. 289–291.
21. M. Tortonese, R.C. Barrett, and C.F. Quate (1993). *Appl. Phys. Lett.*, **62**, pp. 834–836.
22. M. Tortonese, H. Yamada, R.C. Barrett, R.C., and C.F. Quate (1991). Atomic force microscopy using a piezoresistive cantilever. TRANSDUCERS '91. 1991 International Conference on Solid-State Sensors and Actuators. Digest of Technical Papers, 448–451.
23. P.A. Rasmussen, J. Thaysen, O. Hansen, S.C. Eriksen, and A. Boisen (2003). *Ultramicroscopy*, **97**, pp. 371–376.
24. K. W. Wee, G.Y. Kang, J. Park, J.Y. Kang, D.S. Yoon, J.H. Park, and T.S. Kim (2005). *Biosensors and Bioelectronics*, **20**, pp. 1932–1938.
25. P.A. Rasmussen, O. Hansen, and A. Boisen (2005). *Appl. Phys. Lett.*, **86**, p. 203502.

26. P.A. Rasmussen, A.V. Grigorov, and A. Boisen (2005). *J. Micromech. Microeng.*, **15**, pp. 1088–1091.

27. S. Kassegne, J.M. Madou, R. Whitten, J. Zoval, E. Mather, K. Sarkar, H. Dalibor, H., and S. Maity (2002). Design Issues in SOI-based high-sensitivity piezoresistive cantilever devices. *Proc. SPIE Conf. on Smart Structures and Materials*, (San Diego, CA 17–21 March).

28. M. Yang, X. Zhang, K. Vafai, and C.S. Ozkan (2003). *J. Micromech. Microeng.*, **13**, pp. 864–872.

29. E.M. Purcell (1977). *Am. J. Phys.*, **45**, pp. 3–11.

30. M.R. Paul and M.C. Cross (2004). *Phys. Rev. Lett.* **92**, p. 235501.

31. D. Sarid (1991). Scanning Force Microscopy with Applications to Electric, Magnetic, and Atomic Forces (New York), pp. 9–13.

32. F. Gittes and C.F. Schmidt (1998). *Eur. Biophys. J.*, **27**, pp. 75–81.

33. J.E. Sader (1998). *J. Appl. Phys.*, **84**, pp. 64–76.

34. E.O. Tuck (1969). *J. Eng. Math.*, **3**, pp. 29–44.

35. L. Rosenhead (1963). *Laminar Boundary Layers*, Oxford University Press (Oxford, Great Britain), pp. 390–393.

36. K.Y. Yasumura, T.D. Stowe, E.M. Chow, T. Pfafman, T.W. Kenny, B.C. Stipe, and D. Rugar (2000). *J MEMS*, **9**, pp. 117–125.

37. O.N. Tufte and E.L. Stelzer (1963). *J. Appl. Phys.*, **34**, pp. 313–318.

38. W.P. Mason and R.N. Thurston (1957). *J. Acoust. Soc. Am.*, **29**, pp. 1096–1101.

39. M.B. Viani, T.E. Schaffer, A. Chand, M. Rief, H.E. Gaub, and P.K. Hansma (1999). *J. Appl. Phys.*, **86**, pp. 2258–2262.

40. J.L. Arlett (2005). Doctoral Thesis, Department of Physics, California Institute of Technology *(unpublished)*.

41. J.A. Harley and T.W. Kenny (1999). *Appl. Phys. Lett.*, **75**, pp. 289–291.

42. C.Y. Ho, R.W. Powell, and P.E. Liley (1972). *P.E. J. Phys. Chem. Ref. Data*, **1**, pp. 279–421.

43. J.V. Sengers and J.T.R. Watson (1986). *J. Phys. Chem. Ref. Data*, **15**, pp. 1291–1314.

44. J.A. Harley and T.W. Kenny (2000). *J. MEMS*, **9**, pp. 226–235.

45. J.H. Hoh, J.P. Cleveland, C.B. Prater, J.-P. Revel, and P.K. Hansma (1992). *J. Am. Chem. Soc.*, **114**, pp. 4917–4918.

46. M. Grandbois, M. Beyer, M. Rief, H. Clausen-Schaumann, and H.E. Gaub (1999). *Science*, **283**, pp. 1727–1730.

47. E. Evans and K. Ritchie (1999). *Biophys. J.*, **76**, pp. 2439–2447.

48. U. Dammer, M. Hegner, D. Anselmetti, P. Wagner, M. Dreier, W. Huber, and H.-J. Güntherodt (1996). *Biophys. J.*, **70**, pp. 2437–2441.

49. P. Hinterdorfer, W. Baumgartner, H.J. Gruber, K. Schilcher, and H. Schindler (1996). *Proc. Natl. Acad. Sci., USA.*, **93**, pp. 3477–3481.

50. F.N. Hooge (1969). *Phys. Lett.*, A **A29**, pp. 139–140.

51. L.K. Vandamme and S. Oosterhoof (1986). *J. Appl. Phys.*, **59**, pp. 3169–3174.

52. J.W.M. Chon, P. Mulvaney, and J. Sader (2000). *J Appl. Phys.*, **87**, pp. 3978–3988.

53. H.B. Callen and R.F. Greene (1952). *Phys. Rev.*, **86**, pp. 3978–3988.

54. Chandler (1987). *Introduction to Modern Statistical Mechanics.* Oxford University Press.

55. H.Q. Yang and V.B. Makhijani (1994). V.B. A strongly-coupled pressure-based CFD algorithm for fluid structure interaction. *AIAA-94-0179*, pp. 1–10.

56. CFD Research Corporation, 215 Wynn Drive, Huntsville, AL 35805.

57. J.-C. Meiners and S.R. Quake (1999). *Phys. Rev. Lett.*, **82**, pp. 2211–2214.

58. J.L. Arlett, et al. (2005). To be published.

59. C.A. Canaria, J.O. Smith, C.J. Yu, S.E. Fraser, and R. Lansford (2005). *Tetrahedron Letters*, **46**, p. 4813; C.A. Canaria, et al., (2005). To be published.

60. D.E. Segall and R. Phillips (2005). To be published.

61. D.J. Harrison, K. Fluri, K. Seiler, Z.H. Fan, C.S. Effenhauser, and A. Manz (1993). *Science*, **261**, pp. 895–897.

62. H.-P. Chou, C. Spence, A. Scherer, and S. Quake (1999) *PNAS*, **96**, pp. 11–13.

63. M.A. Unger, H.-P. Chou, T. Thorsen, A. Scherer, A. and S. Quake (2000). *Science*, **288**, pp. 113–116.

64. J.E. Solomon and M. Paul (2005). "The Kinetics of Analyte Capture on Nanoscale Sensors", submitted to Biophysical Journal.

65. E. Evans (2001). *Annu. Rev. Biophys. Biomol. Struct.*, **30**, pp. 105–128.

13

Nanodevices for Single Molecule Studies

H.G. Craighead, S.M. Stavis, and K.T. Samiee

Applied and Engineering Physics, Cornell University, Ithaca, NY USA
hgcl@cornell.edu
kts3@cornell.edu
sstavis@gmail.com

13.1 Introduction

During the last two decades, biotechnology research has resulted in progress in fields as diverse as the life sciences, agriculture and healthcare. While existing technology enables the analysis of a variety of biological systems, new tools are needed for increasing the efficiency of current methods, and for developing new ones altogether. Interest has grown in single molecule analysis for these reasons.

The ability to detect single molecules provides a number of advantages in biomolecular analysis [1–10]. One benefit is an increase of quantification accuracy, as analysis occurs at the ultimate resolution limit. Single molecule techniques also consume less reagent than conventional techniques, and reduce analysis times. Mass production of micro-total-analysis-systems with the ability to analyze single molecules could increase the scope of otherwise prohibitively expensive and protracted processes, such as genomic sequencing and drug discovery. In addition to increasing the efficiency of existing technologies, single molecule analysis grants access to information that is otherwise unobtainable. The characteristics of biomolecular reactions are of interest in this regard. Molecular biologists have used conventional methods to study the ensemble characteristics of many systems. While this approach yields important information regarding the average behavior of a system, it tells little about the specific behavior of single molecules. This includes the time evolution and statistical distribution of parameters obscured by traditional techniques.

A variety of nanofabricated structures have emerged as potential tools for single molecule analysis. Several nanostructures have been developed for enhanced optical detection, including quantum dots [11–13], metallic nanobarcodes [14], and nanometric slits [15]. Two optical structures in particular have demonstrated their utility for single molecule analysis – fluidic channels with submicrometer and nanometer dimensions, and optical nanostructures known as zero mode waveguides. Fluidic channels provide controlled transport of analytes through a subfemtoliter focal volume, while zero mode waveguides have

H.G. Craighead et al.: *Nanodevices for Single Molecule Studies*, Lect. Notes Phys. **711**, 271–301 (2007)
DOI 10.1007/3-540-49522-3_13

a focal volume in the zeptoliter range that enables single molecule detection at micromolar concentrations. Consequently, the two structures have distinct applications in single molecule analysis. Nanofabricated structures have also been designed to physically confine and manipulate biomolecules at molecular length scales. Entropic traps, for example, use alternating regions of high and low molecular confinement as a molecular separation mechanism. A variety of other mechanisms have been utilized, including solid state nanopores, diffusion arrays [16], post arrays [17, 18] and nanochannels for confinement of DNA [19–21].

In many of these cases, fluorescence microscopy is used for single molecule detection. Continuous wave lasers are typically used as spatially confined and spectrally isolated excitation sources [22]. Because organic fluorophores can emit many photons in a separate band of the visible spectrum, high signal to noise ratios are achievable. As fluorescent molecules move through the device focal volume, they are excited and emit fluorescence. Some fraction of the emitted fluorescence is collected with a high numerical aperture microscope objective. The fluorescent signal is filtered, and in the case of a multicolor experiment, split into two color channels. The signal can then be focused on one or more optical fibers, used simultaneously as pinholes apertures and to transmit the light to photodiodes for transduction and amplification. ICCD cameras are also frequently used for device imaging.

13.2 Nanostructures for Optical Confinement

13.2.1 Fluidic Channels

A principal advantage of performing single molecule spectroscopy in nanofabricated structures is the small focal volume provided by the device. A small focal volume reduces the amount of extraneous fluorescent material observed, such as buffer solution and associated impurities, in addition to the analyte of interest. This increases the signal to noise ratio of single molecule detection. A small focal volume also extends the range of solution concentrations at which single molecule detection is possible. The instantaneous occupation probability of the focal volume is determined by a Poisson distribution:

$$P_k(x) = \frac{k^x e^{-k}}{x!}$$

Where k is the mean number of particles per unit focal volume, given by the solution concentration, and x is the instantaneous occupancy of the focal volume. As the size of the focal volume is reduced, the probability of more than one analyte simultaneously occupying it becomes small, even at high solution concentrations. This is an important consideration for single molecule spectroscopy of biological reactions for which a high concentration is required, typically in the micromolar regime.

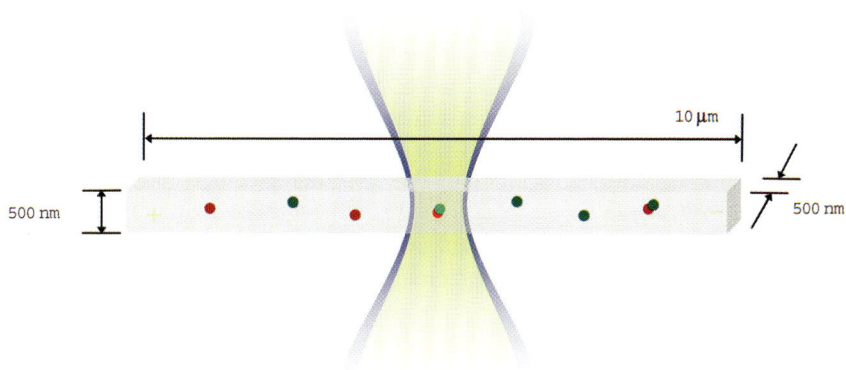

Fig. 13.1. A schematic of a typical submicrometer fluidic channel with a laser beam focused on the center for excitation of flowing analytes. The channel defines the width and depth of the resulting focal volume, while the length is defined by the excitation profile and optical configuration

A natural focal volume to consider for the sake of comparison is that of a diffraction limited spot produced by a high numerical aperture objective. The full-width at half-maximum of such a spot is defined as [23]:

$$\frac{.61\lambda}{NA}$$

where λ is the wavelength of the light and NA is the numerical aperture of the objective. In the case of light with a wavelength of approximately 500 nm, and a numerical aperture of 1.2, a diffraction limited spot has lateral dimensions of approximately 200 nm. The use of pinhole apertures can limit the depth of focus to approximately 1 μm, yielding a focal volume of approximately 0.4 fL. Nanofabricated structures generally have focal volumes less than or equal to this value, depending on the optical setup. In the case of a fluidic channel, the laser profile and microscope configuration define the length of the focal volume, while the channel dimensions define its width and depth, as shown in Fig. 13.1. Typical focal volumes for submicrometer fluidic channels range from 0.1 to 0.5 fL.

Because the dimensions of nanofabricated structures are less than or on the order of the wavelength of light used for fluorescence excitation, it is possible to define the focal volume as a region of constant intensity in some region of the structure. This enables uniform excitation and analysis of molecules entering the device. This is an important issue when quantifying various aspects of fluorescence emission, such as photon burst intensity and length. If single molecule detection and counting are the only goals, this need not be an issue.

Consider the intensity profile of a laser beam focused by a high numerical aperture microscope objective. The intensity is approximated by a gaussian distribution in the two lateral dimensions, and an elongated gaussian in the axial dimension. If the back aperture of the objective is overfilled with an expanded laser beam, a diffraction-limited spot is created, with lateral dimensions on the order of 200 nm. If the objective back aperture is underfilled with a tightly collimated laser beam, the resultant intensity profile can have a 1/e decay distance of several micrometers across the width of the spot [23]. If a structure such as a fluidic channel with submicrometer or nanometer dimensions is placed at the center of such a spot, the excitation intensity will be approximately uniform across the width and through the depth of the structure. In the case of a diffraction-limited laser spot, a channel with dimensions on the order of 50 nm is necessary to achieve uniform intensity. With a larger spot, 500 nm channels can be uniformly illuminated across the width and depth. In either case, the intensity will vary as a gaussian distribution along the length of the channel. When using a confocal microscope, correct selection of pinhole apertures can limit the length of the laser spot along the channel to approximately one micrometer.

Another advantage specific to fluidic channels with submicrometer and nanometer dimensions is that every molecule that enters the channel is individually analyzed. This is in contrast to diffusion-limited single molecule spectroscopy, where an unknown and potentially small fraction of the sample is analyzed. Furthermore, when diffusion is relied upon to deliver analytes to the focal volume, large molecules with small diffusion coefficients may never be observed.

In addition to detecting each molecule that enters the device once, fluidic channels enable much higher throughput than diffusion-limited single molecule spectroscopy. Because analytes can be driven rapidly through the channel, it is possible to analyze a statistically significant number of analytes in a relatively short period of time. Furthermore, because fluidic channels are amenable to massively parallel integration, they are well-suited for ultrahigh throughput single molecule spectroscopy applications.

Precise control over fluid flow and particle motion in fluidic channels has a number of benefits for single molecule spectroscopy. One result is a more uniform distribution of analysis times than is otherwise achieved in a diffusion-limited experiment. This is an important factor to consider when quantification of fluorescence emission is an objective of the experiment, and analysis time uniformity is an important control. Flow control also allows the use of channels in a variety of experiments with different requirements. In an experiment limited by the signal-to-noise ratio, such as the detection of single fluorophores, the flow rate can be decreased to achieve longer analysis times. This results in an increased number of collected photons from each molecule and a corresponding increase in signal-to-noise ratio. In an experiment limited by detection statistics, such as the quantification of a low concentration species, the flow rate can be increased for rapid analysis of a greater number of particles. In practice, a compromise between the two is chosen based on the needs of the particular experiment.

There are a number of techniques to transport analytes through the channel and focal volume. A commonly used method is electrokinetic [24,25] drive, in which an electrical bias is applied across the two sample reservoirs on either end of the channel. This results in two primary physical phenomena, including electrophoresis [26–30] of charged particles and electroosmosis [31,32] of bulk fluid. In the case of electrophoresis, charged particles are attracted to their counter-electrodes with a velocity given by:

$$v_{ep} = \frac{qE}{6\pi\eta r}$$

where q is the effective screened charge of the particle, E is the electric field, η is the viscosity of the solution, and r is the hydrodynamic radius of the particle. The second main effect, electroosmosis, occurs when the glass surfaces of the fluidic channel are wetted and become negatively charged by ionization and adsorption. The negative channel surfaces attract a layer of cations from solution. This double-layer [33–35] of cations is then attracted towards its counter-electrode, and via shear forces drags the bulk fluid contained within the channel along with it. This results in a uniform fluid velocity given by:

$$v_{eo} = \frac{\zeta\varepsilon E}{\eta}$$

where ζ is the Zeta potential of the wall and ε is the dielectric constant of the liquid. Electroosmosis can be used to drive neutral analytes through the channels, since they are dragged along with the moving fluid. Because particle velocity scales linearly in both cases, control over particle motion in the case of electrophoresis, and fluid flow in the case of electroosmosis, is a simple process.

Application of hydrostatic pressure is another method of transporting analytes through fluidic channels with submicrometer and nanometer dimensions. Pressure drive has the advantage of exhibiting no dependence on solution chemistry and analyte charge. However, a detrimental effect of this technique is the parabolic velocity profile across the channel width and depth, resulting in a broad range of analyte speeds. If single molecule detection and counting is the sole purpose of the experiment, this range of particle speeds need not be an issue. If quantification of fluorescent emission is sought, however, this large velocity distribution in turn broadens the distributions of analysis times. This can hinder analysis of photon burst histograms. Additionally, pressure drive results in an average velocity that scales with the square of the channel diameter, making it unfavorable for submicrometer and nanofluidic channels.

Because of the large surface-to-volume ratio of nanofabricated structures, surface interactions are a frequent occurrence and are an additional cause of a variety of problems in single molecule analysis.

One issue is when fluorescent molecules adsorb to the surfaces of the structure in close proximity to the focal volume. This can increase fluorescent background and consequently decreases the signal to noise ratio of single molecule detection. Surface interactions can also generate a population of slow moving or stuck analytes near the device surfaces, in addition to those flowing or diffusing unhindered through the rest of the device. This broadens the distribution of observation times and hinders analysis of autocorrelation curves. Adsorption of analytes can also be a factor in clogging, which is particularly problematic when a microfluidic channel narrows to a nanofluidic constriction. Adsorption may also result in the loss of low concentration analytes.

Interactions between surfaces and analytes in solution can be diminished via an assortment of surface treatments [36]. The surface charge and chemistry of the device can be altered by plasma and chemical exposure. Permanent coatings, such as a silane monolayer [37, 38], are applied before the channels are wetted. Dynamic coatings, including a variety of polymers [39, 40], surfactants [41, 42] and biological agents [43, 44], are added to the running buffer solution and continually coat and refresh the channel walls.

13.2.2 Fabrication

A variety of techniques are used for the fabrication of structures with submicrometer and nanometer dimensions. Conventional photolithography is an attractive choice for larger structures, while electron beam lithography remains the popular for smaller nanoscale structures. A number of methods bridge the gap between the two.

Lithography techniques adopted from the semiconductor industry are commonly used to fabricate a variety of submicrometer and nanometer sized structures. Lithography allows the precise definition of features and integration of structural elements across a wide range of size scales. Photolithography is used to pattern features ranging in size from millimeters to hundreds of nanometers, while electron beam lithography is capable of defining features as small as 10 nm.

Perhaps the simplest fabrication method for a variety of structures is to define the pattern negative into a thin film for pattern transfer [45, 46], and subsequently etch [47, 48] the structure into the bulk substrate. The etch depth is easily controllable from the nanometer to the micrometer scale. To generate an enclosed structure, a cover wafer can be bonded to the substrate to seal the device. This technique is rapid, simple and has the advantage of incorporating a cover wafer compatible with standard high numerical aperture, coverslip-corrected microscope objectives.

A more complex photolithography technique is the use of a sacrificial layer [17, 49, 50], such as polysilicon, to create enclosed fluidic structures on top of a substrate. A positive device pattern is defined via a raised polysilicon structure on a fused silica substrate, for example. This raised structure is then oxidized and selectively removed, resulting in an enclosed fluidic channel on top of the substrate with an extremely thin ceiling and walls. The main advantage of sacrificial layer fabrication is that as the outside of the polysilicon sacrificial layer is oxidized, its interior dimensions shrink. This allows for a significant reduction in size of photolithographically defined features. Fluidic channels with dimensions of approximately 250 nm by 350 nm have been fabricated using this technique.

For any structure fabricated in or on a bulk substrate, material selection of the substrate is an important parameter. High purity synthetic amorphous silicon dioxide, or fused silica [51–53], is preferred for its low fluorescent background. Additionally, fused silica is resistant to thermal stress and is insensitive to solution chemistry. For applications where signal to noise ratio is not a limiting factor, or where high throughput production and device disposability is a consideration, a variety of other fabrication techniques and materials may also be used to form nanostructures [54]. One such method is NanoImprint Lithography (NIL), in which a master mold is used to imprint multiple device copies in soft matter, combining nanoscale resolution with rapid fabrication.

Another recently developed fabrication technique is the use a microfabricated tip to electrospray a nanometer-scale fiber onto a substrate [55]. The microfabricated electrode is coated with a polymer solution and brought in close proximity to a spinning substrate, acting simultaneously as a counter-electrode. A Taylor cone forms at the microfabricated tip and a jet of fluid is driven towards the spinning counter-electrode, upon which the solution is deposited. As the solvent evaporates, a polymer fiber is left behind, which can be used as a sacrificial template for nanofluidic channels.

In one iteration of this technique [56], a polymer nanofiber is deposited on a substrate. The fiber is then coated with a thin film of silicon dioxide and heated to the vaporization temperature of the polymer. As the sacrificial polymer fiber vaporizes, a glass nanofluidic channel is left behind. The primary advantage of this type of fabrication is that electrospun fibers are easily integrated with predefined microfabricated structures. Nanofluidic channels, for example, have been suspended over microfabricated trenches and used for single molecule detection [57]. In this situation, the nanofluidic channel is not surrounded by bulk material, which can otherwise be a source of reflection and scattering. This can increase the signal to noise ratio of single molecule detection, and nanofluidic channels fabricated with this method do indeed provide a low-noise environment for single molecule spectroscopy.

13.3 Applications of Optical Confinement Nanostructures

13.3.1 Fluorescence Correlation Spectroscopy

Fluorescence Correlation Spectroscopy (FCS) [58–60] is a technique in which fluctuations in fluorescence emission resulting from particles moving through a region of high excitation intensity, typically a focused laser beam, are autocorrelated. The autocorrelation function provides information on the temporal decay of fluorescence fluctuations due to a variety of physical mechanisms, including diffusion, photophysical effects, fluid flow, chemical reactions and other phenomena. The concentration of the fluorescent species is also evident from this function, which is fit to an analytical model to extract the desired parameters.

Because nanofabricated structures have dimensions less than or on the order of a focal volume generated by a focused laser beam, the effects of the structure on the observed autocorrelation function must be considered [61]. For example, when a focal volume is situated in free solution, fluorescent molecules are free to diffuse through it in all three dimensions. This is reflected in the autocorrelation function, which accounts for three-dimensional diffusion. If the laser spot is focused on a fluidic channel with submicrometer or nanometer dimensions, however, the channel width and depth can be significantly smaller than the related dimensions of the laser focal volume. The channel is therefore uniformly illuminated across its width and through its depth, and fluorescent species are free to diffuse, and flow, in only one dimension through the focal volume. This altered dimensionality must be accounted for when fitting the measured autocorrelation function to the appropriate analytical model [62, 63]:

$$G\left(\tau\right) = \frac{1}{N} \cdot \underbrace{\underbrace{\underbrace{\frac{1}{\sqrt{\left(1 + \frac{\tau}{\tau_{d_1}}\right)}}}_{\text{1D Diffusion}} \cdot \frac{1}{\sqrt{\left(1 + \frac{\tau}{\tau_{d_2}}\right)}}}_{\text{2D Diffusion}} \cdot \frac{1}{\sqrt{\left(1 + \frac{\tau}{\tau_{d_3}}\right)}}}_{\text{3D Diffusion}} \cdot \underbrace{\exp\left(\frac{-\tau^2}{\tau_f^2\left(1 + \frac{\tau}{\tau_{d_1}}\right)}\right)}_{\text{Flow}}$$

In this equation, $G(\tau)$ is the autocorrelation as a function of time, $\tau_d = w_i/4D$, where w_i is the e^{-2} radius of the ith dimension of the focal volume and D is the diffusion coefficient, and $t_f = w_i/v_i$, where v_i is the speed of analytes in that dimension.

Additionally, because the size of the channel and the laser focal volume are similar, the alignment of the two is critical. Lateral positioning is particularly sensitive in this regard.

13.3.2 DNA Fragment Sizing

DNA fragment sizing is an important process in a variety of biotechnology applications, including DNA sequencing [64]. Reduction of analysis time and sample consumption is the primary motivation for using nanofabricated structures for single molecule analysis in this capacity.

Submicrometer fluidic channels have been used to detect and characterize a mixture of DNA molecules [19,21,49,65]. This process was used to determine the distributions and proportions of DNA fragments in solution as well as the overall concentration of the sample used. The use of nanofabricated fluidic channels in conjunction with a confocal microscope consumed approximately 10,000 molecules, or 76 fg of DNA; far less than conventional methods.

Fluidic channels were fabricated on a fused silica substrate using polysilicon as a sacrificial layer, resulting in channels roughly 300 nm deep and one micrometer wide.

A 40×0.8 NA microscope objective was underfilled with a 488 nm laser beam to achieve an approximately uniform excitation profile across the one micrometer width of the channel. DNA fragments were driven electrokinetically through the channels. Using FCS, the flow speed was found to vary linearly with applied voltage. For fragment size analysis, the mixture of DNA molecules was driven at 3.1 kV/cm. This applied voltage resulted in a flow dominated regime with uniform analysis times.

A histogram of the photon burst size was developed and analyzed to determine the composition of the sample, as shown in Fig. 13.2. Larger fragments with more intercalated dye molecules were brighter, which was reflected in peaks in the histogram corresponding to the various lengths of molecules studied. Peaks were fit to gaussian functions, and the relative peak amplitude provided an estimate of the actual molecular distribution of the molecules. For DNA fragments over one thousand base pairs, this method has resolution

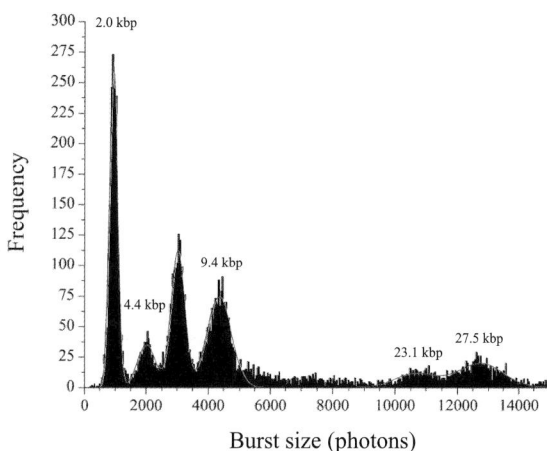

Fig. 13.2. A photon burst size histogram from a mixture of DNA fragments provides information on the sample composition. Because burst size was found to vary linearly with DNA fragment size, the position of each peak is proportional to the size of each DNA fragment, while the peak area is proportional to the relative fragment concentration. Each peak was fit to a gaussian distribution using a least-squares method

greater than or equal to gel electrophoresis, while being faster and consuming less sample. Additionally, the use of submicrometer fluidic channels enables integration with other microfluidic devices for Lab-On-A-Chip and μTotal-Analysis-System applications.

13.3.3 Detecting Labeled Single Molecules

Semiconductor nanocrystals, or quantum dots, have attracted attention as an emerging technology in the field of single molecule spectroscopy. Because of their nanoscale dimensions, quantum dots exhibit unique optical properties that make them ideal for use as fluorescent labels in high throughput single molecule assays.

Signal to noise ratio is often the limiting factor in single molecule spectroscopy. While nanostructures can be used to reduce background noise, fluorescent labels must also be bright enough to detect and analyze in a wide variety of environments. Because of their high quantum yields and extinction coefficients, quantum dots are extraordinarily bright, making them attractive for use in high throughput, low analysis time experiments.

There are a number of other considerations when selecting fluorescent labels for multicolor spectroscopy. In order to detect fluorescence emission from a single fluorescent species, the Stokes shift must be large enough to resolve the emission and excitation peaks. When multiple species of fluorophore are present simultaneously, the emission and excitation spectra can overlap. This is typically managed by confining the range of collected fluorescence, resulting in rejected signal and reduced detection efficiency. This is mitigated by the narrow and symmetrical emission spectra and large effective Stokes shift of quantum dots. Additionally, multiple quantum dots can also be excited by the same excitation source over a wide range of the visible spectrum. This results in the ability to excite several species of quantum dots, or combinations of quantum dots and other fluorescent labels, with a single light source, while the emission spectra are efficiently resolved.

To demonstrate the use of quantum dots in a multicolor single molecule study, DNA oligomers labeled with a single Alexa Fluor 488 fluorophore were identified at low concentrations by detecting their binding to a functionalized quantum dot [66]. Submicrometer fluidic channels with square 500 nm cross sections were used in this experiment. Characterization of single molecule binding has a variety of applications including high throughput immunoassays and array-free hybridization measurements.

All analytes were driven through the channels electrokinetically at 2.3 kV/cm. A single blue (476 nm) laser was used to excite both quantum dots and organic fluorophores, and the emission spectra were resolved without significant signal rejection. Fluorescence emission was collected simultaneously from green (500–590 nm) and red (610–680 nm) regions of the spectrum. A 60×1.2 NA objective was used both to focus the laser beam and collected

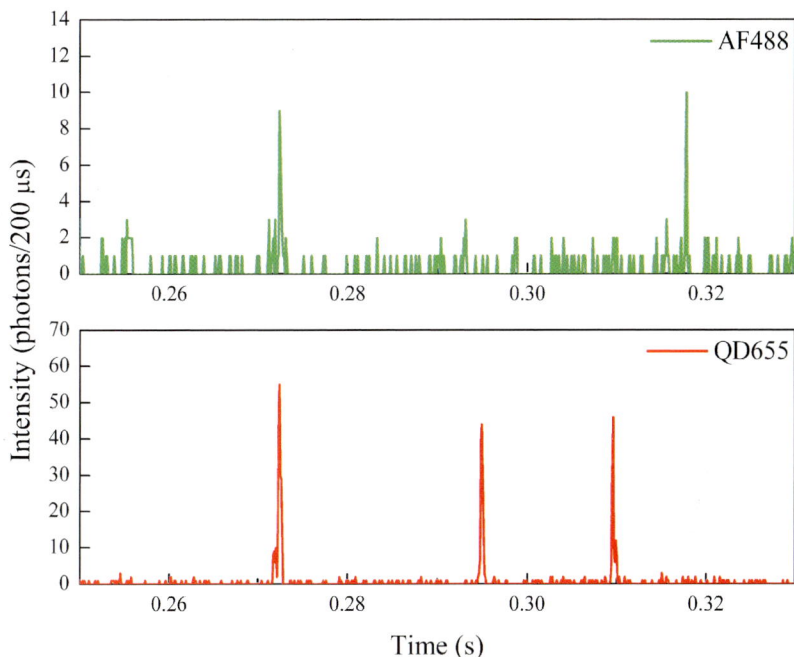

Fig. 13.3. A fluorescence intensity scan shows the detection of individual DNA oligomers labeled with single Alexa Fluor 488 fluorophores in the green color channel (*top*), and individual Quantum Dot 655 Conjugates in the red color channel (*bottom*). The first detection event in both color channels is coincident, indicating that the two species are bound together. Subsequent detection events are temporally separate and are the result of the detection of unbound species

emitted fluorescence. When bound together, the two species were detected simultaneously as they crossed the focal volume, while unbound quantum dots and DNA molecules were detected as non-coincident photon bursts in the two color channels, as shown in Fig. 13.3.

Because Alexa Fluor 488 exhibits a long tail at longer wavelengths, a small amount of spectral cross talk was expected between Alexa Fluor 488 and Qdot 655. False positive signal in the red optical channel due to detection of Alexa Fluor 488 molecules was found to be within background noise. Photon counting histogram analysis was used to quantify coincident detection and to characterize the binding processes of the conjugates, and fluorescence correlation spectroscopy was used to account for possible mobility differences.

13.3.4 Identifying Color-Coded Fluorescent Labels

Nucleic acid engineers have developed fluorescent labels that are uniquely identifiable by the number of conjugated fluorophores, and with binding characteristics that permit recognition of individual specific biomolecules. The viability of this technology for use in single molecule spectroscopy depends on the ability to detect individual labels, and distinguish the fluorescence emission of each label. Submicrometer fluidic channels have been used to rapidly detect individual labels in solution, and labels with small disparities in fluorophore composition have been differentiated with varying degrees of accuracy [67].

Fluorescent labels are synthesized at the molecular level from dendrimer-like DNA. Three oligonucleotide components are annealed to make each Y-shaped DNA fragment, and the complementary ends of corresponding fragments are then ligated to create a specific label. The oliogonucleotide design assures one-way ligation. The outside oligonucleotides of the DNA molecules are covalently conjugated with Alexa Fluor 488 (green) and BODIPY 630/650 (red) fluorescent dyes. The identity of the label is determined by the number of dye molecules incorporated. A variety of labels have been studied for use in single molecule spectroscopy, including one green and one red fluorophore, one green and two red fluorophores, one green and three red fluorophores, three green and one red fluorophores and four green and four red fluorophores.

In order to detect and identify single labels, all analytes were driven electrokinetically through the submicrometer fluidic channels with a bias of 2.3 kV/cm. Blue (488 nm) and yellow-green (568 nm) lasers were used to excite the two fluorophore species. The two laser beams were overlapped to create a single multicolor focal volume. Fluorescence was collected simultaneously from green (500–550 nm) and red (610–680) parts of the spectrum.

To explore the decoding resolution limit, labels with a single fluorophore of each color were detected, and were found to be distinguishable as a group, but not individually, from labels with one additional red fluorophore, as shown in Fig. 13.4. Labels with one green and three red fluorophores were individually distinguishable with greater than 80% accuracy from labels with one red and three green fluorophores. Photon counting histograms were analyzed to differentiate the various labels. The primary means of identification was the spectral intensity ratio, defined as the number of red photons collected divided by the sum of the number of red and green photons collected. Photon burst height and area histograms were also analyzed for label differentiation. Fluorescence correlation spectroscopy was used to measure the mobility and account for differences in analysis time for the labels.

13.3.5 Zero Mode Waveguides

The Zero Mode Waveguide (ZMW) is a nanoscale aperture in a thin metal film on a glass substrate, which is illuminated through the substrate (Fig. 13.5) by a continuous wave laser. The confluence of several physical phenomena results in a high degree of focal volume confinement. First, the 50 nm aperture of the structure is so small that no modes are guided, resulting in a radially confined evanescent field at the bottom of the ZMW. Ohmic losses in the metal

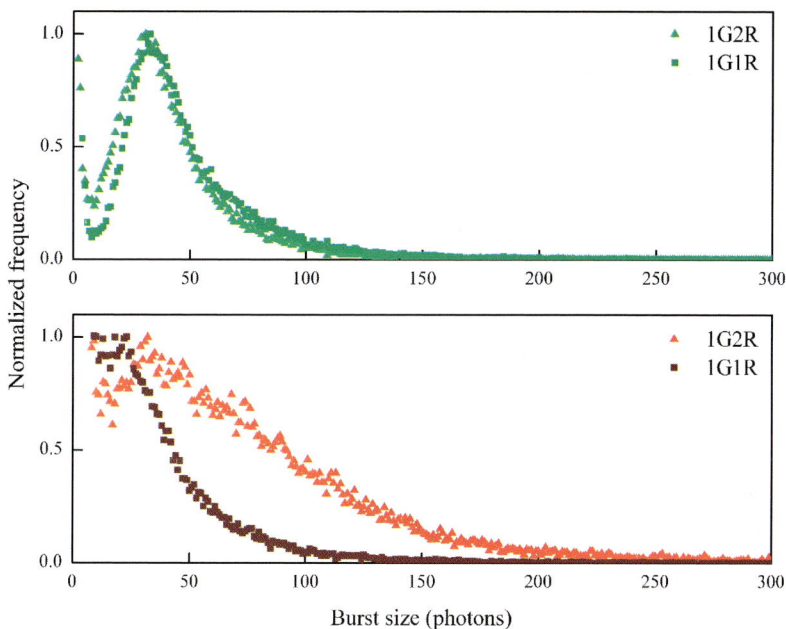

Fig. 13.4. Photon burst size histograms are produced from the detection of two species of nucleic acid engineered fluorescent labels. 1G1R contains one Alexa Fluor 488 fluorophore and one BODIPY 630/650 fluorophore, while 1G2R contains two BODIPY 630/650 fluorophores. While differences in the two populations are evident, in this case individual labels were not distinguishable based on their fluorescence emission. However, labels with greater differences in fluorophore composition were individually identifiable with a higher degree of accuracy

cladding further shorten the evanescent decay of the illumination. Additionally, fluorescence emission must couple back out of the ZMW, which has an efficiency similar to that of the excitation. Finally, fluorophores in a metal cavity are known to have altered emission rates due to changes in the local density of states, and hence emit fewer photons in a ZMW. Taken together, these processes result in a focal volume in the atto- to zeptoliter range [68].

The small ZMW focal volume enables single molecule detection at relatively high, micromolar concentrations, whereas conventional techniques are limited to the nanomolar range. The ZMW is therefore an ideal platform for performing FCS on enzymatic systems requiring high ligand concentrations.

The measured autocorrelation function is closely related to the shape of the ZMW observation volume, which is defined by both the shape and intensity of the excitation illumination and the spatial sensitivity of the detection apparatus [69]. If the fluorophore coupling efficiency and quantum efficiency are assumed to be proportional to the intensity profile then the observation profile can be approximated by an exponential decay in one dimension, $S(z) = e^{-z/L}$,

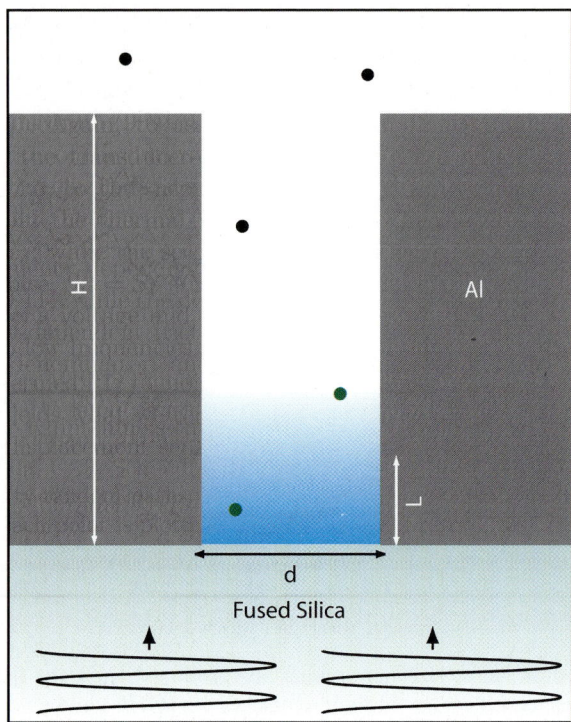

Fig. 13.5. A schematic of a Zero Mode Waveguide is shown, with a diameter d, height H and the evanescent field as a $1/e$ decay length of L. ZMWs are illuminated from the *bottom* by light with a wavelength larger than the diameter of the waveguide. Fluorophores in the evanescent field emit strongly while those closer to the *top* do not

where L is the characteristic decay length. This yields an observation volume given by:

$$V_{Obs} = \frac{\pi d^2}{4} \frac{\left(\int S(z)dz\right)^2}{\int S^2(z)dz} = \frac{\pi d^2 L}{8}$$

where d is the diameter of the ZMW. For a 50 nm hole, L is $\sim$14 nm and V_{Obs} is $\sim$14 zL, corresponding to a maximum solution concentration of $\sim$100 μM while still ensuring a high probability that no more than one molecule occupies the focal volume simultaneously. Because the transverse illumination is approximately uniform in a small ZMW, diffusion of fluorescent species is expected to appear one dimensional. Substituting the observation profile into the general definition of the autocorrelation function in Fourier space yields [58]:

$$G(\tau) \propto \int \frac{L^2 e^{-v^2 \tau D} dv}{(1 + L^2 v^2)^2}$$

This equation has an analytical solution; however, it does not properly describe a ZMW of finite length. Fluorophores diffusing in and out of a ZMW are only reasonably approximated by one dimensional diffusion. Once a fluorescent diffuser leaves the ZMW, the probability of returning to the focal volume becomes exceedingly remote as the diffuser experiences three dimensional diffusion in the free solution outside the ZMW. This results in a depressed return probability at long times, and the long time constant epoch of the autocorrelation curve is consequently depressed when compared to true one dimensional diffusion. To correct the model, Fourier components of the autocorrelation function with wavelengths exceeding the height, H, of the waveguide are excluded from the final integration [68]. No analytical solution was found for the resulting integral, taken from 1/H to infinity, which complicates the fitting procedure.

An approximation yields a suitable fitting function. Let R be the ratio of the observation volume decay constant to the height of the waveguide, L/H, and let the diffusion time, τ_d, be defined as L^2/D. The final fitting function is given by the analytical solution to the 1 dimensional infinite cylinder and a correction term (last term) [70]:

$$G(\tau) = G_0 \left[\frac{\pi}{4} \left(\left(1 - 2\frac{\tau}{\tau_d}\right) e^{\frac{\tau}{\tau_d}} \operatorname{erfc}\left(\sqrt{\frac{\tau}{\tau_D}}\right) - \frac{2}{\sqrt{\pi}} \left(\frac{\tau}{\tau_D}\right)^{1/2} \right) - \sqrt{\frac{\tau_d}{\tau}} \frac{\operatorname{erf}(R)}{(1+R^2)^2} \right],$$

G_0, τ_d, the average residency time in the observation volume and the geometric ratio R are the free parameters for the model. G_0 is related to the average number of molecules in the focal volume, N and the constant background signal, B [71]. A system containing two diffusing species which are statistically independent on the time scale of the experiment can be fit by adding two single species curves together. This yields a total of 5 free parameters: G_0 and τ_d for each of the species and R which is defined by the waveguide only.

In addition to being a powerful tool for FCS, the ZMW provides an ideal platform for single enzyme studies. Several mechanisms exist for fixing a single enzyme to the bottom of a ZMW. Once fixed, the interaction between the enzyme and fluorescent ligands can be characterized by photon burst analysis. Because the observation volume has such a short decay length, the fluorescence emission from a molecule specifically interacting with the enzyme is considerably higher than those freely diffusing in and out of the structure.

Binding the enzyme to the bottom of the ZMW is an important process. The enzyme must be attached to the ZMW in a manner that does not significantly alter its functionality. Simple surface adhesion can be used; however, enzymes may adhere to the surface in an orientation that obscures the active site. Surface adhesion may also alter the structure of the protein, leaving it inactive. An alternative is covalent binding. Silane chemistry and thiol binding to a thin gold film can be used in this capacity. Silanes have linker molecules terminated in a carboxyl or amino group which can be covalently linked to a protein. Because the metal film oxidizes slightly, and the silane linker will attach to metal oxides in addition to glass, enzymes will be bound to all ZMW surfaces. Gold thiol binding does not suffer from this problem, and so a thin layer of gold deposited at the bottom of the ZMW isolates the bound molecules there.

Once the enzyme is fixed in the observation volume, data can be collected in the form of fluorescence intensity traces. Fluorescently labeled ligands that interact with the enzyme will spend more time in the focal volume than those diffuse randomly in and out of the ZMW. Quantitative statements concerning the interaction duration and frequency can then be made by analyzing resulting photon bursts.

13.4 Applications

13.4.1 Lambda Repressor Oligomerization Kinetics

The genetic control system employed by bacteriophage lambda is among the best studied systems in biology [72]. Lambda is a virus that infects bacteria by injecting its ~48 kilobase DNA fragment into the host's genome. A protein encoded on the lambda genome, the repressor, is responsible for deciding the fate of both the bacteria and the virus. Much of the control system is determined by the oligomerization kinetics of the repressor protein. The rate limiting step, going from dimer to tetramer, occurs at μM concentrations, making the ZMW ideal for studying this system.

A fluorescent analog of the repressor, cI, was created by engineering an mRFP-cI fusion. Unlike other members of the fluorescent protein family, the monomeric red fluorescent protein, mRFP, does not form multimers in solution. An approximately 1 micromolar solution of mRFP-cI was prepared and placed in a ZMW array. Fluorescence intensity traces and autocorrelation functions were recorded. As can be seen in Fig. 13.6, two species with differing diffusion constants are present in the system. Fitting the autocorrelation function yields information about the transport properties and concentration of each species. The kinetic constants are decided by the relative concentration of each species which can be calculated from the autocorrelation parameters.

The fitting parameter relevant to the concentration is G_0, which is determined not only by the frequency of fluctuations but also a constant background. When the total concentration of fluorescent proteins is known, the background can be deduced by solving a set of nonlinear equations:

$$4N_{tet} + 2N_{dim} = V_{obs}N_A[CI]_T$$
$$(G_{0,tet}N_{tet})^{\frac{1}{2}} - G_{0,tet} = (G_{0,dim}N_{dim})^{\frac{1}{2}} - G_{0,dim}$$

In this formula, N_{tet} and N_{dim} are the number of tetramers and dimers in the observation volume, V_{obs}; $G_{0,tet}$ and $G_{0,dim}$ are the fitting parameters associated with the two species; NA is Avagadro's number and $[CI]_T$ is the

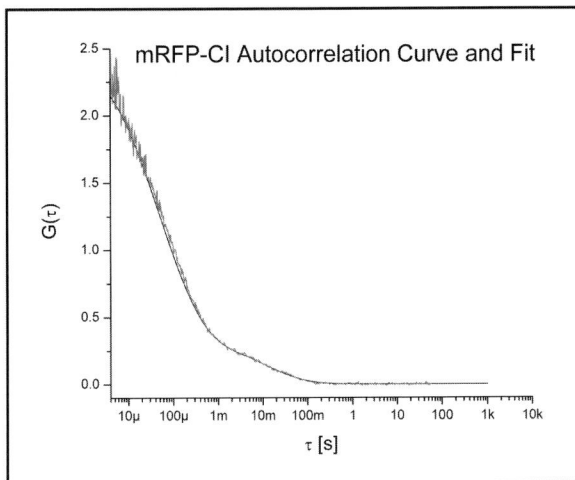

Fig. 13.6. A typical autocorrelation curve from a repressor oligomerization study is shown. The data (*gray*) shows two species in solution with different diffusion constants. The fast component is due to the cI dimer, while the slow component is due to the tetramer. The fit (*black*) yields G_0 for both species and allows the relative concentration to be found

total CI monomer concentration. The first equation follows from conservation of monomer. The second equation simply constrains the constant background to be identical for both species [70]. After finding the relative concentrations, the dimerization constant can be calculated:

$$ K_{D2} = \frac{[CI_2]^2}{[CI_4]} = \frac{1}{N_A V_{obs}} \frac{N_{\dim}^2}{N_{tet}} = \frac{[CI]_T}{4 \left(\frac{N_{tet}}{N_{\dim}} \right)^2 + 2 \left(\frac{N_{tet}}{N_{\dim}} \right)} $$

The results match well with previous biochemical studies. The waveguides yielded $K_{D2} = 4.6 \times 10^{-6} \pm 3 \times 10^{-7}$ M. The free energy change associated with tetramerization was measured by Pray, Burz and Ackers at temperatures between 5°C and 45°C [73]. The corresponding equilibrium constant, related by the well known equation $\Delta G = -RT\ln K$ was $4 \times 10^{-6} \pm 1 \times 10^{-6}$ M at 20°C, 25°C and 30°C.

13.4.2 Real Time Observation of DNA Polymerization

Enzymatic polymerization of DNA is another system well suited to study in the ZMW. DNA strands are copied by a family of enzymes called polymerases. A polymerase molecule will bind a DNA double helix, separate the strands, and proceed to build two new complete DNA helixes from the template strand and nucleotide in solution. Related enzymes transcribe DNA sequences into RNA molecules that are responsible for telling cellular complexes how to construct various proteins. In retroviruses, a relative of polymerase, reverse transcriptase, takes viral RNA and synthesizes a complimentary DNA strand that can be integrated into the host's genome. The equilibrium disassociation constants of nucleotides for most polymerases are above $1\,\mu m$. Efficient polymerization therefore occurs only at high nucleotide concentrations.

A mutant T7 DNA polymerase was immobilized on the glass surface of a ZMW [68]. Because surface adhesion was used, many of the polymerases were inactive. Only a single active enzyme is required, however. The enzyme was incubated with a reaction mixture including all reagents necessary for DNA polymerization including the fluorescent nucleotide analog coumarin-dCTP. Fluorescent bursts were then observed lasting between 1 and 100 ms. No bursts were observed when one or more elements of the reaction mixture were missing (Fig. 13.7). Furthermore, the bursts terminated after roughly 30 minutes which is consistent with the time required for synthesis of the complementary strand of M13 DNA. This strongly suggests that the bursts are due to incorporation events involving the fluorescent dCTP analog.

13.4.3 Nanoscale Optical Observations on Cell Surfaces

ZMWs have been used as small apertures for characterization of diffusion on cell surfaces. RBL cells containing DiI labeled C18 in their cell membranes were incubated on an array of ZMWs [74]. As the cells adhere to the metal surface they appear to protrude into the aperture, as shown in Fig. 13.8. The total fluorescence intensity coupled out of the waveguide increased by an order of magnitude over the course of the 1.5 hr incubation. The adhesion allowed collection of intensity traces that included single fluorophore bursts.

13.5 Nanostructures for Molecular Confinement

13.5.1 Entropic Traps

DNA sorting and separation are important processes in modern biochemistry [65, 75–77]. While these processes are traditionally carried out in gels, there has been a considerable effort to reproduce such effects in microfabricated structures. As DNA strands pass through gel-like media they typically travel in one of two modes. In the first, Ogsten Sieving, the molecules remain relatively relaxed. The DNA molecules are driven with a low electric field and travel through a gel with a pore size larger than the radius of gyration. Roughly speaking, separation occurs because the frequency of interactions with the gel is a function of molecular length. The second transport mode, reptation, occurs at a high electrical field in a matrix with relatively small pore sizes. In this regime, the DNA molecules elongate and "snake" through the gel matrix. In this mode, separation is thought to occur as a result of the continuous drag on the molecule, which is a function of length. Occupying a space between

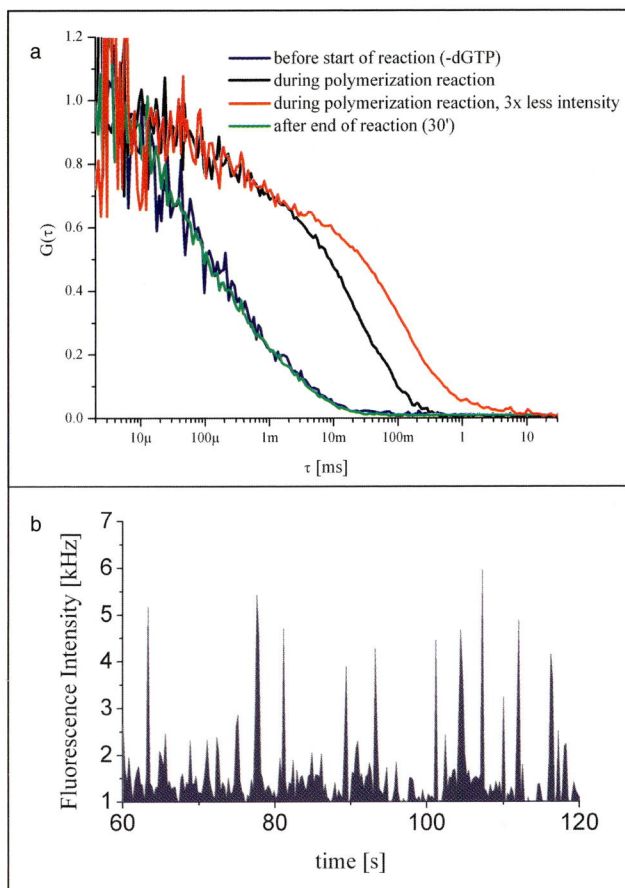

Fig. 13.7. Data is shown from an observation of DNA polymerization. (**a**) FCS curves taken before, during and after DNA polymerization by a polymerase molecule immobilized at the *bottom* of a ZMW. The before and after curves are consistent with diffusion of the fluorescent nucleotide analog. The long residency time of the fluorophore during the reaction is due to its interaction in the active site. (**b**) Photon bursts recorded during polymerization. Each burst corresponds to a base addition

these two separation regimes is entropic trapping. In a medium with small pore sizes, low electric fields will cause DNA molecules to relax in large open regions and then reptate through the small pores. This phenomenon has been used to carry out length-based separations of DNA molecules in the kilobase regime using a microfabricated device. Single molecule techniques are now being applied to characterize the behavior of DNA molecules as they change from a relaxed to an elongated configuration.

To study the behavior of DNA fragments moving in the entropic trapping regime it is necessary to have a nanofabricated structure with controlled

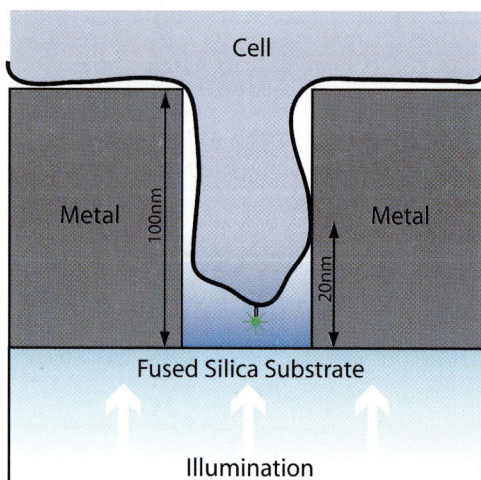

Fig. 13.8. A cell is illustrated protruding into a ZMW. Fluorescence is observed from membrane probes as they diffuse along a protuberance investigating the waveguide

geometry. The entropic trap array, consisting of alternating open and constricted volumes, provides the volume confinement needed for these studies. Entropic traps can be fabricated using two layers of optical lithography, or a single layer of electron beam lithography.

In the absence of an external force, such as an electric field, a DNA molecule will not ordinarily enter the constricted region of an entropic trap [78]. Because the number of possible configurations in an open volume vastly exceeds that of a constricted volume, there is a difference in entropy between the two regions. The resulting free energy barrier prevents DNA fragments from passing through the shallow regions. While the molecule may randomly probe the shallow region, it is energetically unfavorable for it to make any significant entry. In the presence of an electric field, however, the free energy is modified in a way that occasionally allows DNA fragments to fully enter the shallow region. To investigate this behavior, microfabricated channels were built with alternating deep and shallow regions (Fig. 13.9). The deep regions are fabricated to allow the DNA to enter a fully relaxed state where the depth is more than twice the radius of gyration. By varying the geometry of the shallow region, the device can be optimized for different separation procedures [79].

A simple model for DNA escape across an entropic trap was formulated by Han, Turner and Craighead [80]. Assume that the DNA molecule is at the interface between a deep and shallow region in the presence of an electric field E. The molecule will randomly enter a short distance, x, into the shallow region. The total number of bases and hence charge in the shallow region will be proportional to x. The decrease in potential energy due to the entry is given by $x^2 E_S$, where E_s is the field in the shallow region. The increase in free energy is proportional xT. Consequently, the total free energy change

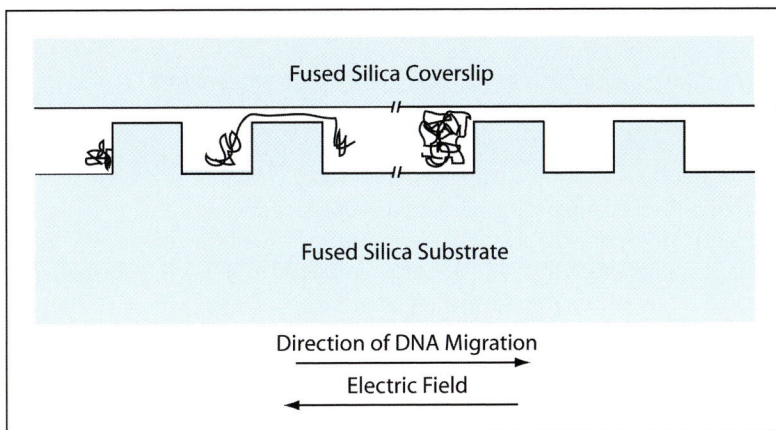

Fig. 13.9. A schematic of the entropic trap array is shown. As DNA migrates from *left* to *right* it repeatedly relaxes in the deep regions. Longer molecules escape through the traps more frequently because they make more escape attempts per unit time than short molecules

caused by introducing monomers by a length x becomes $\Delta F \sim xT - x^2 E_s$. This function first has a positive peak and then decreases as x increases. Thus, there is a transition state, x_c, after which it is energetically favorable for the DNA chain to fully enter the shallow region. The probability of overcoming the free energy barrier is proportional to the Boltzmann factor [80]:

$$P \propto \exp\left(-\frac{\Delta F_{\max}}{k_B T}\right)$$

The trapping time is then given by:

$$\tau = \tau_0 \exp\left(\frac{\Delta F_{\max}}{k_B T}\right) = \tau_0 \exp\left(\frac{\alpha}{E_S k_B T}\right)$$

where α is a constant. When the $\Delta F_{\max} \gg k_B T$ the DNA fragments will be trapped permanently in the deep region. In contrast, when $\Delta F_{\max} \ll k_B T$, the DNA will pass through the shallow region uninhibited. When $\Delta F_{\max} \sim k_B T$, however, molecules will escape the trap only after spending some time trapped.

This formulation is ostensibly length independent. In fact, it has been shown that for T2 and T7 DNA (T2 is 4 times the length of T7), the escape probability is identical. Because the escape mechanism involves random incursions into the shallow region, the probability of escape is proportional to the number of escape attempts made. Larger molecules will explore more configurations than smaller ones. Consequently, larger molecules tend to escape more frequently than shorter fragments. It is this property that makes entropic traps suitable for length based separations.

13.5.2 DNA Separation in Entropic Trap Arrays

The entropic trap array device has been used to separate DNA molecules based on length. Because the trapping time has been found to be a function of length, the mobility of a molecule in the trap array is also length dependent. Length dependent mobility suggests that long arrays of entropic traps could provide an efficient media for molecular separations.

In electrophoresis, the mobility of a molecule is defined as $v = \mu_0 E$, where E is the applied electric field, v is the velocity and μ_0 is the free draining mobility. The mobility provides the critical information about the separation because it contains all length dependant terms. In an entropic trap array the mobility is given by:

$$\frac{\mu}{\mu_0} = \frac{t_{\text{travel}}}{t_{\text{travel}} + \tau}$$

where t_{travel} is the time a DNA molecule takes to move from one trap to the next, τ is the trapping time and μ is the mobility in the trap array. The travel time, t_{travel}, is proportional to $L/\mu_0 E_{av}$, where L is the distance between traps. E_{av} is the average electric field in the device and is related to E_S by a geometrical factor describing the relative volume of the deep and shallow regions. Taken with the definition of the trapping time given above, the mobility can be written:

$$\mu = \frac{\mu_0}{1 + \alpha_1 E_{av} \exp\left(\frac{\alpha_2}{E_{av}}\right)}$$

where α_1 and α_2 are constants. The length dependence is contained within the constant α_1 in this equation. It has been shown that it is the relative size of a DNA fragment that governs how many escape attempts it makes. The critical parameter is R, the radius of gyration, which according to Flory varies as $N^{3/5}$. Hence we can write the mobility with the length dependence explicit as:

$$\mu = \frac{\mu_0}{1 + \alpha_1 N^{-\frac{3}{5}} E_{av} \exp\left(\frac{\alpha_2}{E_{av}}\right)}$$

where N is the number of monomers. This simple model performs very well for DNA fragments larger then the shallow region. However, as the molecules become shorter, the model begins to fail.

Separations were performed with several samples. Initially, a mixture of T2 and T7 DNA was used. The results can be seen in Fig. 13.10 [81]. Further experiments were done showing separation of complicated lambda digests. Figure 13.11 shows separation results for a lambda mono cut overlayed with a simultaneous separation of a 5 kbp ladder. A theoretical plate number as high as 8,500 was obtained, defined as the ratio of the convective and diffusive rates. As can be seen, the entropic trap array is a robust platform for DNA separations at these lengths [81].

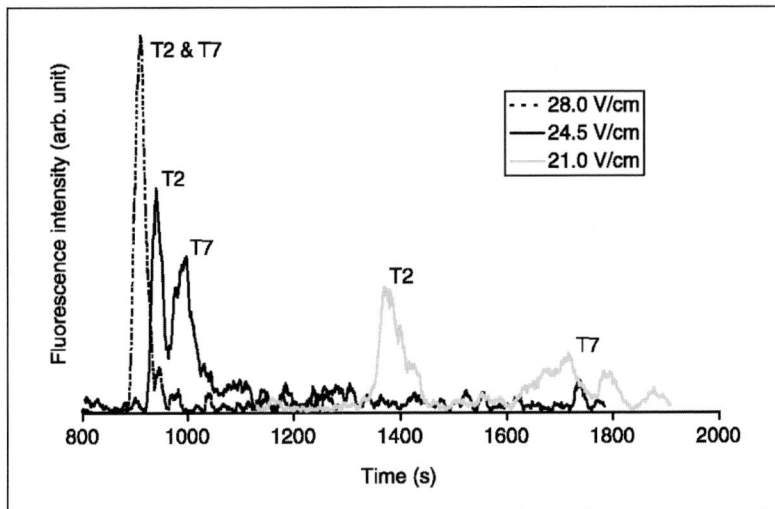

Fig. 13.10. A T2-T7 DNA mixture is separated through an entropic trap channel with 90-nm thin regions, 650-nm thick regions, and a 4-μm channel period. At 21.0 V/cm (*gray line*), the theoretical plate number (N) was 4900 for the T2 peak and 970 for the T7 peak, and the resolution was 1.95. At 24.5 V/cm (*black line*), $N = 8500$ for the T2 peak and 3400 for the T7 peak, and the resolution was 0.89. At 28.0 V/cm, no separation was achieved (*broken line*)

13.5.3 Single Molecule Characterization

DNA trapping times have been characterized as a function of applied electric field and DNA length by measuring the mobility of molecules in the array. This technique is experimentally simple in that one need only measure the transit time of a molecule or group of molecules as they travel along many traps. This analysis produces mean values for the trapping time. However, because the mobility is measured over many traps, single molecule behavior is averaged out.

Single molecule techniques have been applied in an effort to acquire a better understanding of DNA behavior at the traps. Techniques similar to those discussed earlier were used to observe single DNA molecules as they approached the trap interface, became trapped and then ultimately escaped. A confocal microscope in epi-illumination mode was used to focus a laser beam on the trap interface. Fluorescence from YOYO − 1 labeled DNA molecules was recorded via an avalanche photodiode. Fluorescence intensity traces, like those shown in Fig. 13.12, can be analyzed to acquire frequency of DNA arrival and duration of trapping. Not only is the mean trapping time as a function of length, electric field and depth of the shallow region acquired, but also the variance of the trapping time. This variance is critical to explaining band broadening during entropic trap separations. As shown in Fig. 13.13 the

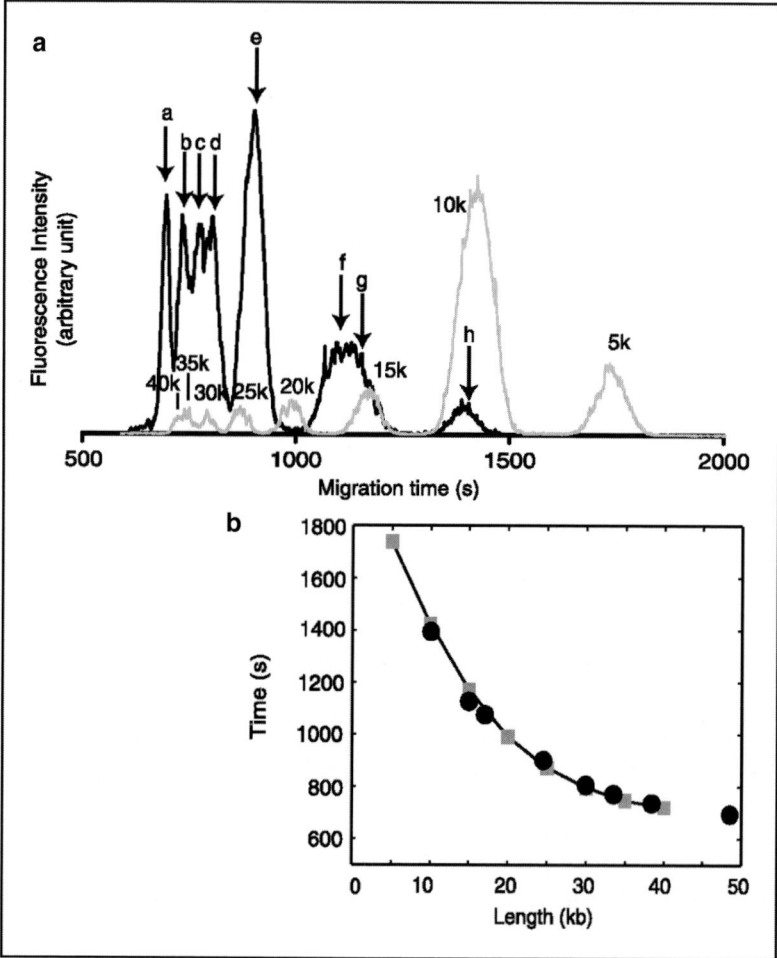

Fig. 13.11. (a) Simultaneous separation of the Mono Cut Mix sample (*black line*) and 5-kbp ladder sample (*gray line*) by the entropic trap array, run at 80 V/cm. The channel has 75-nm thin regions and 1.8-μm thick regions, the channel period is 4 μm and the length is 15 mm. Peak assignment for the Mono Cut Mix sample: (a) 48,502 bp, (b) 38,416 bp, (c) 33,498 bp, (d) 29,946 bp, (e) 24,508/23,994 bp (reference band), (f) 17,053 bp, (g) 15,004 bp, (h) 10,086 bp. For the 5-kbp ladder sample, the 10-kbp peak is the reference peak and brighter than the others. (b) Location of the peaks plotted against DNA length. Peak positions from both samples nicely fall into a single curve

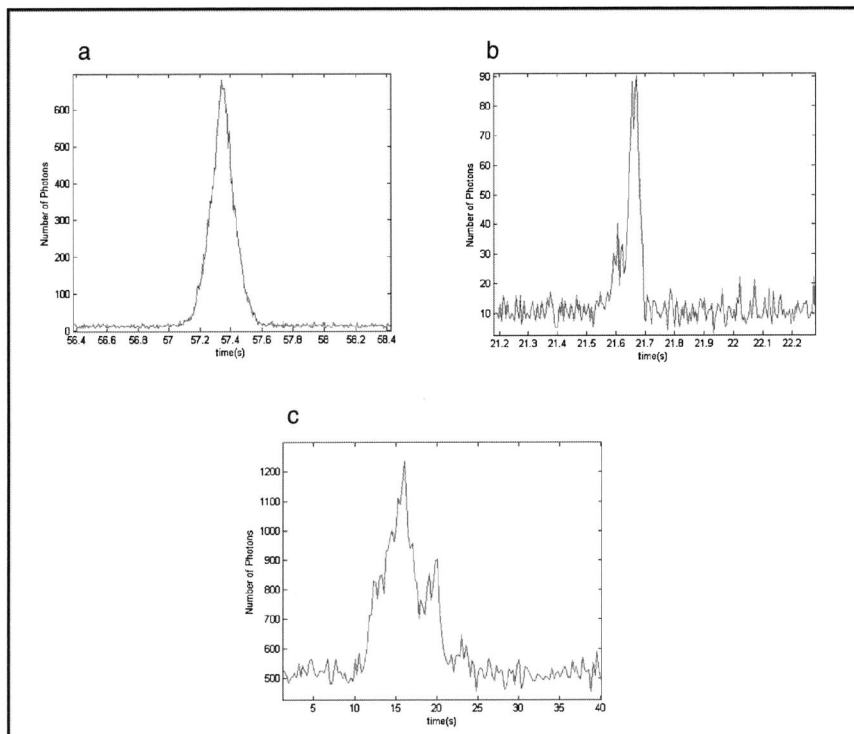

Fig. 13.12. Photon bursts from lambda DNA in entropic traps. (**a**) When no trap is present the burst shape is roughly Gaussian. (**b**) The burst shape becomes asymmetric when DNA fragments are observed at a trap. When the molecule enters the *shallow region*, it travels much faster (the driving electric field is more than 10 times stronger than in the *deep regions*) and so the burst is cut off. (**c**) At *low field*, the DNA can remain at the trap for many seconds. Here a DNA fragment can be seen to migrate around near the trap until it finally escapes more then ten seconds after becoming trapped

variance in trapping time is extremely large at low fields. This large variance suggests that efficient separations cannot be carried out at low electric fields.

13.6 Entropic Recoil

Nanofabricated structures can be used to study the behavior of DNA molecules at the interface between regions of high and low molecular confinement. The physical properties of DNA oligonucleotides can be probed by observing their recoil from regions that highly confine the available molecular conformations to regions of low confinement. Nanostructures that exploit this effect have been termed entropic recoil devices; a schematic is shown in Fig. 13.14.

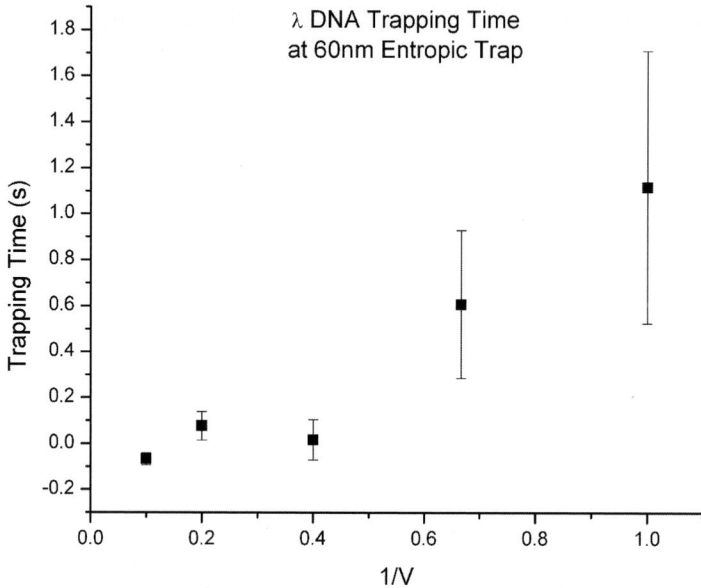

Fig. 13.13. Trapping time is shown as a function of applied voltage. The trapping time is calculated by subtracting the amount of time it takes a typical molecule to approach the trap from the total duration of the burst. The resulting time represents the amount of time spent at the trap interface attempting to escape. The *error bars* represent the variance in the trapping times. Clearly, there is a great deal of variation from molecule to molecule at *low fields*

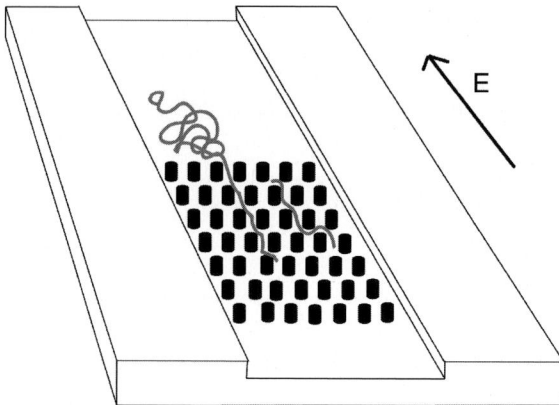

Fig. 13.14. A schematic of an entropic recoil pillar array is shown with two DNA molecules in regions of *high* and *low* confinement. As the electric is turned off, the molecule straddling the interface will recoil out of the pillar array. The molecule inserted into the pillar array will remain there

The first such device, described by Turner et al. [81], consisted of an open region and a confined region containing an array of 35 nm pillars. The pillars in the confined region reduce the number of conformations available to a DNA molecule relative to the open region. At the interface, this difference in entropy leads to an entropic force that draws the molecule out of the pillar region and into the open region. Molecules entirely in the open or pillar regions feel no entropic force and achieve an equilibrium conformation. If an electric field is applied to the device, the DNA molecules can be forced to enter the pillar region. The process is fundamentally identical to DNA molecules jumping from trap to trap in entropic trap devices where entropic changes in free energy are offset by the free energy change due to the applied electric field. The length of a molecule partially inserted into the pillar region can be observed to recoil into the open region according to:

$$L(t) = \sqrt{-\frac{2f(t - t_0)}{\rho}}$$

where f is the entropic force, ρ is the specific drag, t_0 is the time of total recoil and $L(t)$ is the length of molecule within the pillar region.

Entropic recoil devices have been used to separate DNA of different lengths [82]. Bacteriophage T2 and T7 DNA were separated using a pillar device by applying repeated low and high voltages to molecules at the interface. When the electric field pulses are of the correct magnitude and duration, short DNA can be forced to enter the pillar region entirely while longer DNA enter only partially. The partially entered DNA then recoils out while the fully inserted short fragments do not because they feel no entropic force. The procedure can be repeated to achieve separation of DNA based on length.

13.7 Conclusions

Engineered nanostructures can be utilized to isolate and observe the identity and activity of individual biomolecules. This allows one to obtain information otherwise obscured in ensemble averages. Structures can also be created with dimensions sufficiently small that they can exert controlled forces on single molecules. These forces can probe the mechanical properties or deliberately affect the confirmation of the molecules. The selective use of these mechanical effects can be used to sort and analyze molecules on the basis of their mechanical properties.

Only a few examples of nanoscale devices and their uses were described in this chapter, focusing on devices utilizing optical signal transduction. Optical methods are relatively non-invasive and can be used with chemically selective optical probes for exquisitely sensitive detection and recognition of individual molecules. Similar devices can localize molecules for measurement of electronic properties. Ultimately more complex nanofabricated devices and a combination of physical effects may be utilized in molecular analysis systems.

References

1. P. Tamarat, et al. (2000). Ten years of single-molecule spectroscopy. *Journal of Physical Chemistry A*, **104**(1), pp. 1–16.
2. W.P. Ambrose, et al. (1999). Single molecule fluorescence spectroscopy at ambient temperature. *Chemical Reviews*, **99**(10), pp. 2929–2956.
3. A.D. Mehta, et al. (1999). Single-molecule biomechanics with optical methods. *Science*, **283**(5408), pp. 1689–1695.
4. X.S. Xie and J.K. Trautman (1998). Optical studies of single molecules at room temperature. *Annual Review of Physical Chemistry*, **49**, pp. 441–480.
5. T. Plakhotnik, E.A. Donley, and U.P. Wild (1997). Single-molecule spectroscopy. *Annual Review of Physical Chemistry*, **48**, pp. 181–212.
6. S.M. Nie and R.N. Zare (1997). Optical detection of single molecules. *Annual Review of Biophysics and Biomolecular Structure*, **26**, pp. 567–596.
7. P.B. Fernandes (1998). Technological advances in high-throughput screening. *Current Opinion in Chemical Biology*, **2**(5), pp. 597–603.
8. S. Weiss (1999). Fluorescence spectroscopy of single biomolecules. *Science*, **283**(5408), pp. 1676–1683.
9. S. Weiss (2000). Measuring conformational dynamics of biomolecules by single molecule fluorescence spectroscopy. *Nature Structural Biology*, **7**(9), pp. 724–729.
10. M.D. Barnes, W.B. Whitten, and J.M. Ramsey (1995). Detecting Single Molecules in liquids. *Analytical Chemistry*, **67**(13), pp. A418–A423.
11. W.C.W. Chan, et al. (2002). Luminescent quantum dots for multiplexed biological detection and imaging. *Current Opinion in Biotechnology*, **13**(1), pp. 40–46.
12. T. Trindade, P. O'Brien, and N.L. Pickett (2001). Nanocrystalline semiconductors: Synthesis, properties, and perspectives. *Chemistry of Materials*, **13**(11), pp. 3843–3858.
13. A.D., Yoffe (2001). Semiconductor quantum dots and related systems: electronic, optical, luminescence and related properties of low dimensional systems. *Advances in Physics*, **50**(1), pp. 1–208.
14. S.R. Nicewarner-Pena, et al. (2001). Submicrometer metallic barcodes. *Science*, **294**(5540), pp. 137–141.
15. C.H. Wei, et al. (2004) Polarization dependence of light intensity distribution near a nanometric aluminum slit. *Journal of the Optical Society of America B-Optical Physics*, **21**(5), pp. 1005–1012.
16. M. Cabodi, et al. (2002). Continuous separation of biomolecules by the laterally asymmetric diffusion array with out-of-plane sample injection. *Electrophoresis*, **23**(20), pp. 3496–3503.
17. S.W. Turner, et al. (1998). Monolithic nanofluid sieving structures for DNA manipulation. *Journal of Vacuum Science & Technology B*, **16**(6), pp. 3835–3840.
18. S.W.P. Turner, M. Cabodi, and H.G. Craighead (2002). Confinement-induced entropic recoil of single DNA molecules in a nanofluidic structure. *Physical Review Letters*, **88**(12).
19. W. Reisner, et al. (2005). Statics and dynamics of single DNA molecules confined in nanochannels. *Physical Review Letters*, **94**(19).
20. J.O. Tegenfeldt, et al. (2004). Stretching DNA in nanochannels. *Biophysical Journal*, 2004. **8**(1), pp. 596A–596A.

21. J.O. Tegenfeldt, et al. (2004). The dynamics of genomic-length DNA molecules in 100-nm channels. *Proceedings of the National Academy of Sciences of the United States of America*, **101**(30), pp. 10979–10983.

22. P.M. Goodwin, W.P. Ambrose, and R.A. Keller (1996). Single-molecule detection in liquids by laser-induced fluorescence. *Accounts of Chemical Research*, **29**(12), pp. 607–613.

23. J.B. Pawley, ed. 1995. *Handbook of Biological Confocal Microscopy*. 2nd ed. Plenum Press: New York, p. 632.

24. P.K. Wong, et al. (2004). Electrokinetics in micro devices for biotechnology applications. *Ieee-Asme Transactions on Mechatronics*, **9**(2), pp. 366–376.

25. G.J.M. Bruin (2000). Recent developments in electrokinetically driven analysis on microfabricated devices. *Electrophoresis*, **21**(18), pp. 3931–3951.

26. W.G. Kuhr (1990). Capillary Electrophoresis. *Analytical Chemistry*, **62**(12), pp. R403–R414.

27. C.A. Monnig, and R.T. Kennedy (1994). Capillary Electrophoresis. *Analytical Chemistry*, **66**(12), pp. R280–R314.

28. D. Belder, and M. Ludwig (2003). Surface modification in microchip electrophoresis. *Electrophoresis*, **24**(21), pp. 3595–3606.

29. J.L. Viovy (2000). Electrophoresis of DNA and other polyelectrolytes: Physical mechanisms. *Reviews of Modern Physics*, **72**(3), pp. 813–872.

30. K. Swinney and D.J. Bornhop (2000). Detection in capillary electrophoresis. *Electrophoresis*, **21**(7), pp. 1239–1250.

31. S.S. Dukhin (1993). Nonequilibrium Electric Surface Phenomena. *Advances in Colloid and Interface Science*, **44**, pp. 1–134.

32. C. Schwer, and E. Kenndler (1991). Electrophoresis in Fused-Silica Capillaries – the Influence of Organic-Solvents on the Electroosmotic Velocity and the Zeta-Potential. *Analytical Chemistry*, **63**(17), pp. 1801–1807.

33. R. Parsons (1990). Electrical Double-Layer – Recent Experimental and Theoretical Developments. *Chemical Reviews*, **90**(5), pp. 813–826.

34. S.L. Carnie, and G.M. Torrie (1984). The Statistical-Mechanics of the Electrical Double-Layer. *Advances in Chemical Physics*, **56**, pp. 141–253.

35. D.C. Grahame (1947). The Electrical Double Layer and the Theory of Electrocapillarity. *Chemical Reviews*, **41**(3), pp. 441–501.

36. W. Li, D.P. Fries, and A. Malik (2004). Sol-gel stationary phases for capillary electrochromatography. *Journal of Chromatography A*, **1044**(1–2), pp. 23–52.

37. M.A. Shoffner, et al. (1996). Chip PCR.1. Surface passivation of microfabricated silicon-glass chips for PCR. *Nucleic Acids Research*, **24**(2), pp. 375–379.

38. J.B. Brzoska, I. Benazouz, and F. Rondelez (1994). Silanization of Solid Substrates – A Step Toward Reproducibility. *Langmuir*, **10**(11), pp. 4367–4373.

39. C. Manta, et al. (2003). Polyethylene glycol as a spacer for solid-phase enzyme immobilization. *Enzyme and Microbial Technology*, **33**(7), pp. 890–898.

40. J.Q. Li, J. Carlsson, and K. Caldwell (1993). Surface-Properties of Poly- (Ethylene Oxide)-Containing Copolymers on Colloids. *Abstracts of Papers of the American Chemical Society*, **206**, pp. 32–PMSE.

41. Q.S. Huo, et al. (1994). Organization of Organic-Molecules With Inorganic Molecular-Species into Nanocomposite Biphase Arrays. *Chemistry of Materials*, **6**(8), pp. 1176–1191.

42. C. La Mesa (2005). Polymer-surfactant and protein-surfactant interactions. *Journal of Colloid and Interface Science*, **286**(1), pp. 148–157.

43. M. Krieger and J. Herz (1994). Structures and Functions of Multiligand Lipoprotein Receptors – Macrophage Scavenger Receptors and Ldl Receptor-Related Protein (Lrp). *Annual Review of Biochemistry*, **63**, pp. 601–637.
44. S.B. Zimmerman and A.P. Minton (1993). Macromolecular Crowding - Biochemical, Biophysical, and Physiological Consequences. *Annual Review of Biophysics and Biomolecular Structure*, **22**, pp. 27–65.
45. D. Qin, et al. (1998). Microfabrication, microstructures and microsystems, in *Microsystem Technology in Chemistry and Life Science*. pp. 1–20.
46. M. Geissler and Y.N. Xia (2004). Patterning: Principles and some new developments. *Advanced Materials*, **16**(15), pp. 1249–1269.
47. W. Lang (1996). Silicon microstructuring technology. *Materials Science & Engineering R-Reports*, **17**(1), pp. 1–55.
48. S.J. Pearton (1994). Reactive Ion Etching Of Iii-V Semiconductors. *International Journal of Modern Physics B*, **8**(14), pp. 1781–1786.
49. M. Foquet, et al. (2002). DNA fragment sizing by single molecule detection in submicrometer-sized closed fluidic channels. *Analytical Chemistry*, **74**(6), pp. 1415–1422.
50. A. Maciossek, et al. (1995). Galvanoplating and Sacrificial Layers for Surface Micromachining. *Microelectronic Engineering*, **27**(1–4), pp. 503–508.
51. E. Thomson (1925). The mechanical, thermal and optical properties of fused silica. *Journal of the Franklin Institute*, **200**, pp. 313–325.
52. I. Fanderlik (1983). *Optical Properties of Glass.* **Elsevier**.
53. *Melles Griot Optics Guide.* http://www.mellesgriot.com/products/optics/ mp_3 _2.htm.
54. G.M. Whitesides, et al. (2001). Soft lithography in biology and biochemistry. *Annual Review of Biomedical Engineering*, **3**, pp. 335–373.
55. J. Kameoka, et al. (2002). An electrospray ionization source for integration with microfluidics. *Analytical Chemistry*, **74**(22), pp. 5897–5901.
56. J. Kameoka, et al. (2004). Fabrication of suspended silica glass nanofibers from polymeric materials using a scanned electrospinning source. *Nano Letters*, **4**(11), pp. 2105–2108.
57. S.S. Verbridge, et al. (2005). Suspended glass nanochannels coupled with microstructures for single molecule detection. *Journal of Applied Physics*, **97**(12).
58. E.L. Elson and D. Magde (1974). Fluorescence Correlation Spectroscopy.1. Conceptual Basis and Theory. *Biopolymers*, **13**(1), pp. 1–27.
59. D. Magde E.L. Elson, and W.W. Webb (1974). Fluorescence Correlation Spectroscopy.2. Experimental Realization. *Biopolymers*, **13**(1), pp. 29–61.
60. D. Magde, W.W. Webb, and E. Elson (1972). Thermodynamic Fluctuations in a Reacting System – Measurement by Fluorescence Correlation Spectroscopy. *Physical Review Letters*, **29**(11), pp. 705–&.
61. M. Foquet, et al. (2004). Focal volume confinement by submicrometer-sized fluidic channels. *Analytical Chemistry*, **76**(6), pp. 1618–1626.
62. D. Magde and E.L. Elson (1978). Fluorescence Correlation Spectroscopy.3. Uniform Translation and Laminar-Flow. *Biopolymers*, **17**(2), pp. 361–376.
63. R. Rigler, et al. (1993). Fluorescence Correlation Spectroscopy With High Count Rate And Low-Background – Analysis of Translational Diffusion. *European Biophysics Journal With Biophysics Letters*, **22**(3), pp. 169–175.
64. F.S. Collins, et al. (1998). New goals for the US Human Genome Project, 1998-2003. *Science*, **282**(5389), pp. 682–689.

65. H.P. Chou, et al. (1999). A microfabricated device for sizing and sorting DNA molecules. *Proceedings of the National Academy of Sciences of the United States of America*, **96**(1), pp. 11–13.
66. S.M. Stavis, et al. (2005). Single molecule studies of quantum dot conjugates in a submicrometer fluidic channel. *Lab on a Chip*, **5**(3), pp. 337–343.
67. S.M. Stavis et al. (2005). *Detection and identification of nucleic acid engineered fluorescent labels in submicrometre fluidic channels.* Nanotechnology,
68. M.J. Levene, et al. (2003). Zero-mode waveguides for single-molecule analysis at high concentrations. *Science*, **299**(5607), pp. 682–686.
69. S.T. Hess and W.W. Webb (2002). Focal volume optics and experimental artifacts in confocal fluorescence correlation spectroscopy. *Biophysical Journal*, **83**(4), pp. 2300–2317.
70. K.T. Samiee, et al. (2005). Lambda-Repressor Oligomerization Kinetics at High Concentrations Using Fluorescence Correlation Spectroscopy in Zero-Mode Waveguides. *Biophysical Journal*, **88**(3), pp. 2145–2153.
71. D.E. Koppel (1974). Statistical Accuracy in Fluorescence Correlation Spectroscopy. *Physical Review A*, **10**(6), pp. 1938–1945.
72. M. Ptashne (1992). *A Genitic Switch.* 2nd ed. 1992, Cambridge, Cell Press.
73. T.R. Pray, D.S. Burz, and G.K. Ackers (1998). Cooperative non-specific DNA binding by octamerizing lambda cI repressors, A site-specific thermodynamic analysis. *Journal of Molecular Biology*, **282**(5), pp. 947–958.
74. J.B. Edel, et al. (2005). High spatial resolution observation of single molecule dynamics in living cell membranes using zero mode waveguides. *Biophysical Journal*, **88**(1), pp. 195A–195A.
75. E. Carrilho (2000). DNA sequencing by capillary array electrophoresis and microfabricated array systems. *Electrophoresis*, **21**(1), pp. 55–65.
76. N.J. Dovichi (1997). DNA sequencing by capillary electrophoresis. *Electrophoresis*, **18**(12–13), pp. 2393–2399.
77. C. Heller (2001). Principles of DNA separation with capillary electrophoresis. *Electrophoresis*, **22**(4), pp. 629–643.
78. J. Han and H.G. Craighead (1999). Entropic trapping and sieving of long DNA molecules in a nanofluidic channel. *Journal of Vacuum Science & Technology a-Vacuum Surfaces and Films*, **17**(4), pp. 2142–2147.
79. J.Y. Han and H.G. Craighead (2002). Characterization and optimization of an entropic trap for DNA separation. *Analytical Chemistry*, **74**(2), pp. 394–401.
80. J. Han, S.W. Turner, and H.G. Craighead (2001). Entropic trapping and escape of long DNA molecules at submicron size constriction (vol. 83, pp. 1688, 1999). *Physical Review Letters*, **86**(7), pp. 1394–1394.
81. J. Han and H.G. Craighead (2000). Separation of long DNA molecules in a microfabricated entropic trap array. *Science*, **288**(5468), pp. 1026–1029.
82. M. Cabodi, S.W.P. Turner, and H.G. Craighead (2002). Entropic recoil separation of long DNA molecules, *Analytical Chemistry*, **74**(20), pp. 5169–5174.

14

Artificial Dipolar Molecular Rotors

R.D. Horansky,[1] T.F. Magnera,[2] J.C. Price,[1] and J. Michl[2]

[1]Department of Physics and [2]Department of Chemistry and Biochemistry
University of Colorado, Boulder, Colorado, 80309, USA
Robert.Horansky@Colorado.edu
magnera@eefus.colorado.edu
john.price@colorado.edu
michl@eefus.colorado.edu

14.1 Introduction

Rotors are present in almost every macroscopic machine, converting rotational motion into energy of other forms, or converting other forms of energy into rotation. Rotation may be transmitted via belts or gears, converted into linear motion by various linkages, or used to drive propellers to produce fluid motion. Examples of macroscopic rotors include engines which couple to combustible energy sources, windmills which couple to air flows, and most generators of electricity. A key feature of these objects is the presence of a part with rotational freedom relative to a stationary frame. In this chapter we discuss the miniaturization of rotary machines all the way to the molecular scale, where chemical groups form the rotary and stationary parts. For a recent review of molecules with rotary and stationary parts see [1].

When rotor molecules are in the solution or vapor phase, the distinction between the rotating and stationary groups is rather arbitrary. Several such rotors have been designed to be driven by thermal [2] or photochemical [3–6] reactions and demonstrated to rotate unidirectionally, albeit very slowly. If the stationary part of the molecule is mounted to a bulk surface or incorporated in a three dimensional crystal, the rotating group, termed rotator, and the stationary group, termed stator, can be distinguished unambiguously. We are particularly interested in molecules where the group with rotational freedom contains a permanent electric dipole moment that may reorient about the rotational axis. In this case, the rotary motions can be driven and detected by applied electric fields [7–9].

A number of interesting fundamental questions arise in the study of mounted molecular rotors. To what extent can the behavior of molecular rotors be understood in terms of few-degrees-of-freedom models in which the environment is represented by phenomenological parameters (such as a rotary friction constant, or parameters of a rotary potential)? If such models

R.D. Horansky et al.: *Artificial Dipolar Molecular Rotors*, Lect. Notes Phys. **711**, 303–330 (2007)
DOI 10.1007/3-540-49522-3_14   © Springer-Verlag Berlin Heidelberg 2007

work at all, how can the values of the parameters be understood in terms of molecular degrees of freedom? Can rotors be designed to perform useful work? How do nearby rotors interact when assembled into arrays? What collective excitations can be observed in (or engineered into) interacting rotor arrays? How can the motions of single rotors and rotor arrays best be characterized?

Arrays of rotor molecules may exhibit novel dielectric and optical phenomena. Dilute two and three dimensional rotor arrays can be used to explore single-rotor properties, and interactions between single rotors and their immediate environments. Dense two- and three-dimensional rotor arrays may exhibit interesting collective phenomena, such as ferroelectricity [10] and rotary phonon excitations [11].

It is too early to know what applications might arise in this area of nanotechnology, but there are many possibilities. Individual rotors might be used as motors to convert between different forms of energy, or individually addressable rotations might be used to store information. Dipolar rotor arrays will have ordered ground states and are thus novel ferroelectric materials with possible applications to sensors and actuators, and perhaps as non-linear dielectrics. Propagating polar rotary waves [12] are a fascinating possibility which could have applications to radio-frequency filters and delay lines. In the area of microfluidics, the adhesion of the two-dimensional rotors to the surface of a capillary might influence flow at the surface and allow for pumping of the fluid.

We begin this chapter with an introduction to a few specific dipolar molecular rotor species, including their assembly into regular 2- and 3-dimensional arrays. With these molecules in mind, the basic physics of the molecular rotation will be discussed, followed by some of the interesting phenomena predicted should the correct parameter values be achieved. The source of energy to which the rotors will be coupled will be thermal vibration from the bulk substrate, but we will also introduce the theoretical predictions of driving rotation by coupling to a strong electric field. Finally, we will discuss how dielectric spectroscopy can be used to detect the behavior of these rotating dipolar molecules.

14.2 Examples of Molecular Dipolar Rotors

There are a variety of different types of molecules we term dipolar molecular rotors. Our groups have been particularly interested in collections of rotors. The collections are divided into two categories, two-dimensional and three-dimensional. Two-dimensional arrays are formed by attachment of the stator part of the rotor molecule to a bulk surface, which has been either fused silica or gold, depending on the functionalization of the stator. Some dipolar rotors naturally form crystals where the molecular rotors are arranged periodically in three dimensions; the same molecules may form disordered structures. Ex-

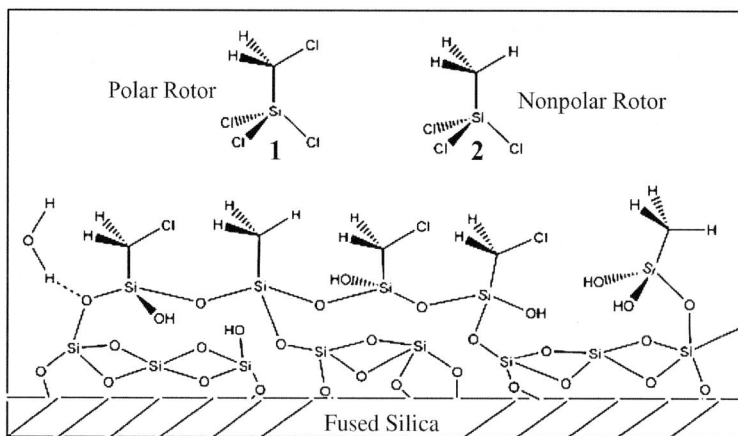

Fig. 14.1. Attachment of chloromethylsilyl and methylsilyl groups to a fused silica surface

amples of dipolar molecular rotors we have studied are presented here both for disordered and regular arrays in both two and three dimensions.

Disordered Two-Dimensional Arrays. *Azimuthal Rotors*. We begin with an example of an azimuthal rotor, or a rotor with the rotation axis perpendicular to the two-dimensional bulk surface. Such rotors have been examined computationally in considerable detail by the methods of classical molecular dynamics [13, 14]. The molecule studied experimentally was chloromethyltrichlorosilane (**1**) mounted on a bulk fused silica surface [7], either neat or diluted with methyltrichlorosilane (**2**), its nonpolar analog. Structures **1** and **2** are shown at the top of Fig. 14.1. In addition to serving as a spacer separating the dipolar molecules, neat **2** has also served as a control sample.

The three chlorine atoms attached to the silicon of the molecules react with water adsorbed on the surface, are replaced with hydroxyl groups and release HCl. Either the remaining chlorine atoms or the hydroxyl groups produced in their hydrolysis then react with hydroxyl groups on the surface of the fused silica to covalently attach to the surface as a self-assembled monolayer. The product is a disordered, two-dimensional collection of rotating dipole groups on the bulk fused silica surface where the rotation axes are approximately perpendicular to the said surface and the rotors are thus termed azimuthal. The chlorine present on the rotating chloromethyl group imparts a permanent electric dipole moment of ~2 Debye (1 debye (D) = 3.336×10^{-30} C-m) in the direction of the C-Cl bond axis, making the molecule a dipolar rotor.

A method for determining the characteristic dynamics of molecular dipolar systems will be addressed in Sect. 4, but some of the properties of these systems will be presented here. When **1** is in the gas phase or on an ideally flat substrate, it is predicted [7] by molecular mechanics [15] simulations to exhibit

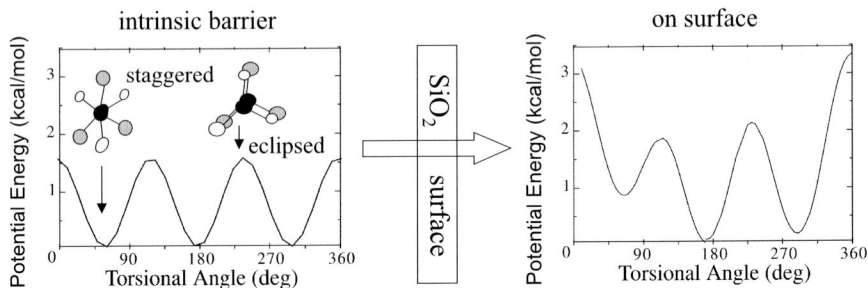

Fig. 14.2. *Left:* The degenerate three-fold potential experienced by rotor **1** as it rotates between staggered and eclipsed conformations. *Right:* upon bonding to fused silica, interactions with the rough surface remove the degeneracy in the wells

a three-well degenerate rotational potential. The stable potential minima, or wells, which are shown in Fig. 14.2, correspond to the three staggered orientations of the chloromethyl group relative to the Si-O bonds. The rotator will orient itself such that it sits in a potential minimum until it acquires enough thermal energy to hop over a barrier into an adjacent potential minimum. In practice, however, the disorder of the silica surface removes the degeneracy of the wells. The barrier energy for hopping out of one well and into another in this system was measured by dielectric spectroscopy (Sect. 14.4) to range between 1 and 4 kcal/mol (between 500 and 2000 Kelvin). This range of barriers was also corroborated using molecular mechanics simulations [7]. Calculations show that the inter-rotor dipole-dipole interactions can be neglected due to the weakness of the dipole moments, the angle the dipoles form to the surface, and the spacing between the rotors, especially in the dilute samples. Since the properties of the sample reflect the behavior of non-interacting rotors, the irregular nature of the two-dimensional array is immaterial.

In an attempt to reduce the effect of disorder in the bulk attachment surface and reduce the inhomogeneity in the rotational potential, compounds similar to **1** have been synthesized with varying groups for attachment to a fused silica surface [16]. A few of them are shown in Fig. 14.3. All of the samples shown have the same rotator as **1**, but they differ in their surface attachment groups. To counteract the roughness of the bulk surface and the resulting inhomogeneity in the rotational potential, longer attachments from the rotator to the surface are used to distance the dipole from the rough surface, leaving only interactions with its own stator as the source of the rotational barrier. With this synthetic control, the rotary dipole behavior is customizable for large or small barriers and more or less interaction with the bulk substrate.

Altitudinal Rotors. Another class of surface-mounted molecular rotors that have been examined in disordered two-dimensional arrays are altitudinal rotors, whose rotational axis is parallel to the gold surface on which they are mounted. Once more, the rotors were located far enough apart to make their

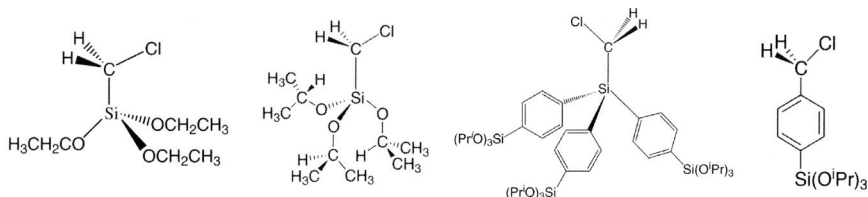

Fig. 14.3. Examples of azimuthal dipolar molecular rotors. By changing the shape and size of the group used for the bonding to bulk surface, the characteristics of the rotational potential may be altered

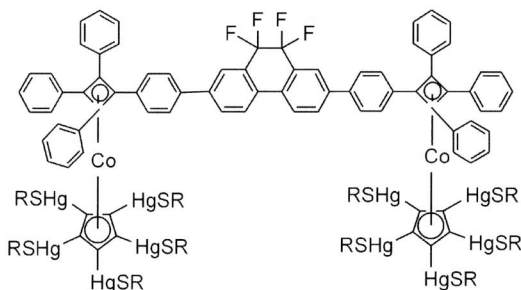

Fig. 14.4. Surface-mountable altitudinal rotor **3** (R = HgS(CH$_2$)$_2$SMe)

mutual interaction negligible. The chemical structure of the most closely examined altitudinal dipolar rotor **3** is shown in Fig. 14.4; a non-polar analog has also been made and examined for comparison [9]. The dipolar rotor has four C-F bonds that produce a net dipole of ~4 Debye perpendicular to the axle. The axle is supported by two cobalt sandwich stands that are attached to the Au surface through ten sulfur containing tentacles. Calculations [9,17] show that when mounted on a gold surface, there is enough space between the surface and the rotator part to permit unhindered rotation, except for those conformations that place one of the tentacles directly beneath the axle.

Whether the surface-mounted rotors are actually free to turn is an interesting issue that was examined [9] with the aid of scanning tunneling microscopy (STM) [18]. This powerful imaging method is capable of resolving individual molecules on atomically flat conducting surfaces. We have used a variant of STM known as barrier height imaging (BHI) [19, 20] to detect whether any particular altitudinal molecular rotor on the Au (111) surface can turn in response to electric field. To understand how this procedure works requires some background information. In STM, a carefully controlled and prepared metal tip is brought to within a few nanometers of a conducting surface. For such small distances a tunneling current is detectable between the tip and the surface and is given by the exponential relationship,

$$i = \rho V \exp(-\beta \phi^{1/2} z), \tag{14.1}$$

where ρ is the integrated density of states, V is the applied potential, z is the tip-surface distance, $\beta = 2(2m)^{1/2}\hbar^{-1} = 1.025\,\mathrm{eV}^{-1/2}\,\text{Å}^{-1}$ and ϕ is the average work function of the tip-surface system. The work function of a metal is the amount of energy needed to remove an electron from its surface to infinity. In a tunneling system where the electron moves between two conducting surfaces the potential barrier to the tunneling process is given roughly by the average work function of the two metals. The work function is very sensitive to the topology of the surface as well as to the presence of surface absorbates.

There are two common modes for collecting STM images. The first is the constant-height mode and is a map of i as a function of the tip position with z held constant except for topographical variations in its magnitude. The second is the constant-current mode and it requires the use of a feedback-servo system that maintains a constant i by varying z as the tip position is scanned. In this second mode, the variation of ϕ and that of ρ are responsible for the contrast in the image. The physics behind variations in ϕ relies on the fact that the easier it is to remove an electron from the surface, the farther out into vacuum will its wave function extend, and the slower will be the fall off or the tunneling current to the tip as z is increased. Adsorbed molecules are sources of surface dipoles and therefore alter the extension of the wavefunction. Atomic corrugations, step edges, and defects will alter the local work function ϕ of clean, crystalline, metal surfaces. The presence of a permanent dipole in an absorbate will increase the local ϕ if its negative end is farther from the surface and decrease ϕ if it is closer. Local surface dipoles are formed when charge is transferred between the absorbate and the surface. The magnitude and direction of these charge transfers is determined by the metal-absorbate system and can diminish or augment any permanent dipoles. Similarly, the strong electric field between the tip and the surface can induce a dipole in an absorbate; in this case the magnitude and direction are proportional to the field strength. The strong dependence of the local work function on surface dipoles allows us to probe molecular rotation induced by an electric field.

BHI is a map of the local work function variation of a surface. The local work function is measured, in our system, by modulating rapidly at 5 kHz by a small amount (0.3 Å) the distance z of the tip from the surface. This causes the tunneling current i to be modulated at the same frequency and with lock-in detection the derivative $\mathrm{d}i/\mathrm{d}z$ is measured. The BHI scan is done in the constant-current mode with the modulation frequency chosen to be much faster than the feedback response time. From equation (14.1),

$$\frac{\mathrm{d}i}{\mathrm{d}z} = -\beta\phi^{1/2}\rho V \exp(-\beta\phi^{1/2}z)\,, \qquad (14.2)$$

$$\frac{\mathrm{d}i}{\mathrm{d}z} = -\beta\phi^{1/2}i\,. \qquad (14.3)$$

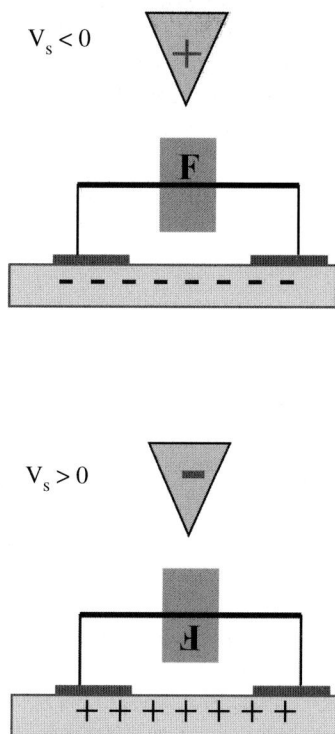

Fig. 14.5. Response of rotator dipole to the sign of the potential difference between the substrate and the STM tip

Equation (14.3) gives the proportionality constant that relates i and $\mathrm{d}i/\mathrm{d}z$ as equal to $-\beta\phi^{1/2}$. Thus, plotting $\mathrm{d}i/\mathrm{d}z$ against tip position is equivalent to mapping the local work function ϕ.

In Fig. 14.5 we schematically represent this dipolar altitudinal rotor in the STM. When voltage difference between the tip and the surface is on the order of volts and the tip-surface distance about a nm, the field strength is on the order of GV/m. A field this strong is sufficient to line up a dipole of about 4 D, even at room temperature, and keep it oriented over 99% of the time. Depending on the direction of the field, we can hold the negative end of the rotator dipole, which carries the fluorine atoms, either close to the surface or far from the surface, as long as it is free to flip. These two situations represent two different conformations of the rotor, and because they differ in the orientation of a surface dipole, they should have detectably different local work functions ϕ_+ and ϕ_-.

In our differential BHI measurement, we raster the tip in the constant-current mode, but go over each line twice, once with each field polarity. When the field holds the negative end of the rotator dipole close to the surface, we

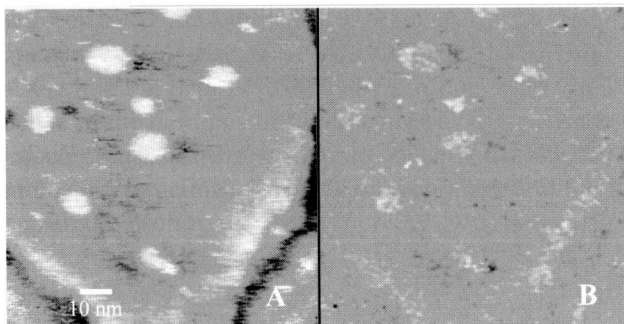

Fig. 14.6. Constant current mode (200 pA, 200 mV) STM (**A**) and differential BHI (**B**) images of the polar altitudinal rotor **3** on Au(111). In the STM image, the molecules appear as bright spots

measure $\phi_+^{1/2}$, and when it is far from the surface, we measure $\phi_-^{1/2}$. This will lead to

$$\left(\frac{di}{dz}\right)_- - \left(\frac{di}{dz}\right)_+ = -i\beta(\phi_-^{1/2} - \phi_+^{1/2}). \tag{14.4}$$

Figure 14.6 shows an ordinary STM image of the dipolar rotors on Au(111) along with a differential plot equivalent to the right.

Figure 14.7 is similar to Fig. 14.6 but shows a control experiment for a similar rotor whose rotator contains no fluorines and carries no permanent dipole. In the differential BHI plot, when the tip is located over clean gold whose surface dipole does not change its normal component when the electric field direction is reversed, there are only weak BHI features, since the local work function differences are very small and the resulting contrast is weak. Faint molecular images are often observed in differential BHI scans since all molecules are polarizable to some degree and have induced dipoles as the

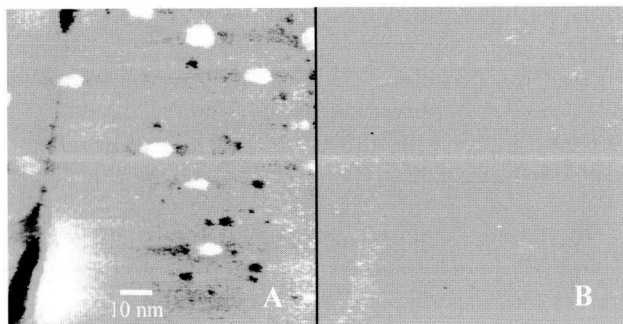

Fig. 14.7. Constant current mode (200 pA, 200 mV) STM (**A**) and differential BHI (**B**) images of a non-polar altitudinal rotor on Au(111). In the STM image, the molecules appear as bright spots

tip passes over them, whose magnitude will depend on the direction of the outside field. However, those molecules that respond to the change in field direction by rotating a substantial dipole, such as that of our dipolar rotator, are expected to give high contrast differential images, and this is indeed seen in Fig. 14.6. The faint BHI images for the control molecule imaged in Fig. 14.7 are consistent with contrast developed from an induced dipole. About one-third of the dipolar rotors give strong, high contrast images. The weak images from the remaining two-thirds are thought to result from those molecules that are unable to turn their dipole when the field reverses direction, due to blockage by surface contaminants or by one of their own tentacles. This blockage is occasionally reversible over the course of tens of minutes, as some molecular images blink on in the differential scan, while some blink off. We attribute this to slow diffusion of the contaminants or the tentacles on the surface.

In the experiment just described, the electric field applied to the rotor from the outside is essentially static, and it changes its direction about twice in a second. An interesting issue regarding the sense of rotation comes up when this kind of single-molecule dipolar altitudinal rotor is mounted on a metal surface and is driven by an electric field that oscillates periodically: can the rotor be driven unidirectionally? This would be easy if the driving electric field were strong enough and rotated. However, above a conductor, electric field is necessarily normal to the surface at all times. At first sight, it might appear impossible to drive unidirectional motion with a field that oscillates in a linear fashion. On second thought, however, this ought to be possible if the rotor does not rotate freely but is subject to a rotational potential that is asymmetric relative to the plane that passes through the axle and is perpendicular to the surface. According to our calculations [9,17], this actually is the case for rotor **3** (Fig. 14.4), at least in one of its stereochemical conformations. In the two potential energy minima in which its dipolar rotator can reside, the dipole is oriented at about 30° away from the surface normal.

In the calculation of the total rotational potential one has to include not only the sinusoidally varying driving field and the intrinsic molecular rotational potential, but also the mirror image of the molecular charge distribution in the metal that mimics the effect of the charges induced by the molecule on the metal surface [21,22]. Since the metal is not a perfect conductor, the molecular dynamics calculations are no longer strictly Newtonian but use Langevin dynamics to include the friction due to the electronic excitation induced in the metal as the charges move [23–25]. The computations predict quite unambiguously that unidirectional rotation can be achieved for many choices of field amplitude and frequency, but at present we have no means to verify this result experimentally.

According to the calculations, the rotor can be driven unidirectionally in two limiting regimes. In the first instance ("driven rotor") the rotor operates best at low temperatures at which thermal hops are nearly impossible and the rotor is essentially forever confined to a single potential minimum in the rota-

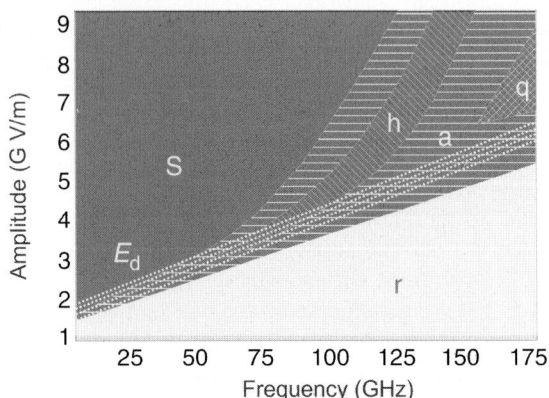

Fig. 14.8. Computed response of one of the stereoisomers of altitudinal rotor **3** mounted on gold surface to oscillating electric field normal to the surface (s: synchronous. a: asynchronous. h: half-synchronous. q: quarter-synchronous. r: random) at 10 K. In the region below the inclined straight bar, an increase in temperature promotes unidirectional rotation and in the region above, it hinders it

tional potential. The time-dependent driving electric field needs to be strong enough to modify this intrinsic potential sufficiently to move the minimum along the rotational angle coordinate in time. Because of electronic friction in the metal substrate, the rotational motion of the rotator lags somewhat behind and its trajectory does not quite follow the path of lowest energy in the time-dependent potential. This difference is accentuated as the frequency of the field increases, and stronger fields are needed to maintain synchronous motion at higher frequencies. Thermal fluctuations are detrimental in that they provide the rotor with an opportunity to hop out of the minimum that carries it along and to move in the opposite sense.

In the second instance (thermal rotor), the opposite is true. In this limiting regime, the operation of the rotor requires a temperature high enough to provide rapid thermally activated hops between the minima in the potential located at different rotational angles, one with the dipole pointing more or less toward the surface, and the other with the dipole pointing in the opposite direction. The directionality of the rotation then is a result of the modulation of the depths of the minima and the heights of the barriers that separate them by the electric field, according to the known principles of "thermal ratchets" [26–29]. At the high frequencies of interest to us, the efficiency of the thermal rotor is not very high and it skips many of the cycles of the electric field.

A phase diagram calculated for rotor **3** in the low-temperature limit, in which only the driven rotation is possible, is shown in Fig. 14.8 [17]. At a given frequency, as the amplitude of the electric field is increased, the rotor motion changes from the random regime, in which it essentially ignores the field, to the synchronous regime, in which it follows the oscillations of the driving field slavishly. A similar change takes place when the electric field amplitude is held constant and its frequency is gradually decreased. At frequencies below about 75 GHz, the change from the random to the synchronous regime is relatively abrupt and proceeds through a brief range of irregular asynchronous motion. At higher frequencies, however, the manner in which the change occurs is much more interesting. There is a range of field amplitudes at which the rotor rotates at half the frequency of the field, skipping every other cycle of the field. Closer analysis showed that this is related to the electronic friction in the metal, which forces the rotor trajectory to deviate excessively from the lowest energy path. Such subharmonic resonances are well known from studies of non-linear mechanics [30,31] but to our knowledge have not been reported before in molecular dynamics studies of realistic models for actual molecules. At higher frequencies, the phase diagram in Fig. 14.8 even contains a region of quarter-synchronous behavior.

Ordered Two-Dimensional Arrays. The creation of ordered and interacting 2-D rotor arrays is an interesting challenge. The only system of this type we are aware of is CO on rock salt [12]. One-dimensional systems can also be envisioned [11].

Although one might find natural crystal surfaces that could serve the purpose, it seems preferable to use interfacial liquid surfaces that permit complete control over the choice of the chemical nature of the rotor attachment point. A liquid surface has no permanent structure and hence offers maximum freedom for an arbitrary choice of a lattice constant; it has no defects such as terrace edges and hence offers the best chances for the formation of large single-crystal domains. It combines a capacity for strong adsorption, and therefore very restricted motion in the vertical direction, with high mobility in the horizontal direction, offering optimal conditions for coupling reactions. Many applications will require the use of trigonal or hexagonal arrays of molecular rotors, and well established Langmuir-Blodgett (LB) techniques [32,32] are useful for the formation of closed-packed trigonal 2-D arrays at interfacial surfaces with trivial simplicity.

To implement the LB method, a rotor molecule is attached to one or more fatty acid chains and spread on a water surface. After compression by a sweeping barrier, the fatty acid chains form a two-dimensional crystalline array of pedestals (Fig. 14.9, bottom) with rotors protruding from the top surface. Since a fatty acid chain occupies a surface area of about 22–23 $Å^2$, inter-rotor spacing are controllable by changing the number of fatty acid pedestals that a rotor rests on from one (∼2.5 Å) to three (∼5 Å). Variable rotor spacing, suitable for keeping very large rotors separated, can be attained with the LB method by diluting the rotor fatty acid molecules with ordinary fatty acids that do not carry a rotor (Fig. 14.9, top). These two-dimensional crystalline films are easily transferred either by traditional vertical or horizontal transfer onto a solid support, or by unconventional transfer from below onto a sieve.

If a sturdier array is desired, covalent linking of the rotor pedestals is necessary [34]. Performing a linking synthesis on interfacially adsorbed species

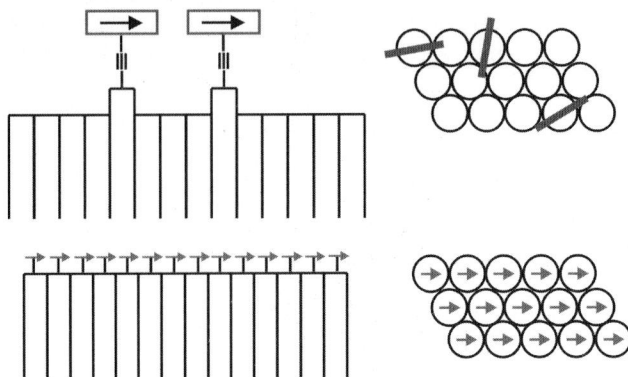

Fig. 14.9. (*Bottom*) Closed-packed Langmuir-Blodgett film of small polar rotors on fatty acid pedestals. (*Top*) Random array of large polar rotors spaced by shorter fatty acid pedestals

avoids reactions in the supernatant solution with the concomitant three-dimensional cross-linking. We use Hg/solvent or Hg/gas interfaces in a home constructed electrochemical Langmuir-Blodgett trough (Fig. 14.10) [35]. This system has many advantages: The surface of mercury is readily cleaned by wiping with a barrier and stays clean for hours under inert atmosphere (N_2) in an ordinary glove-box, which simultaneously prevents its vapor from escaping into the laboratory environment (we have commercial monitors in the vicinity). The metallic surface permits easy ellipsometry, grazing incidence IR, and Raman spectroscopy. The facile definition of the surface potential against a conducting overlayer such as $CH_3CN/LiClO_4$, permitting control of the oxidation state of the adsorbate, controlled generation of metal ions, and electrochemical measurements such as cyclic voltammetry and coulometry (after a small portion of the surface is insulated from the rest in a "ladle"). An example of a metal-ion linked hexagonal array that represents a realistic extrapolation of a synthesis already performed [35] on the electrochemical Hg-LB trough is shown in Fig. 14.11.

Three-Dimensional Arrays. In addition to achieving ordered arrays through film manipulations, another possibility is to use three-dimensional crystals of dipolar molecular rotors to study well-defined collections [8, 36–38]. The molecules we have studied are shown in Fig. 14.12 and are 1,4-bis(3,3,3-triphenylpropynyl)-2-fluorobenzene, **4**, and -difluorobenzene, **5**. Molecules of this type with various substituents X and Y form a crystal lattice that allows rotation of the central benzene ring about the triple bond axes, where the triphenylmethyl groups on either end remain stationary within the crystal framework [37]. By altering the choice of X and Y, we are able to control the rotational dynamics. The crystal structure is exhibited in Fig. 14.13. The fluorine atom on the central benzene ring imparts a 1.5 Debye dipole moment component perpendicular to the rotation axis while the addition of the second

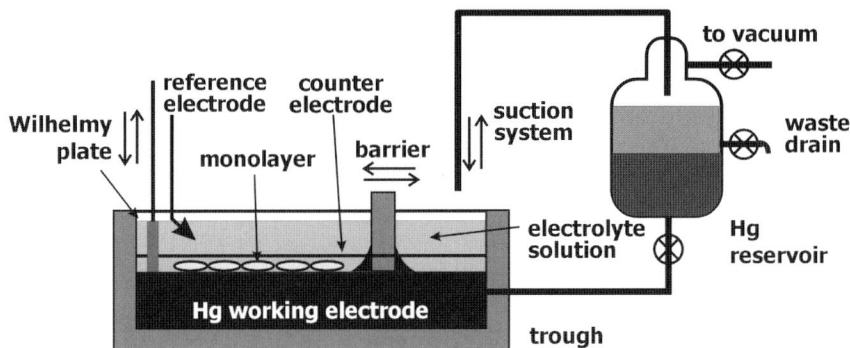

Fig. 14.10. Electrochemical Langmuir-Blodgett trough

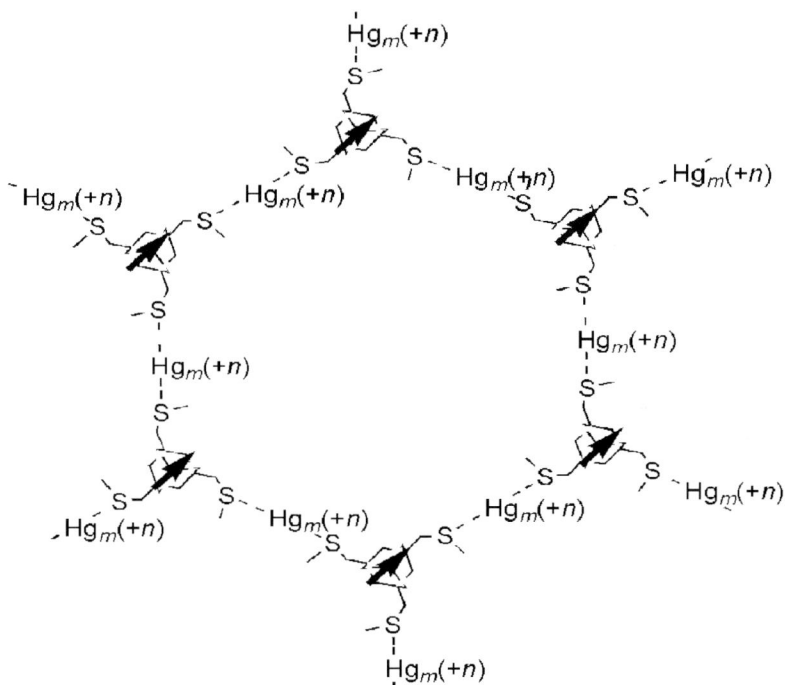

Fig. 14.11. A target structure for a metal-ion bound hexagonal array assembled on an electrochemical LB trough from sulfide-carrying trigonal connectors

fluorine yields 3 Debye. It has been deduced by molecular modeling and ^{2}H NMR techniques that these molecular rotors are essentially free to rotate in the vapor and solution phase, but within the crystal structure, steric interactions with the triphenylmethyl groups of neighboring rotors create a two-well rotational potential with the minima separated by 180° [36]. A schematic drawing of a two-well rotational potential curve is shown in Fig. 14.14. The barrier to

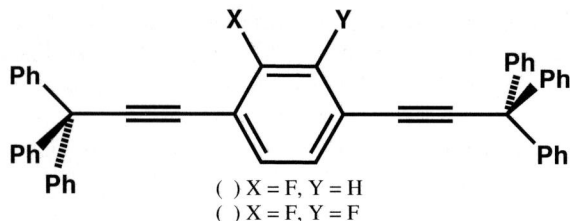

Fig. 14.12. Chemical structure of two crystalline dipolar molecular rotors. The central benzene ring is free to rotate about the triple bonds while the triphenylmethyl groups form the crystal lattice

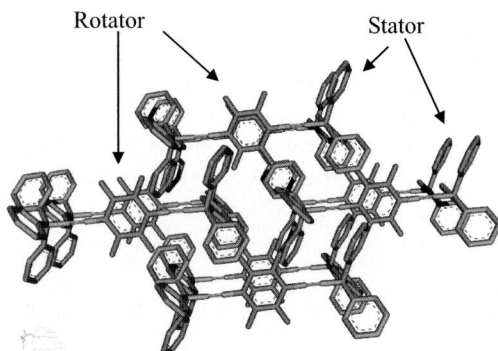

Fig. 14.13. Crystal structure of **5**, isomorphous with **4**. Two of the dipolar rotary groups are labeled Rotator, while two of the stationary triphenylmethyl groups are labeled Stator

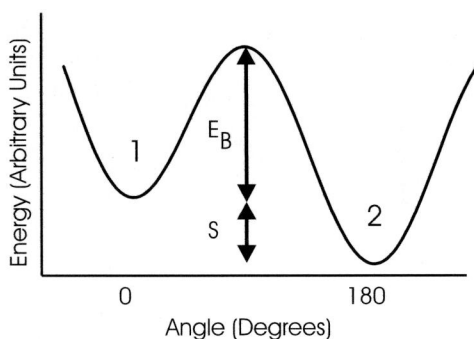

Fig. 14.14. Schematic of a two-well potential energy curve representing the potential experienced by the molecular rotor

rotation E_B for molecules of this type has been measured through dielectric spectroscopy (Sect. 14.4). The regular structure provided by the crystal puts each rotor in an identical rotational potential, in contrast to the disordered surface mounted rotors where each rotor experiences a different environment.

Because of this uniformity of rotor environments, dielectric spectroscopy yields a single barrier to rotation instead of a broad distribution. The dielectric data for **5** are shown in Fig. 14.15. The peaks are fitted well to a theoretical solid line representing a uniform rotor environment. For a spread of barriers, the experimental peak would be broader than the theoretical curve. Rotors **4** and **5** have barriers to rotation of 13.6 kcal/mol (6850 K) and 14.1 kcal/mol (7100 K) respectively. From x-ray studies of these crystals and consideration of the molecular and crystal symmetries, conclusions can be drawn about possible contributions to the asymmetry parameter S, defined as the energy difference between the two potential energy wells. If we focus on one rotor in the array and imagine averaging over the orientations of all other rotors, the S value for **4** may be non-zero (due to steric and intramolecular contributions) while for **5**, S must be zero. From the dielectric spectroscopy data we find $S = 1.5$ kcal/mol for **4** and 0.9 kcal/mol for **5**. In the absence of structural disorder, the observed non-zero value for **5** can only be due to rotor-rotor interactions. Examination of samples with different degrees of disorder suggests that the contributions of disorder are small, implying that the observed asymmetry in **5** is indeed due to interactions [38].

14.3 Behavior of Non-interacting Dipolar Molecular Rotors

There are a number of ways one may choose to model and describe the behavior of dipolar molecular rotors. Numerical molecular models can potentially

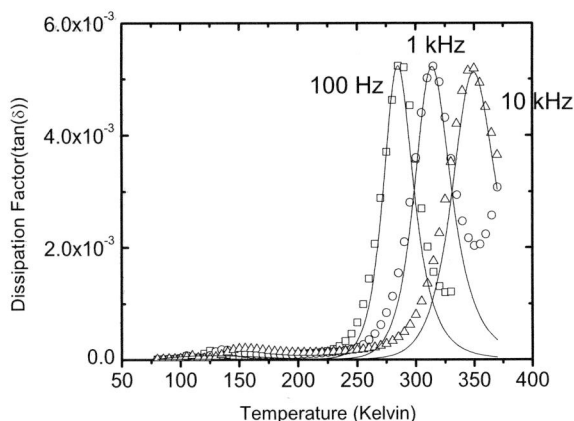

Fig. 14.15. Plot of $\tan(\delta)$ versus temperature for a crystal of **3** with a blank sample subtracted. The solid lines are single Debye peaks with the barrier and attempt frequency dictated by an Arrhenius plot. The overall magnitude is fitted to the 100 Hz peak, and an asymmetry S of 0.9 kcal/mol is chosen to fit the dispersed magnitudes

include all of the important dynamical degrees of freedom. However, these models can be computationally expensive and may not, by themselves, provide a useful summary of the most important physical effects. Another approach is the use of a macro model in which only one or several of the coordinates are treated as dynamical degrees of freedom and the effects of all others appear as friction constants or constraints describing effective potentials.

We will begin by discussing the rotors in the dilute case, where the influences of dipoles upon one another are ignored. In Sect. 14.4, we will introduce dipolar interactions and show how they contribute to the measured well asymmetry. Even in the dilute case, the rotors may be treated in one of four basic ways. They may be treated quantum mechanically, with the barriers to rotation being either high or low, or classically, again with high or low barriers. Systems reported to date appear to be well described by a classical description with high barriers leading to thermally activated hopping. Therefore, after briefly describing each of the approximate regimes, thermally activated rotational hopping will be treated in detail. We will describe analytically the expectations for a two-well rotational potential since it describes most of the studied rotors to date and serves as a tutorial to thermal hopping behavior.

The motion of the rotor is dictated by the total torque Λ_{net}, consisting of a static term and a fluctuating term [39],

$$\Lambda_{net} = \Lambda_{st}(\theta) + \Lambda_{fl}(\theta, t), \tag{14.5}$$

where Λ_{st} is the negative gradient of the rotational potential. The rotational potential arises from interactions of the rotor with the surrounding atoms of the surface, crystal, and/or rotor molecule itself. As was mentioned in the description of the actual molecular rotors, the rotational potential contains wells and barriers between the wells as seen in Fig. 14.14. The fluctuating term, Λ_{fl}, represents interactions of the rotor with the thermal agitation of the surrounding atoms. Averaged over long times, the value of the random direction of the fluctuating forces will be zero. The magnitude of these fluctuations determines the random motion of the rotor, which from equipartition must have an average kinetic energy of $kT/2$, where k is Boltzmann's constant and T is temperature [40].

The appropriate description of the rotor's dynamics will vary depending on the magnitudes of Λ_{fl}, Λ_{st}, and the dissipative coupling to the surrounding bath. If the temperature and dissipation are low enough, then a one-particle Schrödinger equation in the variable θ may give a sufficient description. A characteristic energy scale to consider is,

$$B = \frac{\hbar^2}{2I}, \tag{14.6}$$

which is the ground-state level spacing of the quantum mechanical rigid rotor with a moment of inertia I. At one extreme, if the static rotational potential is low compared to B, then the solution to the dynamics of the rotor are the quantum mechanical rigid rotor wave functions or small perturbations to them. If the potential is larger and may be approximated by a sinusoid, then the dynamics are dictated by solutions to the Schrödinger equation

$$\left(-B\frac{\partial^2}{\partial\theta^2} + \frac{1}{2}E_B\cos n\theta\right)\psi = E\psi\,,\qquad(14.7)$$

where n is the number of wells in the rotational potential. For $n = 2$, this equation may be recast as the Mathieu equation and is exactly solvable [41]. The generalization to n wells is also solvable [39] and has been used to describe hydroxyl "rotors" [12]. The question of how low of a temperature is low enough is an interesting problem and involves the detailed nature of the thermal bath and how it is coupled to the angular coordinate [42], but we may observe that B for the molecular rotors under discussion is about 0.2 K, while kT for the experimental measurements is orders of magnitude higher. Below we attempt to treat the rotors classically.

We may write Newton's law and decompose the fluctuating potential into two forces, one slowly varying, the other rapid [39, 40], giving

$$I\frac{\partial^2\theta}{\partial t^2} = \Lambda_{st}(\theta) - \eta\frac{\partial\theta}{\partial t} + \xi(\theta, t)\,.\qquad(14.8)$$

The terms on the right are just Λ_{net} in the rotational coordinate. The dissipative force $\eta\frac{\partial\theta}{\partial t}$ is the first term of an expansion of the slowly varying fluctuating force with η representing a friction constant, and $\xi(\theta, t)$ is the rapidly fluctuating torque, which averages to zero. If $E_B \ll kT$, Λ_{st} may be ignored and the rotors will exhibit rotational diffusion. Here we are concerned with the situation where kT is about 300 K and E_B, shown in Fig. 14.14, is 5–15 kcal/mol or 2500–7500 K, orders of magnitude larger. With these parameters, we are in the thermally activated or rotational hopping regime.

In this situation the rotor will sit in a particular potential energy minimum and oscillate with a characteristic frequency ω_0, called the libration or attempt frequency. We may consider the probability of finding a single rotor with a given energy at a given time or equivalently, the probability distribution of energy of a large ensemble of rotors. The thermal energy of the ensemble of rotors will distribute in Boltzmann fashion with some rotors having energy well above kT and some well below, but averaging to the given thermal energy. Being trapped in a well, the only mechanism for a rotor to reorient into another minimum is to thermally hop whenever it gains enough thermal energy. This situation is shown in Fig. 14.16 for a three-well scenario where the dipolar rotor is defined to be in a given well if the dipole moment of the rotary group points in the direction dictated by the well. We want to calculate the change in population, N_i, found in the i-th well with time, where i labels the well. This rate of population change will be the rate of rotors hopping into well i minus those that hop out and is given by

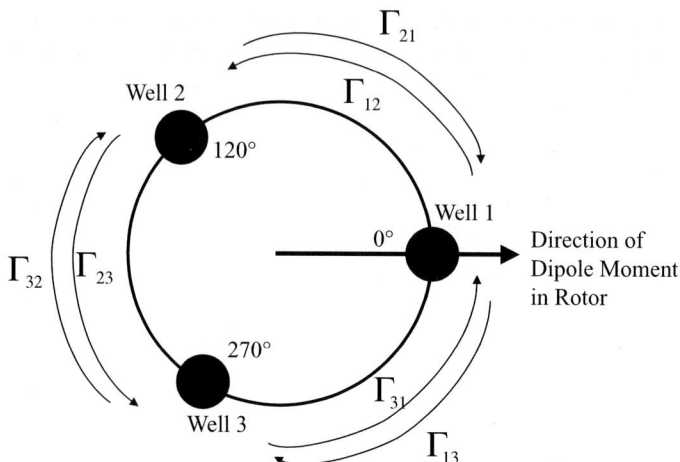

Fig. 14.16. Schematic of thermal hopping for a three-well potential. The dipole moment of the rotator has energy kT and is librating in well 1. If it gains enough energy it may hop over E_B into well 2 or 3. The rate of hopping is shown by *arrows* and Γ_{ij}, where i is the initial and j is the final well involved in the hop

$$\frac{\mathrm{d}N_i}{\mathrm{d}t} = \sum_{j \neq i}^{n} (N_j \Gamma_{ji} - N_i \Gamma_{ij}) . \tag{14.9}$$

Γ_{ij} is the transition rate from well i to well j and the sum is up to n wells. We will attempt to model this transition by a thermally activated hop defined by

$$\Gamma_{ij} = \omega_0 \exp\left(-\frac{W}{kT}\right) , \tag{14.10}$$

where W is the activation or barrier energy between wells [43, 44]. In the example in Fig. 14.14, the barrier from well 1 to 2 is E_B, while the reverse hop has barrier $E_B + S$. To treat only the simplest case, we will suppose that the barrier at 270° is large so that there is only one path between the two minima.

The precise determination of the attempt frequency ω_0 depends upon the shape and depth of the particular potential minimum in which the rotor is trapped. However, for $kT \ll E_B$, the rotor executes small angle oscillations within the well. The rotational potential is approximated as a sinusoidal curve and a typical estimate of the frequency yields

$$\omega_0 \approx \sqrt{\frac{2E_B}{2I}} \approx 10^{12} \ \mathrm{s}^{-1} . \tag{14.11}$$

The attempt frequency could be different in each well depending on energy differences in barriers. We will suppose that these differences are not significant compared to the larger variations of the exponential factor.

Returning to the determination of population changes in the wells, we may use the hopping rates to determine the equilibrium populations of rotor orientations and the response to being pushed out of equilibrium by external fields. Using the two-well potential of Fig. 14.14, we must first calculate the equilibrium populations in each well. There may be an asymmetry in the minimum energy wells, S, that may be due to any number of physical phenomena. For instance, as was mentioned in the two-dimensional rotor case, the wells may be non-degenerate because of interactions with the mounting surface. In the three-dimensional case, crystal and molecular symmetries may impose an asymmetry. In addition, an electric field interacting with the dipole moment will provide a directional preference. When it comes to detecting rotor behavior, the source of S will be important, but for this explanation of rotor behavior, it is just a parameter of the potential.

When the system reaches equilibrium, the populations in each well are no longer changing, so from equation (14.9),

$$\frac{dN_1}{dt} = 0 = \overline{N}_2 \Gamma_{21} - \overline{N}_1 \Gamma_{12} = \overline{N}_2 \omega_0 \exp\left(-\frac{(E_B + S)}{kT}\right) - \overline{N}_1 \omega_0 \exp\left(-\frac{E_B}{kT}\right),$$

(14.12)

where the overbar indicates equilibrium populations. Using the fact that the total number of rotors remains constant,

$$N_1 + N_2 = N,$$

(14.13)

where N is the total number of rotors, we find

$$\overline{N}_1 = \frac{N}{e^{+\frac{S}{kT}} + 1} \quad \text{and} \quad \overline{N}_2 = \frac{N}{e^{-\frac{S}{kT}} + 1}.$$

(14.14)

In any measurements of dynamics, we will be interested in the net difference in populations between wells:

$$\bar{N}_2 - \bar{N}_1 = N \tanh\left(\frac{S}{2kT}\right).$$

(14.15)

Equation (14.15) shows that if the asymmetry is zero, rotors will equally populate both wells, but for cases where one well is lower, the rotors will preferentially point in the direction associated with that well. Notice that as the difference in well energies gets very large, the difference in populations saturates so that all the rotors are in the lower well. Also, as the temperature approaches zero, the rotors will all populate the lower well, but as the temperature becomes very high, the wells tend toward equal populations.

The next step is the calculation of the dynamics when the rotor is forced out of equilibrium. Namely, what is the rate to reestablish a new equilibrium? We know from equation (14.9),

$$\frac{d(N_2 - N_1)}{dt} = 2[N_1 \Gamma_{12} - N_2 \Gamma_{21}] = 2\left[N_1 \omega_0 e^{-\frac{E_B}{kT}} - N_2 \omega_0 e^{-\frac{(E_B + S)}{kT}}\right].$$

(14.16)

Equation (14.16) may be recast into a relaxation equation [44] of the form,

$$\tau \frac{\mathrm{d}A(t)}{\mathrm{d}t} = A_{eq} - A(t) \,. \tag{14.17}$$

$A(t)$ represents any parameter which at some time, t, is displaced from its equilibrium value A_{eq}. Combining equations (14.13) and (14.16) yields

$$\tau \frac{\mathrm{d}(N_2 - N_1)}{\mathrm{d}t} = N \tanh\left(\frac{S}{2kT}\right) - (N_2 - N_1) \,, \tag{14.18}$$

where relaxation back to equilibrium occurs in a characteristic time,

$$\tau \equiv \frac{1}{\omega_0 e^{-\frac{E_B}{kT}} + \omega_0 e^{-\frac{(E_B + S)}{kT}}} = \frac{1}{\Gamma_{12} + \Gamma_{21}} \,, \tag{14.19}$$

or at a rate, $1/\tau$, equal to the sum of the hopping rates in the system. Since the rates contain exponentials, a small difference in well energies corresponds to orders of magnitude difference in hopping rates, with the smallest hopping barrier dominating the behavior. Thus, even though the total relaxation rate of the rotor depends on hopping in and out of both wells, the most significant parameter is often the smaller barrier energy, thus the nomenclature of E_B for that barrier in Fig. 14.14.

The dynamics derived from the two-minimum potential case are instructive toward treatment of systems with an increased number of stable rotor positions. The same general approach used so far may be used to solve the relaxation dynamics for **1** where there are three stable positions that may or may not be degenerate in energy [45]. However, notice that for a pair of potential wells, there appeared one characteristic relaxation time τ involving the rates of hopping in and out of the two wells. In general, the dynamics will involve a sum of exponential relaxations. Group theory techniques can be used to show that the number of distinct relaxation times will be reduced when the potential has symmetries. For example, Willams and Cook [45] show that a four well potential with C_2 symmetry has one relaxation time, but when the symmetry is reduced to D_{2h}, there are two relaxation times.

14.4 Detection of Rotation by Dielectric Spectroscopy

A capacitor has a capacitance C defined by $C = Q/V$ where Q is the charge stored on the capacitor plates and V is the applied voltage. If we place moving dipoles into the capacitor as a dielectric, the dipoles will respond to the electric field, and we can use this response as a probe of the rotor dynamics. This method is known as dielectric spectroscopy. An introduction to this measurement technique will be presented for the two-well case again, followed by comments on higher number well systems. In addition to measuring barrier to rotation and attempt frequency, dielectric spectroscopy gives access to well asymmetries which can provide a measure of rotor-rotor interactions. In special circumstances, the weak a.c. electric field used for dielectric spectroscopy yields unidirectional motion in two-well rotors similar to the driven rotation of Sect. 14.2 [28,29]. A brief explanation of this phenomenon will be provided.

We first calculate the equilibrium polarization, P_{eq}, of a given rotor sample which is the average dipole moment per unit volume. Let's assume the rotational potential is that of Fig. 14.14. The two wells may differ in energy due to an intrinsic asymmetry s and a small electric field, E. Then, the difference in well energies is $S = s + 2p_0E$. We have already calculated the difference in population caused by a disturbance in equation (14.15) and the overall dipole moment of the sample per unit volume at equilibrium is

$$P_{eq} = \frac{p_0}{V_T}\left(\bar{N}_2 - \bar{N}_1\right) = \frac{p_0}{V_T}N\tanh\left(\frac{s + 2p_0E}{2kT}\right), \qquad (14.20)$$

where V_T is the total volume of the sample. As was mentioned, the intrinsic asymmetry, s, may be caused by the mounting surface or crystal symmetries. The applied electric field, however, will be a small perturbation such that $p_0E \ll kT$. A Taylor expansion around s, keeping terms linear in p_0E, leads to

$$P_{eq} \approx \frac{p_0}{V_T}N\tanh\left(\frac{s}{2kT}\right) + \frac{p_0^2EN}{V_TkT}\text{sech}^2\left(\frac{s}{2kT}\right). \qquad (14.21)$$

The first term is the spontaneous polarization due to the intrinsic well asymmetry. The second term is the field dependent polarization. The sech factor is suppression to the rotor polarization since the well difference s will freeze some proportion of rotors into the lower well, not allowing them to respond to the applied field. We want to know the polarization response, $P(t)$, to an applied a.c. electric field. Only the field dependent term in equation (14.21) will contribute. We already examined this problem with equation (14.18). If the a.c. field has the form $E = E_0e^{i\omega t}$, we may assume $P(t)$ has the form $P_0e^{i\omega t}$ and solve equation (14.18). The solution is

$$P_0 = \frac{p_0^2E_0N}{V_TkT}\text{sech}^2\left(\frac{s}{2kT}\right)\frac{1}{1 + i\omega\tau} = P_s\frac{1}{1 + i\omega\tau}. \qquad (14.22)$$

where the relaxation time τ was defined by equation (14.19). At low frequencies ($\omega\tau \ll 1$) this formula is simply the field dependent part of the equilibrium polarization in equation (14.21).

The polarization P_0 is a complex number. In order to understand this, we should think about the change in charge on the capacitor with time, or the current I_C . For a lossless capacitor $Q(t) = CV = CV_0e^{i\omega t}$,

$$I_C = \frac{dQ}{dt} = C\frac{dV}{dt} = i\omega CV_0\exp(i\omega t) = \omega CV_0\exp\left[i\left(\omega t + \frac{\pi}{2}\right)\right]. \qquad (14.23)$$

There is a current due to the charging of the capacitor that is 90° out of phase with the applied voltage. The current I_C represents energy flow that is stored in the capacitor and is completely recoverable from the system. In addition to I_C, there may be current I_L that looks like charge flowing across the capacitor plates through a resistive element leading to energy loss and this is known as a loss current. If I_L is due to conducting charges moving between the plates, $I_L = V/R$, where R is the resistance of the transport mechanism. The loss current may be due to other effects, such as rotor motion, and not proportional to $1/R$, but it will be proportional to V. Thus, the capacitive and loss currents may be measured by detecting electrical signal that is out of phase and in phase with the applied voltage respectively [46].

These results are usually expressed through a complex dielectric permittivity ε^*, a material property. Recall that the capacitance of a parallel plate capacitor in vacuum is $C_0 = A\varepsilon_0/d$, where A is the area of the capacitor plate, d is the distance between them, and ε_0 is the permittivity of free space. If a dielectric with permittivity ε is inserted between the plates, $C = C_0\varepsilon/\varepsilon_0$. The total current amplitude I_0 is,

$$I_0 = I_C + I_L = \left(i\omega C + \frac{1}{R}\right) V_0 = i\omega C_0 \frac{\varepsilon^*}{\varepsilon_0} V_0 = (i\omega\varepsilon' + \omega\varepsilon'') \frac{C_0}{\varepsilon_0} V_0 , \quad (14.24)$$

where

$$\varepsilon^* = \varepsilon' - i\varepsilon'' \quad (14.25)$$

is the complex dielectric permittivity [46]. Comparing equation (14.25) to equation (14.24) we see that the real component represents current stored capacitively while the imaginary part corresponds to energy lost through a resistive mechanism.

A standard way to express a measurement of the complex permittivity is through the loss tangent, $\tan(\delta)$, which is the ratio of the imaginary to real part of the complex permittivity. Alternatively, the loss tangent is the ratio of the real to imaginary part of the current. The reason it is referred to as an angle is that we may represent the current in the complex plane with δ as the angle between the current vector and the imaginary axis. If the system is just a resistor, then the imaginary part of the current is zero, and the loss tangent goes to infinity. If the system is a perfect capacitor, the real part goes to zero, there is no loss, and the loss tangent is zero.

Finally, in order to relate all of this back to our polarization, we have to remember that the total polarization amplitude P_T is related to the electric field amplitude by

$$P_T = (\varepsilon^* - \varepsilon_0) E_0 . \quad (14.26)$$

The total polarization $P_T = P_\infty + P_0$ is composed of two parts, the high frequency electronic polarization P_∞ and the slower rotor polarization P_0 from equation (14.22). Solving for ε^* we find

$$\varepsilon^* = \frac{P_T}{E_0} + \varepsilon_0 = \frac{P_\infty}{E_0} + \frac{P_0}{E_0} + \varepsilon_0 = \frac{P_\infty}{E_0} + \frac{P_s}{E_0} \frac{1}{1 + i\omega\tau} + \varepsilon_0 . \quad (14.27)$$

It is conventional now to define the high frequency dielectric constant

$$\varepsilon_\infty \equiv \frac{P_\infty}{E_0} + \varepsilon_0 \tag{14.28}$$

and the static (low frequency) dielectric constant

$$\varepsilon_s \equiv \frac{P_\infty}{E_0} + \frac{P_s}{E_0} + \varepsilon_0 = \varepsilon_\infty + \frac{P_s}{E_0}. \tag{14.29}$$

With these definition we find

$$\varepsilon^* = \varepsilon_\infty + \frac{\varepsilon_s - \varepsilon_\infty}{1 + i\omega\tau}, \tag{14.30}$$

and

$$\tan(\delta) = \frac{(\varepsilon_s - \varepsilon_\infty)\,\omega\tau}{\varepsilon_\infty \left(\frac{\varepsilon_s}{\varepsilon_\infty} + \omega^2\tau^2\right)}, \tag{14.31}$$

which is referred to as the Debye response [47]. We associate $(\varepsilon_s - \varepsilon_\infty)$ with the capacitance C_R due to the rotor motion, while ε_∞ is associated with the capacitance C_0 that would be measured if the rotors could not move. The capacitance due to the rotors is normally much smaller than C_0 so that $\varepsilon_s/\varepsilon_\infty \approx 1$ and

$$\tan(\delta) = \frac{C_R}{C_0}\frac{\omega\tau}{(1 + \omega^2\tau^2)}. \tag{14.32}$$

The last step is to express C_R in terms of the known value of the polarization from equation (14.22). However, before we do, there are three additional factors that need to be addressed. The first is consideration of the geometry of the capacitor. If the rotors are in a parallel plate capacitor,

$$C_R = \frac{(\varepsilon_s - \varepsilon_\infty)\,A}{d} = \frac{A}{d}\frac{P_s}{E_0}, \tag{14.33}$$

where A and d are the area and gap, but for other geometries there will be a numerical factor. The second issue is the orientation of the dipolar sample. In most cases the sample is polycrystalline and, rather than the electric field interacting with the dipole moment as p_0E, it must be replaced by $p_0E\cos\theta$ and integrated over the appropriate angles. This is addressed in many sources [1, 46, 48] so only the result will be quoted here which is to multiply the resulting polarization by $1/2$ in two dimensions and $1/3$ in three. Finally, if the rotors are dilute, we must account for the dielectric nature of the environment surrounding the rotors. This is done by using the Clausius-Mossotti method where the dipole is assumed to be in a spherical cavity surrounded by an infinite medium with dielectric constant ε_∞ [48]. This induces an additional field in the cavity and multiplies the preceding polarization by $(\varepsilon_\infty + 2)/3$, where ε_∞ is estimated to be approximately 2 for a typical organic substance. Now one can express $\tan(\delta)$ as

$$\tan(\delta) = \frac{K}{\varepsilon_\infty} \frac{p_0^2 N}{V_T kT} \mathrm{sech}^2 \left(\frac{\mathrm{s}}{2kT}\right) \frac{\omega\tau}{(1 + \omega^2\tau^2)}, \tag{14.34}$$

where K represents the numerical prefactor due to angular integrations, internal field corrections, and capacitor geometry. The important aspect of equation (14.34) is the behavior with respect to temperature. For a given applied field frequency ω, at low temperatures, the relaxation time, $\tau = \omega_0^{-1}\exp(E_B/kT)$, will be large, but the denominator of equation (14.34) will get large at a faster rate so that the loss tangent approaches zero. This corresponds to the rotors being unable to hop at the lower temperature in response to the quickly oscillating field and therefore having no loss of energy. On the other hand, at high temperatures, the relaxation time becomes very small, so that the numerator approaches zero while the denominator approaches one. Therefore, the loss tangent approaches zero once again. This time, the applied field is oscillating at a much slower rate than the rotors are hopping, thus the rotor are responding immediately to the applied field with no loss of energy contributing to a lag. However, when the temperature approaches the point where $\omega\tau = 1$, the loss tangent will peak. Thus, by measuring the in-phase and out of phase components of the current with voltage, calculating the loss tangent, and observing where it peaks with respect to temperature, we obtain a measure of the relaxation time as the reciprocal of the applied electric field frequency. One may also hold the temperature constant and observe the frequency where the loss tangent peaks for the same results, but the temperature experiment allows access to a larger range of relaxation times.

An experiment generally entails scanning temperature and measuring $\tan(\delta)$ for a number of fixed frequencies. Figure 14.17 shows what the data might look like if they follow the Debye response for measurements at the frequencies shown and there is no asymmetry in the potential wells so that the sech factor in equation (14.34) is one. There are several features to note about the data. To begin with, as the applied frequency increases from 100 Hz to 10 kHz, the location of the loss peak moves up in temperature. This corresponds to the molecular rotors requiring more thermal energy to keep up with the measurement frequency. Also, as the peaks move up in temperature and frequency, the peak magnitude decreases. This is due to the $1/kT$ factor in equation (14.34), also called the Curie factor. It accounts for the growing depolarization of the rotor due to thermal agitation as the temperature increases.

Whether observing sharp or broad peaks, a dominant energy barrier is often present. From the dispersion of peaks with frequency, an Ahrrenius plot may be made. Since a peak occurs when $\omega\tau = 1$, we may write for the peak temperatures,

$$\omega = 1/\tau = \omega_0 \exp\left(\frac{-E_B}{kT_{peak}}\right) \tag{14.35}$$

$$\ln\omega = \ln\omega_0 - \frac{E_B}{k}\frac{1}{T_{peak}}. \tag{14.36}$$

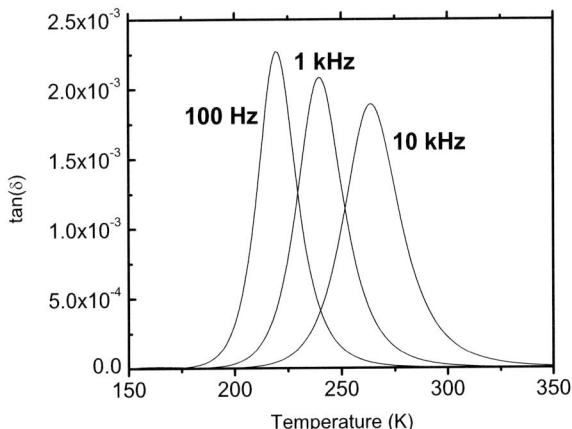

Fig. 14.17. Example of theoretical loss tangent versus temperature with $s = 0$, and a barrier of 12 kcal/mol

A plot of $\ln \omega$ versus $1/T_{peak}$ where T_{peak} is the temperature where the loss peak occurs for a given ω will give a straight line with a slope of E_B and a y-intercept of $\ln \omega_0$. Using equation (14.34) the barriers and attempt frequencies for the dipolar molecular rotors may be extracted from experimental data. See [8] for an Ahrrenius plot for **4** obtained by this method.

The narrow, singular peaks found for three-dimensional rotors also allow characterization of the asymmetry in the two wells also. In Fig. 14.17, where no asymmetry is present, the magnitude of the peaks decreases with increasing frequency due to the Curie factor. However, the observed peaks in Fig. 14.15 increase with increasing frequency. This is due to the sech factor, or the suppression due to rotors freezing into the lower well. The greater the asymmetry, the larger the increase in peak magnitude with increasing frequency or temperature. Thus, by comparing the relative magnitudes of loss at different frequencies, the asymmetry may be ascertained. For **4**, this asymmetry was 1.5 kcal/mol. X-ray studies and calculation showed this to be due to steric hindrance.

Another interesting phenomenon arising with the two-well rotors is net unidirectional motion under the influence of the a.c. electric field. This is significant in the effort to extract useful work from the rotors upon application of an electric field and perhaps for studying electrooptic effects. We may calculate the efficiency of this rotation. Let us imagine the potential of Fig. 14.14 with degenerate rotational potential wells, $S = 0$, and an electric field pointing in an arbitrary angle θ to the wells. The same master equation approach as was applied in Sect. 14.3 may be used for this situation [28, 29]. Makhnovskii et al. performed this calculation for field switching dynamics that may be considered a Markov process, but the result is approximately valid for the case of periodic driving as well. The average rate of rotation J is found to be

$$J = \frac{\omega_0}{2} \exp\left(\frac{-E_B + v}{kT}\right) \sinh\left(\frac{u}{kT}\right)$$

$$\times \frac{1 - \exp\left(\frac{-2v}{kT}\right)}{1 + \left[\frac{\omega_0}{\omega} \exp\left(\frac{-E_B + v}{kT}\right) \cosh\left(\frac{u}{kT}\right)\left(1 + \exp\left(\frac{-2v}{kT}\right)\right)\right]}, \quad (14.37)$$

where $u = p_0 E \cos\theta$ is the interaction of the dipole with the electric field in the potential wells and $v = p_0 E \sin\theta$ is the change in the barrier due to the electric field with p_0 being the dipole moment of the dipolar group in the rotor and E the strength of the applied electric field. θ is the angle between a line connecting the two stable orientations of the rotors and the direction of the electric field. Some typical numbers for the parameters to use for the three-dimensional, symmetric **5** are: $E_B = 14$ kcal/mol, $T = 300$ K, $p = 3$ Debye, $\theta = 45°$, $\omega = 2\pi \times 1$ kHz, and $E = 5 \bullet 10^5$ V/m. This yields a net forward rotation every 69 minutes. For $\omega \ll 1/\tau$, or when the frequency of the applied field is much less than the relaxation rate of the rotor, the rotation rate increases linearly with applied field frequency. When, ω approaches $1/\tau$, the rotation rate saturates. So, at 10 kHz, there is a rotation every 28 minutes and at 1 GHz it takes 23 minutes for a single forward rotation. When the applied field frequency is above the attempt frequency, this hopping rate approach breaks down because the rotor does not have time to equilibrate when the field is switched and equation (14.37) is no longer valid [28]. However, for parameters that are easily accessible experimentally, equation (14.37) should be applicable.

14.5 Summary

Dipolar molecular rotors represent a novel environment for investigation of some interesting physical phenomena. A few of these discussed were the investigation of molecular dynamics and the effects of interactions among collections of rotating molecular dipoles. In addition, the synthetic control over the creation of these molecular systems will allow customization in future work, such as decreasing the rotational barrier and increasing dipole strengths to further interactions at higher thermal temperatures. Some of the synthetic work discussed here has been the creation of disordered, two-dimensional arrays and evidence of their rotation, the efforts toward ordered arrays in this regime, and work in ordered, three-dimensional collections. Each of these provides an interesting framework for observing various parameters of the molecular rotor physics, such as relaxation in varying potential minima configurations. The characteristics of the molecular rotors' environment play a crucial role in determining the description necessary to understand the rotational behavior. Thus far, we have been in a situation apparently described by thermally activated hopping. An introduction to this behavior has been discussed and its detection through dielectric spectroscopy has also been evaluated. It should be mentioned that there are other techniques for measuring the rates for rotary molecular groups, such as ^{2}H NMR and CP-MAS NMR [49]. The possibilities for dipolar molecular rotors are vast. From excited rotary modes, to unidirectional motors, to novel phase behavior, this is a rich area for future investigation.

References

1. G.S. Kottas, L.I. Clarke, D. Horinek, and J. Michl (2005). *Chem. Rev.*, **105**, p. 1281.
2. T.R. Kelly, H. De Silva, and R.A. Silva (1999). *Nature*, **401**, p. 150.
3. N. Koumura, R.W.J. Zijlstra, R.A. van Delden, N. Harada, and B.L. Feringa (1999). *Nature*, **401**, p. 152.
4. D.A. Leigh, J.K.Y. Wong, F. Dehez, and F. Zerbetto (2003). *Nature*, **424**, p. 174.
5. J.V. Hernández, E.R. Kay, and D.A. Leigh (2004). *Science*, **306**, p. 1532.
6. S.P. Fletcher, F. Dumur, M.M. Pollard, and B.L. Feringa (2005). *Science* **310**, p. 80.
7. L.I. Clarke, D. Horinek, G.S. Kottas, N. Varaska, T.F. Magnera, T.P. Hinderer, R.D. Horansky, J. Michl, and J.C. Price (2002). *Nanotechnology*, **13**, p. 533.
8. R.D. Horansky, L.I. Clarke, T.-A.V. Khuong, P.D. Jarowski, M.A. Garcia-Garibay, and J.C. Price (2005). *Phys. Rev. B*, **72**, p. 014302.
9. X. Zheng, M.E. Mulcahy, D. Horinek, F. Galeotti, T.F. Magnera, and J. Michl, (2004). *J. Amer. Chem. Soc.*, **126**, p. 4540.
10. V.M. Rozenbaum (1996). *Phys. Rev. B*, **53**, p. 6240.
11. J. de Jonge, M. Ratner, and R.S.S.W. de Leeuw (2004). *J. Phys. Chem. B*, **108**, p. 2666.
12. V.M. Rozenbaum, V.M. Ogenko, and A.A. Chuiko (1991). *Sov. Phys. Usp.*, **34**, p. 883.
13. J. Vacek and J. Michl (2001). *Proc. Natl. Acad. Sci. USA*, **98**, p. 5481.
14. D. Horinek and J. Michl (2003). *J. Amer. Chem. Soc.*, **125**, p. 11900.
15. U. Burkert and N.L. Allinger (1982). *ACS Monograph, No. 177: Molecular Mechanics* (ACS, Washington D.C.).
16. G.S. Kottas (2004). Ph.D. Dissertation, University of Colorado at Boulder.
17. D. Horinek and J. Michl, *Proc. Natl. Acad. Sci.* USA, in press.
18. D.A. Bonnell (1993). Ed. *Scanning Tunneling Microscopy and Spectroscopy – Theory, Techniques and Applications* (VCH Publishers, New York).
19. A. Sakai (2000). In: *Advances in Materials Reasearch: Advances in Scanning Probe Microscopy* edited by T. Sakurai and Y. Watanabe (Springer-Verlag, New York). p. 143.
20. Y. Hasegawa, J.F. Jia, T. Sakurai, Z.Q. Li, K. Ohno, and Y. Kawazoe (2000). In: *Advances in Materials Reasearch: Advances in Scanning Probe Microscopy* edited by T. Sakurai and Y. Watanabe (Springer-Verlag, New York). p. 167.
21. J.A. Appelbaum and D.R. Hamann (1972). *Phys. Rev. B,* **6**, p. 1122.
22. L. Wang and J. Hermans (1995). *J. Phys. Chem.*, **99**, p. 12001.
23. E.G. d'Agliano, P. Kumar, W. Schaich, and H. Suhl (1975). *Phys. Rev. B*, **11**, p. 2122.

24. K. Schönhammer and O. Gunnarson (1980). *Phys. Rev. B*, **22**, p. 1629.
25. Y. Li and G. Wahnström (1992). *Phys. Rev. Lett.*, **68**, p. 3444.
26. R.D. Astumian (1996). *J. Phys. Chem.*, **100**, p. 19075.
27. P. Reimann (2002). *Phys. Rep.*, **57**, p. 361.
28. Y.A. Makhnovskii, V.M. Rozenbaum, D.-Y. Yang, S.H. Lin, and T.Y. Tsong (2004). *Phys. Rev. E*, **69**, p. 021102.
29. R.D. Astumian (2005). *Proc. Natl. Acad. Sci. USA*, **102**, p. 1843.
30. J.V. José and E.J. Saletan (1998). *Classical dynamics: a contemporary approach* (Cambridge University Press, New York). p. 382 ff.
31. B.V. Chirikov (1979). *Phys. Rep.*, **52**, p. 265.
32. G. Roberts (1990). Ed. *Langmuir-Blodgett Films* (Plenum Press, New York).
33. F. MacRitchie (1990). *Chemistry at Interfaces* (Academic Press, San Diego).
34. T.F. Magnera and J. Michl (2002). *Proc. Nat. Acad. Sci. USA*, **99**, p. 4788.
35. N. Varaksa, L. Pospíšil, T.F. Magnera, and J. Michl (2002). *Proc. Nat. Acad. Sci. USA*, **99**, p. 5012.
36. Z. Dominguez, H. Dang, M.J. Strouse, and M.A. Garcia-Garibay (2002). *J. Amer. Chem. Soc.*, **124**, p. 7719.
37. Z. Dominguez, T.-A. V. Khuong, H. Dang, C.N. Sanrame, J.E. Nuñez, and M.A. Garcia-Garibay (2003). *J. Amer. Chem. Soc.*, **125**, p. 8827.
38. R.D. Horansky, L.I. Clarke, E.B. Winston, S. Karlen, M.A. Garcia-Garibay, and J.C. Price (unpublished).
39. W. Press (1981). *Single-Particle Rotations in Molecular Crystals* (Springer-Verlag, Berlin).
40. F. Reif (1965). *Fundamentals of Statistical and Thermal Physics* (McGraw-Hill, New York).
41. N.W. McLachlan (1947). *Theory and Application of the Mathieu Function* (Clarendon, Oxford).
42. A.J. Legget, in (1986). *Directions in Condensed Matter Physics*, edited by G. Grinstein and G. Mazenko (World Scientific, Singapore). p. 187.
43. K.A. Dill and S. Bromberg (2002). *Molecular Driving Forces* (Garland Science, New York).
44. N.G. McCrum, B.E. Read, and G. Williams (1967). *Anelastic and Dielectric Effects in Polymer Solids* (Dover Publications Inc., New York).
45. G. Williams and M. Cook (1971). *Trans. Faraday Soc.*, **67**, p. 990.
46. A.R. Von Hippel (1966). *Dielectrics and Waves* (M.I.T. Press, Cambridge).
47. F. Kremers and A. Schönhals (2002). *Broadband Dielectric Spectroscopy* (Springer-Verlag, Berlin).
48. J.D. Jackson (1975). *Classical Electrodynamics* (Wiley, New York).
49. M.J. Duer (2004). *Introduction to Solid State NMR Spectroscopy* (Oxford, Malden).

15

Using DNA to Power the Nanoworld

B. Yurke

Bell Laboratories, 600 Mountain Ave., Murray Hill, NJ 07974
yurke@lucent.com

The simplicity of the rules by which DNA strands interact has allowed the construction, out of DNA, of complex nanodevices that can execute motion. These devices possess DNA-based molecular motors that are powered by DNA strands that serve as fuel. Among the variety of such devices constructed are ones that can direct chemical synthesis, that can control the properties of bulk materials, and that can control the binding of chemical species to protein molecules. This suggests that DNA-based nanodevices powered by DNA-based molecular motors may find application in fields such as chemistry, materials science, and medicine. Here we describe the principles by which the motors that power these devices work and survey the range of devices that have been constructed.

15.1 Introduction

A large number of different kinds of molecular motors, molecules that convert chemical energy into mechanical work, can be found within living cells. These motors carry out a variety of tasks, only some of which are listed here. They drive muscle contraction, pseudopod extension, and the beating of cilia and flagella, allowing cells and organisms to move. They also drive DNA replication, the transcription of DNA, and convert energy obtained from food or sunlight into the fuels that power the cell's molecular machinery. Clearly, biological organisms have found molecular motors to be extremely useful. It is likely that we also will find molecular motors very useful, once we have mastered chemistry and nanotechnology sufficiently well.

There has been considerable work on the construction of synthetic molecular motors [1] such as light-driven molecular motors [2–5]. However, the ability to controllably move things on a nanoscale using molecular devices is best developed in DNA-based nanotechnology. Quite a number of DNA-based machines have been constructed [6] that can be cycled through a set of states

B. Yurke: *Using DNA to Power the Nanoworld*, Lect. Notes Phys. **711**, 331–347 (2007)
DOI 10.1007/3-540-49522-3_15   © Springer-Verlag Berlin Heidelberg 2007

driven by DNA strands that serve as fuel [7–19]. The function and application of these machines is the focus of this chapter.

15.2 Structural Properties of DNA

In its native form in biological organisms DNA (deoxyribonucleic acid) consists of two linear molecules that twist around each other to form a double helix. This double-stranded DNA is often referred to as duplex DNA. Each single strand of this double helix is a linear polymer consisting of monomer units which have chemical units called bases attached to them. The monomer units are linked together by a phosphodiester bridge, the "phosphate" units forming the DNA's backbone. There are four kinds of bases labeled A, G, C, and T for adenine, guanine, cytosine, and thymine, respectively. A DNA strand is uniquely specified by listing the sequence with which the bases occur. There is a slight complication in that DNA is a directed polymer. It is thus necessary to specify which end is which. By convention one end is labeled the 5′ end and the other is labeled the 3′ end. When listing DNA sequences it is conventional to begin with the 5′ end and conclude with the 3′ end as was done for the base sequences A, B, and C in Fig. 15.1(a). In the double helix the two strands of DNA align with each other in an antiparallel way so that the 5′ end of one strand matches with the 3′ end of the other strand. Each base A of one DNA strand preferentially binds with a base T from the other strand, while a base G of one strand of DNA preferentially binds with a base C of the other strand. The two strands of a double helix are said to be complementary if every A of one strand is matched with a T of the other strand and every G of one strand is matched with a C of the other strand. Double-stranded DNA in biological organisms is of this form. The paired bases form the rungs of the twisted ladder in the usual iconic representations of duplex DNA. The spacing between bases in double-stranded DNA is 0.34 nm and there are approximately 10 base pairs for one full turn of the double helix. The double helix has a width of 2 nm.

The double helix is maximally stable if there is complementary base paring throughout. If the mismatch density is high enough, the double helix will not be stable and the structure will rapidly fall apart into two single strands. Conversely, two single strands of DNA that happen to have complementary base sequences, when in solution, will rapidly combine to form double-stranded DNA. This process is called hybridization. The binding strength between two DNA strands depends on the base sequences of the strands and can be calculated with reasonable precision. This allows for the design of sets of DNA strands which, when placed in solution, will combine with each other in predetermined ways to self-assemble into complex nanostructures. The linearity of the molecule and the predictability of strand-strand interactions account, to a large extent, for why DNA has proven to be a particularly convenient medium to work with in the construction of nanostructures and nanodevices.

A   5' TGCCTTGTAAGAGCGACCATCAACCTGGAATGCTTCGGAT 3'

B   5' GGTCGCTCTTACAAGGCACTGGTAACAATCACGGTCTATGCG 3'   **(a)**

C   5' GGAGTCCTACTGTCTGAACTAACGATCCGAAGCATTCCAGGT 3'

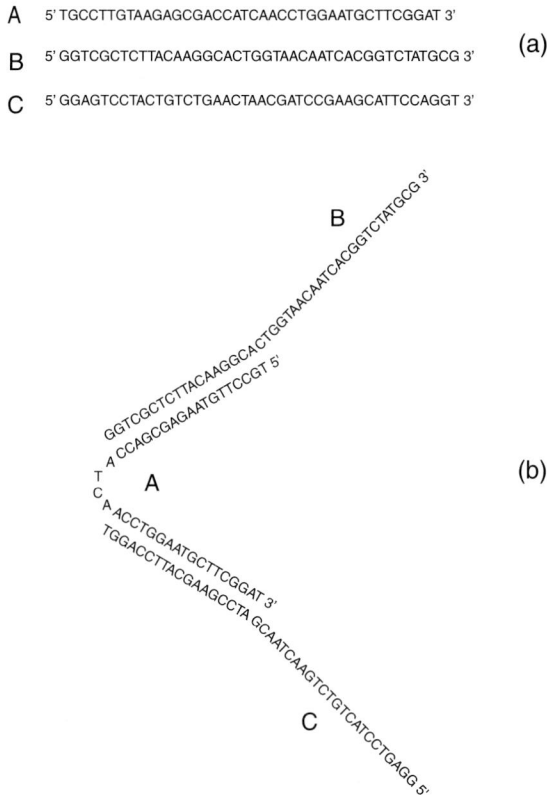

(b)

Fig. 15.1. Self-assembly of a simple nanodevice. The three strands of DNA, A, B, and C, depicted in (**a**), have complementary regions. In solution these strands will diffuse and thereby randomly contact each other. When complementary regions between DNA strands contact each other these will zip together to form double-stranded DNA. The base sequences of A, B, and C were designed to form the structure shown in (**b**). Double-stranded DNA is some 50 times stiffer than single-stranded DNA. Hence, the device in (b) has been shown with bends in some of the single-stranded regions

It also helps that foundries are available that will manufacture DNA strands of arbitrary base sequence up to several hundred nucleotides in length.

Figure 15.1 illustrates the self-assembly of a simple DNA-based nanodevice [7]. In (a) the base sequences for strands A, B, and C are given in standard form, starting with the 5′ end. Both B and C have regions that are complementary to portions of A. When in an aqueous solution these strands randomly migrate through the fluid volume due to the random motion of the molecules making up the fluid. In an aqueous solution the phosphate backbone becomes negatively charged because each phosphate unit donates a proton (a positively charged hydrogen ion) into solution. Compounds that donate protons to solution when in water are referred to as acids. Hence, one has the name deoxyribonucleic *acid* for DNA. Because DNA strands are negatively charged in solution, they tend to electrostatically repel each other. However, salt in solution reduces this electrostatic force and, if the fluid has sufficient salt, occasionally the DNA strands contact each other. When complementary regions touch, base pairs will form and the complementary regions of the two strands will zip together to form a double helix. The base sequences for the strands A, B, and C of Fig. 15.1(a) have been designed so that they will bind together to form the structure shown in (b). This structure consists of two-double stranded arms connected by a single-stranded region which is four bases in length. Each arm possesses a single-stranded extension. Single-stranded DNA in solution is floppy. Hence, the structure has been depicted as folding back onto itself. This structure is rather simple, but it will be used in the next section to describe how DNA-based motors work. Much more complex nanostructures have been made by self-assembly through the design of suitable sets of DNA strands [20]. These include DNA sheets [21–25] and tubes [23–26].

The rigidity of nanostructures constructed of DNA is determined by the stiffness of DNA. Long strands of DNA behave as floppy strings. In solution, such strands writhe due to the Brownian motion of the molecules making up the solvent. At any instant of time the strand configuration is that of a random coil. However, since it costs energy to bend a DNA molecule, there is a limit to the degree of tightness of the random coil. This limit is determined by the thermal energy available in the Brownian motion. The length which characterizes the tightness of the random coiling is called the persistence length. This length is 50 nm for double-stranded DNA [27] and approximately 1 nm for single-stranded DNA [28]. That is, single-stranded DNA is 50 times more floppy than double-stranded DNA. Because it is difficult for the random motion of the solvent molecules to bend DNA over a distance that is shorter than the persistence length, double-stranded DNA shorter than 50 nm behaves more like a rigid rod than a floppy string. Double-stranded DNA is, thus, sufficiently stiff to allow one to build nanostructures with structural integrity.

15.2.1 Motorized DNA Tweezers

To describe how DNA-based motors powered by DNA-stand interactions operate, consider the device shown in Fig. 15.2 which was reported on in [7]. The base sequences for the strands making up the device and a simpler representation of the device have already been shown in Fig. 15.1(a). This device is constructed from three strands of DNA to form a two-armed structure. The arms are 18 base pairs long (about 6 nm) and consist of double-stranded DNA. Since this is much smaller than the persistence length of double-stranded DNA, the arms behave as rigid rods. The arms are linked together by single-stranded DNA four bases in length. Since the spacing between bases in single-stranded DNA is 0.43 nm, the length of the linker between the arms is greater than the

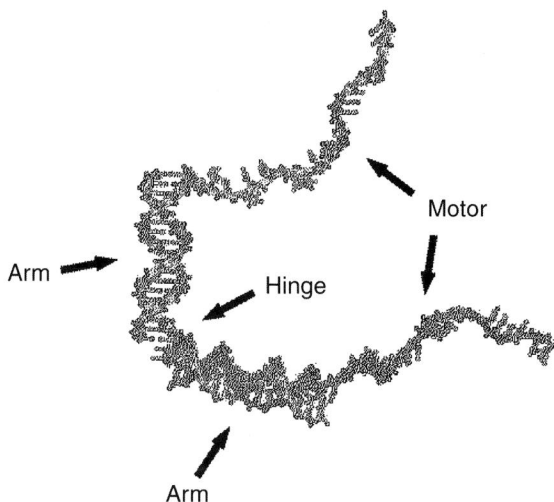

Fig. 15.2. DNA-motor driven tweezers. The arms of the tweezers consist of relatively rigid double-stranded DNA. The arms are connected together by a hinge consisting of single-stranded DNA four bases in length. The motor that is used to open and close these tweezers consists of single-stranded DNA that extends from the arms of the tweezers

persistence length of single-stranded DNA. Consequently, the linker function as a flexible hinge. The two arms can be brought together in a tweezers-like fashion. Since these tweezers are too small to be seen by the human eye and too small to be operated by the human hand, it was necessary to motorize them in order to open and to close them. The motor consists of two strands of single-stranded DNA that extend off of the tips of the arms of the tweezers. The two motor strands were 24 bases in length.

How the tweezers are closed is illustrated in Fig. 15.3. A DNA strand that has regions which are complementary to the single-stranded extensions of the tweezers' arms is introduced into the solution. This strand is labeled F for "fuel" in Fig. 15.3(a). When regions of this strand come into contact with complementary regions of the single-stranded extensions (overhangs) of the arms of the tweezers, double-stranded DNA begins to form, as shown in Fig. 15.3(b). By forming base pairs, the system is going downhill energetically and the tweezer arms are driven shut, as shown in Fig. 15.3(c).

As illustrated by the fact that DNA hybridization can pull the arms of the DNA tweezers together, DNA hybridization is able to develop a force. To address what the maximum force is that DNA hybridization can develop, consider the thought experiment depicted in Fig. 15.4. Here DNA-hybridization is used to lift a weight against the force of gravity. Clearly, there is a maximum size to the weight that can be lifted. The force exerted by this weight is called the stall force. Since work is force times distance, the stall force can be estimated by dividing the energy available to do work when a base pair is

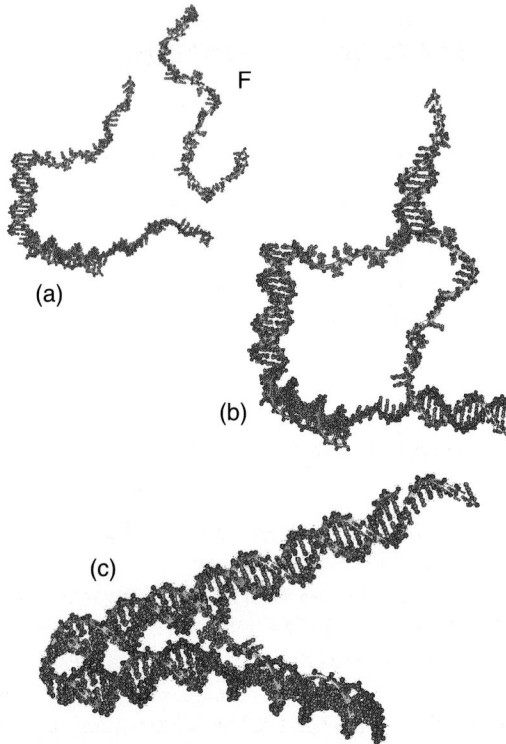

Fig. 15.3. Closing the DNA tweezers. The tweezers are closed by the introduction of a fuel strand F

formed by the distance the weight moves during the formation of a base pair. The energy released in the formation of a base pair depends on whether the base pair formed is between a G and a C or between an A and a T. It also depends on what the neighboring bases happen to be and on the temperature and salt concentration. Averaged over all such combinations the energy ΔG available to do work [29] is 1.8 kcal/mol or 1.25×10^{-20} J at 20°C in 1 molar salt. The distance the weight moves during the formation of a base pair is twice the spacing between neighboring bases on single-stranded DNA. The distance between bases on single-stranded DNA is 0.42 nm. Hence, the distance the weight moves is $\Delta x = 0.84$ nm. The stall force $F = \Delta G/\Delta x$ is thus 15 pN. This number is in line with the forces that have been measured by pulling double-stranded DNA apart [30–32]. The force generated by DNA hybridization is quite respectable when compared with biological molecular motors. For example, kinesin [33] has a stall force of 5 pN. RNA polymerase, one of the more powerful molecular motors to have been characterized [34], has a stall force of 30 pN.

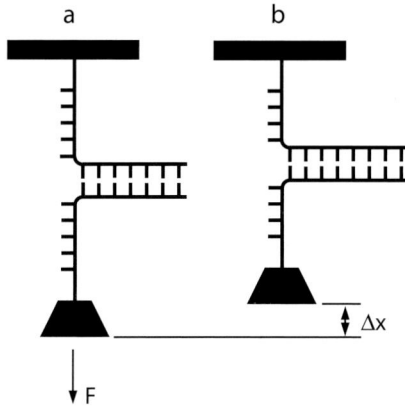

Fig. 15.4. A Gedanken experiment in which a weight is lifted by DNA hybridization

15.3 Reopening the Motorized DNA-based Tweezers

In order to describe how the closed tweezers are reopened, it is necessary to describe a mechanism by which one DNA strand can be removed from a double helix. This process is called strand displacement [35,36] and is depicted in Fig. 15.5. In Fig. 15.5(a) two DNA strands labeled A and B are shown which are hybridized together by virtue of the fact that B possesses a region in which its base sequence is complementary to that of A. The strand B is longer than A so that, when the two are hybridized, one end of B is still single-stranded DNA.

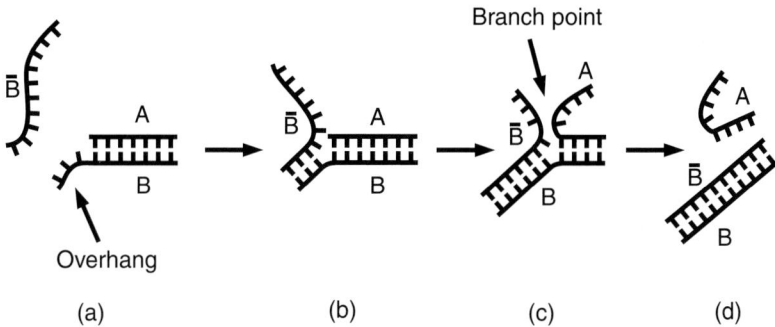

Fig. 15.5. Strand displacement via branch migration. The sequence shows how the strand B can be removed from A by the introduction of the complement $\bar{B}$ of B. In (**a**) A and B are held together via bonding between complementary base pairs. B possesses a single-stranded extention or "toehold." In (**b**) the complement of B binds to the toehold. A branch migration process is initiated. In (**c**) the branch point has moved toward the center of strand B. The branch point moves via a Brownian walk. In (**d**) the branch point has made a first passage to the far end of B. At this point A has been cleared of strand B

Such a single-stranded extension is called an overhang and it facilitates the initiation of the strand displacement process. In particular, as shown in (b), it serves as a site to which a strand of DNA complementary to B, denoted as $\bar{B}$, can bind. Because of its function as an initiation site for strand displacement, the overhang is often referred to as a toehold. To understand what happens next, it is important to realize that the binding between base pairs is relatively weak. In fact, it is sufficiently weak that the random Brownian motion of the molecules making up the solution cause base pairs to temporarily break apart. The disrupted base pairing is usually quickly reestablished. However, in the situation depicted in (b), if the pairing between the base pair of A and B at the toehold end of B is disrupted, it is possible for the base pairing to reform with the base of B now hybridized with its complement on $\bar{B}$. Strand A will now have one free base. The point at which the A and $\bar{B}$ strands meet is called a branch point, as indicated in (c). Because of the random breaking and making of base pairs at the branch point region, the branch point undergoes a random walk along B. The average time between steps [37, 38] is 10 to 100 μsec. At 10 μsec per step it takes on average 16 msec for the branch point to migrate to the far end of B. Once this happens, A is no longer attached to B and B and $\bar{B}$ are fully hybridized. The reverse reaction is also possible. A random fluctuation could cause the end of the double strand in (d) to partially unzip. This would allow A to reattach itself, and the branch point then could migrate back to the state depicted in (b). To go from (b) to (a) would cost energy, because it would require breaking the base pairs between $\bar{B}$ and B in the toehold region. This energy difference between state (a) and state (d) drives the chemical equilibrium toward (d). The toehold is, thus, crucially important. By making the toehold longer than a few bases, one can insure that the forward reaction where $\bar{B}$ displaces A is highly favored over the backward reaction where A displaces $\bar{B}$. This strand displacement process which clears A of B is essentially the same process that is used to clear the tweezers of the fuel strand F.

Figure 15.6 shows the full machine cycle of the motorized DNA tweezers. In addition to regions that are complementary to the single-stranded extensions of the open tweezer arms depicted in (a), the fuel strand F possesses eight extra bases at one end so that when F is fully hybridized with the tweezers there is a single-stranded portion of F that extends off of the end of one of the arms of the tweezers. This is shown in (b) of Fig. 15.6. This overhang functions as a toehold. The tweezers are opened by introducing the complement of the fuel strand. This strand is denoted as $\bar{F}$. The complement attaches itself to the toehold region of F as shown in (c). Branch migration is now initiated in which the tweezers and $\bar{F}$ compete for base pairs with the fuel strand. In (d) the branch point has moved roughly halfway along one of the arms of the tweezers. Eventually, the branch point makes a first passage to the far end of F. At this point the tweezers are released from F and restored to their open configuration (a). At this point waste product W is also produced which consists of the fuel strand fully hybridized with its complement. It should be

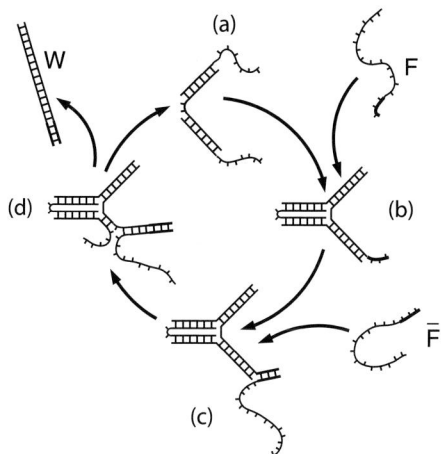

Fig. 15.6. The machine cycle for the DNA tweezers. (**a**) The tweezers in their open state. (**b**) Upon the addition of the fuel strand F the tweezers close. (**c**) The complement $\bar{F}$ of the fuel strand first attaches itself to the fuel strand toehold. (**d**) A branch migration process is initiated in which the complement of the fuel strand competes with the tweezers for base pairing with the fuel strand. Eventually, a first passage of the branch point to the far end of F occurs. At this point the tweezers are cleared of the fuel strand and have been returned to their open configuration (a). In addition, waste product W is created which consists of the fuel strand hybridized with its complement

noted that both the closing and the opening of the tweezers are energetically downhill processes, each resulting in a net increase in base pairs. Thus, steps in the reverse direction are disfavored. The motor that opens and closes the tweezers can be regarded as an engine that is driven by the free energy of hybridization of the fuel strand with its complement.

In the operation of this motor it would be inefficient to simultaneously supply the fuel strand and its complement, because these two strands can hybridize directly. For efficient operation this motor must be stepped through its states by the successive application of the fuel strands and the removal strands, leaving enough time between applications to allow for the reactions to go to completion. Motors for which the advancement through their states is controlled from the outside, such as through the successive application of DNA strands, will be referred to as stepped motors. This is in contrast with free-running motors which have internal timing mechanisms that allow them to autonomously advance through their sequence of states.

The DNA-hybridization motor that opens and closes the tweezers produces back-and-forth motion instead of rotary motion, and, as already stated, it is a stepped motor. In these respects, it is thus quite different in behavior from commonly encountered electrical and gasoline motors. However, a motor is any device that causes motion. The hybridization motor qualifies as a motor because it opens and closes tweezers. Perhaps, an extreme example of what qualifies as a motor is a rocket motor which often consists of a chamber without moving parts in which fuel is burned. Closer in function to the hybridization motors are linear electric motors which produce back-and-forth motion and stepper motors which are stepped through a sequence of states by successively activating various sets of electromagnets. In the case of stepper motors, the switching that activates the electromagnets is done externally to the motor. This process is analogous to the application of a sequence of DNA strands in the case of the hybridization motor.

15.4 A Three-State Machine

The DNA tweezers in their closed state form a robust structure in the sense that the two arms are held close together by the hybridization motor. However, since the four-base single-stranded hinge is flexible, the open tweezers are flexible. The arms jiggle around as a result of the random bombardment by solvent molecules and, consequently do not maintain a well-defined configuration. The device depicted in Fig. 15.7 has two robust states. It is referred to as an actuator [8, 9] to distinguish it from the tweezers. It is very similar to the tweezers in that it possesses two double-stranded arms connected by a hinge, H of Fig. 15.7. The motor domain M of the actuator now consists of a strand of single-stranded DNA running between the tips of the arms.

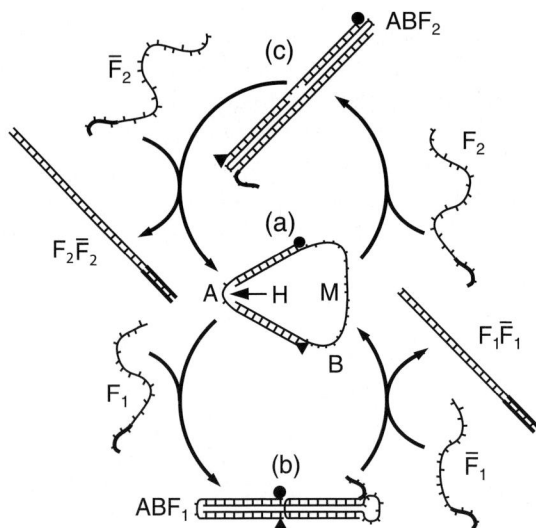

Fig. 15.7. Transition diagram for the actuator. This device can be switched between three distinct states, one of which (**a**) is floppy, the other two states (**b**) and (**c**) are robust. See text for details

The device is made by mixing together two strands of DNA, labeled A and B in the figure. These hybridize together to make a loop in which the double-stranded regions form the actuator's arms. As with the tweezers, the hinge is four bases in length, the arms are eighteen bases in length, and the motor domain is 48 bases in length. The actuator can be closed in the same manner as the tweezers through the introduction of a fuel strand F1 that hybridizes with the motor domain in such a way that the actuator arms are brought close together, as shown in (b). The fuel strand possesses a toehold that allows the actuator to be restored to its open configuration (a), through the application of the complement $\bar{F}_1$ of the fuel strand, in much the same way that the tweezers were opened. By introducing a fuel strand F_2 whose base sequence is complementary to M, the actuator's arms are pushed apart as shown in (c). This configuration is robust due to the rigidity of the double-stranded complex of F_2 hybridized with the motor domain. Because F_2 also possesses a toehold, it also can be restored to state (a), this time by the application of the fuel strand complement $\bar{F}_2$. A detail worth pointing out is that both F_1 and F_2 were designed to be complementary with only the 40 bases central to the motor domain. Hence, when the fuel strands are fully hybridized with the motor domain, four-base single-stranded regions remain at each end of the motor domain. This provides enough flexibility to the structure to enable the complements of the fuel strands to remove the fuel strands from the motor domain.

As shown in Fig. 15.7, a pair of dye molecules, represented by a filled triangle and a filled circle, is attached to the ends of the arms of the actuator. These dyes provide a diagnostic signal used to follow the operation of the device. The physical phenomenon employed is called fluorescence resonance energy transfer (FRET) [39]. One of the dyes, referred to as the donor, is illuminated with monochromatic light at frequencies within the dye's absorption band. The absorbed energy is normally reradiated as fluorescence at lower frequencies lying in the emission band of the dye. The other dye, called the acceptor, is chosen to have its absorption band overlap that of the donor's emission band. When the two dyes are close to each other, the donor, rather than reradiate its absorbed energy as fluorescence, transfers its energy to the acceptor dye. The acceptor dye disposes of the energy either nonradiatively or as florescence within its own emission band which is at lower frequencies than the emission band of the donor dye. The degree to which the donor fluorescence is quenched or the acceptor fluorescence is increased, due to energy transfer from donor to acceptor, is a smooth but strongly varying function of the distance between the two dyes. The degree of fluorescence of the donor or acceptor can thus be used as a "molecular ruler" to measure the distance between the two dyes. The distance at which the behavior of the fluorescence is halfway between the two extremes is called the Förster distance and sets the length scale at which the molecular ruler is useful. The Förster distance is typically in the 1 to 10 nm range, depending on the dye pair used [39]. For the two DNA-based nanodevices so far discussed, whose arms are 7 nm in length, FRET is particularly well-suited for determining the distance separating the arms. The donor and acceptor dyes used in the tweezer and actuator experiments are generally referred to by their acronyms TET and TAMRA, respectively, and this pair has a Förster distance of 6 nm [7].

15.5 Towards Applications

A number of DNA-based devices driven by hybridization motors have been constructed since the publication of the first reports about the DNA-tweezers [7] and the DNA-actuators [8]. Increasingly complex and sophisticated devices have been constructed and work is in progress directed towards the practical application of such devices. Here we discuss some examples to indicate how the field is progressing.

Yan, Zhang, Shen, and Seeman [11] have constructed a sophisticated DNA-hybridization motor that is considerably more complex than the tweezers or actuator. It can be considered as a device in which two shafts come out from opposite ends of the motor. Each shaft consists of two strands of duplex DNA in which the two duplex molecules are held together by virtue of the fact that the single strands making up the duplex strands at several points cross over from one duplex to the other. Such structures are referred to as double-crossover structures. The motor consists of two single strands of DNA running between the two shafts. Pairs of set strands (equivalent to what I have referred to as fuel strands) hybridize to these motor domains. One pair of set strands results in a structure consisting of two parallel duplex molecules held together by the crossovers in the motor shafts. The other pair of set strands results in a structure in which the shafts are rotated, axially 180 degrees with respect to each other. Complements of the set strands (called reset strands) were used to remove the set strands from the motor. A linear chain of these motors was formed. Inserted between the motors were flags constructed from DNA that served as indicators of degree of rotation between the shafts. Thus, the chain of motors could be put into one of two states: In one state all the flags appear on one side of the chain of motors and in the other state the flags alternate. H. Yan et al. [11] were able to observe these two states with atomic force microscopy.

Liao and Seeman [17] have used three such motors, each addressable with a different pair of set or reset strands, to make a DNA-based device that has much of the functionality of a ribozyme, the enzyme that translates messenger RNA into peptide sequences in biological organisms. In the present case, instead of manufacturing a peptide sequence, the device manufactures DNA strands whose base sequence depends on the state of the three motors, which in turn depends on which pairs of set strands have been applied. Therefore, the device translates sets of set strands into DNA sequences. This work is directed towards translating sets of set strands into monomer sequences for polymers other than DNA. When mature, this may provide chemists with a powerful new tool, allowing them to process synthetic polymers having particular monomer sequences with the same ease that RNA can be translated into peptides. This may, for example, allow one to harness the power of in vitro evolution for making synthetic polymers with specific functions [40].

Sherman and Seeman [18] and Shin and Pierce [19] have constructed DNA walkers, nanodevices that walk on a DNA-based track. These devices typically have a pair of feet whose binding with single strands of DNA attached to the substrate is controlled by the addition of set strands and reset strands in a sequence that steps the walker along the substrate. Feng, Park, Reif, and Yan [14], by incorporating motor domains in DNA lattices, have made lattices whose lattice spacing can be changed through the application of DNA strands. Single-stranded DNA, having a base sequence that is guanine rich, can fold on itself to form quadruplex DNA, rather than the more usual duplex DNA. Li and Tan [13] and Alberti and Mergny [15] have constructed hybridization motors based on quadruplex DNA. In such motors, the fuel strand unfolds the compact quadruplex structure into a long linear molecule. The fuel complement by means of strand displacement restores the motor to its compact quadruplex configuration.

Lin, Yurke, and Langrana have carried out work on DNA-crosslinked gels [41] that incorporate single-stranded regions to which fuel strands can attach. The fuel strands are introduced into the gel either by diffusion or, since DNA is negatively charged, by electrophoresis. Binding of the fuel strand with the crosslink causes the crosslink to lengthen and stiffen. The polyacrylamide strands attached to the crosslink are pushed outward. As a result, the polyacrylamide strands are stretched and come under tension. This causes the gel to stiffen. By providing the fuel strand with a toehold, the gel can be returned to its relaxed configuration by strand displacement through the introduction the fuel strand's complement into the gel. Hence, one has a gel whose stiffness can be changed as a function of time through the addition of DNA strands. Lin et al. have observed the stiffness of a gel to change by a factor of three. These materials, in which the stiffness can be changed on a schedule in a biocompatible manner, may be useful as substrates for cell and tissue culture, because biological cells are sensitive to the mechanical properties of the extracellular matrix. In particular, their shape, the way they migrate, and the way they differentiate all depend on the stiffness of the extracellular matrix [42,43].

Dittmer, Reuter, and Simmel [16] have constructed a motorized DNA-based device for which the motor controls the binding of the device to thrombin, a protein molecule that is involved in blood clotting. Such a DNA strand that is capable of binding to a protein molecule is referred to as an aptamer. By binding to the motor domain of the motorized aptamer, the fuel strand pulls the aptamer off of the thrombin molecule. By toehold-mediated strand exchange, the fuel strand can be removed from the aptamer, allowing the aptamer to insert itself back into the thrombin molecule. This work suggests means by which hybridization motors could control biochemical reactions within living organisms and may lead to medical applications.

The motors described so far are stepped motors, as defined in Sect. 15.3. The advancement of such motors through the successive application of fuel strands and their complements allows for control of the timing of the state transitions which may be advantageous in certain applications. For other applications these motors lack the autonomy and convenience of free-running motors. Free-running motors possess internal mechanisms which automatically guide the motor through the sequence of states that constitute the machine cycle. There has been considerable advance in the construction of free-running DNA-based motors. Two approaches have been used to construct free-running molecular motors. In one method, explored by Turberfield, Mitchell, Yurke, Mills, Jr., Blakey, and Simmel [12], the fuel and fuel complement, through hybridization with auxiliary strands, are folded into complexes that hinder the direct hybridization of the fuel strand with its complement. Ideally, only after binding with the motor domain does the fuel complex become reactive with the fuel complement. When reactive, the fuel complement removes the fuel complex from the motor domain by strand displacement. Such free-running motors thus function as a catalyst that increases the rate with which the fuel complex and its complement react.

The second approach that has been explored for making free-running hybridization motors is the use of some means other than strand displacement to remove the fuel strand from the motor domain. Through a judicious choice of base sequences, motor domains and fuel strands can be designed for which there are enzymes that will cut the fuel strand into smaller pieces after it has been bound to the motor strand. Such enzymes exist that are incapable of cutting DNA strands which are single-stranded. The machine cycle of such a motor would consist of one in which the fuel strand first binds to the motor domain, inducing a state change. The fuel strand is then attacked by the cutting enzyme. The resulting shorter DNA strands are not bound as strongly to the motor domain as the full-length fuel strand was. The shorter DNA strands subsequently detach from the motor domain, and the motor returns to its initial state. A version of such a motor was constructed by Chen, Wang, and Mao [44] in which the motor domain served as the cutting enzyme. DNA strands, having a particular DNA sequence, are able to cut RNA molecules. Such a strand was used in this motor and its fuel consisted of an RNA molecule that could be cleaved by the DNA strand. This motor was used to power DNA tweezers. More recently, Stojanovic [45] has constructed RNA-based "spiders" whose legs are RNA molecules that are capable of cutting the DNA molecules to which the RNA molecules can bind. Such spiders can crawl over a lawn of single-stranded DNA molecules. The spiders' feet attach to blades of the DNA grass. After a short amount of time a given leg cuts the grass blade and dissociates from the resulting grass fragments. Then, by Brownian motion, the leg reaches over to attach itself to the next nearest blade of grass. The spider thus walks over the carpet leaving behind a trail of cut DNA grass. Such spiders walking through a DNA-crosslinked gel will cut the crosslinks and cause the gel to disintegrate. Since drugs are routinely embedded in gels,

DNA-crosslinked gels with such spiders may find application in drug delivery. An autonomous walker, powered by a biologically derived restriction enzyme that nicks double-stranded DNA, has been recently devised [46]. An autonomous walker has also been reported that is powered by two restriction enzymes that cleave double-stranded DNA and ligase, an enzyme that splices together nicks in the backbone of double-stranded DNA [47].

What I have presented here is not an exhaustive survey of the field, so not every work has been mentioned. In particlar, it should be noted that hybridization motors are only a subset of DNA-based nanodevices which are capable of inducing motion [48]. Nevertheless, it has been my intention to show that the field is advancing very rapidly and that it has progressed to the point where one can now seriously contemplate applications in chemical synthesis, therapeutics, drug delivery, and tissue engineering.

References

1. V. Balzani, A. Credi, F.M. Raymo, and J.F. Stoddart (2000). "Artifical molecular machines," *Angew. Chem. Int. Ed.*, **39**, p. 3348.
2. T.R. Kelly, H. De Silva, and R.A. Silva (1999). "Unidirectional rotary motion in a molecular system," *Nature*, **401**, p. 150.
3. N. Koumura, R.W.J. Zijlstra, R.A. van Delden, N. Harada, and B.L. Feringa (1999). "Light-driven monodirectional molecular rotor," *Nature*, **401**, p. 152.
4. A.M. Brouwer, C. Frochot, F.G. Gatti, D.A. Leigh, L. Mottier, F. Paolucci, S. Roffia, G.W.H. Wurpel (2001). "Photoinduction of fast reversible translational motion in a hydrogen-bonded molecular shuttle," *Science*, **291**, p. 2124.
5. J.V. Hernández, E.R. Kay, and D.A. Leigh (2004). "A Reversible Synthetic Rotary Molecular Motor," *Science*, **306**, p. 1532.
6. C.M. Niemeyer and M. Adler (2002). "Nanomechanical devices based on DNA," *Angew. Chem. Int. Ed.*, **41**, p. 3779.
7. B. Yurke, A.J. Turberfield, A.P. Mills, Jr., F.C. Simmel, and J.L. Neumann (2000). "A DNA-fuelled molecular machine made of DNA," *Nature*, **406**, p. 605.
8. F.C. Simmel and B. Yurke (2001). "Using DNA to construct and power a nanoactuator," *Phys. Rev. E*, **63**, p. 041913.
9. F.C. Simmel and B. Yurke (2002). "A DNA-based molecular device switchable between three distinct mechanical states," *Appl. Phys. Lett.*, **80**, p. 883.
10. J.C. Mitchell and B. Yurke, "DNA scissors," in *DNA Based Computers VII*, No. 2340 in *LNCS*, edited by N. Jonoska and N.C. Seeman (Springer Verlag, Heidelberg, 2002).
11. H. Yan, X. Zhang, Z. Shen, and N.C. Seeman (2002). "A robust DNA mechanical device controlled by hybridization topology," *Nature*, **415**, p. 62.
12. A.J. Turberfield, J.C. Mitchell, B. Yurke, A.P. Mills, Jr., M.I. Blakey, and F.C. Simmel (2003). "DNA fuel for free-running nanomachines," *Phys. Rev. Lett.*, **90**, p. 118102.
13. J.J. Li and W. Tan (2002). "A single DNA molecule nanomotor," *Nano Lett.*, **2**, p. 315.
14. L. Feng, S.H. Park, J.H. Reif, and H. Yan (2003). "A two-state DNA lattice switched by DNA nanoactuator," *Angew. Chem. Int. Ed.*, **42**, p. 4342.

15. P. Alberti and J.L. Mergny (2003). "DNA duplex-quadruplex exchange as the basis for a nanomolecular machine," *Proc. Natl. Acad. Sci. U.S.A.*, **100**, p. 1569.
16. W.U. Dittmer, A. Reuter, and F.C. Simmel (2004). "A DNA-based machine that can cyclically bind and release thrombin," *Angew. Chem. Int. Ed.*, **43**, p. 3549.
17. S.P. Liao and N.C. Seeman (2004). "Translation of DNA signals into polymer assembly instructions," *Science*, **306**, p. 2072.
18. W.B. Sherman and N.C. Seeman (2004). "A precisely controlled DNA biped walking device," *Nano Letters*, **4**, p. 1203.
19. J.S. Shin and N.A. Pierce (2004). "A synthetic DNA walker for molecular transport," *J. Am. Chem. Soc.*, **126**, P. 10834.
20. N.C. Seeman (2003). "DNA in a material world," *Nature*, **421**, p. 427.
21. E. Winfree, F. Liu, L.A. Wenzler, and N. Seeman (1998). "Design and self-assembly of two-dimensional DNA crystals," *Nature*, **394**, p. 539.
22. C. Mao, W. Sun, and N.C. Seeman (1999). "Designed two-dimensional DNA Holliday junction arrays visualized by atomic force microscopy," *J. Am. Chem. Soc.*, **121**, p. 5437.
23. H. Yan, S.H. Park, G. Finkelstein, J.H. Reif, and T.H. LaBean (2003). "DNA-templated self-assembly of protein arrays and highly conductive nanowires," *Science*, **301**, p. 1882.
24. A. Chworos, I. Severcan, A.Y. Koyfman, P. Weinkam, E. Oroudjev, H.G. Hansma, and L. Jaeger (2004). "Building Programmable Jigsaw Puzzles with RNA," *Science*, **306**, p. 2068.
25. P. W. K. Rothemund, N. Papadakis, and E. Winfree (2004). "Algorithmic self-assembly of DNA Sierpinski triangles," *PLoS Biology*, **2**, p. 2041.
26. P.W.K. Rothemund, A. Ekani-Nkodo, N. Papadakis, A. Kumar, D.K. Fygenson, and E. Winfree (2004). "Design and characterization of programmable DNA nanotubes," *J. Am. Chem. Soc.*, **126**, p. 16344.
27. C. Bustamante, Z. Bryant, and S. B. Smith (2003). "Ten years of tension: single-molecule DNA mechanics," *Nature*, **421**, p. 423.
28. B. Tinland, A. Pluen, J. Sturm, and G. Weill (1997). "Persistence length of single-stranded DNA," *Macromolecules*, **30**, p. 5763.
29. J. SantaLucia, Jr. (1998). "A unified view of polymer, dumbbell, and oligonucleotide DNA nearest-neighbor thermodynamics," *Proc. Natl. Acad. Sci. USA*, **95**, p. 1460.
30. E. Essevaz-Roulet, U. Bockelmann, and F. Heslot (1997). "Mechanical separation of the complementary strands of DNA," *Proc. Nat. Acad. Sci. USA*, **94**, p. 11935.
31. M. Rief, H. Clausen-Schaumann, and H.E. Gaub (1999). "Sequence-dependent mechanics of single DNA molecules," *Nature Struct. Biol.*, **6**, p. 346.
32. J. Liphardt, B. Onoa, S. B. Smith, I. Tinoco, Jr., and C. Bustamante (2001). "Reversible unfolding of single RNA molecules by mechanical force," *Science*, **292**, p. 733.
33. K. Visscher, M.J. Schnitzer, and S.M. Block (1999). "Single kinesin molecules studied with a molecular force clamp," *Nature*, **400**, p. 184.
34. M.D. Wang, M.J. Schnitzer, H. Yin, R. Landick, J. Gelles, and S.M. Block, (1998). "Force and velocity measured for single molecules of RNA polymerase," *Science*, **282**, p. 902.
35. L.P. Reynaldo, A.V. Vologodskii, B.P. Neri, and V.L. Lyamichev (2000). "The kinetics of oligonucleotide replacements," *J. Mol. Biol.*, **297**, p. 511.

36. B. Yurke and A.P. Mills, Jr. (2003). "Using DNA to power nanostructures," *Genet. Program. Evol. Mach.*, **4**, p. 111.

37. C.M. Radding, K.L. Beattie, W.K. Holloman, and R.C. Wiegand (1977). "Uptake of homologous single-stranded fragments by superhelical DNA. IV. Branch migration," *J. Mol. Biol.*, **116**, p. 825.

38. C. Green and C. Tibbetts (1981). "Reassociation rate limited displacement of DNA," *Nucl. Acids Res.*, **9**, p. 1905.

39. E.A. Jares-Erijman and T.M. Jovin (2003). "FRET imaging," *Nature Biotechnology*, **21**, p. 1387.

40. D.R. Halpin and P.B. Harbury (2004). "DNA display I. Sequence-encoded routing of DNA populations," *PLoS Biology*, **2**, p. 1015.

41. D. C. Lin, B. Yurke, and N. A. Langrana (2005). "Inducing reversible stiffness changes in DNA-crosslinked gels," *J. Materials Research*, **20**, p. 1456.

42. E.J. Semler and P.V. Moghe (2001). "Engineering hepatocyte functional fate through growth factor dynamics: the role of cell morphologic priming," *Biotechnol. Bioeng.*, **75**, p. 510.

43. E.J. Semler, C.S. Ranucci, and P.V. Moghe (2000). "Mechanochemical manipulation of hepatocyte aggregation can selectively induce or repress liver-specific function," *Biotechnol. Bioeng.*, **69**, p. 359.

44. Y. Chen, M.S. Wang, and C.D. Mao (2004). "An autonomous DNA nanomotor powered by a DNA enzyme," *Angew. Chem. Int. Ed.*, **43**, p. 3554.

45. M.N. Stojanovic (2005). "Agile and Intelligent DNA Molecules," presented at the "Engineering a DNA World" workshop at the California Institute of Technology, *January*, pp. 6–8.

46. J. Bath, S.J. Green, and A.J. Turberfield (2005). "A free-running DNA motor powered by a nicking enzyme," *Andgew. Chem. Int. Ed.*, **44**, p. 4358.

47. P. Yin, H. Yan, X.G. Daniell, A.J. Turberfield, and J.H. Reif (2004). "A unidirectional DNA walker that moves autonomusly along a track," *Andgew. Chem. Int. Ed.*, **43**, p. 4906.

48. C. Mao, W. Sun, Z. Shen, and N.C. Seeman (1999). "A nanomechanical device based on the B-Z transition of DNA," *Nature*, **397**, p. 144.

Tuning Ion Current Rectification in Synthetic Nanotubes

Z.S. Siwy[1,2] and C.R. Martin[3]

[1] Department of Physics and Astronomy, University of California,
Irvine, Irvine CA 92697
zsiwy@uci.edu
[2] Department of Chemistry, Silesian University of Technology, 44-100 Gliwice,
Poland
[3] Department of Chemistry, University of Florida, Gainesville FL 32611

Abstract. We prepared and studied ion current rectifiers consisting of single asymmetric nanotubes in polymer films. The small opening is as small as several nanometers, while the big opening is in the micrometer range. We fabricated two nanotube systems, which exhibit ion current rectification through two distinct mechanisms (i) electrostatic interactions, based on asymmetric shape of electrostatic potential inside the pore, and (ii) electro-mechanical gate placed at the entrance of a conical pore, responsive to the external field applied across the membrane. Biosensors consisting of single conical nanotubes are discussed as well.

16.1 Introduction

Membranes and porous materials have found various applications in filtration and separation processes. The modern biotechnology has posed, however, new challenges to the producers of membranes and has required pores with diameters similar to those of molecules under study, therefore as small as several nanometers. The nanometer scale of the pores is necessary e.g. for developing sensors for single molecule detection [1–4]. Transport properties of so extremely narrow pores are however not well understood yet. The restricted geometry of nanopores creates a system with very strong interactions between translocating ions and the pore walls, which brings about new transport properties [5–7]. The hint that the transport properties of nanopores are different from the behavior of micropores was given by Nature. Ion channels and pores are nanometer-scale protein structures in biological membranes that mediate communication of a cell with other cells, and they are the basis of almost all physiological processes of a living organism [8,9]. Biological ion channels and pores exhibit transport properties not observed in microscopic nanoporous systems, for example (i) selectivity for ions or molecules [8], (ii) ionic current rectification, which indicates that there is a preferential direction of ionic flow

Z.S. Siwy and C.R. Martin: *Tuning Ion Current Rectification in Synthetic Nanotubes*, Lect.
Notes Phys. **711**, 349–365 (2007)
DOI 10.1007/3-540-49522-3_16

[e.g. [10]], (iii) ion current fluctuations for a constant voltage applied across the membrane [9], (iv) enhanced diffusion i.e. facilitated transport of molecules through nanopores, (v) pumping of ions and molecules against their electro-chemical potential gradient [8,9].

Our research has been focused on designing engineered single nanopores with the aim to elucidate basic physical and chemical phenomena underlying transport properties at the nanoscale. Our special interest was to mimic the behavior of the family of voltage-gated biochannels. Ion current rectification and voltage dependence of the pattern of ion current fluctuations in time are fingerprints of this type of channels [9]. We wanted to prepare the simplest abiotic system, which would exhibit similar transport properties to these of voltage-gated biochannels. In this way we could grasp the basic physics responsible for this fascinating voltage-dependent behavior.

The pores we have been working with are prepared by the track etching technique based on irradiating a polymer foil with swift heavy ions and subsequent chemical development of the latent tracks [11]. What differs the track etching technique from conventional lithographic methods is the single-particle type of recording. It is one swift heavy ion, which penetrates the foil and produces one latent track [11]. Subsequently, one latent track after chemical development results in formation of one pore. Controlling irradiation down to one ion enables fabricating a macroscopic sample containing just one nanopore. Department of Materials Research, Gesellschaft fuer Schwerionenforschung (GSI) Darmstadt, possesses world wide unique facilities suitable for single-ion irradiation [12]. A membrane with a single pore creates an optimal system for fundamental studies of ion transport through nanopores, because we avoid averaging effects resulting from ion transport through many pores.

We developed etching techniques, suitable for preparation of nanopores with openings as small as 2 nm. We examined transport properties of cylindrical and tapered-cone shaped nanopores and found that conical nanopores exhibit transport properties similar to these of biological voltage-gated channels [13–17]. These conical nanopores were shown to rectify ion current and exhibit ion current fluctuations of similar statistical properties as the ion current through biological voltage-gated channels.

In this report we have focused mainly on ion current rectification, presenting both experimental data and theoretical considerations on the rectification mechanism. To elucidate the mechanism of ion current rectification, we designed a system of single conical metal nanotubes of controllable surface charge. The tubes were obtained by electroless plating with gold of single nanopore polymer templates [18].

The manuscript is organized as follows. Preparation of templates with single asymmetric nanopores has been briefly reviewed in Sect. 16.2. Section 16.3 shows rectifying transport properties of conically shaped nanopores, studied by recording current-voltage curves. Section 16.4 discusses two possible mechanisms of ion current rectification based on (i) electrostatic interactions of passing ions with surface charge of nanopores, and (ii) an electro-mechanical

gate placed inside the pore, respectively. Sections 16.5 and 16.6 describe two
systems of metal nanotubes, designed specifically to rectify ion current according to only one of the above-mentioned mechanisms. Section 16.7 shows
application of asymmetric nanotubes in designing sensors for single-molecule
detection.

16.2 System of Single Conical Nanopores in Polymer Films

16.2.1 Fabrication of Single Conical Nanopores

We have used foils of polyethylene terephthalate (PET) (Hostaphan RN12,
Hoechst) and polyimide (Kapton HN50, Du Pont) of 12 micrometers thickness. The foils were irradiated with single heavy ions of total kinetic energy
of $\sim$2 GeV at normal incidence. The irradiation was performed at the linear accelerator UNILAC (GSI, Darmstadt). We also used polycarbonate foils
irradiated with $\sim$50 ions per cm^2, purchased from Osmonics (Bryan, TX).

Chemical etching of single-ion irradiated foils was performed in a conductivity cell, connected to a voltage source and picoammeter (Keithley 6487 and
Axopatch 200B Molecular Devices). To obtain conical pores, the etching was
performed only from one side. The other side of the membrane was protected
against etching by a stopping medium, which neutralizes the etchant [19–22].
For etching of PET we used 9 M NaOH, and as stopping medium we used
acidic solution of formic acid and KCl. Ion tracks in Kapton were developed
in sodium hypochlorite with 13% active chlorine content. Potassium iodide,
which reduces the chlorine ions to Cl$^-$ [21, 22], was the stopping medium.

The big opening of the pores D was determined by scanning electron
microscopy. For conical pores in PET, $D \sim 600$ nm, and for pores in Kapton,
$D \sim 2$ μm. The value of D for a given polymer results from the so-called bulk
etch rate, which is a rate of a non-specific etching of a polymer material [22].
The small opening of the conical pores is below scanning electron microscopy
resolution and its diameter d can be estimated by measuring a current-voltage
curve of a single nanopore at a standard solution of 1 M KCl. Assuming an
ideal conical shape of the pore its small opening can be calculated as [19, 21]

$$d = \frac{4LI}{\kappa \pi DU} \tag{16.1}$$

where L is the length of the pore, κ stands for the specific conductivity of
the electrolyte, U denotes the voltage applied across the membrane and I is
the ion current. However, as discussed below, these conical nanopores produce
nonlinear current-voltage curves. For the size determination we use therefore
the linear, low-voltage part of the I-V characteristics, which typically is between –200 mV and +200 mV. This etching process gives the possibility of
producing pores with an effective diameter d as small as 2 nm.

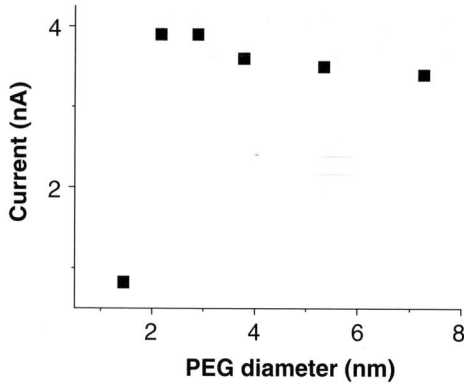

Fig. 16.1. Ion current through a single conically shaped nanopore in Kapton at presence of 20% PEG molecules in 1 M KCl solution. Note the cut-off with PEG molecules of hydrodynamic radius ~2 nm that is a measure of the pore size

To confirm the nanometer opening of the pore, we additionally measured its diameter by the size-exclusion technique [23]. This technique is based on recording conductance of a pore at presence of macromolecules of a well-defined size e.g. polyethylene glycol (PEG). PEG molecules assume a shape of a random coil with a hydrodynamic diameter D_{PEG} related to its molecular weight M_{PEG} by the so-called Kuga relation [24]:

$$D_{PEG} = 0.24 \ M_{PEG}{}^{1/3} - 0.58 \quad [\text{nm}] \tag{16.2}$$

When the PEG molecules are larger than the opening of the pore, they cannot enter the pore and, consequently, they do not alter ionic conductivity of the system. Only when the size of PEG molecules becomes comparable to the pore diameter, PEG fills the pore and hinders the ion current flow. This size of PEG molecules is called a cut-off and it is treated as a measure of the pore's opening diameter. Figure 16.1 shows an example of conductance measurements through a single Kapton nanopore at presence of 20% PEG solution in 1 M KCl.

16.2.2 Surface Characteristics of Irradiated and Etched PET and Polyimide Foils

Irradiation with heavy ions creates a zone of chemically changed material, the so-called latent track, which is more susceptible to chemical development than the bulk material [11]. The irradiation process can be visualized as scission of polymer chains along the trajectory of a swift heavy ion. In this way one gets a directional developing of the track during chemical etching with velocity of etching v_t. At the same time, a not-specific etching occurs on the surface that is called bulk etch rate v_b. The ratio between the two velocities determines the shape of the pore [25].

The surface chemistry of the pores depends on the type of polymer foil as well as on the applied etchant. Polyethylene terephthalate is a linear polyester and ion tracks in this material are developed in sodium hydroxide. NaOH causes hydrolysis of ester bonds resulting in formation of carboxylate and hydroxyl groups on the surface of the membrane and inside the pores [26]. Etching of polyimide foils also leads to formation of carboxylate groups by hydrolysis of imide bonds [27]. Immersing the membranes in an electrolyte of neutral or basic pH renders therefore the polymer surfaces negatively charged. The maximum surface charge density has been estimated to be app. 1.5 e/nm^2 [28]. Lowering pH results in protonation of carboxylate groups, and eventually at pH close to pK$_a$ of the carboxylates, the net surface charge is zero. For both polymers the surface becomes neutral at pH $\sim$3.

Chemical composition and structure of the two polymers determines different properties of the surfaces of the membranes and the pore walls. Scanning force microscopy revealed a much smoother surface of the etched Kapton in comparison to the etched PET surface [21]. We presume that the smoothness of Kapton foils results from the amorphous structure of the polymer [29]. PET is semi-crystalline [30] and chemical etching in NaOH was shown to result in a high surface roughness (tens of nanometers on the etched side). Also on the nanoscale the two materials differ from each other. The cut polymer chains in PET, due to the presence of the flexible ethylene group, can freely perform movement when in a solution, and they are called dangling ends. Kapton chains contain aromatic rings that make the cut polymer chains much more rigid [20].

16.3 Transport Properties of Single Conical Pores

Membrane with a single conical nanopore was placed between two chambers of a conductivity cell connected to a picoammeter/voltage source (Keithley 6487) [16, 17, 19, 20]. Both chambers were filled with a solution of potassium chloride buffered to a given pH. For ion current recordings we used Ag/AgCl electrodes. Current-voltage (I-V) characteristics were determined by stepping the voltage between –1 V and +1V with 50 mV steps.

Figure 16.2 shows current-voltage curves for a single conical nanopore in PET recorded at 0.1 M KCl, buffered to pH 8 and pH 3, respectively. At neutral and basic pH values the surface of the pores is negatively charged and the pore rectifies the ion current. At pH 3, which is close to the isoelectric point of the track-etched polymer surface, the overall surface charge is zero, resulting in a linear current-voltage characteristic [31]. Ion current rectification has also been found very strongly KCl concentration dependent: at higher electrolyte concentrations the pores rectify less. We were however surprised to observe that very small pores rectified even at 3 M KCl [32].

Ion current rectification indicates existence of a preferential direction of ion flow. In order to determine the rectification direction we had to check which

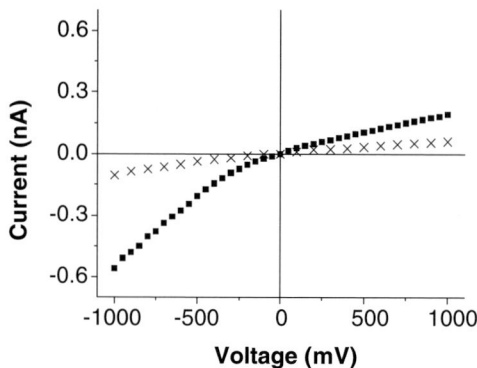

Fig. 16.2. Transport properties of a single conical nanopore in a PET membrane recorded at 0.1 M KCl buffered to pH 8 (■), and pH 3 (×). The diameters of the pore openings are ∼3 nm and 0.6 μm, respectively [32]. Copyright 2006, Wiley-VCH Verlag

ions contribute to the measured ion current: cations (K^+), anions (Cl^-) or both, therefore to measure transference numbers for cations t^+ and anions t^- [33]. The transference numbers fulfil the relation $t^+ + t^- = 1$, and indicate fraction of the ion current carried by cations and anions, respectively. For example, for $t^+ = 1$ cations are the only charge carriers, therefore the system is perfectly cation selective. The transference numbers can be determined by measuring the potential difference, the so-called reversal potential E_r, which is established across a membrane separating two chambers of a conductivity cell, filled with the same electrolyte but of different concentrations c_1 and c_2. E_r indicates a potential difference across the membrane at which the ion current is zero. For $1:1$ electrolytes, E_r can be approximated by the following formula:

$$E_r = (2.303\,RT/nF)(t_+ - t_-)\log(c_1/c_2) \tag{16.3}$$

where F is a Faraday constant and n is the charge of ions. Equation (16.3) was derived assuming independent contribution to ion current from potassium and chloride ions. Ion currents of potassium and chloride ions fulfill the Ohm's law and the membrane exhibits a given conductance for potassium and chloride ions. The flow of a given ion occurs under an influence of potential difference equal to a difference of membrane potential and the Nernst potential of that ion. Membrane potential at which the net ion current is zero equals E_r. Figure 16.3 presents a current-voltage curve of a single conical nanopore recorded at 0.1 M KCl, pH 8 and 1.0 M KCl, pH 8, at the side of the narrow and wide opening of the pore, respectively. As can be seen from the figure, the reversal potential is ∼50 mV, which indicates $t^+ \sim 0.9$.

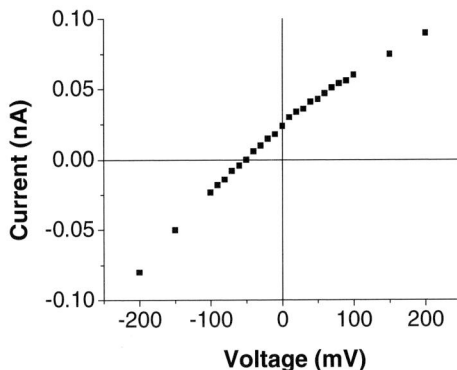

Fig. 16.3. Current-voltage characteristic of a conical pore in PET with opening diameters 3 nm and 0.6 μm, recorded at asymmetric electrolyte concentrations of 0.1 M KCl, pH 8 /1.0 M KCl, pH 8, with a higher concentration on the big opening of the pore. The transmembrane potential at which the current is zero determines the reversal potential related to the transference numbers for cations and anions

16.4 Mechanism of Ion Current Rectification

In order to analyze the origin of ion current rectification, let us summarize the conditions, which were found necessary for ion current rectification to occur. Our experiments showed that the rectifying pores have to be asymmetric in shape i.e. the pore openings on the two sides of the membrane are different from each other. As we have shown before, cylindrical very small pores at symmetric electrolyte conditions exhibit a linear current-voltage characteristic [34]. Another prerequisite to observe ion current rectification is a non-zero surface charge of the pore walls. Eventually, with increase of the pore diameter the rectification property becomes weaker.

We have suggested two mechanisms for ion current rectification, which we want to present here. A full review of various synthetic rectifying systems and developed theoretical approaches has been given in [32]. The first mechanism for ion current rectification, which has been developed for PET pores, points to the existence of an electro-mechanical gate inside a polymer pore. As discussed in the previous section, the pore walls are covered with polymer 'dangling ends', which are created in the process of irradiation with heavy ions and chemical etching. These dangling ends have carboxylate groups, therefore, they can respond to the external electric field, changing the effective diameter of the pore opening [17]. Due to the conical shape of the pore, the same voltage of opposite polarities will make the pore larger or smaller, respectively [17].

We realized however that the effect of ion current rectification could also be connected with distribution of electric potential inside an asymmetric pore with excess surface charge. Ions move through a pore in the direction determined by the external potential difference. The translocation of ions is however strongly influenced by the internal electric field, created due to the presence of surface charges on the pore walls. Interactions of ions with the pore walls are also the basis of pore selectivity, e.g. excess negative surface charge allows mainly cations to enter the pore while anions are rejected [18]. The first hint on the importance of the internal electrostatic potential in the rectification phenomenon came from the studies on Kapton pores. As we discussed in the previous section, nanopores in Kapton rectify the ion current although most probably they are deprived of flexible dangling ends due to the polymer chemical structure based on planar arrangement of aromatic groups [21].

In order to model the profile of electrostatic potential inside a conical pore, we considered a cation moving along the pore axis and interacting with negative charges on the pore wall. We assumed Debye type of interactions between ions. Note that the interactions of passing ions with the surface charge of the pore walls occur only if the pore is sufficiently narrow, with the diameter comparable to the thickness of the double-electrical layer [35]. This is caused by a short-range character of electrostatic interactions in an electrolyte solution, resulting from a strong screening induced by presence of other ions. For conical nanopores, without any voltage applied from the outside, the shape of the internal potential has been shown to be asymmetric-tooth like, reminding the shape of a ratchet potential (see Fig. 16.4) [36, 37]. The potential minimum is situated at the tip of the cone, which can be visualized as an electrostatic trap for cations [38].

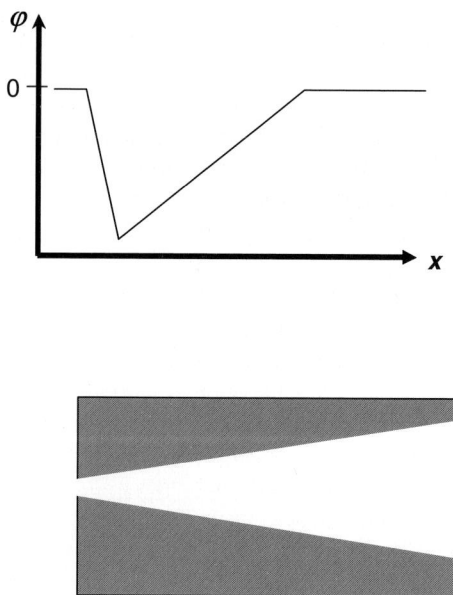

Fig. 16.4. Schematic representation of a profile of electrostatic potential φ inside a conical pore [16] along the cone axis x. The potential is most negative at the tip of a conical nanopore

When we apply electric field to the system, this external potential difference is superimposed with the internal electric field of the pore. As a result, a positive potential difference deepens the electrostatic trap for cations and creates an "off" state of the pore (Fig. 16.4) [38]. A negative transmembrane potential results in the "on" state for the ion flow, because the trap is much shallower [32, 38]. The mechanism by which an asymmetric channel rectifies the ion current is therefore similar to the so-called rocking ratchets [37].

Polymer pores, however, because of the presence of the dangling ends, do not allow a reliable checking which of the mechanisms of rectification dominates and whether each of the mechanisms alone can bring about this rectification effect.

As the next step, we designed two nanoporous systems in which only one rectification mechanism was in operation. In order to explore in a bigger detail the importance of electrostatic interactions of translocating ions with the excess surface charge of the pore walls, we prepared rigid metal nanotubes of controllable surface charge. Metal nanotubes were also the starting point for designing a system that rectifies the ion current due to a presence of an electromechanical gate of known charge and length.

16.5 Gold Tubes with Tailored Surface Charge – An Ionic "Rocking Ratchet"

The polymer PET membranes with single conical pores were templates for preparation of gold tubes. The polymer templates were covered by gold in the process of electroless plating [18, 39] that consists of three steps, (i) sensitization of the pore surface with tin, (ii) subsequent reduction of silver, and (iii) reduction of gold with formaldehyde as a reducing agent. We plated conical nanopores in PET with diameter of the small opening $\sim$20 nm, closing them down with gold to 5–10 nm. The gold tubes alone allow for checking the importance of surface charge in the ion current rectification process. It is known that Cl^- strongly adsorbs to Au surface and that the electrosorption valence is -1 [18]. When a gold tube is immersed in KCl, its net surface charge is therefore negative. This system allows checking whether a rigid asymmetric nanoporous system rectifies the current. On the other hand, if KF is used as an electrolyte, the net surface charge is zero, because F^- ions do not adsorb to Au surface. According to the surface charge model, an asymmetric tube in KF should exhibit ohmic = linear I-V behavior.

Figure 16.5 shows a series of current-voltage curves for a single conical gold nanotube recorded at symmetric electrolyte conditions, for various KCl concentrations. As we can see from the figure, the rectification effect is much weaker at higher electrolyte concentrations. This observation is in accordance with the electrostatic model for ion current rectification: at higher electrolyte concentrations, the interactions between ions are strongly screened by other ions present in a solution, which leads to diminishing of the asymmetry of the

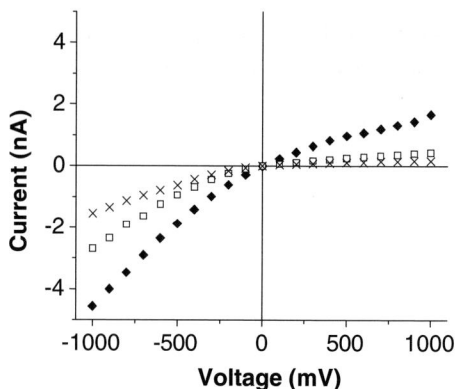

Fig. 16.5. Current-voltage characteristics of a single gold tube of opening diameters 10 nm and 0.6 μm, respectively, at symmetric electrolyte conditions (♦) 1 M KCl, pH 6.6; (□) 0.1 M KCl, pH 6.0; (×) 0.1 M KCl, pH 3.0 [38]

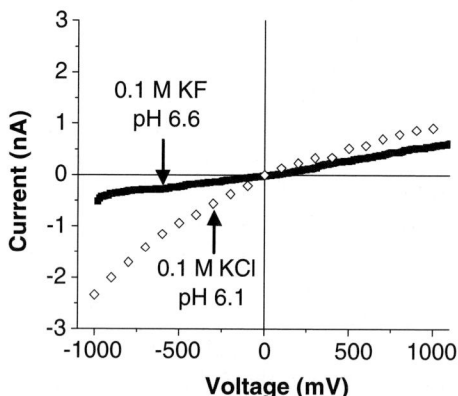

Fig. 16.6. Current-voltage characteristic of a single gold tube recorded at 0.1 M KF and 0.1 M KCl, as indicated in the figure. The small opening of the Au tube is ∼10 nm [38]. Copyright 2004, American Chemical Society

potential (Fig. 16.4) and consequently, to a weaker ion current rectification [40].

Figure 16.6 shows an I-V curve for a conical Au tube recorded at 0.1 M KF, therefore when the net surface charge of the tube is zero. As expected, at these conditions the tube behaves like a simple resistor and it exhibits a linear current-voltage curve.

The gold tubes also present a very convenient system for further chemical modification. It is well known that thiols – molecules containing the –SH group - spontaneously chemisorb to gold, and this provides a route to systematically and predictably changing the surface chemistry [33, 41]. To modify the excess surface charge of the gold tubes, we used thiols with negatively and positively charged groups, respectively [38]. We used short-chain thiols to prevent a possible influence on the rectification phenomenon from the thiol "dangling ends". 2-mercaptopropionic acid and mercaptoethylammonium cation were applied in the surface modification. We showed that chemisorption of these negatively and positively charged thiols resulted in systems rectifying the ion current in opposite directions [38].

Our studies also demonstrated that rectification properties of gold nanotubes are strongly influenced by the opening diameter of the tubes. With the small opening larger than ~20 nm the tubes stop rectifying ion current.

16.6 DNA-Au Tubes Rectify Because of Presence of Electrochemical Gate

Au nanotubes with openings larger than 20 nm were the starting point for designing a system into which we wanted to introduce an electromechanical gate of known length and charge [42]. Due to the large opening of these pores, electrostatic interactions of passing ions with the surface charge of the nanotubes were not significant for transport properties of these nanotubes. Single stranded DNA (ssDNA) with thiol groups [42] (purchased from alphaDNA, Inc and HPLC purified by the supplier), was the optimal choice for the electromechanical gate. The length of ssDNA can be calculated by multiplying the number of nucleotides by the mean distance between nucleotides that is equal in ssDNA to 0.4 nm. It is also known that each nucleotide has at least one negative charge due to a phosphate group [43]. Figure 16.7A schematically presents principles of operation of this type of rectifier. Placing an anode close to the small opening of the pore deflects the DNA strands so that an "on" state of the pore is created. In this case the effective diameter of the pore is larger than in the situation with the anode placed on the big side of the pore.

Figure 16.7B shows current-voltage curves of a single Au tube of 40 nm diameter before and after modification with ssDNA. Longer DNA strands change the pore diameter to a larger extend, which leads to stronger ion current rectification. It is exactly what we observed experimentally.

We would like to emphasize that operation of biological voltage-gated channel is based on an electromechanical gate [44–46].

16.7 Application of Conical Nanopores in Building Single Molecule Sensors

Our experiments with conical nanopores demonstrated that their transport properties are very sensitive to surface characteristics of the pore walls. We used this feature to design sensors based on single conical nanopores whose

(a)

"off"   "on"

(b)

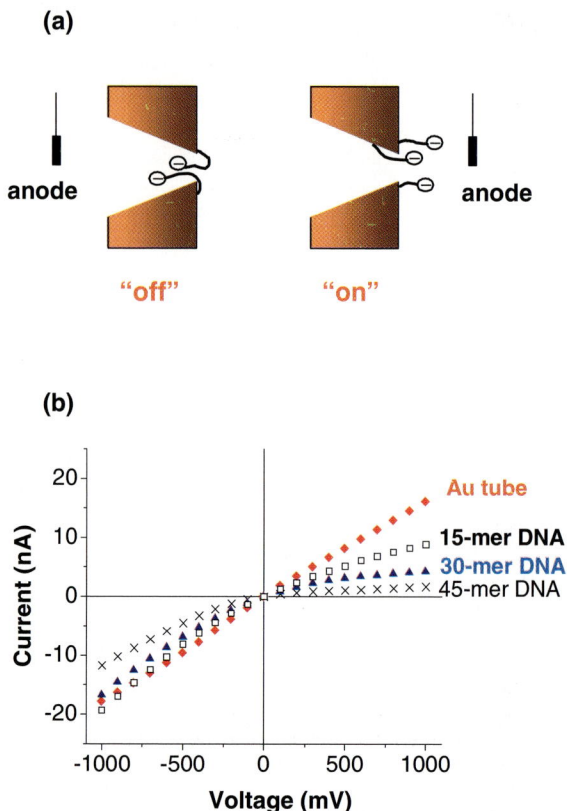

Fig. 16.7. Ion current rectifier with an electromechanical gate [42]. (**a**) Schematic representation of operation of the rectifier (**b**) Current-voltage curves for a single gold tube before and after modification with SH-ssDNA of various lengths, as indicated in the figure. The small opening of the gold tube was ~40 nm, the big opening 5 μm. Copyright 2004, American Chemical Society

pore walls were "decorated" with recognition sites for a specific molecule [47]. In contrast to other single-nanopore based sensors, where the signal of ion current in time is the basis for detection [e.g [1,2,4]], we have used an I-V curve as a main detection signal. The principles of operation of the sensor are very simple. We would start with recording an I-V curve of a pore with attached recognition sites. This measurement would be performed at a standard buffer solution e.g. 1 M KCl. This pore would subsequently be exposed to a solution containing a molecule that binds to the recognition sites in the pore and a new I-V curve would be recorded at 1 M KCl. If a molecule is present in the solution and binds to the sites, we will see it as a change in the current-voltage curve of the nanotube. If additionally, the pore diameter is comparable to the size of the molecule, a total blockage of the current should be observed. In this

(a)

(b)

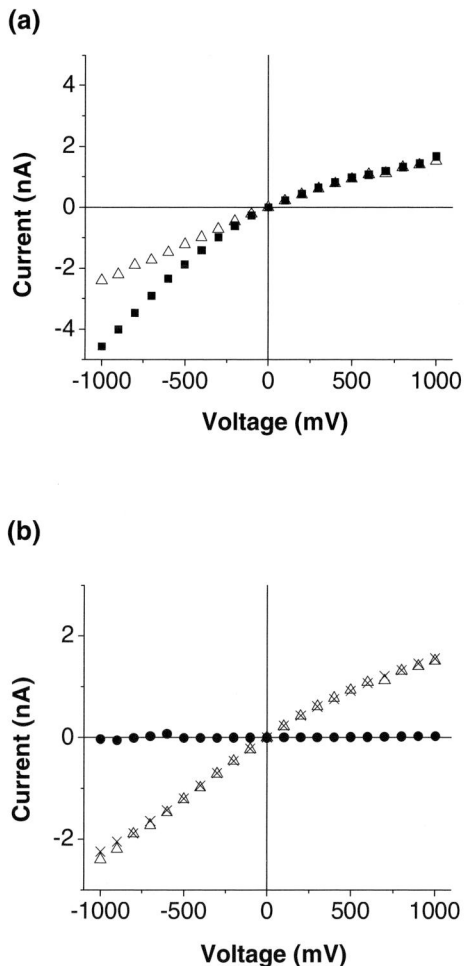

Fig. 16.8. Functioning of a single nanopore biosensor targeted at streptavidin. (**a**) Current-voltage curve of a single gold nanotube at 1 M KCl, pH 8 before (■) and after modification with biotin (△). (**b**) Current-voltage curve of a nanotube (the same as in (A) before (△) and after exposure to 50 nM solution of BSA in 1 M KCl (×) and 0.2 nM streptavidin solution in 1 M KCl (●) [47]. Copyright 2005, American Chemical Society

case, although we have many recognition sites attached to the pore walls, just one bound molecule can be detected. We prepared three different nanoporous systems targeted at detecting streptavidin, immunoglobins and ricin [47].

Figure 16.8 shows example of operation of this type of sensor. A gold nanotube of diameter 5 nm was modified with biotin molecules containing thiol groups, as described in [47] Chemisorption of biotin to the gold surface did not change the effective diameter of the pore (molecular weight of biotin is only

244), however the pore exhibits weaker rectification properties (Fig. 16.8A). This is because biotin is an uncharged molecule therefore its chemisorption reduced the excess surface charge on the pore walls. This biotin modified nanotube was first exposed to a solution of protein which does not bind to biotin, namely lysozyme and BSA [47]. Figure 16.8B shows that no change in the I-V curve of a gold tube was observed with 50 nM solution of BSA. Exposure of this biotin-modified tube to streptavidin, which is known to bind very strongly to biotin [48], resulted in a total shutting off of the current. This total blockage of the current occurred because the diameter of this pore (5 nm) was comparable to the size of a streptavidin molecule [49].

Similar experiments were performed with sensors designed to detect various immunoglobins and ricin [47].

16.8 Conclusions

We have presented single conical polymer nanopores and conical metal nanotubes that rectify ion current. We have shown that the ion current rectification can result from two effects (i) asymmetry of internal electric field inside a nanotube, and (ii) presence of an electromechanical gate, which deflects in external electric field and changes the effective pore opening. The latter system resembles in its operation biological rectifying voltage-gated channels, where charged structural elements create a voltage gate [14, 46].

The first type of an ionic rectifier is based on very narrow pores, where the pore opening is comparable to the thickness of electrical double-layer. Our results show the importance of interactions of ions that pass through the pore with surface charge on the pore walls. The surface charge produces internal electric field, which in a conical nanotube has a shape of an asymmetric ratchet tooth. For one polarity of the applied voltage an electrostatic trap for ions is created, which results in hindering the ion flow. For the opposite polarity of applied voltage on the other hand, the electrostatic trap is not formed and higher values of ion current are measured [16, 32, 38]. Changing the sign of excess surface charge enabled us achieving a control over the direction of rectification of asymmetric nanotubes.

We have also presented another type of rectifier consisting of Au tubes with diameters larger than 20 nm, in which electrostatic interactions do not play a significant role for ion transport properties. An electromechanical gate was introduced inside the pore in a form of ssDNA chains, which at external electric field change the effective diameter of the nanotubes [42].

Single nanopores and nanotubes also create a very convenient system for building sensors for detecting biomolecules as well as studying binding affinities between proteins [47].

Acknowledgments

Irradiation with swift heavy ions was performed at the Gesellschaft fuer Schwerionenforschung (GSI), Darmstadt, Germany. Discussions with members of the Department of Materials Research of GSI, Dr. Pavel Apel, Prof. Andrzej Fulinski and Dr. Ilona Kosinska are greatly acknowledged.

References

1. J.J. Kasianowicz, E. Brandin, D. Branton, D.W. Deamer (1996). Characterization of individual polynucleotide molecules using a membrane channel. *Proc. Natl. Acad. Sci. USA*, **93**, pp. 13770–13773.
2. M. Akeson, D. Branton, J.J. Kasianowicz, E. Brandin, D.W. Deamer (1999). Microsecond Time-Scale Discrimination Among Polycytidylic Acid, Polyadenylic Acid, and Polyuridylic Acid as Homopolymers or as Segments Within Single RNA Molecules. *Biophys. J.*, **77**, pp. 3227–3233.
3. J. Li, D. Stein, C. McMullan, D. Branton, M.J. Aziz, J.A. Golovchenko (2001). Ion-beam sculpting at nanometer length scales. *Nature*, **412**, pp. 166–169.
4. J. Li, M. Gershow, D. Stein, E. Brandin, J.A. Golovchenko (2003). DNA molecules and configurations in a solidstate nanopore microscope. *Nature Materials*, **2**, pp. 611–615.
5. S. Kuyucak, O.S. Andersen, S.-H. Chung (2001). Model of permeation in ion channels. *Rep. Prog. Phys.*, **64**, pp. 1427–1472.
6. D. Stein, M. Kruithof, C. Dekker (2004). Surface-Charge-Governed Ion Transport in Nanofluidic Channels. *Phys. Rev. Lett.*, **93**, pp. 035901/1–035901/4.
7. W. Nonner, D.P. Chen, B. Eisenberg (1999). Progress and prospects in permeation. *J. Gen. Physiol.*, **113**, pp. 773–782.
8. *The Cell – A Molecular Approach*, 2nd ed. G.M. Cooper (Sunderland (MA): Sinauer Associates, Inc. 2000) pp. 81–84, 476–491.
9. B. Hille (1992). *Ionic Channels of Excitable Membranes* (Sinauer, Sunderland, MA), 2nd ed.
10. M. Nishida, R. MacKinnon (2002). Structural basis of inward rectification: cytoplasmic pore of the G protein-gated inward rectifier GIRK1 at 1.8 Å resolution. *Cell*, **111**, pp. 957–965.
11. R.L. Fleischer, P.B. Price, R.M. Walker (1975). *Nuclear Tracks in Solids. Principles and Applications* (Univ. of California Press, Berkeley).
12. R. Spohr. Methods and device to generate a predetermined number of ion tracks, German Patent DE 2951376 C2 (1983); United States Patent No. 4369370 (1983).
13. D.A. Doyle, J.M. Cabral, R.A. Pfuetzner, A. Kuo, J.M. Gulbis, S.L. Cohen, B.T. Chait, R. MacKinnon (1998). The Structure of the Potassium Channel: Molecular Basis of K1 Conduction and Selectivity. *Science*, **280**, pp. 69–77.
14. Y. Jiang, A. Lee, J. Chen, M. Cadene, B.T. Chait, R. MacKinnon (2002). Crystal structure and mechanism of a calcium-gated potassium channel. *Nature*, **417**, pp. 515–522.
15. Y. Jiang, V. Ruta, J. Chen, A. Lee, R. MacKinnon (2003). The principle of gating charge movement in a voltage-dependent K^+ channel. *Nature*, **423**, pp. 42–48.

16. Z. Siwy, A. Fulinski (2002). Fabrication of a Synthetic Nanopore Ion Pump. *Phys. Rev. Lett.*, **89**, pp. 158101/1–158101/4.

17. Z. Siwy, Y. Gu, H.A. Spohr, D. Baur, Wolf-Reber A, R. Spohr, P. Apel, Y.E. Korchev (2002). Rectification and voltage gating of ion currents in a nanofabricated pore. *Europhys. Lett.*, **60**, pp. 349–355.

18. M. Nishizawa, V.P. Menon, C.R. Martin (1995). Metal Nanotube Membrane with Electrochemically Switchable Ion-Transport Selectivity. *Science*, **268**, pp. 700–702.

19. P. Apel, Y.E. Korchev, Z. Siwy, R. Spohr, M. Yoshida (2001). Diode-like single-ion track membrane prepared by electro-stopping. *Nucl. Instr. Meth. B*, **184**, pp. 337–346.

20. Z. Siwy, P. Apel, D. Baur, D. Dobrev, Y.E. Korchev, R. Neumann, R. Spohr, C. Trautmann, K. Voss (2003). Preparation of synthetic nanopores with transport properties analogues to biological channels. **Surface Science** 532–535: 1061–1066.

21. Z. Siwy, D.D. Dobrev, R. Neumann, C. Trautmann, K. Voss (2003). Electro-responsive asymmetric nanopores in polyimide with stable ion-current signal. *Applied Physics A*, **76**, pp. 781–785.

22. Z. Siwy, A. Apel, D.D. Dobrev, R. Neumann, R. Spohr, C. Trautmann, K. Voss (2003). Ion transport through asymmetric nanopores prepared by ion track etching. *Nucl. Instr. Meth. B*, **208**, pp. 143–148.

23. S.M. Bezrukov, I. Vodyanoy, A.V. Parsegian (1994). Counting polymers moving through a single ion channel. *Nature*, **370**, pp. 279–281.

24. S.J. Kuga (1981). Pore size distribution analysis of gel substances by size exclusion chromatography. *J. Chromatogr.*, **206**, pp. 449–461.

25. Spohr R. (1990). Ion Tracks and Microtechnology. Principles Applications and (Friedr. Vieweg & Sohn Verlagsgesellschaft mbH, Braunschweig).

26. A. Wolf, N. Reber, P.Y. Apel, B.E. Fischer, R. Spohr (1995). Electrolyte transport in charged single ion track capillaries. *Nucl. Instr. Meth. B*, **105**, pp. 291–293.

27. J. March (1968). Advanced Organic Chemistry: Reactions, Mechanisms, Structure and (McGraw-Hill Book Company, New York), p. 313.

28. Wolf-Reber A (2002). Aufbau eines Rasterionenleitwertmikroskops. Stromfluktuationen in Nanoporen, PhD dissertation, ISBN 3-89825-490-9, dissertation.de.

29. http://www.dupont.com/kapton/general/capabilities-broch.html.

30. http://www.dupontteijinfilms.com/datasheets/mylar/overview/h67160.pdf.

31. L.E. Ermakova, M.P. Sidorova, M.E. Bezrukova (1998). Filtration and electrokinetic characteristics of track membranes. *J. Colloid.*, **52**, pp. 705–712.

32. Z. Siwy (2006). Ion-Current Rectification in Nanopores and Nanotubes with Broken Symmetry. *Adv. Func. Mat.* **16**, pp. 735–746.

33. C.R. Martin, M. Nishizawa, K. Jirage, M. Kang, S.B. Lee (2001). Controlling Ion-Transport Selectivity in Gold Nanotubule Membranes. *Adv. Mat.*, **13**, pp. 1351–1362.

34. C.C. Harrell, S.B. Lee, C.R. Martin (2003). Synthetic Single-Nanopore and Nanotube Membranes. *Anal. Chem.*, **75**, pp. 6861–6867.

35. J. Israelachvili (1991) *Intermolecular and Surface Forces* (2nd Ed., Academic Press, London).

36. P. Hänggi and R. Bartussek. Brownian rectifiers: How to convert Brownian motion into directed transport. In *Nonlinear Physics of Complex Systems*, edited by

J. Parisi, S.C. Müller, W. Zimmermann (Springer, Berlin, 1997), **476**, pp. 294–308.

37. R.D. Astumian (1997). Thermodynamics and Kinetics of a Brownian Motor. *Science* **276**, pp. 917–922.

38. Z. Siwy, E. Heins, C.C. Harrell, P. Kohli, C.R. Martin (2004). Conical-Nanotube Ion-Current Rectifiers: The Role of Surface Charge. *J. Am. Chem. Soc.*, **126**, pp. 10850–10851.

39. K.B. Jirage, J.C. Hulteen, C.R. Martin (1997). Nanotubule-Based Molecular-Filtration Membranes. *Science*, **278**, pp. 655–658.

40. I.D. Kosińska and A. Fuliński (2005). Asymmetric nanodifussion. *Phys. Rev. E*, **72**, p. 011201 (1–7).

41. A. Ulman (1996). Formation and Structure of Self-Assembled Monolayers. *Chem. Rev.*, **96**, pp. 1533–1554.

42. C.C. Harrell, P. Kohli, Z. Siwy, C.R. Martin (2004). DNA – Nanotube Artificial Ion Channels. *J. Am. Chem. Soc.*, **126**, pp. 15646–15647.

43. G.M. Cooper, Hausman R.E. The Cell. A Molecular Approach. 3rd Ed. (Sinauer Associates, Inc. Sunderland, Massachusetts, 2004).

44. A. Yarnell (2004). Conflicting research findings on the mechanism of voltage-gating in K$^+$ channels has caused controversy. *Chem. Eng. News*, **82**, pp. 35–36.

45. R.O. Blaustein, C. Miller (2004). Ion channels: Shake, rattle or roll? *Nature*, **427**, pp. 499–501.

46. T. Hessa, S.H. White, von Heijne G. (2005). Membrane Insertion of a Potassium-Channel Voltage Sensor. *Science*, **307**, p. 1427.

47. Z. Siwy, Q. Trofin, P. Kohli, L.A. Baker, C. Trautmann, C.R. Martin (2005) Protein Biosensors Based on Biofunctionalized Conical Gold Nanotubes. *J. Am. Chem. Soc.*, **127**, p. 5000.

48. L. Movileanu, S. Howorka, O. Braha, H. Bayley (2000) Detecting protein analytes that modulate transmembrane movement of a polymer chain within a single protein pore Nat. *Biotechnol.*, **18**, pp. 1091–1095.

49. W.A. Hendrickson, A. Pahler, J.L. Smith, Y. Satow, E.A. Merritt, R.P. Phizackerly (1989). Crystal structure of core streptdavidin determined from multi-wavelength anomalous diffraction of synchrotron radiation. *Proc. Natl. Acad. Sci. USA*, **86**, pp. 2190–2194.

17

NanoShuttles: Harnessing Motor Proteins to Transport Cargo in Synthetic Environments

V. Vogel[1] and H. Hess[2]

[1] Department of Materials, Swiss Federal Institute of Technology (ETH), Zürich, Switzerland
viola.vogel@mat.ethz.ch

[2] Department of Materials Science and Engineering, University of Florida, 160 Rhines Hall Gainesville, FL 32611-6400, USA
hhess@mse.ufl.edu

Abstract. Motors have become a crucial commodity in our daily lives, from transportation to driving conveyor belts that enable the sequential assembly of cars and other industrial machines. For the sequential assembly of building blocks at the nanoscale that would not assemble spontaneously into larger functional systems, however, active transport systems are not yet available. In contrast, cells have evolved sophisticated molecular machinery that drives movement and active transport. Driven by the conversion of chemical into mechanical energy, namely through hydrolysis of the biological fuel ATP, molecular motors enable cells to operate far away from equilibrium by transporting organelles and molecules to designated locations within the cell, often against concentration gradients. Inspired by the biological concept of active transport, major efforts are underway to learn how to build nanoscale transport systems that are driven by molecular motors. Emerging engineering principles are discussed of how to build tracks and junctions to guide such nanoshuttles, how to load them with cargo and control their speed, how to use active transport to assemble mesoscopic structures that would otherwise not assemble spontaneously and what polymeric materials to choose to integrate motors into MEMS and other biohybrid devices. Finally, two applications that exploit the physical properties of microtubules are discussed, surface imaging by a swarm of microtubules and a self-assembled picoNewton force meter to probe receptor-ligand interactions.

17.1 Introduction

Technological revolutions often involve synthesis and processing of new materials; the mastery of a new material is so fundamental to mankind that historic ages are defined by the state-of-the art material, hence the "Stone Age" or "Bronze Age". However, some technological revolutions are characterized by the newfound ability to convert energy into mechanical work, based, for example, on the invention of the steam engine, which powered the

V. Vogel and H. Hess: *NanoShuttles: Harnessing Motor Proteins to Transport Cargo In Synthetic Environments*, Lect. Notes Phys. **711**, 367–383 (2007)
DOI 10.1007/3-540-49522-3_17

industrial revolution. Will nanotechnology revolutionize the way in which we employ biological and synthetic nanoscale machines to convert energy, from one form into the other? For example, will this revolution in part be driven by a nanomotor that will power tomorrow's nanodevices? Currently, we lack man-made nanomotors that are capable of having an impact on technology in a manner similar to the steam engine. However, while the first prototypes of synthetic nanomotors are studied [1–4], nature already provides us with a wide range of biological motors which have evolved to perform specific functions at high efficiency.

Cells, for example, have evolved sophisticated molecular machinery that drives movement and active transport. Driven by the conversion of chemical into mechanical energy, namely through hydrolysis of the biological fuel ATP, biological motors enable cells to operate far away from equilibrium by transporting organelles and molecules to designated locations within the cell, often against concentration gradients. Beyond muscle contraction, motor proteins drive a range of complex cellular tasks including shuttling of molecular cargo, cell motion and division, pumping of ions against concentration gradients, reading and duplicating genetic information, and the packaging of viral DNA. Motor protein-driven transport processes are thus essential for cell survival, and significant progress has been made in the last decade in learning how various biological motors work. This has enabled recent efforts in bionanotechnology aimed at applying biological motors for technological applications. Choosing motor proteins as engines to drive active transport has far-reaching consequences for the design of nanoshuttles, but is dictated by the lack of alternatives. While motor proteins have impressive characteristics in some respects, their need for a defined environment, which closely resembles the conditions in cells, limits their applications. In the same way as the transition from horses to cars did not change all transportation concepts, we believe that the eventual replacement of motor proteins by synthetic motors can be done at a later stage and will profit from the experiences in nanoscale transport gained with motor proteins.

What technological needs can be addressed with active transporters driven by biomolecular motors? Most of the biological machinery is assembled and disassembled in a sequential manner, often involving motor proteins. Motors are also crucial to our macroscopic constructions whether we assemble cars on conveyor belts or build sky scrapers. Yet for the sequential assembly of nanoscale components that would not assemble spontaneously into larger functional systems, active transport systems are not available. Nanotechnology requires active transport in the same way as parts of complex macroscopic machines are assembled in a modern factory by the use of conveyor belts. Taking advantage of molecular motors is challenging, as they have to be interfaced with and integrated into synthetic materials and devices. What is required to translate the biological concept of active transport of intracellular cargo by motor proteins into a technological context? After discussing the central elements that have to be in place to exploit motor proteins in synthetic environment to carry cargo, we will provide several unanticipated proof-of principle demonstrations for future applications including self-propelled nanoprobes for surface imaging, and a self-assembled picoNewton force meter to measure receptor-ligand interactions.

17.2 Engineering Concepts to Realize Motor Protein Driven Nanoshuttles

Engineering a shuttle system for applications in nanotechnology requires a nanoscale motor as central element. Motor proteins whose task it is to transport cargo within cells already possess the necessary structural features: they move along the cytoskeleton in a directed manner, they have the ability to bind cargo, and mechanisms regulating their speed exist, as illustrated in the contribution by Hirokawa and Takemura in this volume. Consequently, all studies aimed at building a nanoscale transport system on the basis of motor proteins have chosen as engines motors from the well-known kinesin or myosin families [5]. These motor proteins are fueled by ATP and convert its chemical energy into linear motion with an efficiency exceeding 50%. Kinesin for example moves in discrete steps of 8 nm against an opposing force of 5 pN, while hydrolyzing one ATP molecule per step [6].

Inverting the biological concept where the rod-like microtubules serve as tracks along which the kinesin motors carry their cargo, we investigated how to engineer tracks that can guide the movement of microtubules. In our systems, microtubules serve as shuttles which are pushed forward by surface-anchored motors, namely kinesin adsorbed to engineered tracks on synthetic surfaces (Fig. 17.1). This setup roughly resembles a conveyor belt, with the stationary motors moving a filament onto which the cargo is loaded.

The conceptual challenge in building a transport system is that solutions have to be developed to (a) control the movement of the shuttle between user-controlled locations, (b) load and unload cargo, and (c) control the speed by which the shuttles move [7]. As outlined in detail below, we have explored a variety of strategies to improve the guidance of molecular shuttles along predetermined paths, such as chemical modification, guiding channels, and combinations of surface chemistry and topography. In order to permit selective loading with cargo, we have functionalized microtubules with biotin linkers, as well as with fluorescent dyes to facilitate observation of these nanoscale filaments. We control the shuttle velocity by triggering the photolytic cleavage of caged ATP in the solution, which provides a varying amount of ATP fuel to the motors.

In the following, we will also discuss new insights into the mechanism by which microtubules are guided along engineered tracks, as well as how they traffic across junctions of engineered networks, how they can be loaded with cargo and finally how to control their speed on user-demand. Finally, we will

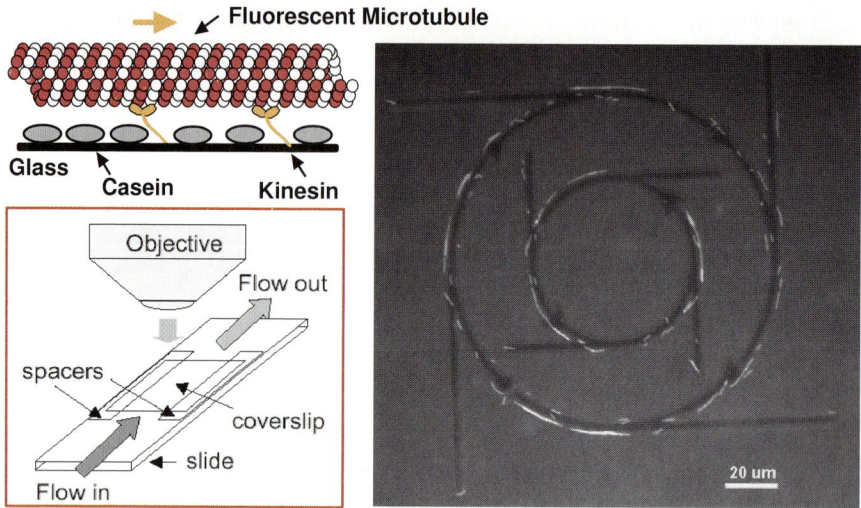

Fig. 17.1. In the applications discussed here, the motor protein kinesin is adsorbed to engineered tracks and pushes forward the microtubules which serve as shuttles. Kinesin is embedded in a monolayer of casein. Fluorescence labeling of the microtubule permits the direct observation of its movement with a fluorescence microscope. A basic flow cell, consisting of a slide, spacers and a coverslip patterned with microfabricated tracks, serves as a versatile experimental platform. The direction of motion of microtubules in open microfabricated channels can be biased through pinwheel arms that force microtubules to reverse direction and the triangular structures that act as rectifiers as shown on the *right*

give two additional proof-of-principle demonstrations of how nanoshuttles can be utilized in novel ways.

Track design for unidirectional guidance: Efficient guidance of the shuttles along predetermined tracks is a critical issue in developing biomolecular motor-based systems. The probability of loosing a microtubule when hitting the track edge has to be minimized and methods are needed to bias the motion of microtubules in one preferred direction. While initial studies relied on the deposition of polytetrafluoroethylene (Teflon) films with parallel nanoscale grooves to guide the movement of microtubules [8] or actin filaments [9], current methodologies employed by other laboratories and us rely on fabrication of complex track patterns using electron-beam lithography, photolithography or soft-lithography techniques [10–18]. To quantify the efficiency of guidance, we tested the following track designs: (1) planar chemical tracks where lanes of surface-bound kinesin border a non-adhesive surface chemistry [8, 19] (Fig. 17.2a), (2) topographic tracks with steep side-walls where all surfaces are functionalized with kinesin [20, 21], (Fig. 17.2b), and (3) the combination of both approaches [22, 23] where only the bottom of the guiding channel adsorbs motor proteins while the wall is non-adhesive (Fig. 17.2c).

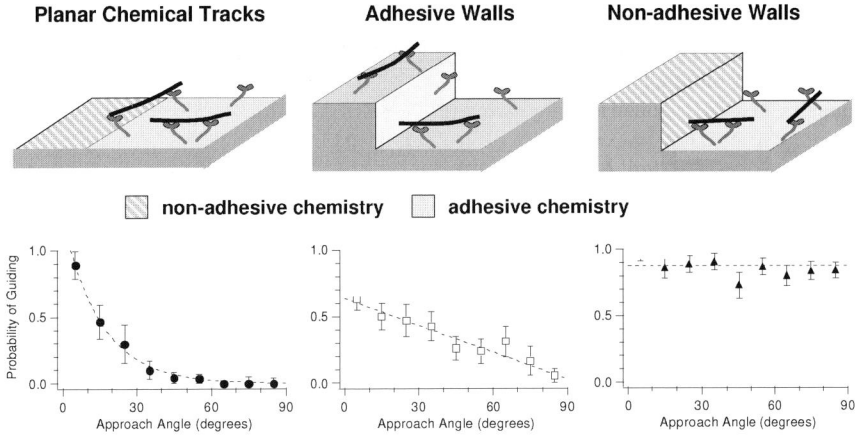

Fig. 17.2. Technological approaches to guide the movement of microtubules on engineered surfaces functionalized with motor proteins. The first approach (**a**) is of limited utility for the guiding of stiff filaments, such as microtubules. Due to their high bending rigidity, microtubules are pushed across the boundary to the non-adhesive surface by the kinesins on the track and eventually detach. The second approach (**b**) has been hampered by the ability of the microtubules to climb the kinesin-coated side-walls and subsequently escape from the track. The implementation of the third approach (**c**) demonstrated highly efficient guiding. The technique relies on selective adsorption of motor proteins to glass exposed at the bottom surface of a guiding channel patterned in photoresist

Due to the high bending rigidity of microtubules, steep walls are far more effective in guiding microtubules than planar chemical surface patterns of kinesin [3], provided they are of sufficient height [24]. The reason is that a microtubule colliding with a steep wall is forced physically to bent along the track, whereas a microtubule that is pushed across a flat chemical edge derails if the thermally swiveling tip does not bent backward far enough to catch on to a next kinesin sitting on the track. Best guidance, however, is not obtained with straight but with overhanging walls [25, 26] as shown in Fig. 17.3 where we prepared 1 µm high walls with a 200 nm high and 1 µm deep undercut at the bottom by photolithography. Both the resist and glass surface will adsorb kinesin motor protein and support microtubule binding if the photoresist surface is rendered hydrophilic by oxygen plasma treatment. However, microtubules moving on the bottom surface are unable to climb the sidewall and thus remain on the bottom surface, preferentially moving in the undercut section of the channel. This result is significant in several ways: (1) their ability to climb the wall is suppressed by physical means and thus is independent of the surface chemistry of the wall; (2) they are restrained to move at the bottom of the open channel, while ATP and cargo can be freely exchanged with the solution; and (3) it is the first step towards

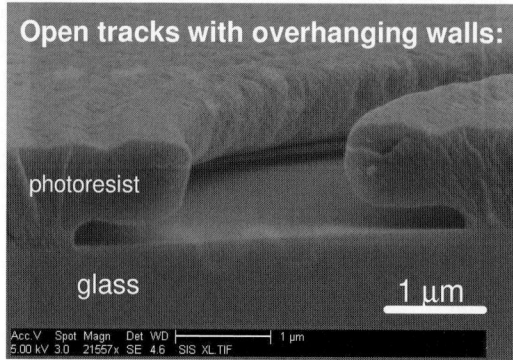

Fig. 17.3. Novel wall geometry for efficient guiding of microtubules on motor protein-coated surfaces imaged by scanning electron microscopy. The undercut prevents microtubules that move in the channel from climbing the sidewall, even if all surfaces are coated with motor proteins. Reproduced with permission from Nano Letters 2003, 3, 1651–1655. Copyright 2003 Am. Chem. Soc.

three-dimensional architectures, since the bottom of the channel and the top surfaces can support different functionalities.

Directing the cargo transport to defined locations requires that the motion of the microtubules in the tracks is biased. Since kinesin is adsorbed randomly on the track surfaces, biasing the direction of motion is most easily done by geometric means where asymmetric channel features turn around all those microtubules that move into the wrong direction (Fig. 17.1, right, see also [13, 22, 27, 28].

How wide can the channels be without loosing the ability to support directional movement? The probability of a MT turning around in channels is given by the channel width and the persistence length of the filament (or more exactly the persistence length of the filament path [29]). Thus the maximum width of the channel is linked to the physical properties of the filament. For microtubules which have a persistence length of a few millimeters, turnarounds are rarely seen in channels that are a few microns wide. In contrast, actin filaments require nanoscale channels for guidance [11] since their persistence length is less than 20 micrometer [30].

Junction Design: While we understand quantitatively how the shuttle system will perform on straight track segments, junctions are crucial in the transition of fabricating track networks that perform specific tasks. To what extent can the probability of a microtubule passing a junction be predicted and tuned through the geometric design of the junction? Our goal therefore was to measure quantitatively how motor proteins move through junctions of different geometries. We tested three new types of junctions that have significance for track networks in devices [22], namely crossing junctions (both

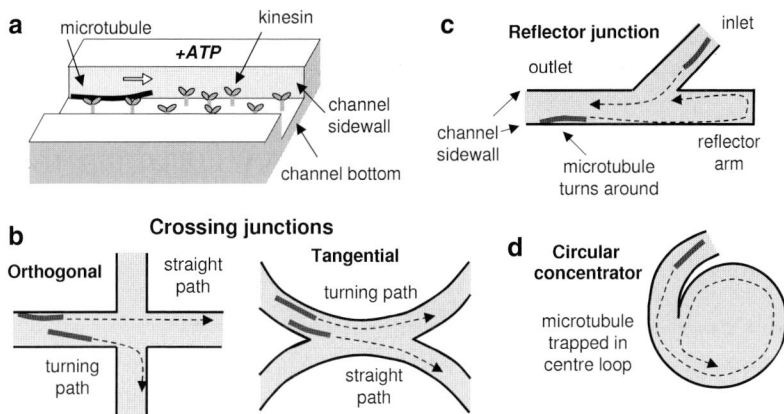

Fig. 17.4. Experimental detail of various track junctions. (**a**) The motor protein ki-nesin adsorbs at the bottom of microfabricated channels and translates microtubules. The track junctions investigated in this paper are (**b**) a crossing junction that al-lows microtubule tracks to cross one another at right angles (orthogonal crossing junction) or at shallow angles (tangential crossing junction), (**c**) the unidirectional reflector junction that traps and turns around microtubule in its arm, and (**d**) a circular concentrator that traps and collects microtubule in a central loop

orthogonal and tangential), a unidirectional reflector, and a circular concen-trator as illustrated in Fig. 17.4.

Cargo loading: One reason why so many different motors have evolved within given motor protein families seems to be that the tails of each motor connect via different scaffolding proteins to selective cellular cargo [31]. In our assays, though, the kinesin tail is anchored to the track surface while the mi-crotubules serve as shuttles. Consequently, chemical modifications of tubulin are needed to specifically bind cargo. The strong and selective binding between biotin and streptavidin, for example, was successfully utilized as a coupling element to connect cargo to the shuttles [21]. Biotinylated microtubules were shown to bind and transport streptavidin-coated beads (Fig. 17.5). We could further observe events where beads transported by microtubules fused with larger immobile aggregates of beads showing that this system could be capable of assembling complex structures in a LEGO-like fashion. Biotin-streptavidin linkages were later utilized by others to load DNA [32] and quantum dots [33, 34] and the relevant parameters affecting cargo transport have been assessed [35]. Alternative approaches include the ligation of DNA oligos to microtubules [36] or of cyclodextrin [37], which enables reversible binding of molecular cargo.

In addition to cargo loading, the same approach can be used to crosslink moving microtubules to create mesoscopic structures that only form in the presence of active transport [35, 38]. As seen in Fig. 17.5, "sticky" micro-tubules (biotinylated microtubules partially coated with streptavidin), are

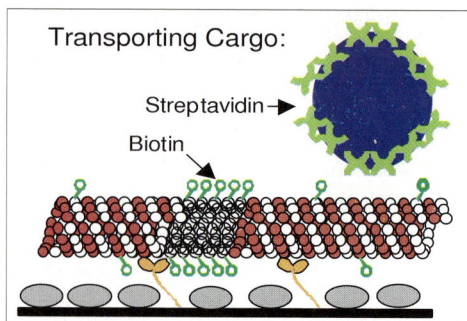

Fig. 17.5. Biotinylated tubulin can be mixed with plain or fluorescently labeled tubulin and assembled into microtubules which selectively bind cargo coated with streptavidin

transported across a motor-coated surface, and collide with each other. Collisions frequently lead to cross-linking, but since the motors generate sufficient force to bend the microtubules or break the biotin-streptavidin bonds, only aggregates where the microtubules point in the same direction are formed. These initially linear aggregates grow in length until the tip collides with the center or tail of the linear structure. At this point a "spool" begins to form. The spool collects additional microtubules colliding with it, but can spontaneously unwind when its tip binds again to motors on the surface and is pulled away from the rotating structure.

The structures assembled in this experiment differ from other self-assembled supramolecular structures fundamentally: (1) The internal order (unidirectional alignment of the microtubules) is solely a consequence of the active motion where parallel moving microtubules bind whereas the bonds among antiparallel moving microtubules are ruptured. (2) Strain is exerted by the motors only on mismatched (misaligned) parts. Once proper binding has occurred the motors transport the microtubules in the same direction without experiencing a net force between them. In contrast, in thermally activated self-assembly, thermal forces continue to strain the assembly. (3) The "spool" structures ultimately formed are under tremendous bending strain (>10,000 kT per turn), and thus in a configuration not accessible to thermally activated self-assembly. (4) The self-assembly process can be externally controlled by controlling the supply of fuel to the motors, so intermediate and final structures can be stabilized by deactivating the motors.

Controlling the shuttle speed: Controlling the speed of motion and especially stopping the nanoshuttles at defined locations is desirable for many applications. In a cargo loading station, for example, a pause can increase the probability of a collision event between the shuttle and cargo. Restricting the fuel access on user demand is an efficient way to reduce the speed by which microtubules are gliding on the kinesin tracks. Using caged ATP,

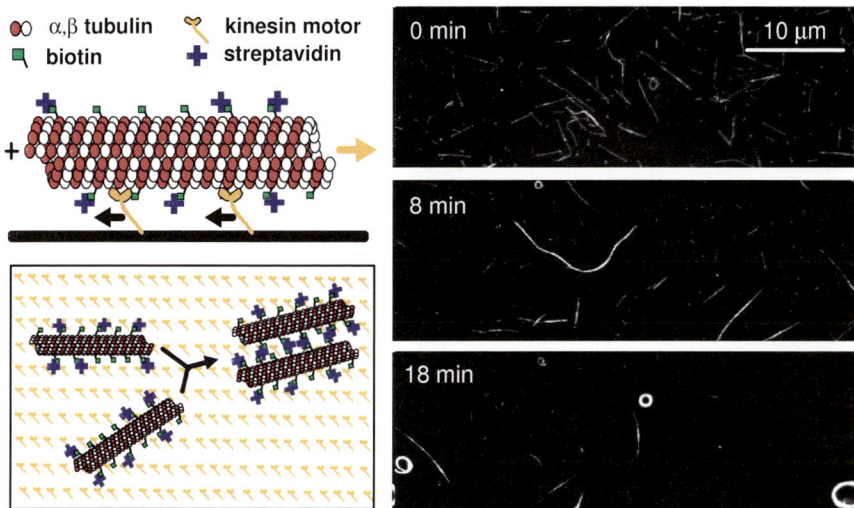

Fig. 17.6. Self-assembly of wires and spools. Biotinylated microtubules partially coated with the tetravalent streptavidin are cross-linked when colliding. In contrast to a solution-based process which results in random aggregates, microtubules gliding on a kinesin-coated surface form first oriented wires and then spools within 30 min. Reproduced with permission from Nano Letters 2005, 5, 629–633. Copyright 2005 Am. Chem. Soc.

a UV light pulse was used to release ATP resulting in a fast acceleration of the microtubules (Fig. 17.7) [21]. Since the ATP consumption of the motors alone was too small to deplete the solution of ATP in a reasonable time frame, the ATP-sequestering enzyme hexokinase was added to reduce the free ATP concentration to background levels on a timescale between several seconds and minutes. Using light, the microtubules can be moved in several discrete micrometer steps. The step size is defined by the UV-exposure time, concentration of the caged ATP supply, and the hexokinase concentration. An exciting new avenue is the in-situ production of ATP through an enzymatic reaction localized to the microtubules [39].

Choice of polymeric materials: What polymeric materials are best suited to build micro- and nanofluidic devices with integrated motor proteins? Is the overall lifetime of the device with all of its components affected by the polymeric environment? The lifetime of a device is determined by the lifetime of the most fragile component. While a number of polymers can replace glass as packaging material to integrate kinesin motors and microtubules, special care has to be taken in the choice of polymers when intense light illumination is needed, for example, for fluorescence imaging. To test the time over which all the biological constituents remain active in flow cells made from different polymers [40], the motor proteins were adhered to the glass bottoms of flow cells, and the flow cell covers were fabricated from glass, poly(urethane)

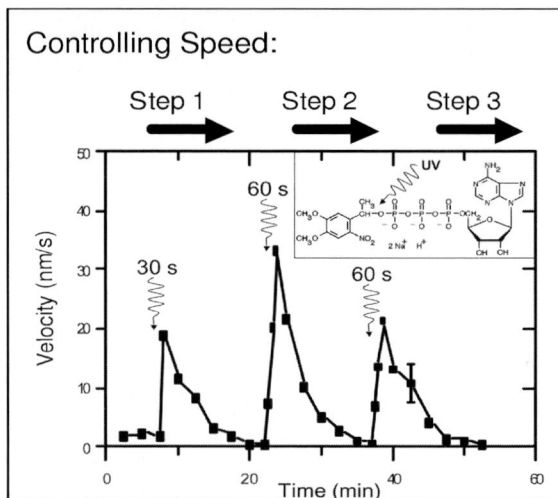

Fig. 17.7. Control of shuttle movement can be achieved by user-controlled release of ATP from a reservoir of caged ATP. Reproduced with permission from Nano Letters 2001, 1, 235–239. Copyright 2001 Am. Chem. Soc.

(PU), poly(methyl-methacrylate) (PMMA), poly(dimethylsiloxane) (PDMS) and ethylene-vinyl alcohol copolymer (EVOH) (Fig. 17.8). Under dark storage, 50% of the microtubules were still moving after 8 hours in all cells, with the exception of PU, in which they moved for two hours. Under light illumination and in the presence of oxygen scavengers, however, the number of moving microtubules had already decayed by 50% within the first few seconds in PDMS cells, while they lasted for a few minutes in PMMA, and their number was little altered even after 20 minutes in glass, EVOH and PU cells. Microtubules degraded rapidly in the presence of PDMS or PMMA. A similar fast degradation as seen in PDMS cells was otherwise observed with glass as cover only if oxygen scavengers were not added to the medium. Thus PDMS and PMMA, two widely used materials in micro- and nanofabrication, cause rapid disintegration of microtubules under exposure to light, even in the presence of oxygen quenchers, and the accelerated microtubule degradation coincide with rapid photobleaching. The oxygen permeability of PDMS is several orders of magnitude higher than for alternative polymers, including EVOH. The comparative analysis thus reveals that the lifetime of the entire device is not limited by the time for which the kinesin motors stay functional, but that the microtubules are the most fragile biological structure in the device. Strategies to further stabilize microtubules, using non-hydrolysable GTP analogs and chemical cross-linkers exist, and can now be compared against the taxol stabilization method.

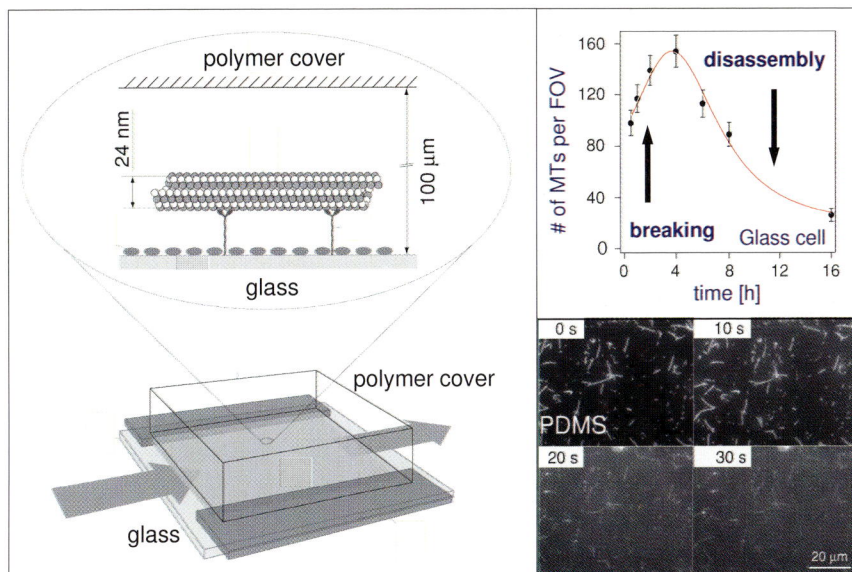

Fig. 17.8. *Left* – The lifetime of kinesin motor-based nanodevices and the compatibility of polymeric materials is assayed in flow cells by fluorescence microscopy. *Right top* – Degradation of microtubules under dark conditions: The depolymerization of microtubules is the primary mechanism of degradation for kinesin-powered bionanodevices. By tracking the number of microtubules per field-of-view (FOV), we can distinguish degradation pathways such as breaking and disassembly. Currently the lifetime of a device is on the order of several hours. *Right bottom* – In flow cells with covers fabricated from PDMS, microtubules photobleach rapidly and disassemble when the fluorescence excitation light is on [40]

17.3 First Applications of NanoShuttles

After gaining insights into how to exploit the physical properties of microtubules, we designed two devices: first, a swarm of randomly moving microtubules was traced to image the topography of a surface (Fig. 17.9) [41] and second a nanoscale force meter to probe the lifetime of receptor-ligand under load (Fig. 17.9) [42]. A recent theoretical study has investigated the feasibility of a biomolecular motor-powered pump [43].

Surface imaging: New approaches to surface imaging are very desirable which exploit active movement of a large number of simultaneously operating probes, no matter whether at the macroscopic scale, for example when exploring the surface of the planet Mars using a swarm of robots, or the microscopic topography of unknown surfaces. Molecular motors can be exploited as self-propelled nanoscale probes, which independently roam in large numbers across the surface and provide information about surface conditions. As illustrated in Fig. 17.9, the movement of hundreds of microtubules moving on a kinesin-

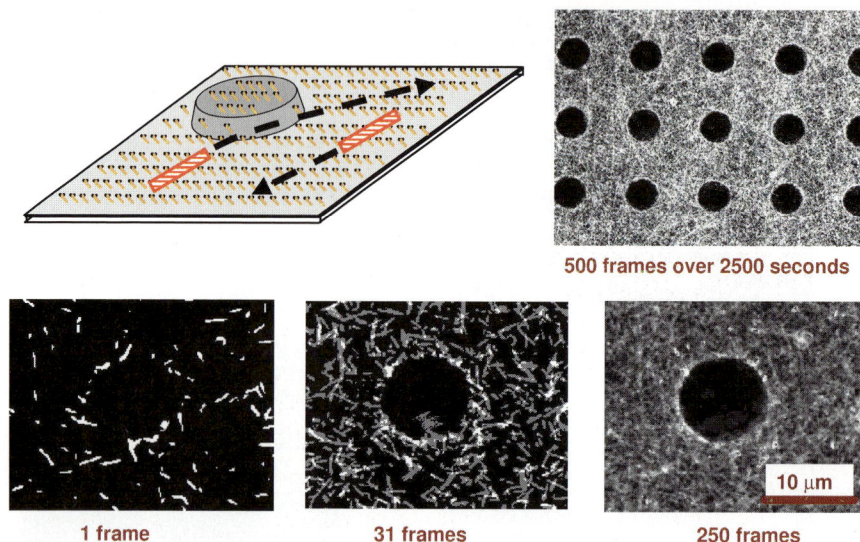

500 frames over 2500 seconds

1 frame   31 frames   250 frames

Fig. 17.9. A novel surface imaging method based on microtubules gliding on surface-bound kinesins: The trajectories of the microtubules are superimposed to create an image where the grayscale is a measure of how frequently a certain pixel was visited by a microtubule. The position of an array of posts with micrometer dimensions was imaged based on the preference of microtubules to reorient and move along rather than to climb the walls. Reproduced with permission from Nano Letters 2002, 2, 113–116. Copyright 2002 Am. Chem. Soc.

coated surface has been recorded over time. As the microtubules collide with microfabricated posts on the surface, their direction of movement changes and they weave their way around this obstacle course. The accumulated information of all microtubule paths can be used to visualize the positions of the obstacles. Although this imaging technique is limited by the fragility of the proteins and the optical detection of microtubule position, it demonstrates how self-propelled nanoprobes enable an approach to surface imaging, which is very different from classic scanning probe microscopy. It could find useful applications in the imaging of buried surfaces.

Self-assembled "force meter" to probe single ligand–receptor interactions: Force measurements on the molecular scale typically require a device capable of exerting picoNewton forces, such as optical tweezers or the atomic force microscope [44]. Recently, we have demonstrated a miniaturized force meter for the measurement of intermolecular bond strengths [42]. In this self-assembled force meter (Fig. 17.10), kinesin motors were used to provide the force to strain and rupture the bond between a receptor-ligand pair. The magnitude of the strain is reflected in the bending of a microtubule attached to the ligand. The microtubule thus serves as a nanoscale-cantilevered beam of known stiffness. Inherently, such a device is particularly suited to

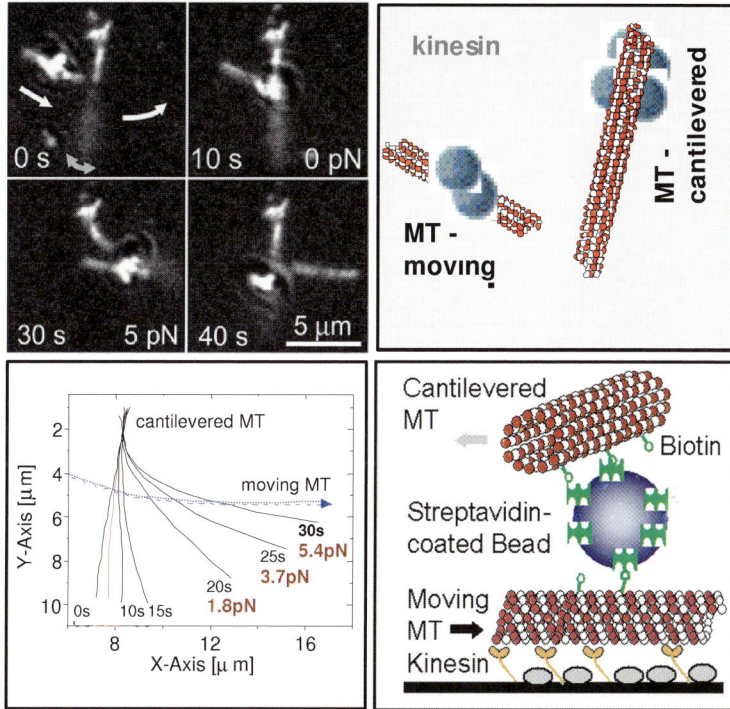

Fig. 17.10. A microscopic forcemeter for the measurement of intermolecular forces
on the order of a few piconewton can be assembled from microtubules functionalized with ligands, kinesins, and beads coated with receptors (0 s, 10 s). It consists of a cantilevered microtubule that binds to a streptavidin-coated bead loaded
onto a microtubule moved by kinesins. The kinesins push the moving microtubule
(30 s) straining the bond between streptavidin and biotin until it ruptures (40 s).
Observation of the concurrent bending of the cantilevered microtubule allows the
determination of the strain forces based on the known stiffness of microtubules in
a frame-by-frame analysis. Reproduced with permission from Nano Letters 2002, 2,
1113–1115. Copyright 2002 Am. Chem. Soc.

the study of rupture events between biological receptors and their ligands under physiologically significant loading conditions, which are characterized by
the application of picoNewton forces on a timescale of seconds [44]. Its small
dimensions also permit the assembly of an array of such force meters, which
would facilitate the observation of a statistically significant number of rupture
events. Notice that the lifetime of the streptavidin-biotin bond of several days
under static conditions is reduced to just 30 seconds when it is loaded with
5 pN, as expected [45, 46].

17.4 Conclusions

Twenty years after the discovery of kinesin [47, 48] the time has come to exploit the accumulated knowledge of how biomolecular motors work for technical applications. Integrating molecular motors into synthetic systems adds a new dimension to nanotechnology since active motion can be utilized to enable directed assembly processes in contrast to random assembly in solution. Molecular motors complement our ability to fabricate static structures with nanoscale dimensions with the ability to translocate, interrogate, assemble and disassemble building blocks in potentially highly parallel processes. As pointed out, many engineering challenges have to be overcome to ultimately realize concepts where molecular motors, for example, would drive an assembly line at the nanoscale. It is furthermore our hope, that advances in the design of hybrid nanofluidic and nanomechanical systems based on biomolecular motors will also contribute to our understanding of biological systems which have inspired us in the first place.

Acknowledgements

We would like to thank the many students and colleagues who have contributed to our team effort over the past years, and NASA (grant NAG5-8784), the DOE/BES (grant DE-FG03-03ER46024 and DE-FG02-05ER46193), the DARPA Biomolecular Motors Program, the ETH Zuerich, and the Alexander-von-Humboldt foundation (H.H.) for financial support.

References

1. V.V. Balzani, A. Credi, F. M. Raymo, J.F. Stoddart (2000). Artificial Molecular Machines. *Angew Chem Int Ed Engl*, **39**, pp. 3348–3391
2. A.M. Fennimore, T.D. Yuzvinsky, W.Q. Han, M.S. Fuhrer, J. Cumings, A. Zettl (2003). Rotational actuators based on carbon nanotubes. *Nature*, **424**, pp. 408–410
3. B.L. Feringa (2001). In control of motion: from molecular switches to molecular motors. *Acc Chem Res*, **34**, pp. 504–513
4. G.S. Kottas, L.I. Clarke, D. Horinek, J. Michl (2005). Artificial molecular rotors. *Chemical Reviews*, **105**, pp. 1281–1376
5. R.D. Vale, R.A. Milligan (2000). The way things move: looking under the hood of molecular motor proteins. *Science*, **288**, pp. 88–95
6. J. Howard (2001). Mechanics of Motor Proteins and the Cytoskeleton. Sindauer, Sunderland, MA, p. 367
7. H. Hess, V. Vogel (2001). Molecular shuttles based on motor proteins: Active transport in synthetic environments. *Reviews in Molecular Biotechnology*, **82**, pp. 67–85
8. J.R. Dennis, J. Howard, V. Vogel (1999). Molecular shuttles: directed motion of microtubules along nanoscale kinesin tracks. *Nanotechnology*, **10**, pp. 232–236

9. H. Suzuki, K. Oiwa, A. Yamada, H. Sakakibara, H. Nakayama, S. Mashiko (1995). Linear arrangement of motor protein on a mechanically deposited fluoropolymer thin film. Jap. *J. Appl. Phys.* Part 1 **34**, pp. 3937–3941

10. R. Bunk, J. Klinth, L. Montelius, I.A. Nicholls, P. Omling, S. Tagerud, A. Mansson (2003). Actomyosin motility on nanostructured surfaces. *Biochem Biophys Res Commun*, **301**, pp. 783–788

11. R. Bunk, M. Sundberg, A. Mansson, I.A. Nicholls, P. Omling, S. Tagerud, L. Montelius (2005). Guiding motor-propelled molecules with nanoscale precision through silanized bi-channel structures. *Nanotechnology*, **16**, pp. 710–717

12. L.J. Cheng, M.T. Kao, E. Meyhofer, L.J. Guo (2005). Highly efficient guiding of microtubule transport with imprinted CYTOP nanotracks. *Small*, **1**, pp. 409–414

13. Y. Hiratsuka, T. Tada, K. Oiwa, T. Kanayama, T.Q. Uyeda (2001). Controlling the Direction of Kinesin-Driven Microtubule Movements along Microlithographic Tracks. *Biophys J*, **81**, pp. 1555–61

14. J.A. Jaber, P.B. Chase, J.B. Schlenoff (2003). Actomyosin-Driven Motility on Patterned Polyelectrolyte Mono- and Multilayers. *Nano Letters*, **3**, pp. 1505–1509

15. C. Mahanivong, J.P. Wright, M. Kekic, D.K. Pham, C. dos Remedios, D.V. Nicolau (2002). Manipulation of the motility of protein molecular motors on microfabricated substrates. *Biomedical Microdevices*, **4**, pp. 111–116

16. P. Manandhar, L. Huang, J.R. Grubich, J.W. Hutchinson, P.B. Chase, S.H. Hong (2005). Highly selective directed assembly of functional actomyosin on Au surfaces. *Langmuir*, **21**, pp. 3213–3216

17. S.G. Moorjani, L. Jia, T.N. Jackson, W.O. Hancock (2003). Lithographically Patterned Channels Spatially Segregate Kinesin Motor Activity and Effectively Guide Microtubule Movements. *Nano Letters*, **3**, pp. 633–637

18. D.V. Nicolau, H. Suzuki, S. Mashiko, T. Taguchi, S. Yoshikawa (1999). Actin motion on microlithographically functionalized myosin surfaces and tracks. *Biophys J*, **77**, pp. 1126–34

19. R.C. Lipscomb, J. Clemmens, Y. Hanein, M.R. Holl, V. Vogel, B.D. Ratner, D.D. Denton, K.F. Böhringer (2002). Controlled Microtubules Transport on Patterned Non-adhesive Surfaces Second International IEEE-EMBS Special Topic Conference on Microtechnologies in Medicine & Biology. IEEE, Madison, Wisconsin, pp. 21–26

20. J. Clemmens, H. Hess, J. Howard, V. Vogel (2003a). Analysis of Microtubule Guidance in Open Microfabricated Channels Coated with the Motor Protein Kinesin. *Langmuir*, **19**, pp. 1738–1744

21. H. Hess, J. Clemmens, D. Qin, J. Howard, V. Vogel (2001). Light-Controlled Molecular Shuttles Made from Motor Proteins Carrying Cargo on Engineered Surfaces. *Nano Letters*, **1**, pp. 235–239

22. J. Clemmens, H. Hess, R. Doot, C.M. Matzke, G.D. Bachand, V. Vogel (2004). Motor-protein "roundabouts": Microtubules moving on kinesin-coated tracks through engineered networks. *Lab on a Chip*, **4**, pp. 83–86

23. J. Clemmens, H. Hess, R. Lipscomb, Y. Hanein, K.F. Boehringer, C.M. Matzke, G.D. Bachand, B.C. Bunker, V. Vogel (2003b). Principles of Microtubule Guiding on Microfabricated Kinesin-coated Surfaces: Chemical and Topographic Surface Patterns. *Langmuir*, **19**, pp. 10967–10974

24. P. Stracke, K.J. Bohm, J. Burgold, H.J. Schacht, E. Unger (2000). Physical and technical parameters determining the functioning of a kinesin-based cell-free motor system. *Nanotechnology*, **11**, pp. 52–56

25. S. Diez, J.H. Hellenius, J. Howard (2004). Biomolecular Motors operating in Engineered Environments. In: Niemeyer CM, Mirkin CA (eds.) *Nanobiotechnology*. Wiley-VCH, Weinheim

26. H. Hess, C.M. Matzke, R.K. Doot, J. Clemmens, G.D. Bachand, B.C. Bunker, V. Vogel (2003). Molecular Shuttles Operating Undercover: A New Photolithographic Approach for the Fabrication of Structured Surfaces Supporting Directed Motility. *Nano Letters*, **3**, pp. 1651–1655

27. H. Hess, J. Clemmens, C.M. Matzke, G.D. Bachand, B.C. Bunker, V. Vogel (2002b). Ratchet patterns sort molecular shuttles. *Appl Phys A*, **75**, pp. 309–313

28. M.G. van den Heuvel, C.T. Butcher, R.M. Smeets, S. Diez, C. Dekker (2005). High rectifying efficiencies of microtubule motility on kinesin-coated gold nanostructures. *Nano Lett*, **5**, pp. 1117–1122

29. T. Nitta, H. Hess (2005). Dispersion in Active Transport by Kinesin-Powered Molecular Shuttles. Nano Letters web-released 10-Jun-05

30. F. Gittes, B. Mickey, J. Nettleton, J. Howard (1993). Flexural rigidity of microtubules and actin filaments measured from thermal fluctuations in shape. *J. Cell Biol.* **120**, pp. 923–934

31. N. Hirokawa, R. Takemura (2005). Molecular motors and mechanisms of directional transport in neurons. *Nat Rev Neurosci*, **6**, pp. 201–14

32. S. Diez, C. Reuther, C. Dinu, R. Seidel, M. Mertig, W. Pompe, J. Howard (2003). Stretching and Transporting DNA Molecules Using Motor Proteins. *Nano Letters*, **3**, pp. 1251–1254

33. G.D. Bachand, S.B. Rivera, A.K. Boal, J. Gaudioso, J. Liu, B.C. Bunker (2004). Assembly and transport of nanocrystal CdSe quantum dot nanocomposites using microtubules and kinesin motor proteins. *Nano Letters*, **4**, pp. 817–821

34. A. Mansson, M. Sundberg, M. Balaz, R. Bunk, I.A. Nicholls, P. Omling, S. Tagerud, L. Montelius (2004). In vitro sliding of actin filaments labelled with single quantum dots. *Biochemical and Biophysical Research Communications*, **314**, pp. 529–534

35. M. Bachand, A.M. Trent, B.C. Bunker, G.D. Bachand (2005). Physical factors affecting kinesin-based transport of synthetic nanoparticle cargo. *Journal of Nanoscience and Nanotechnology*, **5**, pp. 718–722

36. G. Muthukrishnan, C.A. Roberts, Y.C. Chen, J.D. Zahn, W.O. Hancock (2004). Patterning surface-bound microtubules through reversible DNA hybridization. *Nano Letters*, **4**, pp. 2127–2132

37. K.A. Kato, R. Goto, K. Katoh, M. Shibakami (2005). Microtubule-cyclodextrin conjugate: Functionalization of motile filament with molecular inclusion ability. *Bioscience Biotechnology and Biochemistry*, **69**, pp. 646–648

38. H. Hess, J. Clemmens, C. Brunner, R. Doot, S. Luna, K.-H. Ernst, V. Vogel (2005)Molecular self-assembly of "Nanowires" and "Nanospools" using active transport. *Nano Letters*, **5**, pp. 629–633

39. Y.-Z. Du, Y. Hiratsuka, S. Taira, M. Eguchi, T.Q.P. Uyeda, N. Yumoto, M. Kodaka (2005). Motor protein nano-biomachine powered by self-supplying ATP. *Chemical Communications*, **16**, pp. 2080–2082

40. C. Brunner, H. Hess, K.-H. Ernst, V. Vogel (2004). Lifetime of biomolecules in hybrid nanodevices. *Nanotechnology*, **15**, pp. S540–S548

41. H. Hess, J. Clemmens, J. Howard, V. Vogel (2002a). Surface Imaging by Self-Propelled Nanoscale Probes. *Nano Letters*, **2**, pp. 113–116

42. H. Hess, J. Howard, V. Vogel (2002c). A Piconewton Forcemeter assembled from Microtubules and Kinesins. *Nano Letters*, **2**, pp. 1113–1115

43. J.L. Bull, A.J. Hunt, E. Meyhofer (2005). A theoretical model of a molecular-motor-powered pump. *Biomedical Microdevices*, **7**, 21–33

44. H. Clausen-Schaumann, M. Seitz, R. Krautbauer, H.E. Gaub (2000). Force spectroscopy with single bio-molecules. *Curr Opin Chem Biol*, **4**, pp. 524–530

45. E. Evans (2001). Probing the relation between force–lifetime–and chemistry in single molecular bonds. *Annu Rev Biophys Biomol Struct*, **30**, pp. 105–128

46. R. Merkel, P. Nassoy, A. Leung, K. Ritchie, E. Evans (1999). Energy landscapes of receptor-ligand bonds explored with dynamic force spectroscopy. *Nature*, **397**, pp. 50–3.

47. S.T. Brady (1985). A novel brain ATPase with properties expected for the fast axonal transport motor. *Nature*, **317**, pp. 73–75

48. R.D. Vale, T.S. Reese, M.P. Sheetz (1985). Identification of a novel force-generating protein, kinesin, involved in microtubule-based motility. *Cell*, **42**, pp. 39–50

18

Nanotechnology Enhanced Functional Assays of Actomyosin Motility – Potentials and Challenges

A. Månsson[1], I.A. Nicholls[1] P. Omling[2], S. Tågerud[1], and L. Montelius[2]

[1] School of Pure and Applied Natural Sciences, University of Kalmar,
391 82 Kalmar, Sweden
[2] Division of Solid State Physics and The Nanometer Consortium, University of
Lund, Box 118, 221 00 Lund, Sweden

Abstract. Muscle contraction occurs as a result of force-producing interactions between the contractile proteins myosin II and actin with the two proteins highly ordered in the filament lattice of the muscle sarcomere. In contrast to this well-ordered structure, most in vitro studies are performed with the contractile proteins in a disordered arrangement. Here we first review the existing in vitro motility assays and then consider how they can be improved by the use of nanotechnology. As a basis for such improvement we describe our recent work where we used chemically and topographically patterned surfaces to achieve selective localization of actomyosin motor function to predetermined areas of sub-micrometer dimensions. We also describe guidance and unidirectional actin filament sliding on nanosized tracks and suggest how such tracks can be combined with 1. microfluidics-based rapid solution exchange and 2. application of electromagnetic forces of well-defined orientation, thus simulating the lifting of a weight by actomyosin. As a related issue we discuss the usefulness of nanotechnology based assay systems for miniaturized high-throughput drug screening systems with molecular motors as drug targets. Finally, we consider the potentials and challenges in using nanotechnology to reconstruct the most essential aspects of cellular order within the muscle sarcomere.

18.1 General Introduction

In order to achieve efficient movement and force-generation, molecular motors often function in localized and oriented supra-molecular assemblies. One extremely refined example of such an assembly is the muscle sarcomere where the actomyosin motor system is organized in a nearly crystalline lattice (e.g. [1]). Force generation and shortening in this system is accomplished by cyclic interactions between the myosin II motor molecules and actin filaments.

A. Månsson et al.: *Nanotechnology Enhanced Functional Assays of Actomyosin Motility – Potentials and Challenges*, Lect. Notes Phys. **711**, 385–406 (2007)
DOI 10.1007/3-540-49522-3_18   © Springer-Verlag Berlin Heidelberg 2007

Important information about the force-generating mechanism of single myosin motors has been obtained by single molecule mechanical and bio-chemical studies [2, 3] for review, see [4, 5]. These studies, in turn, have been aided by the structural information available from X-ray crystallography [6–8] and electron microscopy [9] and through mutational studies designed to alter the detailed function of the motors (for review see [10]). However, one aspect not usually taken into account in in vitro single molecule studies is the high degree of order in the muscle sarcomere. The very fact that single molecules are used means that it is impossible to detect cooperative effects involving numbers of motors and actin filaments in an ordered arrangement. This problem with the loss of internal cellular order applies not only to single molecule studies. Rather it is a general problem for in vitro experimental systems. Muscle cells, on the other hand, exhibit high degrees of order and various biophysical experiments on cellular preparations (e.g. [11–13]) have contributed substantially to the understanding of actomyosin function in the ordered lattice of actin and myosin filaments in the sarcomere. However, the large ensemble of myosin heads and actin filaments, makes interpretation of some experimental data difficult. Isolated myofibrils [14, 15] although being most useful experimental systems for some purposes, are as limited as muscle fibres in this respect, with thousands of myosin motor molecules and actin filaments. In order to study cooperative effects it is therefore of interest to develop a new experimental system designed to involve the minimal number of protein components required for the most essential features of the cellular order. In recent years the dimensions of man-made nanostructures have rapidly approached a length scale in the 1–10 nm range (http://public.itrs.net) relevant to the organization of contractile machinery in the muscle sarcomere. The interesting possibility therefore exists that nanotechnology, in the near future, will provide methods for in vitro reconstruction of important aspects of the cellular order of actomyosin and other biomolecules.

This present paper will, on the basis of our recent work, introduce the potentials and challenges involved in such developments. We will also consider more short-term nanotechnology assisted improvements of in vitro test-systems for studies of actomyosin function. First, however, we will briefly review the hierarchial organization of a muscle cell and discuss some existing in vitro motility and single molecule experimental systems.

18.2 Actomyosin Interactions in the Muscle Cell

The elongated skeletal muscle cell has its contractile machinery organized in parallel threadlike myofibrils. These have a diameter of about 1 μm and run the entire length of the cell (up to tens of centimetres). The myofibrils exhibit a repeating pattern of striations (Fig. 18.1a) where each repetitive

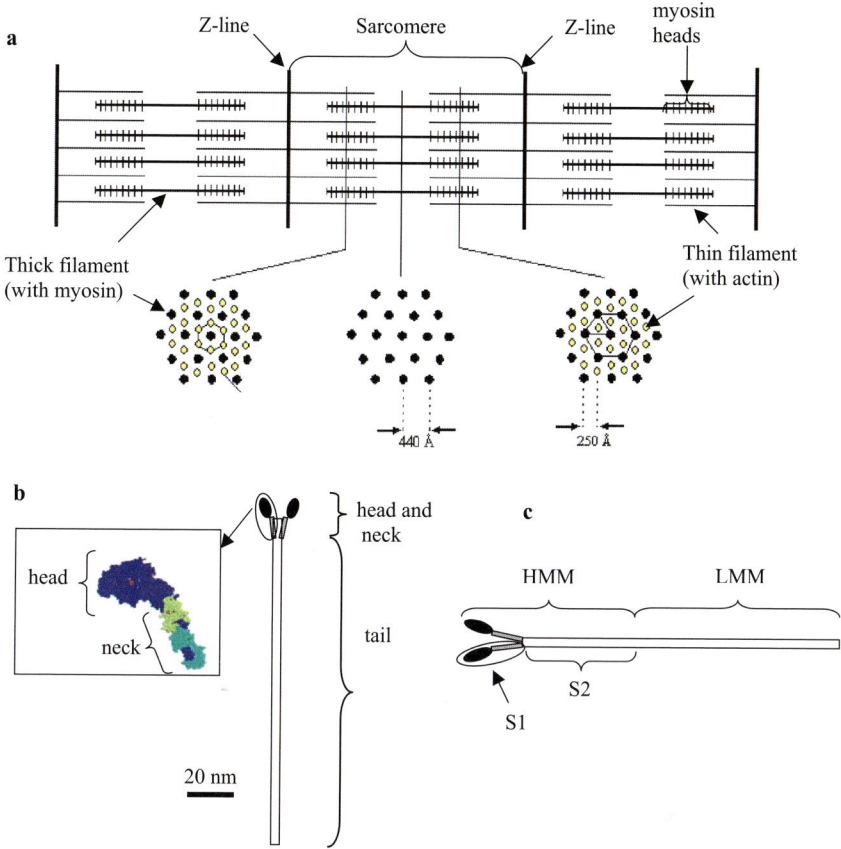

Fig. 18.1. (a) Schematic illustration of the sarcomere with thick and thin filaments organized in a hexagonal lattice. Distances between actin and myosin refer to the physiological resting state in vertebrate muscle [1]. The length of a sarcomere is about 2 μm. (b) Schematic drawing of a myosin molecule (approximately to scale) with two globular head domains containing actin and ATP-binding sites. The heads are connected via α-helical neck regions to a tail (*rod*) where the two myosin heavy chains are intertwined in a coiled coil. In addition to the two heavy chains there are two sets of light chains attached to the neck domain of each heavy chain. Inset: the crystallographic structure of myosin II head and neck printed from "The Protein Explorer" software (http://molvis.sdsc.edu/protexpl/). (c) Schematic illustration of different proteolytic fragments of myosin: heavy meromyosin (HMM), light meromyosin (LMM), subfragment 1 (S1) and subfragment 2 (S2). The fragments are obtained by limited proteolysis as described in the text

unit is denoted a sarcomere (length 2–3 μm in vertebrates). The basis for the striated appearance is the arrangement of the contractile proteins, actin and myosin and associated proteins into thin and thick filaments, respectively. These two interdigitating sets of filaments partly overlap in the sarcomere centre (Fig. 18.1a; [1]). Here globular myosin heads extend from the thick filament backbone allowing them to cyclically interact with binding sites on the actin filaments. The interactions ([16]; e.g. reviewed in [4, 10]) produce a sliding motion of the thin and thick filaments relative to each other leading to shortening of the sarcomere and hence of the entire muscle cell. The degree to which actin, in addition to myosin, is an active player in the force-generating process is still a matter of debate [5]). However, there is strong evidence (e.g. [10, 17]); see also Rayment and Manstein and Oiwa, this issue) that different steps in the ATP hydrolysis by myosin motor domains are coupled to structural changes that could account for the translation of the actin filaments.

The thin actin-containing filaments are anchored at the Z-lines. These structures separate the sarcomeres from each other and each contains several actin binding and cross-linking proteins. A transverse section of the sarcomere further illustrates the high degree of order. It can thus be seen in Fig. 18.1a that the thick and thin filaments are arranged in a hexagonal lattice with an actin to myosin distance (centre to centre) of approximately 250 Å. This almost crystalline arrangement of the contractile proteins is the basis for the highly informative X-ray diffraction pattern (cf. [13, 18, 19]) that can be recorded from living muscle cells.

18.3 Actin and the Thin Filaments

F-actin and the regulatory proteins troponin and tropomyosin constitute the thin filaments of the sarcomere [20]. The building blocks of the filamentous F-actin are G-actin monomers. These are organized in two proto-filaments that form a right-handed double-helix with a cross-over every 36 nm. The regulatory tropomyosin molecules of 36 nm length are situated in two chains over the myosin binding sites on the F-actin filament and one troponin protein complex is attached to each tropomyosin molecule [20].

18.4 Myosin II and the Thick Filaments

Myosin II (for review see Sellers [21]) is a hexameric polypeptide composed of two heavy chains (200 kDa each in skeletal muscle) and two pairs of light chains (15–27 kDa). There are several isoforms of myosin II, both muscular and non-muscular, but here we focus entirely on the isoforms from muscle. Below, these isoforms are often referred to simply as myosin. The myosin II molecule is highly asymmetrical (Fig. 18.1b) consisting of a long tail and two globular head-neck regions. The carboxy-terminal sections of the two heavy chains are mostly α-helical forming the coiled-coil tail under physiological conditions. This tail (or rod) structure aggregates with other myosin molecules to form the thick filaments of the sarcomere. The amino-terminal section of the heavy chain constitutes the globular motor domain (head) of myosin with the nucleotide and actin binding sites. This motor domain is connected to the rod portion of myosin via the α-helical neck that is stabilised by two light chains, the essential and regulatory light chains. The thick filaments are bipolar with the myosin heads always pointing in different directions on either side of the filament midpoint. The myosin molecules are helically ordered on the filament surface with a repeat period of 42.9 nm [1] and myosin heads extend pair-wise in opposite directions every 14.3 nm. The neighbouring pairs of heads along the filament are rotated relative to each other by 120 degrees.

By use of controlled proteolysis, myosin can be cleaved into different functional domains (Fig. 18.1c) that in many respects (e.g. solubility), are more suitable for in vitro studies than the myosin molecule *per se*. Digestion of myosin with either trypsin or chymotrypsin results in two fragments, heavy meromyosin (HMM; $\sim$350 kDa) and light meromyosin (LMM; $\sim$150 kDa). The HMM fragment contains small α-helical parts of the myosin heavy chain tail which allows the formation of a structure consisting of both motor domains and a short coiled-coil tail. Further limited proteolysis of HMM by papain or chymotrypsin results in two additional fragments (Fig. 18.1c), referred to as subfragment-1 (S-1; motor domain + neck domain) and subfragment-2 (S-2).

18.5 The in vitro Motility Assay and Single Molecule Mechanics

The in vitro motility assay [22, 23] has been immensely valuable for the study of molecular motor function [24–28]. In an initial version of the assay, movement of myosin coated beads was followed as they slided along oriented actin filaments [29, 30]. In the version that is most frequently used today the reverse geometry is employed. Thus, fluorescence labeled actin filaments are observed as they slide on a surface to which randomly oriented molecular motors are adsorbed. This version was developed by Spudich and co-workers [22, 23] following the demonstration by Yanagida and coworkers [31] that individual fluorescence labeled actin filaments could be readily observed in the fluorescence microscope. In most studies of the myosin II motor of muscle the proteolytic heavy meroyosin (HMM) motor fragment is used rather than the entire myosin molecule. One reason for this is that the LMM tail of myosin appears to interact with the actin filaments and exert a braking force that counteracts filament sliding [32].

The in vitro assay, in contrast to in vivo studies, enables detailed control of the chemical environment and the use of protein engineering [26] and single molecule force measurements [2, 33, 34]. However, the motors are generally disordered and often bound to the surface by non-specific adsorption, possibly resulting in several different binding configurations [35]. The disorder and possible surface-protein interactions are likely to have profound effects on the mechanisms of force generation and regulation [3, 36]. In accordance with this idea there are numerous examples where actomyosin motor function is different in vivo and in vitro. First, a lower sliding velocity in vitro than in vivo has been reported by several authors [36, 37] although there are conflicting results [38]. Other examples involve differences in temperature dependence and a different pH dependence of the sliding velocity [25].

Some of the differences between fibers and in vitro motility assays can be attributed to the lack of order of the actomyosin system in vitro. Evidence for this view is provided by modified in vitro motility assay studies (e.g. [3,36,39]) where entire myosin filaments (either native or synthetic) were adsorbed to a surface. Measurement of actin sliding velocity and myosin induced force development in these systems suggest that the sliding speed and the amplitude of the force generating structural change in myosin is strongly dependent on the relative orientation of the actin filaments and the myosin motors.

In addition to the importance of the order of the motors, it is not surprising that motor function is seriously affected by interactions between the surface and the myosin motor domains [3,35] or the actin filaments [36]. Interactions of the myosin head with the underlying substrate may lead to structural changes in the motor domain or steric hindrance of actin binding and power-stroke production. This is in good agreement with difference in actin translating capability of HMM when adsorbed to different surface chemistries [40] as further discussed below.

In addition to the degree of order and surface-protein interactions there are other differences between the cellular and in vitro function of actomyosin. One of these is the removal of protein components in the purification of actin and myosin for the in vitro experiments. That this may have significant effects on function has been clearly demonstrated for the case of troponin and tropomyosin [20, 24, 41]. However, there may also be effects of other protein components of the sarcomere. Finally, in this context, it is of interest to mention some other factors that have large effects on the in vitro sliding speed. Such factors include buffer composition (e.g. Mg^{2+}-concentration; [42]), type of motor fragment (myosin, chymotryptic HMM, tryptic HMM or S1; [32,43]) and motor surface density [27]. These factors, if not well understood and carefully controlled, will contribute to differences between sliding speeds in vivo and in vitro.

Some of the above limitations of the in vitro motility assay extend to many single molecule experiments where the force production of single motors is studied, e.g. by the use of optical traps [2]. In these experiments the motors are often attached by non-specific adsorption to the surface without detailed control of their orientation (e.g. [2,34,44]). Furthermore, the possibility of the motor interacting with more than one actin filament has not been considered in these experiments and the actin filaments are free to rotate in optical traps destroying the ordering of the binding site for myosin.

18.6 An Ideal Ordered in vitro Motility Assay System

In spite of some of the limitations considered above, in vitro motility assays and single molecule studies of actomyosin have increased our knowledge of the molecular-level events underlying muscle function much beyond what was learnt from studies of muscle cells. However, there are several reasons (cf. [3, 45, 46]) for the development of assay systems at an intermediate level of organization. Such assay systems should bridge the gap between the single molecule studies and experiments on whole cells by allowing in vitro studies of few (rather than single) molecules interacting in ordered arrangements. In such assays one may envisage an ordered array of myosin motors interacting with two actin filaments attached to each other at an interfilament distance similar to that in vivo. The ideal assay system should also allow modification of the protein components by genetic or biochemical means (e.g. to produce point mutations or single-headed myosin). Finally, it is important to be able to perform the studies with or without regulatory proteins (tropomyosin and troponin) on the actin filaments [36, 47].

In an ideal assay system the surface-motor interactions should be minimal or at least well understood. A test of the success by which the in vitro system reproduces the properties of the living system is the extent to which a set of experimental results correspond to similar results in cell based assay systems, e.g. through studies of maximum sliding velocity and the relationship between velocity and temperature, ionic strength, pH, ATP concentration etc.

The use of myosin filaments rather than randomly adsorbed myosin motors on a surface was considered above. A more complex in vitro system is the recently developed A-band in vitro motility assay [46]. Here myofibrils are isolated followed by removal of the actin filaments by treatment with the actin severing protein gelsolin. This supposedly leaves the thick filament lattice of the sarcomere intact, and thereby the ordered arrangement of the myosin motors. Force generation between actin and myosin is initiated by bringing an actin filament, held by an optical tweezer (via a polystyrene bead), into the thick filament lattice. Although the A-band motility assay is an interesting system with a level of organization, in between that of single molecules and a myofibril, we would like to develop an even simpler ordered system. Here it should be possible to limit the number of protein components (e.g. only myosin and actin) and allow studies of one-headed and two-headed myosin motors at appropriate orientations interacting with one or two actin filaments. Unlike such a system the A-band motility assay is likely to contain several more proteins than actin and myosin (with limited possibility for control) and the myosin molecules in the intact thick filaments may also be difficult to manipulate biochemically or genetically (i.e. to create one-headed myosin or special functionalities). Furthermore, the A-band motility assay, in its present form, does not consider the interaction of a given myosin molecule with two actin filaments. The interaction with two actin filaments may be important for cooperative interactions between the two heads of myosin II in force generation.

Below we consider potentials and challenges of using nanotechnology to create an in vitro motility assay with a minimal number of protein components in ordered arrangements.

18.7 Recent Developments of Nanotechnology Enhanced in vitro Motility Assays

Surface Chemistry and Actomyosin Function

As mentioned above the surface chemistry used for HMM adsorption is likely to be an important determinant of actomyosin function. In accordance with this view we found in recent work [40, 50] differences in the sliding velocities and in the fraction of motile filaments on different resist polymers and on different silanized surfaces, nitrocellulose and glass/SiO$_2$ (Fig. 18.2). For the experiments in Fig. 18.2 methylcellulose was included in the assay solution to prevent the actin filaments from detaching from surfaces where only a fraction of the HMM molecules take part in actin binding. Importantly, if the methylcellulose was removed from the assay solution there was complete selective localization of function to TMCS with no actin filaments sliding

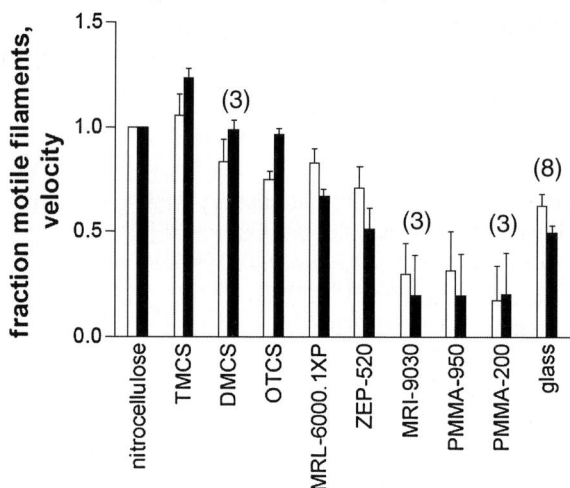

Fig. 18.2. HMM induced actin sliding velocity (*filled bars*) and fraction of motile filaments (*open bars*) with HMM adsorbed either to different silanized surfaces (TMCS: trimethylchlorosilane; DMCS: dimethyldichlorosilane and OTCS: octade-cyltrichlorosilane), to different resist polymers (MRL-6000.1XP, ZEP-520, MRI-9030, PMMA-950 and PMMA-200) or to glass. Data normalized to results obtained on a control nitrocellulose surface in each given experiment. Four replicate surfaces tested if not otherwise stated within parenthesis. Results based on data in [50] and [40]

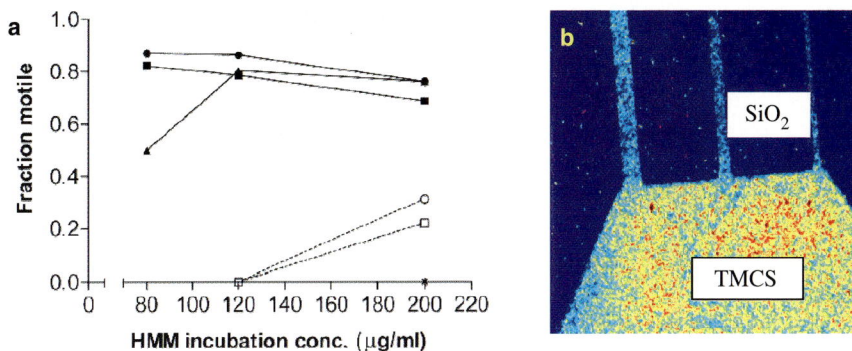

Fig. 18.3. The quality of HMM induced actin filament motility (fraction of motile actin filaments) on TMCS-treated SiO$_2$ (*filled symbols*) and unsilanized SiO$_2$ (*open symbols* and *star* on horizontal axis at 200 µg/ml) as a function of HMM incubation concentration. The actomyosin function was tested at ionic strengths 20 mM (*circles*), 30 mM (*squares*) and 50 mM (*triangles*) on both TMCS (*full symbols*) and SiO$_2$ (*open symbols*) in a given experiment. *Star* refers to ionic strength 20 mM tested in an additional experiment for SiO$_2$ without TMCS. No methylcellulose present in the assay solutions. Temperature: 32.0–32.6°C. (**b**) Fully selective binding of Alexa488-phalloidin (APh) actin filaments to HMM on TMCS derivatized part of SiO$_2$/silicon wafer. Image color-coded from zero (*dark blue*; low actin concentration) to 256 gray scale levels (*red*; highest actin concentration) to indicate the actin filament density on the surface. The filaments on TMCS exhibited high-quality motility. HMM added at 120 µg ml^{-1}, APh-actin added at 1 µM (actin monomer concentration). Assay solution without methylcellulose and ionic strength 40 mM (18°C). Total image size 230 × 230 µm^2. Figure 18.3a reproduced from [40] with permission from Elsevier

on SiO$_2$ provided that the HMM incubation concentration was lower than 200 µg/ml (Fig. 18.3). The motility selectivity between TMCS and SiO$_2$ was observed in spite of similar densities of catalytically active HMM molecules on the two surface chemistries [77]. These experiments, as well as data from Toyoshima [35] suggest that there exists a spectrum of HMM configurations on artificial surfaces and that different configurations may predominate on different surface chemistries.

The differences in motility quality (speed, fraction of motile filaments) between TMCS and SiO$_2$/glass was quite marked. However, as illustrated in Fig. 18.2, more subtle differences in protein function exist between some other surface chemistries with only small differences in sliding velocity and fraction of motile filaments. In addition, two different surface chemistries may only give similar motility quality in some assay solutions. For instance at physiological ionic strength we observed a slightly different pH dependence of HMM induced actin sliding velocities on TMCS and nitrocellulose in spite of very similar sliding velocity at physiological pH (Fig. 18.4). Furthermore, Marston [41], on basis of in vitro motility assay experiments at different levels of Ca^{2+} and troponin reconstitution of the actin filaments, provided argu-

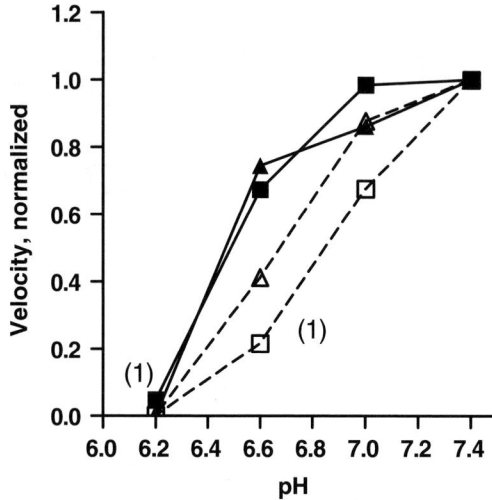

Fig. 18.4. HMM induced actin filament sliding velocity as a function of pH with HMM adsorbed either to nitrocellulose (*filled symbols*) or to TMCS derivatized surface (*open symbols*). *Triangles* and *squares* refer to two different experiments with different batches of assay solutions, HMM and surface preparations. Sliding velocity normalized to velocity on each surface at pH 7.4. Data are given as mean ± standard error of the mean but the error bars are smaller than the size of the symbols. Number of filaments observed at each pH-level given within parenthesis if less than 24. For all but one filament on nitrocellulose the sliding velocity was 0 µm/s at pH 6.2

ments for a greater drag force on filament sliding on nitrocellulose than on silanized surfaces [41]. It is clear from the above results and discussion that understanding the effect of surface-protein interaction on actomyosin function is essential for correct interpretation of in vitro motility assays. However, as further considered below, such understanding is also important in the development of nanotechnology assisted motility assays.

Nanostructured Surfaces for Spatial Control
of HMM Induced Actin Filament Sliding

Recent developments of nanotechnology enhanced in vitro motility assay systems [48–60] have focused on the selective localization of motor function to certain micrometer or nanometer sized areas on a surface and studies of the conditions for guidance and rectification of actin filament or microtubule sliding along such areas. In most of this earlier work a main motivation has been the use of motor proteins for cargo-transportation in nanotechnology (reviewed in [61, 62] see also Vogel and Hess and Manstein and Oiwa, this volume). Below we consider the recent achievements and their short-term and long-term potential in functional studies of actomyosin.

(a)

280 nm

50 nm
100 nm
200 nm
400 nm
600 nm

42 µm

42 µm

50 µm

100 µm

R=5µm 4µm 3µm 2µm 1µm

1200 µm

(b) 100 nm

PMMA

LOR

SiO₂

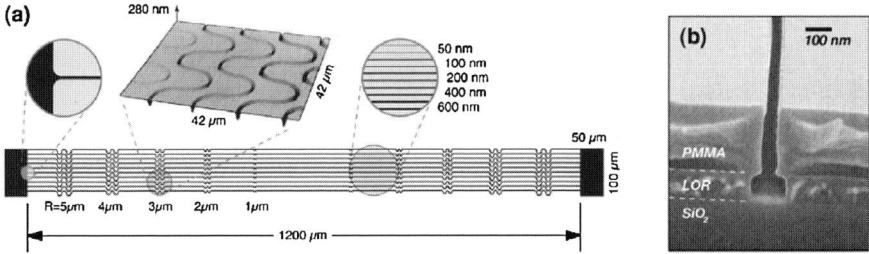

Fig. 18.5. Schematic diagram of electron beam lithography pattern for spatial confinement of actomyosin function. Dark areas functionalized with TMCS. Other parts of the surface are covered with PMMA layer that was made hydrophilic by oxygen plasma treatment. The *insets* show from *left* to *right*: tapered line opening towards actin filament loading zone, atomic force microscopy image of meander structure in PMMA forming the basis for TMCS derivatized track and straight section with tracks of different widths. (**b**) Scanning electron microscopy image showing the cross-section of the sample with bi-layer channels in two resist layers (LOR and PMMA) spin-coated onto the surface before patterning with electron beam. Borders between resist layers indicated by *dashed lines*. Figure reproduced from [53] with permission from IOP Publishing Limited

Selective localization of actomyosin function has been realized using a variety of micro-, and nanopatterning techniques to create chemically defined patterns [52,53,63,64] e.g. of different surface charge [65] or different degree of surface hydrophobicity [40,64]. The results suggest that there exist a range of useful strategies to selectively localize actomyosin function to predetermined areas. As regards efficient guidance of motor propelled cytoskeletal filaments without U-turns of the filaments on the tracks this is more difficult to realize for actin filaments than for microtubules. The main reason is the low flexural rigidity of the actin filaments. We have shown that tracks of width less than about 400 nm are required to ensure that U-turns of actin filaments are prevented at high myosin head densities [52,53]. This width is considerably lower than the persistence length of the actin filaments (18 nm; [66]) due to energy input from action of the myosin motors. In our most recent work [53] with nanosized tracks we used a silicon dioxide surface derivatized with trimethylchlorosilane (TMCS) for the adsorption of functional HMM molecules. This substrate had previously been observed to reproducibly allow actomyosin function of equal or better quality than on standard nitrcoellulose substrates (Fig. 18.2). Electron beam lithography was used in combination with chemical vapor deposition to create nanosized TMCS-tracks surrounded by a polymethylmethacrylate (PMMA) layer. The tracks, or rather partly roofed channels (Fig. 18.5), were designed to have a width down to 100 nm and a length of more than 1 mm. By oxygen plasma treatment the PMMA layer was made hydrophilic and as a consequence completely motility suppressing. Furthermore, the PMMA layer also served as a topographical barrier

396   A. Månsson et al.

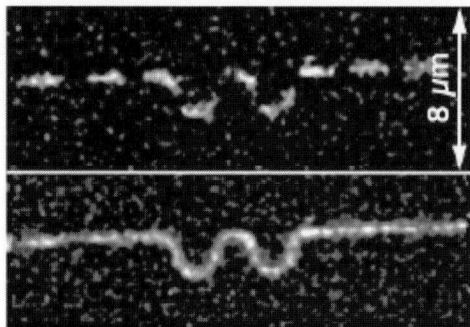

Fig. 18.6. Fluorescence micrograph showing part of bilayer resist channel (cf. Fig. 18.5) with one actin filament being propelled by HMM from the *right* to the *left*. The *upper panel* shows stroboscopic image with equal time lapse between frames. The *lower panel* shows integration of all filament images during sliding. The radius of curvature of the meandering part of the bilayer resist channel was 1 μm and the channel width was about 100 nm. Figure reproduced from [53] with permission from IOP Publishing Limited

that partly covered the TMCS derivatized area. This was possible since the PMMA layer was spin coated on top of another resist polymer (LOR) which was specifically etched to produce an undercut below the PMMA layer. The approach was very successful with high quality actin motility on the TMCS tracks and complete actin filament guidance without U-turns for all track widths in the range of 100–500 nm. Guidance, without escape events, was observed in the presence of curvatures of radius down to 1 μm (Fig. 18.6). However, the sliding velocity along a track was similar for filaments moving in both directions suggesting that there was no preferential orientation of the myosin motors on the track (cf. [3, 39]).

Usefulness of Quantum Dots

As summarized above top-down nanofabrication techniques, allow unprecedented spatial localization of actomyosin function. Another aspect of nanotechnology, the development of fluorescent nanocrystals (quantum dots) also has considerable potentials for studies of molecular motors. This was recently pointed out in [67] where properties of quantum dots such as negligible photobleaching and the emission of polarized light were mentioned as some of the useful properties. Månsson et al. further demonstrated that actin filaments could be readily labeled with single fluorescent quantum dots via streptavidin-biotin-phalloidin linkages and this labeling could be achieved without loss of myosin induced sliding of actin filaments (Fig. 18.7). Below we discuss the usefulness of quantum dots for achieving the cross-linking of two actin filaments at similar interfilament distances to those in the muscle sarcomere.

18.8 Nanotechnology Enhanced in vitro Motility Assays – Future Directions

One possible way to achieve an ideal ordered assay system for the study of actomyosin function would be to use nanotechnology to produce scaffolds to guide the positioning of myosin motors and actin filaments in a similar way as in the muscle sarcomere. This is one of the long-term goals of our recent efforts. However, so far, our work has necessarily focused on the much simpler tasks as outlined above. It is clear that the challenges in achieving both exact positioning and ordering of the actomyosin system by nanotechnology are enormous. Both the small dimensions of the actomyosin lattice in the sarcomere and the possibly adverse effects of surface protein interactions contribute to the difficulties. However, the potential advantages of nanotechnology based assay systems in functional studies are numerous. As further discussed below these also extend beyond the long-term goal of reconstructing the ordered arrangement between actin and myosin.

Electromagnetic Manipulation of Actomyosin
in Nanotechnology Based Assays

One example of a development viable within the near future, concerns the possibility to affect actin filament sliding with electrical or magnetic fields that have well defined orientation with respect to the sliding direction. This principle has already been illustrated by work of Riveline et al. [68] using micrometer wide PMMA grooves for immobilization of myosin. Such grooves with PMMA walls were required to achieve straight filament paths and prevent changes in actin filament sliding direction when electric fields were applied. In the absence of the PMMA grid the filaments were only partly aligned with the field causing variation in the electric force affecting the actin filaments in the sliding direction. However, in spite of straight sliding paths Riveline et al. demonstrated some lateral fluctuations of the filaments in the grooves. Such fluctuations may be attributable to Brownian motion of the actin filaments in regions that are temporarily without attached myosin heads. Here, our recent demonstration of completely rectified sliding along nanosized tracks suggests important improvements. Thus, the nanosized tracks have a width that is more than two orders of magnitude smaller than the persistence length of the

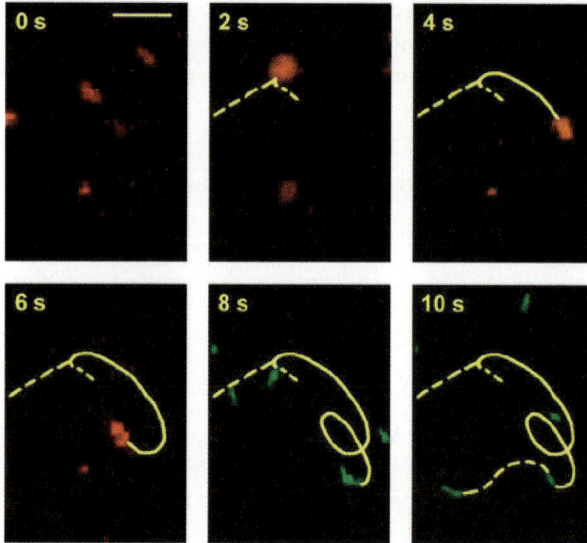

Fig. 18.7. Transportation of CdSe quantum dot labelled actin filaments. Image sequence illustrating events where two actin filaments first meet and cross-link (0– 2 s) presumably via the streptavidin coated quantum dot. Then the two filaments move together (2–8 s) and finally (8–10 s) they separate again. One of them stops moving whereas the other filament continues its sliding. *Dashed yellow lines* show the single filament paths, *solid line* represents sliding of double filaments. Switch between TRITC and FITC filter sets at time point between 6 and 8 s. Images enhanced for clarity, scale bar 5 μm. Reproduced from [67] with permission from Elsevier

actin filaments (18 μm; [66]). This would significantly reduce the amplitude of any thermal fluctuations in a plane parallel to the surface. Furthermore, by using the type of nanostructures in Fig. 18.5 the thermal fluctuations would also be significantly reduced for a plane that is perpendicular to the surface provided that the filament is located below one of the PMMA overhangs of the bilayer resist channel.

By pursuing the developments of nanostructuring techniques, in combination with the application of electric and magnetic forces (e.g. via magnetic beads), we hope to arrive at conditions where it is possible to observe force-generating events by individual myosin II motors under different experimental conditions than in optical tweezer systems.

Miniaturized High Throughput in vitro Motility Assay

The controlled localization of actomyosin function also allows straightforward incorporation of microfluidics systems. Such combinations should allow for highly controlled and localized changes in assay conditions. Such changes would result as a filament moves from one region of laminar flow to another region where the assay solution has a different composition. This approach may enable the detection of changes in sliding velocity (e.g. caused by a drug) considerably lower than 10%, a current limit set by the variation due to the use of different flow cells for different assay conditions.

Fully predictable sliding paths with unidirectional myosin induced sliding of actin filaments on nanosized tracks open for developments of greatly miniaturized high-throughput in vitro motility assays. Most importantly, in this regard, we observed efficient guidance along tracks that were highly curved with radius as low as 1 μm. This suggests that myosin induced actin filament sliding around circular tracks with such radius is also possible. Assuming slightly larger circles e.g. 2 μm radius, would give a path of a total length of approximately 12 μm covering a circular area of about 12 μm^2. With such a minimal test area very small amounts of test substances (e.g. inhibitors of motor protein function) are required provided that the area is appropriately delineated. With a sliding velocity of 5 μm s^{-1} the cycle time for an actin filament sliding around the track would be a little more than 2 s. One easy way to monitor the velocity of such a filament in a high-throughput assay is to simply calculate the number of full cycles around the circular path over, say 10–20 s. This would be the time required for testing the effect of one particular chemical substance whether this was a foreign chemical (e.g. a drug) or a intracellular regulatory protein.

One can question the potential usefulness of nanopatterned surface in high-throughput testing due to the limited lifetime of the proteins and low-throughput fabrication of the test surfaces by means of electron beam lithography. However, we have previously demonstrated [40] that actomyosin function of high quality can be maintained for considerable time on a TMCS surface (up to a week). This suggests that rather infrequent exchange of motor protein binding surfaces may be required. Furthermore, we have recently demonstrated the potential to use nanoimprint lithography for production of the test surfaces with preserved actomyosin function [52]. By turning to such a high-throughput nanofabrication procedure, large series of disposable nanopatterned surfaces can be produced allowing exchange to a new set of proteins and a new test chip through a simple and relatively cheap procedure. This will enable us to develop the nanotechnology enhanced motility assay into a routine biological testing procedure with important future implications for drug testing etc. Such systems are of interest in view of the emerging role of different motor proteins as potential drug targets [69, 70].

Reconstruction of Important Features of the Sarcomere

Nanosized tracks (Figs. 18.5–18.6) with rectified sliding of actin filaments may be useful for cross-linking the filaments at well-defined orientations and inter-filament distance similar to that in the muscle sarcomere. Such cross-linking may be realized by adding a cross-linker of suitable length as two filaments slide with the same direction along a track. Alternatively, the cross-linking molecule could be attached to one of the filaments. The possibility to achieve interfilament cross-linking with appropriate interfilament distance by the latter method was demonstrated previously (see Fig. 18.7). Here, two actin filaments were cross-linked to each other during sliding via quantum dots and biotin-streptavidin links. Since sliding direction was not controlled in the experiment in Fig. 18.7 neither was the orientation of the two actin filaments with respect to each other. We expect that nanostructured surfaces may be adapted to readily allow cross-linking similar to that in Fig. 18.7 but with appropriate interfilament orientation and distances.

Whereas the selective localization of motor function to certain areas is relatively straightforward it is highly challenging to produce such localization with all motors in the same orientation. In principle it should be possible to orient HMM molecules by electric fields during adsorption since both HMM and its S2-part are permanent electric dipoles [71]. However, calculations show that very large field strengths (>1 MV/m) are needed to counteract thermal fluctuations. This is quite awkward even if small inter-electrode distances are used to keep down the voltages. Furthermore, even if this method of orientation is successful it is not entirely clear that the adsorbed myosin motors (e.g. HMM) behave as the myosin motors in the sarcomere. The disposition of these motors is likely to be affected by the surface chemistry of the highly ordered thick filament backbone. Future nanostructures for in vitro motility assays would have to take this aspect into account. Possibly, hybrid devices may be constructed where parts of the myosin filament backbone are integrated with nanolithographically produced tracks.

Clearly, the challenges involved in achieving oriented arrays of myosin motors by artificial methods are significant and we realize that this may be a very long-term project. However, there are several advantages in pursuing this task instead of using myosin filaments to achieve oriented arrays of motors. The advantages include the possibility to create extended tracks of motors with the same orientation. The bipolar nature of synthetic myosin filaments or native thick filaments only allow maximum lengths of 10–25 μm [36,39] for arrays of myosin motors pointing in the same direction. The longest arrays are obtained with invertebrate myosin or vertebrate myosin copolymerized with invertebrate myosin rod [39].

The advantages of precise positioning of the motors (e.g. simplified tracking, combination with microfluidics, precise application of electric forces) and topographical barriers around nanosized tracks (reduced thermal fluctuations) cannot be readily reproduced with myosin filaments. Finally, it would be of interest to add the individual motors (and other protein components) to the ordered array in a more controlled way than realized by the self-assembly underlying formation of myosin filaments.

18.9 Conclusions

Nanotechnology enhanced assays for basic biophysical studies of actomyosin, as well as for miniaturized high-throughput drug screening have several attractive features, even though tough challenges remain to be overcome. These challenges involve both developments of nanotechnological methods and better understanding of surface-protein interactions in different experimental systems. We believe that the efforts to meet the challenges will lead to new developments and insights both in nanotechnology and in the life sciences.

One problem that has been the inspiration for much of our work has been the long-standing issue [45, 72, 73], in muscle physiology concerning the degree of cooperativity between the two heads of myosin II in the development of force and shortening. We have no guarantee that our efforts will resolve this issue. However, even if we will not be successful in this respect the multidisciplinary engineering approach taken in many of our studies will give insights into surface-protein interactions and actomyosin function different from those achieved by conventional biophysical methods. The results may also pave the way for the use of nanotechnology to reconstruct the cellular order of other macromolecular assemblies than the muscle sarcomere.

Acknowledgements

Mark Sundberg and Richard Bunk are acknowledged for performing the actual experiments for Fig. 18.3b and Mark Sundberg, Martina Balaz and Richard Bunk for work on earlier original papers upon which parts of the current review is based. The work was supported by grants from The Carl Trygger Foundation, The Swedish Research Council (Project # 621-2004-3449), The Swedish Foundation for Strategic Research, The Knowledge Foundation (KK-stiftelsen), The Faculty of Natural Sciences and Engineering, University of Kalmar and The Nanometer Consortium at Lund University.

References

1. H.E. Huxley (1969). The mechanism of muscular contraction. *Science*, **164**, pp. 1356–1365.
2. J.T. Finer, R.M. Simmons, and J.A. Spudich (1994). Single myosin molecule mechanics: piconewton forces and nanometre steps. *Nature*, **368**, pp. 113–119.
3. H. Tanaka, A. Ishijima, M. Honda, K. Saito, and T. Yanagida (1998). Orientation dependence of displacements by a single one-headed myosin relative to the actin filament. *Biophys. J.*, **75**, pp. 1886–1894.
4. M.J. Tyska and D.M. Warshaw (2002). The myosin power stroke. *Cell Motil Cytoskeleton*, **51**, pp. 1–15.
5. T. Yanagida, K. Kitamura, H. Tanaka, H.A. Iwane, and S. Esaki (2000). Single molecule analysis of the actomyosin motor. *Curr. Opin. Cell. Biol.*, **12**, pp. 20–25.

6. A. Houdusse, A.G. Szent-Gyorgyi, and C. Cohen (2000). Three conformational states of scallop myosin S1. *Proc. Natl. Acad. Sci. USA*, **97**, pp. 11238–11243.

7. W. Kabsch, H.G. Mannherz, D. Suck, E.F. Pai, and K.C. Holmes (1990). Atomic structure of the actin : DNase I complex. *Nature*, **347**, pp. 37–44.

8. I. Rayment, W.R. Rypniewski, K. Schmidt-Base, R. Smith, D.R. Tomchick, M.M. Benning, D.A. Winkelmann, G. Wesenberg, and H.M. Holden (1993b). Three-dimensional structure of myosin subfragment-1: A molecular motor. *Science*, **261**, pp. 50–58.

9. J. Liu, M.C. Reedy, Y.E. Goldman, C. Franzini-Armstrong, H. Sasaki, R.T. Tregear, C. Lucaveche, H. Winkler, B.A. Baumann, J.M. Squire, T.C. Irving, M.K. Reedy, and K.A. Taylor (2004). Electron tomography of fast frozen, stretched rigor fibers reveals elastic distortions in the myosin crossbridges. *J. Struct. Biol.*, **147**, pp. 268–282.

10. M.A. Geeves, R. Fedorov, and D.J. Manstein (2005). Molecular mechanism of actomyosin-based motility. *Cell Mol. Life Sci.*, **62**, pp. 1462–1477.

11. K.A. Edman, A. Mansson, and C. Caputo (1997). The biphasic force-velocity relationship in frog muscle fibres and its evaluation in terms of cross-bridge function. *J. Physiol.*, **503**, pp. 141–156.

12. A.F. Huxley and R.M. Simmons (1971). Proposed mechanism of force generation in striated muscle. *Nature*, **233**, pp. 533–538.

13. V. Lombardi, G. Piazzesi, M.A. Ferenczi, H. Thirlwell, I. Dobbie, and M. Irving (1995). Elastic distortion of myosin heads and repriming of the working stroke in muscle. *Nature*, **374**, pp. 553–555.

14. C. Lionne, R. Stehle, F. Travers, and T. Barman (1999). Cryoenzymic studies on an organized system: myofibrillar ATPases and shortening. *Biochemistry*, **38**, pp. 8512–8520.

15. C. Tesi, F. Colomo, S. Nencini, N. Piroddi, and C. Poggesi (2000). The effect of inorganic phosphate on force generation in single myofibrils from rabbit skeletal muscle. *Biophys. J.*, **78**, pp. 3081–3092.

16. A.F. Huxley (1957). Muscle structure and theories of contraction. *Prog. Biophys. Biophys. Chem.*, **7**, pp. 255–318.

17. J. Howard (2001). *Mechanics of motor proteins and the cytoskeleton*. Sinauer Associates Inc., Sunderland, MA.

18. H.E. Huxley (1984). Time-resolved X-ray diffraction studies of cross-bridge movement and their interpretation. *Adv. Exp. Med. Biol.*, **170**, pp. 161–175.

19. M. Reconditi, N. Koubassova, M. Linari, I. Dobbie, T. Narayanan, O. Diat, G. Piazzesi, V. Lombardi and M. Irving (2003). The conformation of myosin head domains in rigor muscle determined by X-ray interference. *Biophys. J.*, **85**, pp. 1098–1110.

20. A.M. Gordon, E. Homsher, and M. Regnier (2000). Regulation of contraction in striated muscle. *Physiol. Rev.*, **80**, pp. 853–924.

21. J.R. Sellers (1999). *Myosins*. Oxford University Press, Oxford.

22. S.J. Kron and J.A. Spudich (1986). Fluorescent actin filaments move on myosin fixed to a glass surface. *Proc. Natl. Acad. Sci. USA*, **83**, pp. 6272–6276.

23. S.J. Kron, Y.Y. Toyoshima, T.Q. Uyeda, and J.A. Spudich (1991). Assays for actin sliding movement over myosin-coated surfaces. *Methods Enzymol.* **196**, pp. 399–416.

24. Y. Harada, K. Sakurada, T. Aoki, D.D. Thomas, and T. Yanagida (1990). Mechanochemical coupling in actomyosin energy transduction studied by in vitro movement assay. *J. Mol. Biol.*, **216**, pp. 49–68.

25. E. Homsher, F. Wang, and J.R. Sellers (1992). Factors affecting movement of F-actin filaments propelled by skeletal muscle heavy meromyosin. *Am. J. Physiol.*, **262**, pp. C714–723.
26. G. Tsiavaliaris, S. Fujita-Becker, and D.J. Manstein (2004). Molecular engineering of a backwards-moving myosin motor. *Nature*, **427**, pp. 558–561.
27. T.Q. Uyeda, S.J. Kron, and J.A. Spudich (1990). Myosin step size. Estimation from slow sliding movement of actin over low densities of heavy meromyosin. *J. Mol. Biol.*, **214**, pp. 699–710.
28. A.L. Wells, A.W. Lin, L.Q. Chen, D. Safer, S.M. Cain, T. Hasson, B.O. Carragher, R.A. Milligan, and H.L. Sweeney (1999). Myosin VI is an actin-based motor that moves backwards. *Nature*, **401**, pp. 505–508.
29. M.P. Sheetz and J.A. Spudich (1983). Movement of myosin–coated structures on actin cables. *Cell Motil.*, **3**, pp. 485–489.
30. J.A. Spudich, S.J. Kron, and M.P. Sheetz (1985). Movement of myosin-coated beads on oriented filaments reconstituted from purified actin. *Nature*, **315**, pp. 584–586.
31. T. Yanagida, M. Nakase, K. Nishiyama, and F. Oosawa (1984). Direct observation of motion of single F-actin filaments in the presence of myosin. *Nature*, **307**, pp. 58–60.
32. B. Guo and W.H. Guilford (2004). The tail of myosin reduces actin filament velocity in the in vitro motility assay. *Cell Motil Cytoskeleton*, **59**, pp. 264–272.
33. A. Kishino and T. Yanagida (1988). Force measurements by micromanipulation of a single actin filament by glass needles. *Nature*, **334**, pp. 74–76.
34. J.E. Molloy, J.E. Burns, K.-J. Jones, R.T. Tregear, and D.C. White (1995). Movement and force produced by a single myosin head. *Nature*, **378**, pp. 209–212.
35. Y.Y. Toyoshima (1993). How are myosin fragments bound to nitrocellulose film? *Adv. Exp. Med. Biol.*, **332**, pp. 259–265.
36. T. Scholz and B. Brenner (2003). Actin sliding on reconstituted myosin filaments containing only one myosin heavy chain isoform. *J. Muscle Res. Cell Motil.*, **24**, pp. 77–86.
37. M. Canepari, R. Rossi, M.A. Pellegrino, C. Reggiani, and R. Bottinelli (1999). Speeds of actin translocation in vitro by myosins extracted from single rat muscle fibres of different types. *Exp. Physiol.*, **84**, pp. 803–806.
38. P. Hook and L. Larsson (2000). Actomyosin interactions in a novel single muscle fiber in vitro motility assay. *J. Muscle Res. Cell. Motil.*, **21**, pp. 357–365.
39. A. Yamada, M. Yoshio, and H. Nakayama (1997). Bi-directional movement of actin filaments along long bipolar tracks of oriented rabbit skeletal muscle myosin molecules. *FEBS Lett.*, **409**, pp. 380–384.
40. M. Sundberg, J.P. Rosengren, R. Bunk, J. Lindahl, I.A. Nicholls, S. Tagerud, P. Omling, L. Montelius, and A. Mansson (2003). Silanized surfaces for in vitro studies of actomyosin function and nanotechnology applications. *Anal. Biochem.*, **323**, pp. 127–138.
41. S. Marston (2003). Random walks with thin filaments: application of in vitro motility assay to the study of actomyosin regulation. *J. Muscle Res. Cell Motil.*, **24**, pp. 149–156.
42. S. Fujita-Becker, U. Durrwang, M. Erent, R.J. Clark, M.A. Geeves, and D.J. Manstein (2005). Changes in Mg2+ ion concentration and heavy chain phosphorylation regulate the motor activity of a class I myosin. *J. Biol. Chem.*, **280**, pp. 6064–6071.

43. J.R. Sellers, G. Cuda, F. Wang, and E. Homsher (1993). Myosin-specific adaptations of the motility assay. *Methods Cell Biol.* **39**, pp. 23–49.

44. M.J. Tyska, D.E. Dupuis, W.H. Guilford, J.B. Patlak, G.S. Waller, K.M. Trybus, D.M. Warshaw, and S. Lowey (1999). Two heads of myosin are better than one for generating force and motion. *Proc. Natl. Acad. Sci. USA*, **96**, pp. 4402–4407.

45. J.E. Molloy (2005). Muscle contraction: actin filaments enter the fray. *Biophys. J.*, **89**, pp. 1–2.

46. M. Suzuki, H. Fujita, and S. Ishiwata (2005). A new muscle contractile system composed of a thick filament lattice and a single actin filament. *Biophys. J.*, **89**, pp. 321–328.

47. E. Homsher, D.M. Lee, C. Morris, D. Pavlov, and L.S. Tobacman (2000). Regulation of force and unloaded sliding speed in single thin filaments: effects of regulatory proteins and calcium. *J. Physiol.*, **524 Pt 1**, pp. 233–243.

48. K.J. Bohm, R. Stracke, P. Muhlig, and E. Unger (2001). Motor protein-driven unidirectional transport of micrometer-sized cargoes across isopolar microtubule arrays. *Nanotechnol.*, **12**, pp. 238–244.

49. T.B. Brown and W.O. Hancock (2002). A polarized microtubule array for kinesin-powered nanoscale assembly and force generation. *Nano Letters*, **2**, pp. 1131–1135.

50. R. Bunk, J. Klinth, L. Montelius, I.A. Nicholls, P. Omling, S. Tagerud, and A. Mansson (2003a). Actomyosin motility on nanostructured surfaces. *Biochem. Biophys. Res. Commun.*, **301**, pp. 783–788.

51. R. Bunk, J. Klinth, J. Rosengren, I. Nicholls, S. Tagerud, P. Omling, A. Mansson, and L. Montelius (2003b). Towards a "nano-traffic" system powered by molecular motors. *Microelectron. eng.*, **67–8**, pp. 899–904.

52. R. Bunk, A. Månsson, I.A. Nicholls, P. Omling, M. Sundberg, S. Tågerud, P. Carlberg, and L. Montelius (2005a). Guiding molecular motors by nanoimprinted structures. *Jap. J. Appl. Phys.*, **44**, pp. 3337–3340.

53. R. Bunk, M. Sundberg, I.A. Nicholls, P. Omling, Tågerud, S.A. Månsson, and L. Montelius (2005b). Guiding motor-propelled molecules with nanoscale precision through silanized bi-channel structures. *Nanotechnol*, **16**, pp. 710–717.

54. J.R. Dennis, J. Howard, and V. Vogel (1999). Molecular shuttles: directed motion of microtubules slang nanoscale kinesin tracks. *Nanotechnol.*, **10**, pp. 232–236.

55. H. Hess, J. Clemmens, D. Qin, J. Howard, and V. Vogel (2001). Light-controlled molecular shuttles made from motor proteins carrying cargo on engineered surfaces. *Nano Letters*, **1**, pp. 235–239.

56. H. Hess, C.M. Matzke, R.K. Doot, J. Clemmens, G.D. Bachand, B.C. Bunker, and V. Vogel (2003). Molecular shuttles operating undercover: A new photolithographic approach for the fabrication of structured surfaces supporting directed motility. *Nano Letters*, **3**, pp. 1651–1655.

57. Y. Hiratsuka, T. Tada, K. Oiwa, T. Kanayama, and T.Q. Uyeda (2001). Controlling the direction of kinesin-driven microtubule movements along microlithographic tracks. *Biophys. J.*, **81**, pp. 1555–1561.

58. L. Limberis and R.J. Stewart (2000). Toward kinesin-powered microdevices. *Nanotechnol.*, **11**, pp. 47–51.

59. H. Suzuki, A. Yamada, K. Oiwa, H. Nakayama, and S. Mashiko (1997). Control of actin moving trajectory by patterned poly(methylmethacrylate) tracks. *Biophys. J.*, **72**, pp. 1997–2001.

60. D.C. Turner, C. Chang, K. Fang, S.L. Brandow, and D.B. Murphy (1995). Selective adhesion of functional microtubules to patterned silane surfaces. *Biophys. J.*, **69**, pp. 2782–2789.
61. H. Hess, G.D. Bachand, and V. Vogel (2004). Powering nanodevices with biomolecular motors. *Chemistry-a European Journal*, **10**, pp. 2110–2116.
62. A. Månsson, M. Sundberg, R. Bunk, M. Balaz, I.A. Nicholls, P. Omling, J.O. Tegenfeldt, S. Tågerud, and L. Montelius (2005). Actin-based molecular motors for cargo transportation in nanotechnology – potentials and challenges. *IEEE trans. Adv. Pack*, **28**, pp. 547–555.
63. J. Clemmens, H. Hess, R. Lipscomb, Y. Hanein, K.F. Bohringer, C.M. Matzke, G.D. Bachand, B.C. Bunker, and V. Vogel (2003). Mechanisms of microtubule guiding on microfabricated kinesin-coated surfaces: Chemical and topographic surface patterns. *Langmuir*, **19**, pp. 10967–10974.
64. D.V. Nicolau, H. Suzuki, S. Mashiko, T. Taguchi, and S. Yoshikawa (1999). Actin motion on microlithographically functionalized myosin surfaces and tracks. *Biophysical Journal*, **77**, pp. 1126–1134.
65. J.A. Jaber, P.B. Chase, and J.B. Schlenoff (2003). Actomyosin-driven motility on patterned polyelectrolyte mono- and multilayers. *Nano Letters*, **3**, pp. 1505–1509.
66. F. Gittes, B. Mickey, J. Nettleton and J. Howard (1993). Flexural Rigidity of Microtubules and Actin-Filaments Measured from Thermal Fluctuations in Shape. *Journal of Cell Biology*, **120**, pp. 923–934.
67. A. Mansson, M. Sundberg, M. Balaz, R. Bunk, I.A. Nicholls, P. Omling, S. Tagerud, and L. Montelius (2004). In vitro sliding of actin filaments labelled with single quantum dots. *Biochem. Biophys. Res. Commun.*, **314**, pp. 529–534.
68. D. Riveline, A. Ott, F. Julicher, D.A. Winkelmann, O. Cardoso, J.J. Lacapere, S. Magnusdottir, J.L. Viovy, L. Gorre-Talini, and J. Prost (1998). Acting on actin: the electric motility assay. *Eur. Biophys. J.*, **27**, pp. 403–408.
69. M.S. Duxbury, S.W. Ashley, and E.E. Whang (2004). Inhibition of pancreatic adenocarcinoma cellular invasiveness by blebbistatin: A novel myosin II inhibitor. *Biochem. Biophys. Res. Commun.*, **313**, pp. 992–997.
70. K.W. Wood, W.D. Cornwell, and J.R. Jackson (2001). Past and future of the mitotic spindle as an oncology target. *Curr. Opin. Pharmacol.*, **1**, pp. 370–377.
71. S. Highsmith and D. Eden (1985). Transient electrical birefringence characterization of heavy meromyosin. *Biochemistry*, **24**, pp. 4917–4924.
72. A.F. Huxley (1974). Muscular contraction. *J. Physiol. (Lond.)*, **243**, pp. 1–43.
73. R.D. Vale and R.A. Milligan (2000). The way things move: looking under the hood of molecular motor proteins. *Science*, **288**, pp. 88–95.
74. M. Irving, St. Claire Allen, T., C. Sabido-David, J.S. Craik, Brandmeier, B., J. Kendrick-Jones, J.E. Corrie, D.R. Trentham, and Y.E. Goldman (1995). Tilting of the light-chain region of myosin during step length changes and active force generation in skeletal muscle. *Nature*, **375**, pp. 688–691.
75. A. Ishijima, H. Kojima, T. Funatsu, M. Tokunaga, H. Higuchi, H. Tanaka and T. Yanagida (1998). Simultaneous observation of individual ATPase and mechanical events by a single myosin molecule during interaction with actin. *Cell*, **92**, pp. 161–171.
76. I. Rayment, H.M. Holden, M. Whittaker, C.B. Yohn, M. Lorenz, K.C. Holmes, and R.A. Milligan (1993a). Structure of the actin-myosin complex and its implications for muscle contraction. *Science*, **261**, pp. 58–65.

77. M. Sundberg, M. Balaz, R. Bunk, J. Rosengren-Holmberg, L. Montelius, IA. Nicholls, P. Omling, S. Tågerud and A. Månsson (2006). Selective spatial localization of actomyosin motor function by chemical surface patterning. *Langmuir* **22**(17), 7302–7312.

Index